Lecture Notes in Computer Science　　　14348

Founding Editors

Gerhard Goos
Juris Hartmanis

Editorial Board Members

The series Lecture Notes in Computer Science (LNCS), including its subseries Lecture Notes in Artificial Intelligence (LNAI) and Lecture Notes in Bioinformatics (LNBI), has established itself as a medium for the publication of new developments in computer science and information technology research, teaching, and education.

LNCS enjoys close cooperation with the computer science R & D community, the series counts many renowned academics among its volume editors and paper authors, and collaborates with prestigious societies. Its mission is to serve this international community by providing an invaluable service, mainly focused on the publication of conference and workshop proceedings and postproceedings. LNCS commenced publication in 1973.

Xiaohuan Cao · Xuanang Xu · Islem Rekik ·
Zhiming Cui · Xi Ouyang
Editors

Machine Learning in Medical Imaging

14th International Workshop, MLMI 2023
Held in Conjunction with MICCAI 2023
Vancouver, BC, Canada, October 8, 2023
Proceedings, Part I

Editors
Xiaohuan Cao ⓘ
Shanghai United Imaging Intelligence Co.,
Ltd.
Shanghai, China

Islem Rekik ⓘ
Imperial College London
London, UK

Xi Ouyang ⓘ
Shanghai United Imaging Intelligence Co.,
Ltd.
Shanghai, China

Xuanang Xu ⓘ
Rensselaer Polytechnic Institute
Troy, NY, USA

Zhiming Cui ⓘ
ShanghaiTech University
Shanghai, China

ISSN 0302-9743   ISSN 1611-3349 (electronic)
Lecture Notes in Computer Science
ISBN 978-3-031-45672-5   ISBN 978-3-031-45673-2 (eBook)
https://doi.org/10.1007/978-3-031-45673-2

This Springer imprint is published by the registered company Springer Nature Switzerland AG
The registered company address is: Gewerbestrasse 11, 6330 Cham, Switzerland

Paper in this product is recyclable.

Preface

The 14th International Workshop on Machine Learning in Medical Imaging (MLMI 2023) was held in Vancouver, Canada, on October 8, 2023, in conjunction with the 26th International Conference on Medical Image Computing and Computer Assisted Intervention (MICCAI 2023).

As artificial intelligence (AI) and machine learning (ML) continue to significantly influence both academia and industry, MLMI 2023 aims to facilitate new cutting-edge techniques and their applications in the medical imaging field, including, but not limited to medical image reconstruction, medical image registration, medical image segmentation, computer-aided detection and diagnosis, image fusion, image-guided intervention, image retrieval, etc. MLMI 2023 focused on major trends and challenges in this area and facilitated translating medical imaging research into clinical practice. Topics of interests included deep learning, generative adversarial learning, ensemble learning, transfer learning, multi-task learning, manifold learning, and reinforcement learning, along with their applications to medical image analysis, computer-aided diagnosis, multi-modality fusion, image reconstruction, image retrieval, cellular image analysis, molecular imaging, digital pathology, etc.

The MLMI workshop has attracted original, high-quality submissions on innovative research work in medical imaging using AI and ML. MLMI 2023 received a large number of submissions (139 in total). All the submissions underwent a rigorous double-blind peer-review process, with each paper being reviewed by at least two members of the Program Committee, composed of 89 experts in the field. Based on the reviewing scores and critiques, 93 papers were accepted for presentation at the workshop and chosen to be included in two Springer LNCS volumes, which resulted in an acceptance rate of 66.9%. It was a tough decision and many high-quality papers had to be rejected due to the page limitation.

We are grateful to all Program Committee members for reviewing the submissions and giving constructive comments. We also thank all the authors for making the workshop very fruitful and successful.

October 2023

Xiaohuan Cao
Xuanang Xu
Islem Rekik
Zhiming Cui
Xi Ouyang

Organization

Workshop Organizers

Xiaohuan Cao   Shanghai United Imaging Intelligence Co., Ltd.,
                      China
Xuanang Xu   Rensselaer Polytechnic Institute, USA
Islem Rekik   Imperial College London, UK
Zhiming Cui   ShanghaiTech University, China
Xi Ouyang   Shanghai United Imaging Intelligence Co., Ltd.,
                      China

Steering Committee

Dinggang Shen   ShanghaiTech University, China/Shanghai United
                      Imaging Intelligence Co., Ltd., China
Pingkun Yan   Rensselaer Polytechnic Institute, USA
Kenji Suzuki   Tokyo Institute of Technology, Japan
Fei Wang   Visa Research, USA

Program Committee

Reza Azad   RWTH Aachen University, Germany
Ulas Bagci   Northwestern University, USA
Xiaohuan Cao   Shanghai United Imaging Intelligence, China
Heang-Ping Chan   University of Michigan Medical Center, USA
Jiale Cheng   South China University of Technology, China
Cong Cong   University of New South Wales, Australia
Zhiming Cui   ShanghaiTech University, China
Haixing Dai   University of Georgia, USA
Yulong Dou   ShanghaiTech University, China
Yuqi Fang   University of North Carolina at Chapel Hill, USA
Yuyan Ge   Xi'an Jiaotong University, China
Hao Guan   University of North Carolina at Chapel Hill, USA
Hengtao Guo   Rensselaer Polytechnic Institute, USA
Yu Guo   Tianjin University, China
Minghao Han   Fudan University, China

Shijie Huang	ShanghaiTech University, China
Yongsong Huang	Tohoku University, Japan
Jiayu Huo	King's College London, UK
Xi Jia	University of Birmingham, UK
Caiwen Jiang	ShanghaiTech University, China
Xi Jiang	University of Electronic Science and Technology of China, China
Yanyun Jiang	Shandong Normal University, China
Ze Jin	Tokyo Institute of Technology, Japan
Nathan Lampen	Rensselaer Polytechnic Institute, USA
Junghwan Lee	Georgia Institute of Technology, USA
Gang Li	University of North Carolina at Chapel Hill, USA
Yunxiang Li	UT Southwestern Medical Center, USA
Yuxuan Liang	Rensselaer Polytechnic Institute, USA
Mingquan Lin	Weill Cornell Medicine, USA
Jiameng Liu	ShanghaiTech University, China
Mingxia Liu	University of North Carolina at Chapel Hill, USA
Muran Liu	ShanghaiTech University, China
Siyuan Liu	Dalian Maritime University, China
Tao Liu	Fudan University, China
Xiaoming Liu	United Imaging Research Institute of Intelligent Imaging, China
Yang Liu	King's College London, UK
Yuxiao Liu	ShanghaiTech University, China
Zhentao Liu	ShanghaiTech University, China
Lei Ma	ShanghaiTech University, China
Diego Machado Reyes	Rensselaer Polytechnic Institute, USA
Runqi Meng	ShanghaiTech University, China
Janne Nappi	Massachusetts General Hospital, USA
Mohammadreza Negahdar	Genentech, USA
Chuang Niu	Rensselaer Polytechnic Institute, USA
Xi Ouyang	Shanghai United Imaging Intelligence, China
Caner Ozer	Istanbul Technical University, Turkey
Huazheng Pan	East China Normal University, China
Yongsheng Pan	ShanghaiTech University, China
Linkai Peng	Southern University of Science and Technology, China
Saed Rezayi	University of Georgia, USA
Hongming Shan	Fudan University, China
Siyu Cheng	ShanghaiTech University, China
Xinrui Song	Rensselaer Polytechnic Institute, USA
Yue Sun	University of North Carolina at Chapel Hill, USA

Contents – Part I

Structural MRI Harmonization via Disentangled Latent Energy-Based
Style Translation 1
   Mengqi Wu, Lintao Zhang, Pew-Thian Yap, Weili Lin, Hongtu Zhu,
   and Mingxia Liu

Cross-Domain Iterative Network for Simultaneous Denoising,
Limited-Angle Reconstruction, and Attenuation Correction of Cardiac
SPECT 12
   Xiongchao Chen, Bo Zhou, Huidong Xie, Xueqi Guo, Qiong Liu,
   Albert J. Sinusas, and Chi Liu

Arbitrary Reduction of MRI Inter-slice Spacing Using Hierarchical
Feature Conditional Diffusion 23
   Xin Wang, Zhenrong Shen, Zhiyun Song, Sheng Wang, Mengjun Liu,
   Lichi Zhang, Kai Xuan, and Qian Wang

Reconstruction of 3D Fetal Brain MRI from 2D Cross-Sectional
Acquisitions Using Unsupervised Learning Network 33
   Yimeng Yang, Dongdong Gu, Xukun Zhang, Zhongxiang Ding, Fei Gao,
   Zhong Xue, and Dinggang Shen

Robust Unsupervised Super-Resolution of Infant MRI via Dual-Modal
Deep Image Prior 42
   Cheng Che Tsai, Xiaoyang Chen, Sahar Ahmad, and Pew-Thian Yap

SR4ZCT: Self-supervised Through-Plane Resolution Enhancement for CT
Images with Arbitrary Resolution and Overlap 52
   Jiayang Shi, Daniël M. Pelt, and K. Joost Batenburg

unORANIC: Unsupervised Orthogonalization of Anatomy
and Image-Characteristic Features 62
   Sebastian Doerrich, Francesco Di Salvo, and Christian Ledig

An Investigation of Different Deep Learning Pipelines for GABA-Edited
MRS Reconstruction 72
   Rodrigo Berto, Hanna Bugler, Roberto Souza, and Ashley Harris

Towards Abdominal 3-D Scene Rendering from Laparoscopy Surgical
Videos Using NeRFs 83
Khoa Tuan Nguyen, Francesca Tozzi, Nikdokht Rashidian,
Wouter Willaert, Joris Vankerschaver, and Wesley De Neve

Brain MRI to PET Synthesis and Amyloid Estimation in Alzheimer's
Disease via 3D Multimodal Contrastive GAN 94
Yan Jin, Jonathan DuBois, Chongyue Zhao, Liang Zhan,
Audrey Gabelle, Neda Jahanshad, Paul M. Thompson, Arie Gafson,
and Shibeshih Belachew

Accelerated MRI Reconstruction via Dynamic Deformable Alignment
Based Transformer 104
Wafa Alghallabi, Akshay Dudhane, Waqas Zamir, Salman Khan,
and Fahad Shahbaz Khan

Deformable Cross-Attention Transformer for Medical Image Registration 115
Junyu Chen, Yihao Liu, Yufan He, and Yong Du

Deformable Medical Image Registration Under Distribution Shifts
with Neural Instance Optimization 126
Tony C. W. Mok, Zi Li, Yingda Xia, Jiawen Yao, Ling Zhang,
Jingren Zhou, and Le Lu

Implicitly Solved Regularization for Learning-Based Image Registration 137
Jan Ehrhardt and Heinz Handels

BHSD: A 3D Multi-class Brain Hemorrhage Segmentation Dataset 147
Biao Wu, Yutong Xie, Zeyu Zhang, Jinchao Ge, Kaspar Yaxley,
Suzan Bahadir, Qi Wu, Yifan Liu, and Minh-Son To

Contrastive Learning-Based Breast Tumor Segmentation in DCE-MRI 157
Shanshan Guo, Jiadong Zhang, Dongdong Gu, Fei Gao, Yiqiang Zhan,
Zhong Xue, and Dinggang Shen

FFPN: Fourier Feature Pyramid Network for Ultrasound Image
Segmentation 166
Chaoyu Chen, Xin Yang, Rusi Chen, Junxuan Yu, Liwei Du, Jian Wang,
Xindi Hu, Yan Cao, Yingying Liu, and Dong Ni

Mammo-SAM: Adapting Foundation Segment Anything Model
for Automatic Breast Mass Segmentation in Whole Mammograms 176
Xinyu Xiong, Churan Wang, Wenxue Li, and Guanbin Li

Consistent and Accurate Segmentation for Serial Infant Brain MR Images
with Registration Assistance 186
    Yuhang Sun, Jiameng Liu, Feihong Liu, Kaicong Sun, Han Zhang,
    Feng Shi, Qianjin Feng, and Dinggang Shen

Unifying and Personalizing Weakly-Supervised Federated Medical Image
Segmentation via Adaptive Representation and Aggregation 196
    Li Lin, Jiewei Wu, Yixiang Liu, Kenneth K. Y. Wong, and Xiaoying Tang

Unlocking Fine-Grained Details with Wavelet-Based High-Frequency
Enhancement in Transformers 207
    Reza Azad, Amirhossein Kazerouni, Alaa Sulaiman, Afshin Bozorgpour,
    Ehsan Khodapanah Aghdam, Abin Jose, and Dorit Merhof

Prostate Segmentation Using Multiparametric and Multiplanar Magnetic
Resonance Images 217
    Kuruparan Shanmugalingam, Arcot Sowmya, Daniel Moses,
    and Erik Meijering

SPPNet: A Single-Point Prompt Network for Nuclei Image Segmentation 227
    Qing Xu, Wenwei Kuang, Zeyu Zhang, Xueyao Bao, Haoran Chen,
    and Wenting Duan

Automated Coarse-to-Fine Segmentation of Thoracic Duct Using
Anatomy Priors and Topology-Guided Curved Planar Reformation 237
    Puyang Wang, Panwen Hu, Jiali Liu, Hang Yu, Xianghua Ye,
    Jinliang Zhang, Hui Li, Li Yang, Le Lu, Dakai Jin,
    and Feng-Ming (Spring) Kong

Leveraging Self-attention Mechanism in Vision Transformers
for Unsupervised Segmentation of Optical Coherence Microscopy White
Matter Images 247
    Mohamad Hawchar and Joël Lefebvre

PE-MED: Prompt Enhancement for Interactive Medical Image
Segmentation 257
    Ao Chang, Xing Tao, Xin Yang, Yuhao Huang, Xinrui Zhou,
    Jiajun Zeng, Ruobing Huang, and Dong Ni

A Super Token Vision Transformer and CNN Parallel Branch Network
for mCNV Lesion Segmentation in OCT Images 267
    Xiang Dong, Hai Xie, Yunlong Sun, Zhenquan Wu, Bao Yang,
    Junlong Qu, Guoming Zhang, and Baiying Lei

Boundary-RL: Reinforcement Learning for Weakly-Supervised Prostate
Segmentation in TRUS Images 277
   Weixi Yi, Vasilis Stavrinides, Zachary M. C. Baum, Qianye Yang,
   Dean C. Barratt, Matthew J. Clarkson, Yipeng Hu, and Shaheer U. Saeed

A Domain-Free Semi-supervised Method for Myocardium Segmentation
in 2D Echocardiography Sequences 289
   Wenming Song, Xing An, Ting Liu, Yanbo Liu, Lei Yu, Jian Wang,
   Yuxiao Zhang, Lei Li, Longfei Cong, and Lei Zhu

Self-training with Domain-Mixed Data for Few-Shot Domain Adaptation
in Medical Image Segmentation Tasks 299
   Yongze Wang, Maurice Pagnucco, and Yang Song

Bridging the Task Barriers: Online Knowledge Distillation Across Tasks
for Semi-supervised Mediastinal Segmentation in CT 310
   Muhammad F. A. Chaudhary, Seyed Soheil Hosseini, R. Graham Barr,
   Joseph M. Reinhardt, Eric A. Hoffman, and Sarah E. Gerard

RelationalUNet for Image Segmentation 320
   Ivaxi Sheth, Pedro H. M. Braga, Shivakanth Sujit, Sahar Dastani,
   and Samira Ebrahimi Kahou

Interpretability-Guided Data Augmentation for Robust Segmentation
in Multi-centre Colonoscopy Data 330
   Valentina Corbetta, Regina Beets-Tan, and Wilson Silva

Improving Automated Prostate Cancer Detection and Classification
Accuracy with Multi-scale Cancer Information 341
   Cynthia Xinran Li, Indrani Bhattacharya, Sulaiman Vesal,
   Sara Saunders, Simon John Christoph Soerensen, Richard E. Fan,
   Geoffrey A. Sonn, and Mirabela Rusu

Skin Lesion Segmentation Improved by Transformer-Based Networks
with Inter-scale Dependency Modeling 351
   Sania Eskandari, Janet Lumpp, and Luis Sanchez Giraldo

MagNET: Modality-Agnostic Network for Brain Tumor Segmentation
and Characterization with Missing Modalities 361
   Aishik Konwer, Chao Chen, and Prateek Prasanna

Unsupervised Anomaly Detection in Medical Images Using Masked
Diffusion Model 372
   Hasan Iqbal, Umar Khalid, Chen Chen, and Jing Hua

IA-GCN: Interpretable Attention Based Graph Convolutional Network
for Disease Prediction 382
  Anees Kazi, Soroush Farghadani, Iman Aganj, and Nassir Navab

Multi-modal Adapter for Medical Vision-and-Language Learning 393
  Zheng Yu, Yanyuan Qiao, Yutong Xie, and Qi Wu

Vector Quantized Multi-modal Guidance for Alzheimer's Disease
Diagnosis Based on Feature Imputation 403
  Yuanwang Zhang, Kaicong Sun, Yuxiao Liu, Zaixin Ou,
  and Dinggang Shen

Finding-Aware Anatomical Tokens for Chest X-Ray Automated Reporting 413
  Francesco Dalla Serra, Chaoyang Wang, Fani Deligianni,
  Jeffrey Dalton, and Alison Q. O'Neil

Dual-Stream Model with Brain Metrics and Images for MRI-Based Fetal
Brain Age Estimation 424
  Shengxian Chen, Xin Zhang, Ruiyan Fang, Wenhao Zhang, He Zhang,
  Chaoxiang Yang, and Gang Li

PECon: Contrastive Pretraining to Enhance Feature Alignment Between
CT and EHR Data for Improved Pulmonary Embolism Diagnosis 434
  Santosh Sanjeev, Salwa K. Al Khatib, Mai A. Shaaban,
  Ibrahim Almakky, Vijay Ram Papineni, and Mohammad Yaqub

Exploring the Transfer Learning Capabilities of CLIP in Domain
Generalization for Diabetic Retinopathy 444
  Sanoojan Baliah, Fadillah A. Maani, Santosh Sanjeev,
  and Muhammad Haris Khan

More from Less: Self-supervised Knowledge Distillation for Routine
Histopathology Data 454
  Lucas Farndale, Robert Insall, and Ke Yuan

Tailoring Large Language Models to Radiology: A Preliminary Approach
to LLM Adaptation for a Highly Specialized Domain 464
  Zhengliang Liu, Aoxiao Zhong, Yiwei Li, Longtao Yang, Chao Ju,
  Zihao Wu, Chong Ma, Peng Shu, Cheng Chen, Sekeun Kim,
  Haixing Dai, Lin Zhao, Dajiang Zhu, Jun Liu, Wei Liu, Dinggang Shen,
  Quanzheng Li, Tianming Liu, and Xiang Li

Author Index 475

Contents – Part II

GEMTrans: A General, Echocardiography-Based, Multi-level Transformer
Framework for Cardiovascular Diagnosis 1
   Masoud Mokhtari, Neda Ahmadi, Teresa S. M. Tsang,
   Purang Abolmaesumi, and Renjie Liao

Unsupervised Anomaly Detection in Medical Images
with a Memory-Augmented Multi-level Cross-Attentional Masked
Autoencoder 11
   Yu Tian, Guansong Pang, Yuyuan Liu, Chong Wang, Yuanhong Chen,
   Fengbei Liu, Rajvinder Singh, Johan W. Verjans, Mengyu Wang,
   and Gustavo Carneiro

LMT: Longitudinal Mixing Training, a Framework to Predict Disease
Progression from a Single Image 22
   Rachid Zeghlache, Pierre-Henri Conze, Mostafa El Habib Daho,
   Yihao Li, Hugo Le Boité, Ramin Tadayoni, Pascal Massin,
   Béatrice Cochener, Ikram Brahim, Gwenolé Quellec,
   and Mathieu Lamard

Identifying Alzheimer's Disease-Induced Topology Alterations
in Structural Networks Using Convolutional Neural Networks 33
   Feihong Liu, Yongsheng Pan, Junwei Yang, Fang Xie, Xiaowei He,
   Han Zhang, Feng Shi, Jun Feng, Qihao Guo, and Dinggang Shen

Specificity-Aware Federated Graph Learning for Brain Disorder Analysis
with Functional MRI 43
   Junhao Zhang, Xiaochuan Wang, Qianqian Wang, Lishan Qiao,
   and Mingxia Liu

3D Transformer Based on Deformable Patch Location for Differential
Diagnosis Between Alzheimer's Disease and Frontotemporal Dementia 53
   Huy-Dung Nguyen, Michaël Clément, Boris Mansencal,
   and Pierrick Coupé

Consisaug: A Consistency-Based Augmentation for Polyp Detection
in Endoscopy Image Analysis 64
   Ziyu Zhou, Wenyuan Shen, and Chang Liu

Cross-view Contrastive Mutual Learning Across Masked Autoencoders
for Mammography Diagnosis 74
   Qingxia Wu, Hongna Tan, Zhi Qiao, Pei Dong, Dinggang Shen,
   Meiyun Wang, and Zhong Xue

Modeling Life-Span Brain Age from Large-Scale Dataset Based
on Multi-level Information Fusion 84
   Nan Zhao, Yongsheng Pan, Kaicong Sun, Yuning Gu, Mianxin Liu,
   Zhong Xue, Han Zhang, Qing Yang, Fei Gao, Feng Shi,
   and Dinggang Shen

Boundary-Constrained Graph Network for Tooth Segmentation on 3D
Dental Surfaces 94
   Yuwen Tan and Xiang Xiang

FAST-Net: A Coarse-to-fine Pyramid Network for Face-Skull
Transformation 104
   Lei Zhao, Lei Ma, Zhiming Cui, Jie Zheng, Zhong Xue, Feng Shi,
   and Dinggang Shen

Mixing Histopathology Prototypes into Robust Slide-Level
Representations for Cancer Subtyping 114
   Joshua Butke, Noriaki Hashimoto, Ichiro Takeuchi, Hiroaki Miyoshi,
   Koichi Ohshima, and Jun Sakuma

Consistency Loss for Improved Colonoscopy Landmark Detection
with Vision Transformers 124
   Aniruddha Tamhane, Daniel Dobkin, Ore Shtalrid, Moshe Bouhnik,
   Erez Posner, and Tse'ela Mida

Radiomics Boosts Deep Learning Model for IPMN Classification 134
   Lanhong Yao, Zheyuan Zhang, Ugur Demir, Elif Keles,
   Camila Vendrami, Emil Agarunov, Candice Bolan, Ivo Schoots,
   Marc Bruno, Rajesh Keswani, Frank Miller, Tamas Gonda,
   Cemal Yazici, Temel Tirkes, Michael Wallace, Concetto Spampinato,
   and Ulas Bagci

Class-Balanced Deep Learning with Adaptive Vector Scaling Loss
for Dementia Stage Detection 144
   Boning Tong, Zhuoping Zhou, Davoud Ataee Tarzanagh, Bojian Hou,
   Andrew J. Saykin, Jason Moore, Marylyn Ritchie, and Li Shen

Enhancing Anomaly Detection in Melanoma Diagnosis Through
Self-Supervised Training and Lesion Comparison 155
Jules Collenne, Rabah Iguernaissi, Séverine Dubuisson,
and Djamal Merad

DynBrainGNN: Towards Spatio-Temporal Interpretable Graph Neural
Network Based on Dynamic Brain Connectome for Psychiatric Diagnosis 164
Kaizhong Zheng, Bin Ma, and Badong Chen

Precise Localization Within the GI Tract by Combining Classification
of CNNs and Time-Series Analysis of HMMs 174
Julia Werner, Christoph Gerum, Moritz Reiber, Jörg Nick,
and Oliver Bringmann

Towards Unified Modality Understanding for Alzheimer's Disease
Diagnosis Using Incomplete Multi-modality Data 184
Kangfu Han, Fenqiang Zhao, Dajiang Zhu, Tianming Liu, Feng Yang,
and Gang Li

COVID-19 Diagnosis Based on Swin Transformer Model
with Demographic Information Fusion and Enhanced Multi-head
Attention Mechanism 194
Yunlong Sun, Yiyao Liu, Junlong Qu, Xiang Dong, Xuegang Song,
and Baiying Lei

MoViT: Memorizing Vision Transformers for Medical Image Analysis 205
Yiqing Shen, Pengfei Guo, Jingpu Wu, Qianqi Huang, Nhat Le,
Jinyuan Zhou, Shanshan Jiang, and Mathias Unberath

Fact-Checking of AI-Generated Reports 214
Razi Mahmood, Ge Wang, Mannudeep Kalra, and Pingkun Yan

Is Visual Explanation with Grad-CAM More Reliable for Deeper Neural
Networks? A Case Study with Automatic Pneumothorax Diagnosis 224
Zirui Qiu, Hassan Rivaz, and Yiming Xiao

Group Distributionally Robust Knowledge Distillation 234
Konstantinos Vilouras, Xiao Liu, Pedro Sanchez, Alison Q. O'Neil,
and Sotirios A. Tsaftaris

A Bone Lesion Identification Network (BLIN) in CT Images with Weakly
Supervised Learning 243
Kehao Deng, Bin Wang, Shanshan Ma, Zhong Xue, and Xiaohuan Cao

Post-Deployment Adaptation with Access to Source Data via Federated
Learning and Source-Target Remote Gradient Alignment 253
  Felix Wagner, Zeju Li, Pramit Saha, and Konstantinos Kamnitsas

Data-Driven Classification of Fatty Liver From 3D Unenhanced
Abdominal CT Scans 264
  Jacob S. Leiby, Matthew E. Lee, Eun Kyung Choe, and Dokyoon Kim

Replica-Based Federated Learning with Heterogeneous Architectures
for Graph Super-Resolution 273
  Ramona Ghilea and Islem Rekik

A Multitask Deep Learning Model for Voxel-Level Brain Age Estimation 283
  Neha Gianchandani, Johanna Ospel, Ethan MacDonald,
  and Roberto Souza

Deep Nearest Neighbors for Anomaly Detection in Chest X-Rays 293
  Xixi Liu, Jennifer Alvén, Ida Häggström, and Christopher Zach

CCMix: Curriculum of Class-Wise Mixup for Long-Tailed Medical Image
Classification 303
  Sirui Li, Fuheng Zhang, Tianyunxi Wei, Li Lin, Yijin Huang,
  Pujin Cheng, and Xiaoying Tang

MEDKD: Enhancing Medical Image Classification with Multiple Expert
Decoupled Knowledge Distillation for Long-Tail Data 314
  Fuheng Zhang, Sirui Li, Tianyunxi Wei, Li Lin, Yijin Huang,
  Pujin Cheng, and Xiaoying Tang

Leveraging Ellipsoid Bounding Shapes and Fast R-CNN for Enlarged
Perivascular Spaces Detection and Segmentation 325
  Mariam Zabihi, Chayanin Tangwiriyasakul, Silvia Ingala,
  Luigi Lorenzini, Robin Camarasa, Frederik Barkhof,
  Marleen de Bruijne, M. Jorge Cardoso, and Carole H. Sudre

Non-uniform Sampling-Based Breast Cancer Classification 335
  Santiago Posso Murillo, Oscar Skean, and Luis G. Sanchez Giraldo

A Scaled Denoising Attention-Based Transformer for Breast Cancer
Detection and Classification 346
  Masum Shah Junayed and Sheida Nabavi

Distilling Local Texture Features for Colorectal Tissue Classification
in Low Data Regimes 357
  Dmitry Demidov, Roba Al Majzoub, Amandeep Kumar, and Fahad Khan

Delving into Ipsilateral Mammogram Assessment Under Multi-view
Network 367
  Toan T. N. Truong, Huy T. Nguyen, Thinh B. Lam, Duy V. M. Nguyen,
  and Phuc H. Nguyen

ARHNet: Adaptive Region Harmonization for Lesion-Aware
Augmentation to Improve Segmentation Performance 377
  Jiayu Huo, Yang Liu, Xi Ouyang, Alejandro Granados,
  Sébastien Ourselin, and Rachel Sparks

Normative Aging for an Individual's Full Brain MRI Using Style GANs
to Detect Localized Neurodegeneration 387
  Shruti P. Gadewar, Alyssa H. Zhu, Sunanda Somu, Abhinaav Ramesh,
  Iyad Ba Gari, Sophia I. Thomopoulos, Paul M. Thompson,
  Talia M. Nir, and Neda Jahanshad

Deep Bayesian Quantization for Supervised Neuroimage Search 396
  Erkun Yang, Cheng Deng, and Mingxia Liu

Triplet Learning for Chest X-Ray Image Search in Automated COVID-19
Analysis 407
  Linmin Wang, Qianqian Wang, Xiaochuan Wang, Yunling Ma,
  Lishan Qiao, and Mingxia Liu

Cascaded Cross-Attention Networks for Data-Efficient Whole-Slide
Image Classification Using Transformers 417
  Firas Khader, Jakob Nikolas Kather, Tianyu Han, Sven Nebelung,
  Christiane Kuhl, Johannes Stegmaier, and Daniel Truhn

Enhanced Diagnostic Fidelity in Pathology Whole Slide Image
Compression via Deep Learning 427
  Maximilian Fischer, Peter Neher, Peter Schüffler, Shuhan Xiao,
  Silvia Dias Almeida, Constantin Ulrich, Alexander Muckenhuber,
  Rickmer Braren, Michael Götz, Jens Kleesiek, Marco Nolden,
  and Klaus Maier-Hein

RoFormer for Position Aware Multiple Instance Learning in Whole Slide
Image Classification 437
  Etienne Pochet, Rami Maroun, and Roger Trullo

Structural Cycle GAN for Virtual Immunohistochemistry Staining
of Gland Markers in the Colon 447
  Shikha Dubey, Tushar Kataria, Beatrice Knudsen,
  and Shireen Y. Elhabian

NCIS: Deep Color Gradient Maps Regression and Three-Class Pixel
Classification for Enhanced Neuronal Cell Instance Segmentation
in Nissl-Stained Histological Images 457
    Valentina Vadori, Antonella Peruffo, Jean-Marie Graïc, Livio Finos,
    Livio Corain, and Enrico Grisan

Regionalized Infant Brain Cortical Development Based on Multi-view,
High-Level fMRI Fingerprint 467
    Tianli Tao, Jiawei Huang, Feihong Liu, Mianxin Liu, Lianghu Guo,
    Xinyi Cai, Zhuoyang Gu, Haifeng Tang, Rui Zhou, Siyan Han,
    Lixuan Zhu, Qing Yang, Dinggang Shen, and Han Zhang

Author Index 477

Structural MRI Harmonization via Disentangled Latent Energy-Based Style Translation

Mengqi Wu[1,2], Lintao Zhang[1], Pew-Thian Yap[1], Weili Lin[1], Hongtu Zhu[3], and Mingxia Liu[1(✉)]

[1] Department of Radiology and Biomedical Research Imaging Center, University of North Carolina at Chapel Hill, Chapel Hill, NC 27599, USA
mingxia_liu@med.unc.edu
[2] Joint Department of Biomedical Engineering, University of North Carolina at Chapel Hill and North Carolina State University, Chapel Hill, NC 27599, USA
[3] Department of Biostatistics and Biomedical Research Imaging Center, University of North Carolina at Chapel Hill, Chapel Hill, NC 27599, USA

Abstract. Multi-site brain magnetic resonance imaging (MRI) has been widely used in clinical and research domains, but usually is sensitive to non-biological variations caused by site effects (*e.g.*, field strengths and scanning protocols). Several retrospective data harmonization methods have shown promising results in removing these non-biological variations at feature or whole-image level. Most existing image-level harmonization methods are implemented through generative adversarial networks, which are generally computationally expensive and generalize poorly on independent data. To this end, this paper proposes a disentangled latent energy-based style translation (DLEST) framework for image-level structural MRI harmonization. Specifically, DLEST disentangles *site-invariant image generation* and *site-specific style translation* via a latent autoencoder and an energy-based model. The autoencoder learns to encode images into low-dimensional latent space, and generates faithful images from latent codes. The energy-based model is placed in between the encoding and generation steps, facilitating style translation from a source domain to a target domain implicitly. This allows *highly generalizable image generation and efficient style translation* through the latent space. We train our model on 4,092 T1-weighted MRIs in 3 tasks: histogram comparison, acquisition site classification, and brain tissue segmentation. Qualitative and quantitative results demonstrate the superiority of our approach, which generally outperforms several state-of-the-art methods.

Keywords: MRI harmonization · style translation · energy-based model

Supplementary Information The online version contains supplementary material available at https://doi.org/10.1007/978-3-031-45673-2_1.

1   Introduction

Structural magnetic resonance imaging (MRI), due to its non-invasive nature and high resolution, has been widely used in clinical diagnoses and various research fields. Many retrospective studies propose to train learning-based models on multi-source MRI data, pooled from different acquisition sites, to increase models' statistical power [1–3]. But they usually suffer from poor generalizability to independent data, as MRI is sensitive to non-biological variations caused by *site effects* such as differences in scanner vendors and field strengths [4–7].

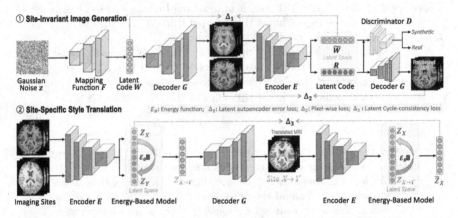

Fig. 1. Illustration of the disentangled latent energy-based style translation (DLEST) framework for structural MRI harmonization at image level.

Existing MRI harmonization methods can be roughly classified into two categories: (1) *feature-level* approach and (2) *image-level* approach. Feature-level methods generally rely on empirical Bayes models with pre-extraction of image, biological, or radiomic features [8,9], limiting their adaptivity to applications that employ different MRI features. In contrast, image-level harmonization targets broader applications and is independent of specific MRI feature representations, which is the focus of this work. Many studies have demonstrated the great potential of generative adversarial networks (GANs) for image-level harmonization [10–15]. Several state-of-the-art methods (*e.g.*, CycleGAN [10], StarGAN [11], and ImUnity [16]) formulate the harmonization as a pixel/voxel-level image-to-image translation task. The objective is to translate the style of one image to another while preserving the content. In this context, *style information* represents high-level features such as contrast, textures, intensity variation, and signal-to-noise ratio, while *content information* represents low-level image features such as contours, edges, orientations, and anatomical structures. However, existing research usually requires substantial time and computational cost for model training due to optimization of a huge number of network parameters [17]. The cost further increases with the addition of cycle-consistency calculation and style translation in high-dimensional image spaces [18]. And they often have to be entirely retrained when applied to new data, limiting their utility in practice.

To this end, we design a disentangled latent energy-based style translation (**DLEST**) framework for image-level structural MRI harmonization. As shown in Fig. 1, DLEST contains *site-invariant image generation* with a latent autoencoder that encodes images into lower-dimensional latent space and generates images from latent codes, and *site-specific style translation* with an energy-based model [19,20] that operates in between the autoencoder, enabling implicit cross-domain style translation via latent space. We evaluate the proposed DLEST on 4,092 T1-weighted (T1-w) MRIs, with results suggesting its superiority over state-of-the-art (SOTA) image-level harmonization methods.

2 Proposed Method

Problem Formulation. The goal of our image harmonization framework is to translate the style (*i.e.*, non-biological variation) from a source domain X to a target domain Y. In the context of a multi-site MRI dataset, scans from one acquisition site/setting will be treated as the *target domain* and the rest will serve as *source domain(s)*. Given a source image $\mathbf{x} \in X$ and a target image $\mathbf{y} \in Y$, we aim to first train a latent autoencoder to map both images to their latent codes (*i.e.*, Z_x and Z_y) through the mapping $\boldsymbol{E}: \{\mathbf{x}, \mathbf{y}\} \rightarrow \{Z_x, Z_y\}$. A latent code $Z = \{S, C\}$ is a lower-dimensional representation of an image, containing implicit *style code S* and *content code C*. By translating the style code S_x of the source domain to the target domain, while maintaining its content code C_x, we can formulate the style translation as a mapping $\boldsymbol{T}: Z_x = \{S_x, C_x\} \rightarrow Z_{x \rightarrow y} = \{S_y, C_x\}$, where $Z_{x \rightarrow y}$ is the latent code of the source image in target domain. Then, we propose to learn another mapping $\boldsymbol{G}: Z_{x \rightarrow y} \rightarrow \tilde{\mathbf{y}}$ via the autoencoder to decode the translated latent code back to the image space, resulting in an image $\tilde{\mathbf{y}}$ in the target domain with the style of $\mathbf{y}$ and the content of $\mathbf{x}$. Thus, the style of source images can be translated into target domain while keeping their content. As shown in Fig. 1, DLEST contains a site-invariant image generation (SIG) module and a site-specific style translation (SST) module.

Site-Invariant Image Generation (SIG). This SIG module is designed to map images into lower-dimensional latent space and generate images from latent codes, which is *independent of site specification*. Specifically, the input of SIG contains a random Gaussian noise z and a real MR image $\mathbf{x}$. A deterministic mapping network $\boldsymbol{F}$ is used to map z into a latent code $\mathcal{W} = F(z)$ [21], followed by a stochastic decoder $\boldsymbol{G}$ that takes the latent code $\mathcal{W}$ with an independent noise $\eta \sim N(0, 1)$ to generate a synthetic image $\tilde{\mathbf{x}} = G(\mathcal{W}, \eta)$. Then, an encoder $\boldsymbol{E}$ is employed to map the synthetic image $\tilde{\mathbf{x}}$ back to latent code $\tilde{\mathcal{W}} = E(\tilde{\mathbf{x}}) = E \circ G \circ F(z)$. The encoder $\boldsymbol{E}$ can also map the real image $\mathbf{x} \sim p_{Data}$ into its latent code $\mathcal{R} = E(\mathbf{x})$. Finally, a discriminator $\boldsymbol{D}$ takes a latent code and outputs a scalar value, indicating if that latent code belongs to a real or synthetic image. We also decode the latent code of the real image back to a recovered image via a decoder $\boldsymbol{G}$, aiming to preserve the content (*e.g.*, anatomic structure) of the real image. To ensure the reciprocity of the autoencoder in latent space, we encourage that $\mathcal{W} = \tilde{\mathcal{W}}$ by minimizing a *latent autoencoder error loss* as:

$$\mathcal{L}_{lae}^{E,G} = \Delta_1(F \| E \circ G \circ F) = \mathbb{E}_{z \sim P_z}[\|F(z) - E \circ G \circ F(z)\|_2^2] \tag{1}$$

through which the pair of networks E and G can be considered as a *latent autoencoder* that autoencodes the latent code $\mathcal{W}$. Instead of functioning in high-dimensional image space like typical autoencoders [22], this helps our method perform image generation in the low-dimensional latent code space, thus improving computational efficiency. To ensure the content preservation of latent autoencoder E and G, we also impose a *pixel-wise loss* by comparing the real image $\mathbf{x}$ with its recovered version $\tilde{\mathbf{x}}$ as:

$$\mathcal{L}_{pix}^{E,G} = \Delta_2(\mathbf{x} \| \tilde{\mathbf{x}}) = \mathbb{E}_{x \sim P_{Data}}[\|\mathbf{x} - G \circ E(\mathbf{x})\|_1] \tag{2}$$

which encourages the model to reconstruct MRI with minimal content loss.

To generate MRI-like images, we optimize an *adversarial loss*:

$$\mathcal{L}_{adv}^{E,D} = \Phi(D \circ E \circ G \circ F(z)) + \Phi(-D \circ E(\mathbf{x})) + R_1 \tag{3}$$

$$\mathcal{L}_{adv}^{F,G} = \Phi(-D \circ E \circ G \circ F(z)) \tag{4}$$

where Φ is the *softplus* function defined as $\Phi(t) = \log(1 + \exp(t))$, R_1 is a zero-centered gradient penalty term defined as $\frac{\gamma}{2}\mathbb{E}_{x \sim P_{Data}}[\|\nabla D \circ E(\mathbf{x})\|^2]$ [21].

Thus, the **full objective functions** of our SIG module can be formulated as:

$$\min_{F,G} \max_{E,D} \mathcal{L}_{adv}^{E,D} + \mathcal{L}_{adv}^{F,G}, \quad \min_{E,G} \mathcal{L}_{lae}^{E,G} + \mathcal{L}_{pix}^{E,G} \tag{5}$$

Site-Specific Style Translation (SST). As illustrated in the bottom panel of Fig. 1, the SST module is composed of (1) the autoencoder (with E and G trained in the proposed SIG module) and (2) an energy-based model (EBM). Given a source image $\mathbf{x} \in X$ and a target image $\mathbf{y} \in Y$, the encoder E first maps them into their corresponding latent codes $Z_x = E(\mathbf{x})$ and $Z_y = E(\mathbf{y})$. The EBM takes these latent codes as input and outputs the translated latent code $Z_{x \to y}$ that will be decoded by G to generate the translated image. We further use the EBM to reversely translate $Z_{x \to y}$ to get a recovered latent code $\tilde{Z}_x$ that is compared to the original Z_x, which helps ensure cycle-consistency to prevent mode collapse during unpaired image translation. The EBM is introduced below.

By observing latent samples from the target site, the EBM aims to model a distribution $P_\theta(Z_y) \sim P_{Data}(Y)$ to represent the data distribution of target samples in latent space. In particular, $P_\theta(Z_y)$ is assumed to follow a Gibbs distribution [23], defined as $P_\theta(Z_y) = \frac{1}{Q(\theta)} \exp(-\mathcal{E}_\theta(Z_y))$, where $\mathcal{E}_\theta(Z_y) : \mathbb{R}^D \to \mathbb{R}$ is a scalar energy function parameterized by θ. Here, $\mathcal{E}_\theta(Z_y)$ is learned to assign low energy values to inputs from the target distribution and high energy values to others. And $Q(\theta) = \int \exp(-\mathcal{E}_\theta(Z_y)) \, dZ_y$ is the intractable partition function.

The EBM can be trained by maximizing the derivative of the negative log-likelihood or minimizing its inverse as follows:

$$\mathcal{L}_{EBM} = -(\frac{\partial}{\partial \theta} - L(\theta)) = -(\mathbb{E}_{Z_y \sim P_{Data}}[\frac{\partial}{\partial \theta}\mathcal{E}_\theta(Z_y)] - \mathbb{E}_{\tilde{Z}_y \sim P_\theta}[\frac{\partial}{\partial \theta}\mathcal{E}_\theta(\tilde{Z}_y)]) \quad (6)$$

where the 2nd expectation term requires direct sampling from intractable model distribution P_θ. It can be approximated using Stochastic Gradient Langevin Dynamics (SGLD) [24] sampling, which iteratively updates the following:

$$\tilde{Z}_y^{t+1} = \tilde{Z}_y^t - \frac{\eta^t}{2}\frac{\partial}{\partial \tilde{Z}_y^t}\mathcal{E}_\theta(\tilde{Z}_y^t) + \sqrt{\eta^t}\epsilon^t, \ \epsilon^t \sim \mathcal{N}(0, I) \quad (7)$$

where $t = [0, \cdots, T]$ is the sampling step, η denotes the step size, and ϵ is an injected Gaussian noise to capture the data uncertainty and to ensure sample convergence [19]. Here, $\tilde{Z}_y^0 = Z_x = E(\mathbf{x})$, which means the initial samples $\tilde{Z}_y^0$ are the latent codes from the source site X. The entire SGLD process aims to iteratively update the latent code using the negative gradients of the energy function $\mathcal{E}_\theta$. Finally, $\tilde{Z}_y^T$ at the end of T SGLD steps can be decoded by G to generate a translated image $\tilde{y} = G(Z_{x \to y}) = G(\tilde{Z}_y^T)$. Given that the EBM does not explicitly separate the content code and the style code during the style translation, we additionally regularize EBM with a **latent content loss**, which compares latent code $Z_{x \to y}$ to the original source latent code Z_x, formulated as:

$$\mathcal{L}_{con} = \mathbb{E}_{Z_x \sim P_{Data}, Z_{x \to y} \sim P_\theta}[\|Z_x - Z_{x \to y}\|_1] \quad (8)$$

To ensure bidirectional latent style translation and prevent mode collapse for unpaired image translation, we define a **latent cycle-consistency loss** as:

$$\mathcal{L}_{cyc} = \mathbb{E}_{Z_x \sim P_{Data}}[\|Z_x - P_\theta^{-1} \circ E \circ G \circ P_\theta(Z_x)\|_1] \quad (9)$$

where P_θ^{-1} denotes the same EBM as P_θ, but with an opposite SGLD sampling process. It updates the latent code using positive energy gradients, rather than negative gradients used by P_θ. The objective function of SST can be written as:

$$\mathcal{L}(\theta) = \mathcal{L}_{EBM} + \alpha\mathcal{L}_{con} + \beta\mathcal{L}_{cyc} \quad (10)$$

where α and β are hyperparameters.

Implementation. The F and D are implemented as multi-layer perceptions (MLPs) with 8 and 3 layers, respectively. The latent autoencoder E and G has 7 encoding/decoding blocks, with each block containing a minimum of 64 and a maximum of 256 filters. The dimension of all latent codes is 512. The SIG is trained using Adam [25] in a *fully unsupervised manner*. After training, the latent autoencoder E and G in SIG can be directly used in SST, without updating their parameters. The training of the SST is also *time and computationally efficient*, since (1) the autoencoder E and G do not require further training, even when applied to unseen datasets; (2) the style translation process via EBM is performed in a low-dimensional (*i.e.*, 512) latent space; and (3) the energy function $\mathcal{E}_\theta$ in EBM is lightweight, implemented as a 2-layer MLP.

3   Experiments

Materials. The Open Big Healthy Brains (OpenBHB) [26] and Strategic Research Program for Brain Science (SRPBS) [27] are used. The OpenBHB contains T1-w MRIs from healthy controls gathered from ≥ 60 centers. It has been split into a training set for SIG training (with 3,227 subjects from 58 acquisition sites/settings) and a validation set (with 757 subjects). The SRPBS contains 108 T1-w MRIs from 9 healthy traveling subjects with 11 sites/settings. More details can be found in the *Supplementary Materials*. All scans were minimally preprocessed using FSL [28], including bias field correction, brain extraction, and registration to the $1mm^3$ MNI-152 template with 9°C of freedom. Images from SRPBS were further segmented by FSL into cerebrospinal fluid (CSF), gray matter (GM), and white matter (WM). We select 10 axial slices from the middle of each MRI volume in the OpenBHB, and 15 axial slices from each volume in the SRPBS. All 2D slices were zero-padded to the size of 256×256.

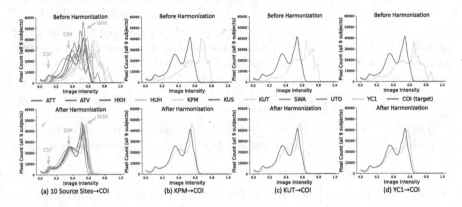

Fig. 2. Histogram comparison of (a) all 10 source sites and (b)-(d) 3 source sites across 9 subjects before and after harmonization by DLEST on SRPBS (COI as target site).

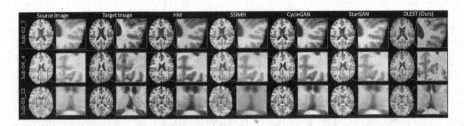

Fig. 3. MRIs of 3 subjects in SRPBS, harmonized from HUH to COI by five methods.

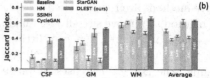

Fig. 4. Segmentation results of three tissues on SRPBS in terms of (a) Dice coefficient and (b) Jaccard index, with COI and HUH as target and source sites.

Table 1. Site classification results on OpenBHB, where '*' denotes the difference between a specific competing method and DLEST is statistically significant ($p < 0.05$).

Method	BACC↓	ACC↓	AUC↓	F1↓	SEN↓
Baseline	$0.676 \pm 0.137^*$	0.722 ± 0.147	0.979 ± 0.022	0.713 ± 0.146	0.722 ± 0.147
HM	$0.619 \pm 0.133^*$	0.696 ± 0.137	0.981 ± 0.019	0.685 ± 0.129	0.696 ± 0.137
SSIMH	$0.673 \pm 0.083^*$	0.659 ± 0.076	0.983 ± 0.008	0.656 ± 0.079	0.659 ± 0.076
CycleGAN	$0.588 \pm 0.081^*$	0.660 ± 0.088	0.974 ± 0.007	0.645 ± 0.088	0.660 ± 0.088
StarGAN	$0.489 \pm 0.201^*$	0.562 ± 0.223	0.951 ± 0.047	0.543 ± 0.215	0.562 ± 0.223
DLEST (Ours)	$\mathbf{0.336 \pm 0.057}$	$\mathbf{0.483 \pm 0.065}$	$\mathbf{0.938 \pm 0.020}$	$\mathbf{0.474 \pm 0.069}$	$\mathbf{0.483 \pm 0.065}$

Experimental Setting. We compare DLEST with 2 non-learning methods: histogram matching (**HM**) [29], **SSIMH** [30], and 2 state-of-the-art (SOTA) methods: **StarGAN** [11] and **CycleGAN** [10]. Three tasks are performed, including histogram comparison, site classification, and brain tissue segmentation.

Task 1: Histogram Comparison. For qualitative evaluation, we compare histograms and MRIs before and after harmonization via DLEST on the SRPBS dataset with traveling subjects that have ground truth scans. We select the site COI as target since it has relatively low within-site variation, and harmonize the rest 10 sites/settings to COI. As shown in Fig. 2, there is increased overlapping between the histogram of each source site and the target after harmonization. In most cases, three prominent histogram peaks (corresponding to CSF, GM, and WM) are better aligned to those of the target, indicating the effectiveness of our DLEST. Also, Fig. 3 suggests that the overall style of harmonized MRIs generated by DLEST is more consistent with target images.

Task 2: Site Classification. We quantitatively assess whether a harmonization method can effectively remove non-biological variations. We first harmonize all data from 58 sites in OpenBHB to a pre-selected target site (ID: 17) using each method. We then train a ResNet18 [31] as site classifier on training data of OpenBHB and validate it on validation data. We record the performance of ResNet18 on raw data as Baseline. Five metrics are used: balanced accuracy (BACC), accuracy (ACC), area under the ROC curve (AUC), F1 score (F1), and sensitivity (SEN). The site classification results based on data harmonized by five methods are reported in Table 1. A lower value indicates a stronger ability of a method to remove site effects, making it more difficult for ResNet18 to learn non-biological features that can distinguish each site. Table 1 suggests that our DLEST consistently outperforms all competing methods by a large margin. For example, the BACC decreases by 0.340 when trained on data harmonized by DLEST compared to raw data, and decreases by 0.153 compared to the second-best method (*i.e.*, StarGAN). Compared with CycleGAN and StarGAN, our DLEST yields very good results in terms of all five metrics. This implies that our method generalizes well to multi-site MRIs. The possible reason could be that our method can capture the underlying distribution of source and target sites in latent space, instead of relying on a single reference/target image.

Task 3: Brain Tissue Segmentation. This experiment aims to validate the impact of DLEST on a brain tissue segmentation task. We first harmonize a selected source site (*i.e.*, HUH) from SRPBS to a target site space (*i.e.*, COI) using each harmonization method, and then train a U-Net [32] segmentation model on images from COI and directly validate the trained U-Net on HUH. The auto-segmentation outputs from FSL [28] are used as the ground truth. The U-Net performance on unharmonized HUH data serves as the baseline. Dice coefficient and Jaccard index are used here. The class-wise and average brain tissue segmentation results are shown in Fig. 4. Overall, our DLEST outperforms all non-learning methods, indicating that fine-grained image-level style translation performed by DLEST leads to a better segmentation result than simply aligning the global intensity distribution. This corroborates the histogram alignments shown in Fig. 2. Compared to SOTA methods, DLEST generally performs better on the average segmentation result and on CSF and GM segmentations.

Ablation Study. We study three degenerated variants of DLEST: (1) DLEST-PIX without pixel-wise loss in Eq. (2) during image generation, (2) DLEST-CON without content loss in Eq. (8), and (3) DLEST-CYC without latent cycle-consistency loss in Eq. (9) during style translation. Figure 5 indicates that, although all variants fall short of the performance, DLEST-PIX in particular struggles to yield satisfactory results when compared to its two counterparts. This implies that pixel-wise loss plays a crucial role in preserving image content in DLEST.

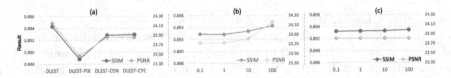

Fig. 5. Results of (a) DLEST variants and DLEST with different (b) α and (c) β.

Parameter Analysis. To investigate the impact of α and β in Eq. (10), we vary one parameter each time while keeping the other unchanged and record the results of DLEST. We use two image-level metrics to assess the harmonization: structural similarity index measurement (SSIM) [33] and peak signal-to-noise ratio (PSNR). The result in Fig. 5 indicates that the model would benefit when the latent content loss carries more weight (*i.e.*, with larger α) in the overall loss function. Besides, the performance of DLEST is less affected by the choice of β.

4   Conclusion

We propose a disentangled latent energy-based style translation framework for MRI harmonization. By disentangling site-invariant image generation and site-specific style translation in latent space, our DLEST can achieve generalizable and efficient style translation on independent data. Experimental results demonstrate that DLEST outperforms several state-of-the-art methods.

Acknowledgement. This work was partly supported by NIH grant AG073297.

References

1. An, L., et al.: Goal-specific brain MRI harmonization. Neuroimage **263**, 119570 (2022)
2. Tofts, P., Collins, D.: Multicentre imaging measurements for oncology and in the brain. Br. J. Radiol. **84**, S213–S226 (2011)
3. Schnack, H.G., et al.: Mapping reliability in multicenter MRI: voxel-based morphometry and cortical thickness. Hum. Brain Mapp. **31**(12), 1967–1982 (2010)
4. Glocker, B., Robinson, R., Castro, D.C., Dou, Q., Konukoglu, E.: Machine learning with multi-site imaging data: an empirical study on the impact of scanner effects. arXiv preprint arXiv:1910.04597 (2019)
5. Wachinger, C., Rieckmann, A., Pölsterl, S.: Detect and correct bias in multi-site neuroimaging datasets. Med. Image Anal. **67**, 101879 (2021)
6. Helmer, K.G., et al.: Multi-site study of diffusion metric variability: characterizing the effects of site, vendor, field strength, and echo time using the histogram distance. In: Medical Imaging 2016: Biomedical Applications in Molecular, Structural, and Functional Imaging, vol. 9788, pp. 363–373. SPIE (2016)
7. Guan, H., Liu, M.: Domain adaptation for medical image analysis: a survey. IEEE Trans. Biomed. Eng. **69**(3), 1173–1185 (2021)
8. Johnson, W.E., Li, C., Rabinovic, A.: Adjusting batch effects in microarray expression data using empirical Bayes methods. Biostatistics **8**(1), 118–127 (2007)
9. Pomponio, R., et al.: Harmonization of large MRI datasets for the analysis of brain imaging patterns throughout the lifespan. Neuroimage **208**, 116450 (2020)
10. Zhu, J.Y., Park, T., Isola, P., Efros, A.A.: Unpaired image-to-image translation using cycle-consistent adversarial networks. In: Proceedings of the IEEE International Conference on Computer Vision, pp. 2223–2232 (2017)
11. Choi, Y., Uh, Y., Yoo, J., Ha, J.W.: StarGAN V2: diverse image synthesis for multiple domains. In: Proceedings of the IEEE/CVF Conference on Computer Vision and Pattern Recognition, pp. 8188–8197 (2020)

12. Liu, M., et al.: Style transfer using generative adversarial networks for multi-site MRI harmonization. In: de Bruijne, M., et al. (eds.) MICCAI 2021. LNCS, vol. 12903, pp. 313–322. Springer, Cham (2021). https://doi.org/10.1007/978-3-030-87199-4_30
13. Sinha, S., Thomopoulos, S.I., Lam, P., Muir, A., Thompson, P.M.: Alzheimer's disease classification accuracy is improved by MRI harmonization based on attention-guided generative adversarial networks. In: International Symposium on Medical Information Processing and Analysis, vol. 12088, pp. 180–189. SPIE (2021)
14. Guan, H., Liu, Y., Yang, E., Yap, P.T., Shen, D., Liu, M.: Multi-site MRI harmonization via attention-guided deep domain adaptation for brain disorder identification. Med. Image Anal. **71**, 102076 (2021)
15. Guan, H., Liu, S., Lin, W., Yap, P.T., Liu, M.: Fast image-level MRI harmonization via spectrum analysis. In: Lian, C., Cao, X., Rekik, I., Xu, X., Cui, Z. (eds.) Machine Learning in Medical Imaging. MLMI 2022. Lecture Notes in Computer Science, vol. 13583, pp. 201–209. Springer, Cham (2022). https://doi.org/10.1007/978-3-031-21014-3_21
16. Cackowski, S., Barbier, E.L., Dojat, M., Christen, T.: ImUnity: a generalizable VAE-GAN solution for multicenter MR image harmonization. Med. Image Anal. **88**, 102799 (2023)
17. Gulrajani, I., Ahmed, F., Arjovsky, M., Dumoulin, V., Courville, A.C.: Improved training of Wasserstein GANs. In: Advances in Neural Information Processing Systems, vol. 30 (2017)
18. Kwon, T., Ye, J.C.: Cycle-free CycleGAN using invertible generator for unsupervised low-dose CT denoising. IEEE Trans. Comput. Imaging **7**, 1354–1368 (2021)
19. LeCun, Y., Chopra, S., Hadsell, R., Ranzato, M., Huang, F.: A tutorial on energy-based learning. Predicting Structured Data, MIT Press (2006)
20. Zhao, Y., Chen, C.: Unpaired image-to-image translation via latent energy transport. In: Proceedings of the IEEE/CVF Conference on Computer Vision and Pattern Recognition, pp. 16418–16427 (2021)
21. Pidhorskyi, S., Adjeroh, D.A., Doretto, G.: Adversarial latent autoencoders. In: Proceedings of the IEEE/CVF Conference on Computer Vision and Pattern Recognition, pp. 14104–14113 (2020)
22. Kingma, D.P., Welling, M.: Auto-encoding variational bayes. arXiv preprint arXiv:1312.6114 (2013)
23. Xiao, Z., Kreis, K., Kautz, J., Vahdat, A.: VAEBM: a symbiosis between variational autoencoders and energy-based models. arXiv preprint arXiv:2010.00654 (2020)
24. Welling, M., Teh, Y.W.: Bayesian learning via stochastic gradient langevin dynamics. In: Proceedings of the 28th International Conference on Machine Learning, pp. 681–688 (2011)
25. Kingma, D.P., Ba, J.: Adam: a method for stochastic optimization. arXiv preprint arXiv:1412.6980 (2014)
26. Dufumier, B., Grigis, A., Victor, J., Ambroise, C., Frouin, V., Duchesnay, E.: OpenBHB: a large-scale multi-site brain MRI data-set for age prediction and debiasing. Neuroimage **263**, 119637 (2022)
27. Tanaka, S., et al.: A multi-site, multi-disorder resting-state magnetic resonance image database. Scientific Data **8**(1), 227 (2021)
28. Smith, S.M., et al.: Advances in functional and structural MR image analysis and implementation as FSL. Neuroimage **23**, S208–S219 (2004)
29. Shinohara, R.T., et al.: Statistical normalization techniques for magnetic resonance imaging. NeuroImage Clin. **6**, 9–19 (2014)

30. Guan, H., Liu, M.: DomainATM: domain adaptation toolbox for medical data analysis. Neuroimage **268**, 119863 (2023)
31. He, K., Zhang, X., Ren, S., Sun, J.: Deep residual learning for image recognition. In: Proceedings of the IEEE Conference on Computer Vision and Pattern Recognition, pp. 770–778 (2016)
32. Ronneberger, O., Fischer, P., Brox, T.: U-Net: convolutional networks for biomedical image segmentation. In: Navab, N., Hornegger, J., Wells, W.M., Frangi, A.F. (eds.) MICCAI 2015. LNCS, vol. 9351, pp. 234–241. Springer, Cham (2015). https://doi.org/10.1007/978-3-319-24574-4_28
33. Wang, Z., Bovik, A.C., Sheikh, H.R., Simoncelli, E.P.: Image quality assessment: from error visibility to structural similarity. IEEE Trans. Image Process. **13**(4), 600–612 (2004)

Cross-Domain Iterative Network for Simultaneous Denoising, Limited-Angle Reconstruction, and Attenuation Correction of Cardiac SPECT

Xiongchao Chen$^{(\boxtimes)}$, Bo Zhou, Huidong Xie, Xueqi Guo, Qiong Liu, Albert J. Sinusas, and Chi Liu

Yale University, New Haven, CT 06511, USA
{xiongchao.chen,chi.liu}@yale.edu

Abstract. Single-Photon Emission Computed Tomography (SPECT) is widely applied for the diagnosis of ischemic heart diseases. Low-dose (LD) SPECT aims to minimize radiation exposure but leads to increased image noise. Limited-angle (LA) SPECT enables faster scanning and reduced hardware costs but results in lower reconstruction accuracy. Additionally, computed tomography (CT)-derived attenuation maps (μ-maps) are commonly used for SPECT attenuation correction (AC), but this will cause extra radiation exposure and SPECT-CT misalignments. Although various deep learning methods have been introduced to separately address these limitations, the solution for simultaneously addressing these challenges still remains highly under-explored and challenging. To this end, we propose a Cross-domain Iterative Network (CDI-Net) for simultaneous denoising, LA reconstruction, and CT-free AC in cardiac SPECT. In CDI-Net, paired projection- and image-domain networks are end-to-end connected to fuse the cross-domain emission and anatomical information in multiple iterations. Adaptive Weight Recalibrators (AWR) adjust the multi-channel input features to further enhance prediction accuracy. Our experiments using clinical data showed that CDI-Net produced more accurate μ-maps, projections, and AC reconstructions compared to existing approaches that addressed each task separately. Ablation studies demonstrated the significance of cross-domain and cross-iteration connections, as well as AWR, in improving the reconstruction performance. The source code of this work is released at https://github.com/XiongchaoChen/CDI-Net.

Keywords: Cardiac SPECT · Cross-domain prediction · Denoising · Limited-angle reconstruction · Attenuation correction

Supplementary Information The online version contains supplementary material available at https://doi.org/10.1007/978-3-031-45673-2_2.

1   Introduction

Cardiac Single-Photon Emission Computed Tomography (SPECT) is the most widely performed non-invasive exam for clinical diagnosis of ischemic heart diseases [13]. Reducing the tracer dose can lower patient radiation exposure, but it will result in increased image noise [25]. Acquiring projections in fewer angles using fewer detectors allows for faster scanning and lower hardware costs, but it also leads to decreased reconstruction accuracy [1,17]. Additionally, in clinical practice, computed tomography (CT)-derived attenuation maps (μ-maps) are commonly used for SPECT attenuation correction (AC) [8,15]. However, most SPECT scanners are stand-alone without the assistance of CT [20]. The CT scan also causes additional radiation exposure and SPECT-CT misalignments [9,10].

Deep learning-based methods have been extensively explored to address the aforementioned limitations individually. To reduce image noise in low-dose (LD) SPECT, convolutional neural networks (CNNs) were employed to process the LD projection, producing the full-dose (FD) projection for SPECT reconstruction [1,24]. Similarly, to perform limited-angle (LA) reconstruction, the LA projection was input to CNNs to predict the full-angle (FA) projection [2,23,26]. In addition, a dual-domain approach, Dual-domain Sinogram Synthesis (DuDoSS), utilized the image-domain output as a prior estimation for the projection domain to predict the FA projection [11]. For the CT-free AC, CNNs were used to generate pseudo attenuation maps (μ-maps) from SPECT emission images [4,5,12,22].

Although various methods have been developed to address these limitations individually, it is of great interest to address all these limitations simultaneously to enable CT-free, low-dose, and accelerated SPECT imaging, which could potentially lead to better performance on those separated but correlated tasks. Dual-domain methods have exhibited superior performance to single-domain methods [6,7]. Thus, we propose a Cross-Domain Iterative Network (CDI-Net) for simultaneous denoising, LA reconstruction, and CT-free AC in cardiac SPECT. In CDI-Net, projection and image-domain networks are end-to-end connected to fuse the predicted cross-domain and cross-modality emission and anatomical features in multiple iterations. Adaptive Weight Recalibrators (AWR) calibrate the fused features to improve prediction accuracy. We tested CDI-Net using clinical data and compared it to existing methods. Ablation studies were conducted to verify the impact of cross-domain, cross-iteration fusions, and AWR on enhancing network performance.

2   Materials and Methods

2.1   Problem Formulation

The aim of this study is to generate the predicted FD & FA projection ($\hat{P}_F$) and μ-map ($\hat{\mu}$) with the LD & LA projection (P_L) as the inputs, formulated as:

$$[\hat{P}_F, \hat{\mu}] = \mathcal{H}(P_L), \tag{1}$$

where $\mathcal{H}(\cdot)$ is the pre-processing and neural network operator. The output labels are the ground-truth FD & FA projection (P_F) and CT-derived μ-map (μ). Then, $\hat{P}_F$ and $\hat{\mu}$ are utilized in an offline maximum-likelihood expectation maximization (ML-EM, 30 iterations) module to reconstruct the target FD & FA SPECT image with AC. Thus, predicting $\hat{P}_F$ performs the denoising and LA reconstruction, while predicting $\hat{\mu}$ enables the CT-free AC.

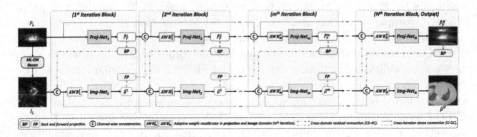

Fig. 1. Cross-Domain Iterative Network (CDI-Net). Projection- (*Proj-Net*) and image-domain networks (*Img-Net*) are end-to-end connected by cross-domain residual connections (CD-RC). Cross-iteration dense connections (CI-DC) fuse features across iterations. Adaptive Weight Recalibrators (AWR) adjust the multi-channel input.

2.2 Dataset and Pre-processing

This work includes 474 anonymized clinical hybrid SPECT/CT myocardial perfusion imaging (MPI) studies. Each study was conducted following the injection of ^{99m}Tc-tetrofosmin on a GE NM/CT 570c system. The GE 530c/570c system has 19 pinhole detectors placed in three columns on a cylindrical surface [3]. The clinical characteristics of the dataset are listed in Supplementary Table S1.

We extracted the 9 angles in the central column to simulate the configurations of the latest cost-effective MyoSPECT ES system [14] as shown in Supplementary Fig. S2, generating the LA projection. The 10% LD projection was produced by randomly decimating the list-mode data using a 10% downsampling rate. P_L was generated by combining the pipelines used for generating the LA and LD projections, and P_F was the original FD & FA projection. The ground-truth CT-derived μ-maps (μ) were well registered with the reconstructed SPECT images.

2.3 Cross-Domain Iterative Network

The overview framework of CDI-Net is shown in Fig. 1. P_L is first fed into an ML-EM reconstruction (30 iterations) module, producing the LD & LA reconstructed SPECT image I_L, which is then employed for the μ-map generation.

Cross-Domain Residual Connection. The projection- (*Proj-Net*) and image-domain networks (*Img-Net*) are both U-Net modules connected through cross-domain residual connections (**CD-RC**) facilitated by forward projection (FP) and back projection (BP) operators. In the 1st iteration, P_L is input to *Proj-Net$_1$* to generate $\hat{P}_F^1$ as:

$$\hat{P}_F^1 = \mathcal{P}_1(P_L), \tag{2}$$

where $\mathcal{P}_1(\cdot)$ is *Proj-Net$_1$* operator. $\hat{P}_F^1$ is then processed by BP and introduced to *Img-Net$_1$* through CD-RC, providing emission information for the μ-map generation. I_L and the BP of $\hat{P}_F^1$ is first fed into the AWR_1^I (described in Sect. 2.4) for multi-channel recalibration, and then input to *Img-Net$_1$* to generate $\hat{\mu}^1$:

$$\hat{\mu}^1 = \mathcal{I}_1(\mathcal{A}_1^I(\{I_L, \mathcal{T}_b(\hat{P}_F^1)\})), \tag{3}$$

where $\mathcal{I}_1(\cdot)$ is the *Img-Net$_1$* operator. $\mathcal{A}_1^I(\cdot)$ refers to the AWR_1^I (superscript I means image-domain). $\{\cdot\}$ is concatenation and $\mathcal{T}_b(\cdot)$ refers to BP. Next, the FP of $\hat{\mu}^1$ is added to *Proj-Net$_2$* of the next iteration by CD-RC, providing anatomical information for the projection prediction. This is the initial attempt to employ anatomical features for the estimation of FD & FA projection in cardiac SPECT.

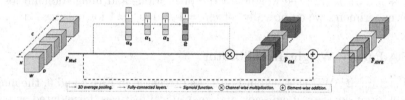

Fig. 2. Adaptive weight recalibrator. The channel weights of the input F_{Mul} is first recalibrated. A global residual connection is then added to retain the original features.

Cross-Iteration Dense Connection. In the $m^{th}(m \geq 2)$ iteration, the predicted projections from previous iterations, $\hat{P}_F^j(j < m)$, are incorporated into *Proj-Net$_m$* through cross-iteration dense connections (**CI-DC**). The FP of $\hat{\mu}^j$ from *Img-Net$_j$*$(j < m)$ are also added to *Proj-Net$_m$* through CD-RC as additional input anatomical information. The multi-channel input of *Proj-Net$_m$* is:

$$U_P^m = \left\{P_L, \hat{P}_F^1, \hat{P}_F^2, \cdots, \hat{P}_F^{(m-1)}, \mathcal{T}_f(\hat{\mu}^1), \mathcal{T}_f(\hat{\mu}^2), \cdots, \mathcal{T}_f(\hat{\mu}^{(m-1)})\right\}, \tag{4}$$

where $\mathcal{T}_f(\cdot)$ refers to the FP. Then, U_P^m is fed into AWR_m^P for recalibration and input to *Proj-Net$_m$* to generate $\hat{P}_F^m$, formulated as:

$$\hat{P}_F^m = \mathcal{P}_m(\mathcal{A}_m^P(U_P^m)), \tag{5}$$

where $\mathcal{P}_m(\cdot)$ refers to *Proj-Net$_m$* and $\mathcal{A}_m^P(\cdot)$ is *AWR$_m^P$*. Similarly, the predicted μ-maps from previous iterations, $\hat{\mu}^j (j < m)$, are integrated into *Img-Net$_m$* by CI-DC. The BP of $\hat{P}_F^j (j \leq m)$ are also added to *Img-Net$_m$* by CD-RC as additional input emission information. The multi-channel input of *Img-Net$_m$* is:

$$U_I^m = \left\{ I_L, \hat{\mu}^1, \hat{\mu}^2, \cdots, \hat{\mu}^{(m-1)}, \mathcal{T}_b(\hat{P}_F^1), \mathcal{T}_b(\hat{P}_F^2), \cdots, \mathcal{T}_b(\hat{P}_F^{(m-1)}), \mathcal{T}_b(\hat{P}_F^m) \right\}. \quad (6)$$

Then, U_I^m is recalibrated by AWR_m^I and input to *Img-Net$_m$* to produce $\hat{\mu}^m$ as:

$$\hat{\mu}^m = \mathcal{I}_m(\mathcal{A}_m^I(U_I^m)), \quad (7)$$

where $\mathcal{I}_m(\cdot)$ is *Img-Net$_m$* and $\mathcal{A}_m^I(\cdot)$ is the AWR_m^I operator.

Loss Function. The network outputs are $\hat{P}_F^N$ and $\hat{\mu}^N$, where N is the number of iterations (default: 5). The overall loss function $\mathcal{L}$ is formulated as:

$$\mathcal{L} = \sum_{i=1}^{N} (w_P \left\| \hat{P}_F^i - P_F \right\|_1 + w_\mu \left\| \hat{\mu}^i - \mu \right\|_1), \quad (8)$$

where w_P and w_μ are the weights of the projection- and image-domain losses. In our experiment, we empirically set $w_P = 1$ and $w_\mu = 1$ for balanced training.

2.4   Adaptive Weight Recalibrator

The diagram of AWR is shown in Fig. 2. As presented in Eqs. 4 and 6, the multi-channel input consists of emission and anatomical features, formulated as:

$$F_{Mul} = [f_1, f_2, \ldots, f_C], \quad (9)$$

where $f_i \in \mathbb{R}^{H \times W \times D}$ indicates the emission or anatomical feature in each individual channel. F_{Mul} is flattened using 3D average pooling, producing α_0 that embeds the channel weights. A recalibration vector $\hat{\alpha}$ is generated using fully-connected layers and a Sigmoid function. $\hat{\alpha}$ is applied to F_{Mul}, described as:

$$\hat{F}_{Chl} = [f_1 \hat{\alpha}_1, f_2 \hat{\alpha}_2, \ldots, f_C \hat{\alpha}_C], \quad (10)$$

where $\hat{\alpha}_i \in [0, 1]$ indicates the channel recalibration factor. Then, a global residual connection is applied to retain the original information, producing the output of AWR as $\hat{F}_{AWR} = \hat{F}_{Chl} + F_{Mul}$. Thus, AWR adaptively adjusts the weight of each input channel to better integrate the emission and anatomical information for higher prediction accuracy.

2.5   Implementation Details

In this study, we tested CDI-Net against many existing methods in terms of the predicted FD & FA projections, μ-maps, and AC SPECT images. U-Net (labeled as Separate-UNet) [21], Attention U-Net (labeled as Separate-AttnUNet) [18], and DuDoSS (labeled as Separate-DuDoSS) [11] were applied to generate $\hat{P}_F$ with P_L as input. U-Net and Attention U-Net were also employed to predict $\hat{\mu}$ with I_L as input. We also tested ablation study groups w/o CI-DC, CD-RC, or AWR. Then, $\hat{P}_F$ and $\hat{\mu}$ were utilized to reconstruct the AC images.

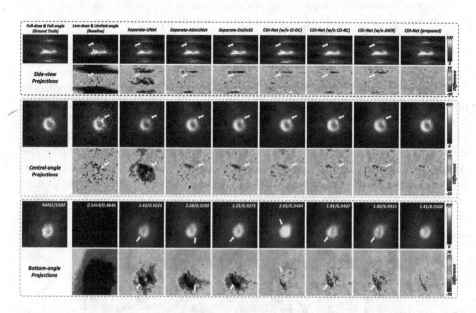

Fig. 3. Predicted FD & FA projections in the side, central-angle, and bottom-angle views with NMSE/SSIM annotated. White arrows point out the prediction inconsistency.

All networks were developed using PyTorch [19] with Adam optimizers [16]. The image- and projection-domain modules were trained with initial learning rates (LR) of 10^{-3} and 10^{-4} with a decay rate of 0.99/epoch [27]. The networks that predict μ-maps or projections separately were trained for 200 epochs, while CDI-Net was trained for 50 epochs. The performance of CDI-Net with different iterations (1 to 6, default 5) is presented in Sect. 3 (Fig. 6), and the impact of multiple LD levels (1 to 80%, default 10%) is shown in Sect. 3 (Fig. 6).

3   Results

Figure 3 shows the predicted FD & FA projections in multiple views. We can observe that CDI-Net outputs more accurate projections than other groups.

Table 1. Evaluation of the predicted projections using normalized mean square error (NMSE), normalized mean absolute error (NMAE), structural similarity (SSIM), and peak signal-to-noise ratio (PSNR). The best results are marked in red.

Methods	NMSE(%)	NMAE(%)	SSIM	PSNR	P-values[a]
Baseline LD & LA	54.56 ± 2.46	62.44 ± 2.53	0.4912 ± 0.0260	19.23 ± 1.68	< 0.001
Separate-UNet [21]	4.21 ± 1.48	16.69 ± 2.24	0.9276 ± 0.0195	30.58 ± 1.79	< 0.001
Separate-AttnUNet [18]	3.45 ± 1.13	15.45 ± 2.56	0.9368 ± 0.0205	31.43 ± 1.65	< 0.001
Separate-DuDoSS [11]	3.19 ± 1.11	14.57 ± 2.29	0.9416 ± 0.0187	31.79 ± 1.65	< 0.001
CDI-Net (w/o CI-DC)	2.56 ± 0.85	13.22 ± 1.81	0.9505 ± 0.0144	32.73 ± 1.65	< 0.001
CDI-Net (w/o CD-RC)	2.39 ± 0.78	13.39 ± 1.94	0.9486 ± 0.0160	33.02 ± 1.65	< 0.001
CDI-Net (w/o AWR)	2.42 ± 0.83	13.40 ± 2.00	0.9478 ± 0.0173	32.98 ± 1.65	< 0.001
CDI-Net (proposed)	2.15 ± 0.69	12.64 ± 1.77	0.9542 ± 0.0142	33.47 ± 1.68	–

[a] P-values of the paired t-tests of NMSE between the current method and CDI-Net (proposed).

Conversely, Separate-UNet, Separate-AttnUNet, and Separate-DuDoSS display underestimations in cardiac regions. This indicates the advantages of fusing emission and anatomical features for simultaneous prediction as in the CDI-Net. Moreover, CDI-Net shows superior performance to the ablation study groups w/o CI-DC, CD-RC, or AWR, confirming the significance of CI-DC, CD-RC, and AWR in enhancing network performance. Table 1 lists the quantitative comparison of the predicted projections. CDI-Net produces more accurate quantitative results than groups conducting separate predictions and ablation study groups ($p < 0.001$).

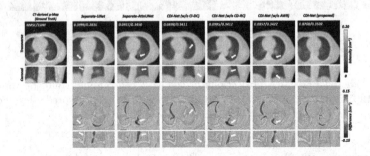

Fig. 4. Predicted μ-maps. White arrows denote the prediction inconsistency.

Figure 4 presents the predicted μ-maps. It can be observed that CDI-Net outputs more accurate μ-maps than other testing groups. The μ-maps predicted by Separate-UNet and Separate-AttnUNet display obvious inconsistency with the ground truth, particularly in the inner boundaries. This indicates that CDI-Net improves the accuracy of generating μ-maps by incorporating emission and anatomical information. Moreover, the μ-map predicted by CDI-Net is more accurate than ablation study groups w/o CI-DC, CD-RC, or AWR. Table 2 lists the quantitative evaluation of the predicted μ-maps. The μ-maps predicted by CDI-Net exhibit lower quantitative errors than other methods ($p < 0.001$).

Table 2. Evaluation of the predicted μ-maps. The best results are marked in red.

Methods	NMSE (%)	NMAE (%)	SSIM	PSNR	P-values[a]
Separate-UNet [21]	12.83 ± 4.85	22.93 ± 5.96	0.2782 ± 0.0617	17.22 ± 1.81	< 0.001
Separate-AttnUNet [18]	12.45 ± 4.49	22.20 ± 5.49	0.2829 ± 0.0582	17.34 ± 1.82	< 0.001
CDI-Net (w/o CI-DC)	11.88 ± 4.18	22.69 ± 5.37	0.2993 ± 0.0624	17.54 ± 1.81	< 0.001
CDI-Net (w/o CD-RC)	11.90 ± 4.69	21.95 ± 5.60	0.3041 ± 0.0660	17.56 ± 1.80	< 0.001
CDI-Net (w/o AWR)	11.84 ± 4.69	21.96 ± 5.48	0.3047 ± 0.0627	17.60 ± 1.89	< 0.001
CDI-Net (proposed)	11.42 ± 4.31	21.54 ± 5.30	0.3066 ± 0.0607	17.83 ± 1.85	–

[a] P-values of the paired t-tests of NMSE between the current method and CDI-Net (proposed).

The predicted projections and μ-maps are then utilized in SPECT reconstruction. As shown in Fig. 5, CDI-Net produces the most accurate AC images. The groups conducting separate predictions or ablation study groups show over- or under-estimations of the myocardial perfusion intensities compared to the ground truth. The quantitative evaluation listed in Table 3 shows that CDI-Net leads to higher reconstruction accuracy than other testing groups ($p < 0.001$). Segment-wise evaluation of the AC images is shown in Supplementary Fig. S1.

Table 3. Evaluation of the reconstructed AC SPECT. The best results are in red.

Methods	NMSE (%)	NMAE (%)	SSIM	PSNR	P-values[a]
Baseline LD & LA	35.80 ± 10.83	54.36 ± 6.13	0.6646 ± 0.0344	24.00 ± 1.80	< 0.001
Separate-UNet [21]	6.63 ± 2.26	23.78 ± 3.53	0.8576 ± 0.0248	31.33 ± 1.72	< 0.001
Separate-AttnUNet [18]	5.85 ± 1.76	22.46 ± 2.96	0.8655 ± 0.0239	31.56 ± 1.67	< 0.001
Separate-DuDoSS [11]	5.68 ± 1.81	22.02 ± 3.11	0.8706 ± 0.0242	32.00 ± 1.70	< 0.001
CDI-Net (w/o CI-DC)	5.45 ± 1.61	21.67 ± 2.92	0.8742 ± 0.0207	32.15 ± 1.69	< 0.001
CDI-Net (w/o CD-RC)	5.55 ± 1.81	21.66 ± 3.13	0.8722 ± 0.0231	32.12 ± 1.69	< 0.001
CDI-Net (w/o AWR)	5.49 ± 1.66	21.59 ± 2.92	0.8729 ± 0.0224	32.13 ± 1.70	< 0.001
CDI-Net (proposed)	4.82 ± 1.44	20.28 ± 2.65	0.8829 ± 0.0194	32.69 ± 1.65	–

[a] P-values of the paired t-tests of NMSE between the current method and CDI-Net (proposed).

Moreover, we tested the performance of CDI-Net with different iterations as shown in Fig. 6 (left). The errors of the predicted projections and μ-maps by CDI-Net decrease as the number of iterations increases, with convergence occurring at 5 iterations. Additionally, we generated more datasets with multiple LD levels to test these methods as shown in Fig. 6 (mid, right). It can be observed that CDI-Net demonstrates consistently higher prediction accuracy of projections and μ-maps than other groups across multiple LD levels.

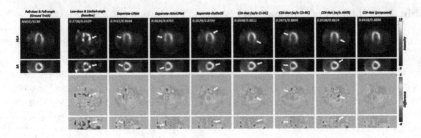

Fig. 5. Reconstructed SPECT images with attenuation correction (AC) using the predicted FD & FA projections and μ-maps. White arrows denote the inconsistency.

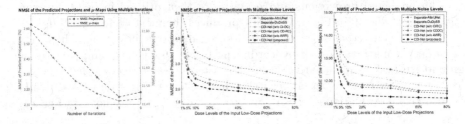

Fig. 6. Evaluation of CDI-Net with different iterations (left). Evaluation of multiple methods based on datasets with multiple low-dose levels (mid, right).

4 Discussion and Conclusion

In this paper, we propose CDI-Net that simultaneously achieves denoising, LA reconstruction, and CT-free AC for low-dose cardiac SPECT. The CD-RC and CI-DC components effectively fuse the cross-domain and cross-modality anatomical and emission features. The fused features are adaptively calibrated by AWR and then jointly employed for the prediction of projections and μ-maps. Thus, CDI-Net effectively combines the cross-domain information that is then used for image estimations in both domains. This approach also marks the initial investigation in employing anatomical features to assist the projection estimation of cardiac SPECT. Experiments using clinical data with different LD levels demonstrated the superiority of CDI-Net over existing methods in predicting projections and μ-maps, as well as in reconstructing AC SPECT images.

For potential clinical impact, CDI-Net enables accurate AC SPECT reconstruction in LD, LA, and CT-less scenarios. This could potentially promote the clinical adoption of the latest cost-effective SPECT scanners with fewer detectors and lower dose levels and without CT. Thus, we can achieve accurate cardiac AC SPECT imaging with reduced hardware expenses and lower radiation exposure.

References

1. Aghakhan Olia, N., et al.: Deep learning-based denoising of low-dose SPECT myocardial perfusion images: quantitative assessment and clinical performance. Eur. J. Nucl. Med. Mol. Imaging **49**, 1508–1522 (2022)
2. Amirrashedi, M., Sarkar, S., Ghadiri, H., Ghafarian, P., Zaidi, H., Ay, M.R.: A deep neural network to recover missing data in small animal pet imaging: comparison between sinogram-and image-domain implementations. In: 2021 IEEE 18th International Symposium on Biomedical Imaging (ISBI), pp. 1365–1368. IEEE (2021)
3. Chan, C., et al.: The impact of system matrix dimension on small FOV SPECT reconstruction with truncated projections. Med. Phys. **43**(1), 213–224 (2016)
4. Chen, X., et al.: Cross-vender, cross-tracer, and cross-protocol deep transfer learning for attenuation map generation of cardiac SPECT. J. Nucl. Cardiol. **29**(6), 3379–3391 (2022)
5. Chen, X., Liu, C.: Deep-learning-based methods of attenuation correction for SPECT and PET. J. Nuclear Cardiol. pp. 1–20 (2022)
6. Chen, X., Peng, Z., Valadez, G.H.: DD-CISENet: dual-domain cross-iteration squeeze and excitation network for accelerated MRI reconstruction. arXiv preprint arXiv:2305.00088 (2023)
7. Chen, X., Shinagawa, Y., Peng, Z., Valadez, G.H.: Dual-domain cross-iteration squeeze-excitation network for sparse reconstruction of brain MRI. arXiv preprint arXiv:2210.02523 (2022)
8. Chen, X., et al.: CT-free attenuation correction for dedicated cardiac SPECT using a 3D dual squeeze-and-excitation residual dense network. J. Nucl. Cardiol. **29**(5), 2235–2250 (2022)
9. Chen, X., et al.: DuSFE: dual-channel squeeze-fusion-excitation co-attention for cross-modality registration of cardiac SPECT and CT. Med. Image Anal. **88**, 102840 (2023)
10. Chen, X., et al.: Dual-branch squeeze-fusion-excitation module for cross-modality registration of cardiac SPECT and CT. In: Wang, L., Dou, Q., Fletcher, P.T., Speidel, S., Li, S. (eds.) Medical Image Computing and Computer Assisted Intervention – MICCAI 2022. MICCAI 2022. Lecture Notes in Computer Science, vol. 13436, pp. 46–55. Springer, Cham (2022). https://doi.org/10.1007/978-3-031-16446-0_5
11. Chen, X., et al.: DuDoSS: deep-learning-based dual-domain sinogram synthesis from sparsely sampled projections of cardiac SPECT. Med. Phys. **50**(1), 89–103 (2022)
12. Chen, X., et al.: Direct and indirect strategies of deep-learning-based attenuation correction for general purpose and dedicated cardiac SPECT. Eur. J. Nucl. Med. Mol. Imaging **49**(9), 3046–3060 (2022)
13. Danad, I., et al.: Comparison of coronary CT angiography, SPECT, PET, and hybrid imaging for diagnosis of ischemic heart disease determined by fractional flow reserve. JAMA Cardiol. **2**(10), 1100–1107 (2017)
14. GE-HealthCare: Ge myospect es: a perfect fit for today's practice of cardiology. https://www.gehealthcare.com/products/molecular-imaging/myospect (2023)
15. Goetze, S., Brown, T.L., Lavely, W.C., Zhang, Z., Bengel, F.M.: Attenuation correction in myocardial perfusion SPECT/CT: effects of misregistration and value of reregistration. J. Nucl. Med. **48**(7), 1090–1095 (2007)
16. Kingma, D.P., Ba, J.: Adam: a method for stochastic optimization. arXiv preprint arXiv:1412.6980 (2014)

17. Niu, S., et al.: Sparse-view x-ray CT reconstruction via total generalized variation regularization. Phys. Med. Biol. **59**(12), 2997 (2014)

18. Oktay, O., et al.: Attention u-Net: learning where to look for the pancreas. arXiv preprint arXiv:1804.03999 (2018)

19. Paszke, A., et al.: PyTorch: an imperative style, high-performance deep learning library. In: Advances in Neural Information Processing Systems, vol. 32 (2019)

20. Rahman, M.A., Zhu, Y., Clarkson, E., Kupinski, M.A., Frey, E.C., Jha, A.K.: Fisher information analysis of list-mode SPECT emission data for joint estimation of activity and attenuation distribution. Inverse Prob. **36**(8), 084002 (2020)

21. Ronneberger, O., Fischer, P., Brox, T.: U-Net: convolutional networks for biomedical image segmentation. In: Navab, N., Hornegger, J., Wells, W.M., Frangi, A.F. (eds.) MICCAI 2015. LNCS, vol. 9351, pp. 234–241. Springer, Cham (2015). https://doi.org/10.1007/978-3-319-24574-4_28

22. Shi, L., Onofrey, J.A., Liu, H., Liu, Y.H., Liu, C.: Deep learning-based attenuation map generation for myocardial perfusion SPECT. Eur. J. Nucl. Med. Mol. Imaging **47**, 2383–2395 (2020)

23. Shiri, I., et al.: Standard SPECT myocardial perfusion estimation from half-time acquisitions using deep convolutional residual neural networks. J. Nucl. Cardiol. **28**(6), 2761–2779 (2020)

24. Sun, J., et al.: Deep learning-based denoising in projection-domain and reconstruction-domain for low-dose myocardial perfusion SPECT. J. Nucl. Cardiol. **30**(3), 970–985 (2022)

25. Wells, R.G.: Dose reduction is good but it is image quality that matters. J. Nucl. Cardiol. **27**(1), 238–240 (2020)

26. Whiteley, W., Gregor, J.: CNN-based pet sinogram repair to mitigate defective block detectors. Phys. Med. Biol. **64**(23), 235017 (2019)

27. You, K., Long, M., Wang, J., Jordan, M.I.: How does learning rate decay help modern neural networks? arXiv preprint arXiv:1908.01878 (2019)

Arbitrary Reduction of MRI Inter-slice Spacing Using Hierarchical Feature Conditional Diffusion

Xin Wang[1], Zhenrong Shen[1], Zhiyun Song[1], Sheng Wang[1], Mengjun Liu[1], Lichi Zhang[1], Kai Xuan[2], and Qian Wang[3(✉)]

[1] School of Biomedical Engineering, Shanghai Jiao Tong University, Shanghai, China
[2] School of Artificial Intelligence, Nanjing University of Information Science and Technology, Nanjing, China
[3] School of Biomedical Engineering, ShanghaiTech University, Shanghai, China
qianwang@shanghaitech.edu.cn

Abstract. Magnetic resonance (MR) images collected in 2D scanning protocols typically have large inter-slice spacing, resulting in high in-plane resolution but reduced through-plane resolution. Super-resolution techniques can reduce the inter-slice spacing of 2D scanned MR images, facilitating the downstream visual experience and computer-aided diagnosis. However, most existing super-resolution methods are trained at a fixed scaling ratio, which is inconvenient in clinical settings where MR scanning may have varying inter-slice spacings. To solve this issue, we propose *Hierarchical Feature Conditional Diffusion (HiFi-Diff)* for arbitrary reduction of MR inter-slice spacing. Given two adjacent MR slices and the relative positional offset, HiFi-Diff can iteratively convert a Gaussian noise map into any desired in-between MR slice. Furthermore, to enable fine-grained conditioning, the Hierarchical Feature Extraction (HiFE) module is proposed to hierarchically extract conditional features and conduct element-wise modulation. Our experimental results on the publicly available HCP-1200 dataset demonstrate the high-fidelity super-resolution capability of HiFi-Diff and its efficacy in enhancing downstream segmentation performance.

Keywords: Magnetic Resonance Imaging · Super-resolution · Diffusion Model · Conditional Image Synthesis

1 Introduction

Magnetic resonance imaging (MRI) is essential for analyzing and diagnosing various diseases, owing to its non-invasive property and superior contrast for soft tissues. In clinical practice, 2D scanning protocols are commonly employed for MR image acquisition due to limitations in scanning time and signal-to-noise

---

X. Wang and Z. Shen—Contributed equally to this work.

X. Cao et al. (Eds.): MLMI 2023, LNCS 14348, pp. 23–32, 2024.
https://doi.org/10.1007/978-3-031-45673-2_3

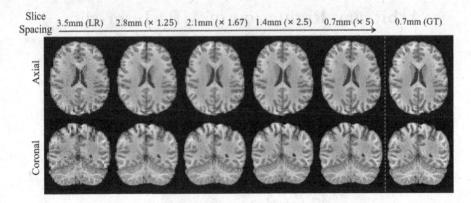

Fig. 1. A test case of applying HiFi-Diff to arbitrary reduction of MR inter-slice (sagittal) spacing. The visual quality of the resulting images in the axial and coronal views is gradually enhanced with increasing SR factors.

ratio. Typically, such scanning protocols produce MR volumes with small intra-slice spacing but much larger inter-slice spacing, which poses a great challenge for many volumetric image processing toolkits [5,13] that require near-isotropic voxel spacing of the input images. Therefore, it is necessary to resample the acquired volumes to align the inter-slice spacing with the intra-slice spacing.

Interpolation methods are widely used to reduce the inter-slice spacing of 2D scanned MR volumes. However, these methods simply calculate missing voxels as a weighted average of the adjacent ones, leading to inevitable blurred results. For better performance, many deep-learning-based super-resolution (SR) studies have been investigated [1,2,7,12,21]. In this paper, we term large slice spacing as *low resolution* (LR) and small slice spacing as *high resolution* (HR). DeepResolve [1] adopts a 3D convolutional network to obtain HR images from LR ones, but it only considers reducing inter-slice spacing at a fixed ratio. Training such an SR model for each scaling ratio is impractical, as it requires significant time and computational resources.

To tackle this issue, local implicit image function (LIIF) [2] and MetaSR [12] are proposed to perform arbitrary-scale SR for natural images. To achieve arbitrary-scale SR of MR images, ArSSR [21] extends LIIF to 3D volumes and utilizes a continuous implicit voxel function for reconstructing HR images at different ratios. Although ArSSR produces competitive quantitative results, it still suffers from image over-smoothing. To solve the aforementioned problem, adversarial learning [9] is usually introduced to synthesize more image details. However, such a training scheme often leads to training instability and is prone to generate artifacts [6,18].

Recently, diffusion models [11,19] have achieved wide success in image synthesis tasks, outperforming other deep generative models in terms of visual fidelity and training stability. Typical denoising diffusion models (*e.g.*, DDPM [11]) use a series of denoising operations to iteratively generate samples from a prior dis-

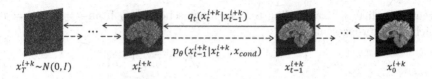

Fig. 2. Conditional diffusion process of HiFi-Diff, where q_t and p_θ denote single-step transitions in forward and reverse processes, respectively.

tribution (*e.g.*, Gaussian) to a desired data distribution. Although there exist several works that apply diffusion models for MR image reconstruction [4] or denoising [3], the application of diffusion models to achieve arbitrary-scale MR image super-resolution has not been studied yet.

In this paper, by leveraging the powerful ability of the diffusion models, we propose **Hi**erarchical **F**eature *Conditional* **Diff**usion *(HiFi-Diff)*, which allows arbitrary reduction of inter-slice spacing for 2D scanned MR images, as shown in Fig. 1. Conditioned on two adjacent LR slices, HiFi-Diff can generate any in-between MR slices. To handle different ratios of inter-slice spacing, we construct continuous representations for the spatial positions between two adjacent LR slices by providing relative positional offsets as additional conditions. Inspired by the core idea of FPN [15], we propose the Hierarchical Feature Extraction (HiFE) module, which applies different-scale feature maps as conditions to perform element-wise feature modulation in each layer. The experimental results demonstrate that HiFi-Diff produces MR slices of excellent quality and effectively enhances downstream image segmentation tasks.

In summary, the main contributions of this paper include: (1) To the best of our knowledge, HiFi-Diff is the first diffusion model for arbitrary-scale SR of MR images. (2) We propose the HiFE module to hierarchically extract conditional features for fine-grained conditioning on MR slice generation.

2 Method

We discuss the conditional diffusion process and the network architecture of HiFi-Diff in Sects. 2.1 and 2.2, respectively.

2.1 Conditional Diffusion for Arbitrary-Scale Super-Resolution

Let $x_0^i \in \mathbb{R}^{H \times W}$ denote a sample from 2D MR slice distribution, where the subscript 0 refers to the initial timestep and the superscript i refers to the slice index. For arbitrary-scale SR, we aim to learn continuous representations for the spatial positions between any two adjacent MR slices in an LR volume. Specifically, we define the generated slice between x_0^i and x_0^{i+1} as x_0^{i+k}, where $k \in [0, 1]$ is a non-integral offset denoting its relative distance to x_0^i.

Similar to DDPM [11], HiFi-Diff learns a Markov chain process to convert the Gaussian distribution into the target data distribution, as demonstrated in Fig. 2. The forward diffusion process gradually adds Gaussian noises ϵ_t to the target MR slice x_0^{i+k} according to a variance schedule β_t from $t = 0$ to $t = T$, which can be represented as:

$$q_t(x_t^{i+k}|x_{t-1}^{i+k}) = \mathcal{N}(x_t^{i+k}; \sqrt{1-\beta_t}x_{t-1}^{i+k}, \beta_t \mathbf{I}), \tag{1}$$

Furthermore, we can directly sample x_t^{i+k} from x_0^{i+k} at an arbitrary timestep t in a closed form using the following accumulated expression:

$$q_t(x_t^{i+k}|x_0^{i+k}) = \mathcal{N}(x_t^{i+k}; \sqrt{\overline{\alpha}_t}x_0^{i+k}, (1-\overline{\alpha}_t)\mathbf{I}) \Rightarrow x_t^{i+k} = \sqrt{\overline{\alpha}_t}x_0^{i+k} + (1-\overline{\alpha}_t)\epsilon, \tag{2}$$

where $\overline{\alpha}_t = \prod_{s=1}^{t}(1 - \beta_s)$ and $\epsilon \sim \mathcal{N}(\mathbf{0}, \mathbf{I})$. To gain the generation ability for MR slices, HiFi-Diff learns the reverse diffusion via a parameterized Gaussian process $p_\theta(x_{t-1}^{i+k}|x_t^{i+k}, x_{cond})$ conditioned on the feature pyramid x_{cond}:

$$p_\theta(x_{t-1}^{i+k}|x_t^{i+k}, x_{cond}) = \mathcal{N}(x_{t-1}^{i+k}; \mu_\theta(x_t^{i+k}, t, x_{cond}), \sigma_t^2\mathbf{I}), \tag{3}$$

where σ_t^2 is a fixed variance and $\mu_\theta(x_t^{i+k}, t, x_{cond})$ is a learned mean defined as:

$$\mu_\theta(x_t^{i+k}, t, x_{cond}) = \frac{1}{\sqrt{1-\beta_t}}\left(x_t^{i+k} - \frac{\beta_t}{\sqrt{1-\overline{\alpha}_t}}\epsilon_\theta(x_t^{i+k}, t, x_{cond})\right) \tag{4}$$

where $\epsilon_\theta(x_t^{i+k}, t, x_{cond})$ denotes the main branch of HiFi-Diff for noise prediction. To generate in-between slices x_0^{i+k}, we iteratively compute the denoising process $x_{t-1}^{i+k} = \mu_\theta(x_t^{i+k}, t, x_{cond}) + \sigma_t z$, where $z \sim \mathcal{N}(\mathbf{0}, \mathbf{I})$.

HiFi-Diff is trained in an end-to-end manner by optimizing the simple variant of the variational lowerbound $\mathcal{L}_{simple}$ with respect to θ and ϕ:

$$\begin{aligned}\mathcal{L}_{simple}(\theta, \phi) &= \mathbb{E}_{x_0^{i+k}, t, x_{cond}}\left[\left\|\epsilon_\theta(x_t^{i+k}, t, x_{cond}) - \epsilon_t\right\|_2^2\right] \\ &= \mathbb{E}_{x_0^{i+k}, t, x_0^i, x_0^{i+1}, k}\left[\left\|\epsilon_\theta(x_t^{i+k}, t, \mathcal{F}_\phi(x_0^i, x_0^{i+1}, k)) - \epsilon_t\right\|_2^2\right],\end{aligned} \tag{5}$$

where $\mathcal{F}_\phi$ parameterizes the proposed HiFE module, ϵ_t is the Gaussian distribution data with $\mathcal{N}(\mathbf{0}, \mathbf{I})$, and t is a timestep uniformly sampled from $[0, T]$.

2.2   Hierarchical Feature Conditioning Framework

Given any pair of x_0^i and x_0^{i+1} from an LR volume with a desired offset k, HiFi-Diff is able to iteratively convert a Gaussian noise map into the target in-between slice x_0^{i+k} through a reversed diffusion process, as described in the last section.

In this section, we introduce the network architecture of HiFi-Diff. As illustrated in Fig. 3(a), the adjacent MR slices x_0^i and x_0^{i+1} are concatenated and input to the proposed HiFE module, which adopts a U-Net [17] architecture consisting of a stack of residual blocks (shown in Fig. 3(b)). The offset k is injected into each residual block to perform channel-wise modulation. Specifically, k is projected by two successive fully connected layers into a 128-dimensional index

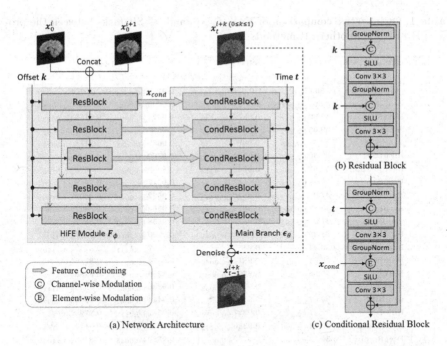

Fig. 3. Overview of *Hierarchical Feature Conditional Diffusion (HiFi-Diff)*.

embedding. Next, for each layer, the index embedding is passed through a learnable affine transformation to obtain the channel-wise scaling and bias parameters (k_s, k_b), which are applied to the feature map h using the expression $\text{ChannelMod}(h, k_s, k_b) = k_s \text{GroupNorm}(h) + k_b$. In this way, the HiFE module yields an hourglass-like feature hierarchy that includes feature maps at different scales, with semantics ranging from low to high levels.

Conditioned on the timestep t and the feature pyramid x_{cond}, the main branch of HiFi-Diff learns to gradually denoise the noise-corrupted input slice x_t^{i+k}. The main branch has the same U-Net architecture as HiFE module and consists of a stack of conditional residual blocks. For each conditional residual block (shown in Fig. 3(c)) in the main branch, the timestep t performs channel-wise modulation in the same way that the offset k does in each residual block of HiFE module. After the channel-wise modulation, the feature map is further modulated by x_{cond} from the lateral connection at the same image level. The conditional feature x_{cond} is transformed into the scaling and bias parameters (x_s, x_b) that have the same spatial sizes as the feature map h, such that x_{cond} can shift group-normalized h in an element-wise manner: $\text{ElementMod}(h, x_s, x_b) = x_s \text{GroupNorm}(h) + x_b$. Through element-wise modulation, the hierarchical feature pyramid x_{cond} provides fine-grained conditioning to guide the MR slice generation.

Table 1. Quantitative comparison of ×4, ×5, ×6, and ×7 SR tasks between the proposed HiFi-Diff and other SR methods.

Task	Method	PSNR	SSIM	Dice	
				WM	GM
×4	Interpolation	$36.05_{\pm 2.394}$	$0.9758_{\pm 0.0070}$	$0.9112_{\pm 0.0084}$	$0.7559_{\pm 0.0177}$
	DeepResolve	$39.65_{\pm 2.281}$	$0.9880_{\pm 0.0039}$	$0.9700_{\pm 0.0021}$	$0.9230_{\pm 0.0055}$
	MetaSR	$39.30_{\pm 2.287}$	$0.9876_{\pm 0.0039}$	$0.9672_{\pm 0.0027}$	$0.9152_{\pm 0.0065}$
	ArSSR	$39.65_{\pm 2.282}$	$0.9884_{\pm 0.0037}$	$0.9687_{\pm 0.0026}$	$0.9171_{\pm 0.0067}$
	w/o HiFE	$38.78_{\pm 2.298}$	$0.9874_{\pm 0.0039}$	$0.9639_{\pm 0.0033}$	$0.9118_{\pm 0.0078}$
	HiFi-Diff	$39.50_{\pm 2.285}$	$0.9890_{\pm 0.0040}$	$0.9700_{\pm 0.0045}$	$0.9229_{\pm 0.0057}$
×5	Interpolation	$31.55_{\pm 2.381}$	$0.9542_{\pm 0.0113}$	$0.8362_{\pm 0.0090}$	$0.6470_{\pm 0.0136}$
	DeepResolve	$38.28_{\pm 2.318}$	$0.9848_{\pm 0.0047}$	$0.9617_{\pm 0.0028}$	$0.9006_{\pm 0.0070}$
	MetaSR	$37.95_{\pm 2.279}$	$0.9840_{\pm 0.0049}$	$0.9522_{\pm 0.0043}$	$0.8808_{\pm 0.0100}$
	ArSSR	$38.27_{\pm 2.261}$	$0.9850_{\pm 0.0045}$	$0.9552_{\pm 0.0039}$	$0.8847_{\pm 0.0100}$
	w/o HiFE	$37.53_{\pm 2.289}$	$0.9832_{\pm 0.0051}$	$0.9538_{\pm 0.0042}$	$0.8827_{\pm 0.0109}$
	HiFi-Diff	$38.25_{\pm 2.260}$	$0.9852_{\pm 0.0047}$	$0.9620_{\pm 0.0060}$	$0.9007_{\pm 0.0104}$
×6	Interpolation	$30.44_{\pm 2.379}$	$0.9457_{\pm 0.0129}$	$0.7967_{\pm 0.0111}$	$0.5749_{\pm 0.0159}$
	DeepResolve	$37.21_{\pm 2.315}$	$0.9814_{\pm 0.0055}$	$0.9521_{\pm 0.0033}$	$0.8786_{\pm 0.0078}$
	MetaSR	$36.55_{\pm 2.286}$	$0.9792_{\pm 0.0061}$	$0.9286_{\pm 0.0066}$	$0.8298_{\pm 0.0142}$
	ArSSR	$36.62_{\pm 2.271}$	$0.9798_{\pm 0.0058}$	$0.9330_{\pm 0.0061}$	$0.8358_{\pm 0.0141}$
	w/o HiFE	$36.67_{\pm 2.292}$	$0.9801_{\pm 0.0058}$	$0.9414_{\pm 0.0042}$	$0.8569_{\pm 0.0097}$
	HiFi-Diff	$37.41_{\pm 2.314}$	$0.9827_{\pm 0.0054}$	$0.9527_{\pm 0.0111}$	$0.8798_{\pm 0.0169}$
×7	Interpolation	$29.61_{\pm 2.379}$	$0.9386_{\pm 0.0142}$	$0.7611_{\pm 0.0130}$	$0.5144_{\pm 0.0185}$
	DeepResolve	$36.35_{\pm 2.303}$	$0.9782_{\pm 0.0063}$	$0.9387_{\pm 0.0046}$	$0.8468_{\pm 0.0104}$
	MetaSR	$35.27_{\pm 2.308}$	$0.9739_{\pm 0.0074}$	$0.8988_{\pm 0.0092}$	$0.7691_{\pm 0.0180}$
	ArSSR	$35.20_{\pm 2.289}$	$0.9741_{\pm 0.0072}$	$0.9034_{\pm 0.0090}$	$0.7739_{\pm 0.0192}$
	w/o HiFE	$35.86_{\pm 2.316}$	$0.9766_{\pm 0.0070}$	$0.9245_{\pm 0.0056}$	$0.8254_{\pm 0.0122}$
	HiFi-Diff	$36.58_{\pm 2.328}$	$0.9797_{\pm 0.0062}$	$0.9401_{\pm 0.0103}$	$0.8550_{\pm 0.0177}$

3   Experimental Results

3.1   Dataset and Experimental Setup

Data Preparation. We collect 1,113 subjects of 3T MR images from the HCP-1200 dataset [8], with all images having an isotropic voxel spacing of 0.7mm×0.7mm×0.7mm. Among these, 891 images are used for training, and the remaining 222 images are used for testing. We perform N4 bias correction and skull-stripping for preprocessing. It is noteworthy that skull-stripping is necessary in order to protect the privacy of the subjects. To simulate the LR images with large slice spacing, we downsample the isotropic HR volumes perpendicular to the sagittal view following [1].

Implementation Details. To achieve a comprehensive evaluation, we compare HiFi-Diff with other methods for MR super-resolution, including trilinear interpolation, DeepResolve [1], MetaSR [12], and ArSSR [21]. DeepResolve is trained and tested for each specific scaling ratio, while HiFi-Diff, MetaSR, and ArSSR are trained using mixed scaling ratios of ×2, ×3, ×4 for arbitrary-scale super-resolution. In each iteration of training, we corrupt the intermediate slice x_0^{i+k} with Gaussian noise according to the randomly sampled timestep t. We set

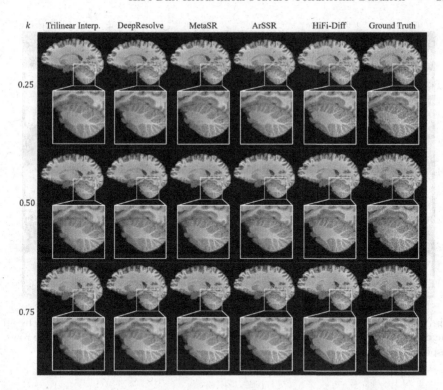

Fig. 4. Qualitative comparison of ×4 SR task between the proposed HiFi-Diff and other SR methods. The cerebellum regions are highlighted and zoomed in.

$T = 1000$ during training, and use DDIM sampler [20] to speed up the reverse process by reducing $T = 1000$ to $T = 100$. All the experiments are conducted using an NVIDIA A100 40G with PyTorch [16]. We use the learning rate of 1.0×10^{-4}, batch size of 1, and Adam optimizer [14] to train our model for 700k iterations.

3.2 Super-Resolution Evaluation

We use Peak Signal-to-Noise Ratio (PSNR) and Structural Similarity Index (SSIM) to evaluate the consistency between the SR results and ground truth. Based on the results in Table 1, the proposed HiFi-Diff method outperforms other state-of-the-art methods, particularly at large scaling ratios. It is worth mentioning that DeepResolve achieves the highest PSNR scores at scaling ratios of ×4 and ×5, which can be attributed to the fact that DeepResolve is specifically trained for each scaling ratio.

In addition, we conduct an ablation study to assess the effectiveness of the HiFE module by removing it and comparing the results. In detail, we directly inject the concatenated slices into the main branch for element-wise modulation, and concatenate the embedding of offset k and timestep t for channel-wise

| Trilinear Interp. | DeepResolve | MetaSR | ArSSR | HiFi-Diff | Ground Truth |

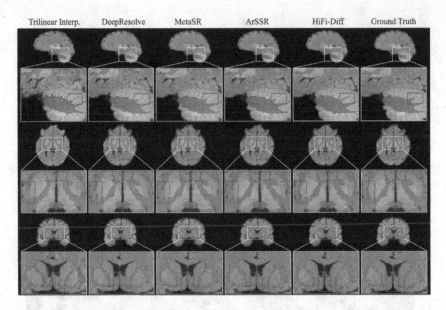

Fig. 5. Visual comparison of the fully automatic segmentation on ×4 SR results by all the comparing models. The sagittal, axial, and coronal views are shown in three rows, respectively. The areas surrounded by the white boxes are zoomed in below.

modulation. The results show a decrease in all metrics when the HiFE module is removed, indicating that HiFi-Diff benefits from the fine-grained conditioning provided by the HiFE module.

The qualitative comparison of the generated in-between MR slices with different offsets is shown in Fig. 4. By inspection of the cerebellum, one can notice that other methods fail to produce complete and clear structures of the white matter for they are optimized using L1/L2 loss, driving their results towards over-smoothing and loss of high-frequency information. In contrast, HiFi-Diff can faithfully reconstruct image details through an iterative diffusion process.

To validate the effectiveness of the proposed HiFi-Diff on downstream tasks, we conduct brain segmentation on different SR results using Fastsurfer [10]. According to Table 1, HiFi-Diff outperforms other methods in terms of Dice score for the white matter (WM) and the gray matter (GM) in most scenarios. The visual comparison in Fig. 5 further demonstrates the superiority of HiFi-Diff, as other methods yield tissue adhesion or discontinuity in their segmented results, while our approach avoids these problems.

4   Conclusion and Discussion

In conclusion, we propose HiFi-Diff to conduct arbitrary reduction of MR inter-slice spacing, outperforming previous methods in both generation capability and downstream task performance by leveraging the power of the diffusion models. To further enhance fine-grained conditioning, we introduce the HiFE module, which hierarchically extracts conditional features and conducts element-wise feature modulations. Despite the superior performance, HiFi-Diff still suffers from slow sampling speed. One possible solution is the implementation of faster sampling algorithms or the utilization of techniques such as knowledge distillation.

References

1. Chaudhari, A.S., et al.: Super-resolution musculoskeletal MRI using deep learning. Magn. Reson. Med. **80**(5), 2139–2154 (2018)
2. Chen, Y., Liu, S., Wang, X.: Learning continuous image representation with local implicit image function. In: 2021 IEEE/CVF Conference on Computer Vision and Pattern Recognition (CVPR), pp. 8624–8634 (2021). DOI: https://doi.org/10.1109/CVPR46437.2021.00852
3. Chung, H., Lee, E.S., Ye, J.C.: MR image denoising and super-resolution using regularized reverse diffusion. IEEE Trans. Med. Imaging **42**, 922–934 (2022)
4. Cui, Z.X., et al.: Self-score: self-supervised learning on score-based models for MRI reconstruction. arXiv preprint arXiv:2209.00835 (2022)
5. Desikan, R.S., et al.: An automated labeling system for subdividing the human cerebral cortex on MRI scans into gyral based regions of interest. Neuroimage **31**(3), 968–980 (2006). https://doi.org/10.1016/j.neuroimage.2006.01.021
6. Dhariwal, P., Nichol, A.: Diffusion models beat GANs on image synthesis. Adv. Neural. Inf. Process. Syst. **34**, 8780–8794 (2021)
7. Dong, C., Loy, C.C., He, K., Tang, X.: Image super-resolution using deep convolutional networks. IEEE Trans. Pattern Anal. Mach. Intell. **38**(2), 295–307 (2016). https://doi.org/10.1109/TPAMI.2015.2439281
8. Glasser, M.F., et al.: The minimal preprocessing pipelines for the human connectome project. NeuroImage **80**, 105–124 (2013). mapping the Connectome
9. Goodfellow, I., et al.: Generative adversarial networks. Commun. ACM **63**(11), 139–144 (2020)
10. Henschel, L., Conjeti, S., Estrada, S., Diers, K., Fischl, B., Reuter, M.: FastSurfer-a fast and accurate deep learning based neuroimaging pipeline. Neuroimage **219**, 117012 (2020)
11. Ho, J., Jain, A., Abbeel, P.: Denoising diffusion probabilistic models. Adv. Neural. Inf. Process. Syst. **33**, 6840–6851 (2020)
12. Hu, X., Mu, H., Zhang, X., Wang, Z., Tan, T., Sun, J.: Meta-SR: a magnification-arbitrary network for super-resolution. In: 2019 IEEE/CVF Conference on Computer Vision and Pattern Recognition (CVPR), pp. 1575–1584 (2019). https://doi.org/10.1109/CVPR.2019.00167
13. Jenkinson, M., Beckmann, C.F., Behrens, T.E., Woolrich, M.W., Smith, S.M.: Fsl. NeuroImage **62**(2), 782–790 (2012)
14. Kingma, D.P., Ba, J.: Adam: A method for stochastic optimization (2014), cite arxiv:1412.6980Comment: Published as a conference paper at the 3rd International Conference for Learning Representations, San Diego, 2015
15. Lin, T.Y., Dollár, P., Girshick, R., He, K., Hariharan, B., Belongie, S.: Feature pyramid networks for object detection. In: Proceedings of the IEEE Conference on Computer Vision and Pattern Recognition, pp. 2117–2125 (2017)
16. Paszke, A., et al.: Automatic differentiation in PyTorch. In: NIPS 2017 Workshop on Autodiff (2017)

17. Ronneberger, O., Fischer, P., Brox, T.: U-Net: convolutional networks for biomedical image segmentation. In: Navab, N., Hornegger, J., Wells, W.M., Frangi, A.F. (eds.) MICCAI 2015. LNCS, vol. 9351, pp. 234–241. Springer, Cham (2015). https://doi.org/10.1007/978-3-319-24574-4_28

18. Salimans, T., Goodfellow, I.J., Zaremba, W., Cheung, V., Radford, A., Chen, X.: Improved techniques for training GANs. CoRR abs/1606.03498, http://arxiv.org/abs/1606.03498 (2016)

19. Sohl-Dickstein, J., Weiss, E., Maheswaranathan, N., Ganguli, S.: Deep unsupervised learning using nonequilibrium thermodynamics. In: International Conference on Machine Learning, pp. 2256–2265. PMLR (2015)

20. Song, J., Meng, C., Ermon, S.: Denoising diffusion implicit models. arXiv preprint arXiv:2010.02502 (2020)

21. Wu, Q., et al.: An arbitrary scale super-resolution approach for 3-Dimensional magnetic resonance image using implicit neural representation (2021)

Reconstruction of 3D Fetal Brain MRI from 2D Cross-Sectional Acquisitions Using Unsupervised Learning Network

Yimeng Yang[1,3], Dongdong Gu[1], Xukun Zhang[5], Zhongxiang Ding[6], Fei Gao[3], Zhong Xue[1(✉)], and Dinggang Shen[1,2,4]

[1] Shanghai United Imaging Intelligence Co., Ltd., Shanghai 200230, China
zhong.xue@uii-ai.com, dgshen@shanghaitech.edu.cn
[2] School of Biomedical Engineering, ShanghaiTech University, Shanghai 201210, China
[3] School of Computer Science and Technology, ShanghaiTech University, Shanghai 201210, China
[4] Shanghai Clinical Research and Trial Center, Shanghai 201210, China
[5] Fudan University, Shanghai 200433, China
[6] The 1st Hospital of Hangzhou, Hangzhou 310006, China

Abstract. Fetal brain magnetic resonance imaging (MRI) is becoming more important for early brain assessment in prenatal examination. Fast acquisition of three cross-sectional series/views is often used to eliminate motion effects using single-shot fast spin-echo sequences. Although stacked in 3D volumes, these slices are essentially 2D images with large slice thickness and distances (4 to 6 mm) resulting blurry multiplanar views. To better visualize and quantify fetal brains, it is desirable to reconstruct 3D images from different 2D cross-sectional series. In this paper, we present a super-resolution CNN-based network for 3D image reconstruction using unsupervised learning, referred to as cross-sectional image reconstruction (C-SIR). The key idea is that different cross-sectional images can help each other for training the C-SIR model. Additionally, existing high resolution data can also be used for pre-training the network in a supervised manner. In experiments, we show that such a network can be trained to reconstruct 3D images using simulated down-sampled adult images with much better image quality and image segmentation accuracy. Then, we illustrate that the proposed C-SIR approach generates relatively clear 3D fetal images than other algorithms.

Keywords: Unsupervised learning · 3D reconstruction from cross-sectional images

1 Introduction

Although ultrasound is the most commonly used modality in prenatal imaging, it cannot depict detailed brain structures of the fetus for detailed quantification

X. Cao et al. (Eds.): MLMI 2023, LNCS 14348, pp. 33–41, 2024.
https://doi.org/10.1007/978-3-031-45673-2_4

Fig. 1. Typical cross-sectional views of an axial sequence of the fetal brain.

due to low echo signals, low tissue contrast, and limited field of view. Magnetic resonance imaging (MRI) [10] can be used as a supplementary technique to provide more detailed structural information and to quantify the brain structures. The advantage of fetal brain MRI is that it has good soft tissue contrast and covers the entire brain with high signal-to-noise ratio and can be collected from any section/direction [7]. Thus, fetal brain MRI is gaining increasing importance in early assessment of the fetal brain during prenatal examinations.

However, MRI acquisition usually takes longer time, and the fetus may move during scanning, causing motion artifacts [2]. To eliminate such effects, fast imaging methods such as single-shot fast spin-echo (SSFSE) that can obtain one high-resolution 2D slice in less than a second is used in clinical examination [8]. A volume can then be formed by stacking the cross-sectional 2D images [9], thereby providing multiplanar axial, sagittal, and coronal views via different acquisitions. Such images are blurry in opposite multiplanar views because the slice spacing is relatively large (4 to 6mm) than normal 3D MRI acquisitions. Moreover, there could be motion among slices and among different cross-sectional images. Thus, although these images are sufficient clinically for examining pathological cases, they could render difficulty for quantitative morphological analysis.

Figure 1 shows three orthogonal multiplanar views of a fetal brain MRI captured from the axial (of the fetal brain) direction. Besides that the axial plane has high resolution, the other two views are with low resolution due to large slice thickness. Conventionally, fetal MRIs are collected from three different planes including as axial, coronal, and sagittal sections, and physicians can review detailed structures from them, which provide complementary information. However, these images are not in 3D, and they (and their slices) are not aligned well due to fetal or mother movement during the scan.

To better quantify fetal brains we study 3D high-resolution image reconstruction from multiple 2D sequences. For such a typical super resolution (SR) problem, there are supervised and unsupervised methods available in the literature [6]. Supervised models can recover missing information through learning from low and high resolution (LR-HR) image pairs, and unsupervised SR can use some image regularization or generative adversarial networks (GAN) [5] to help training the SR networks. Additionally, to handle motion among slices, iterative manipulation of motion correction and super-resolution [3] can be adopted, in which previously reconstructed images can help motion correction, and then motion corrected images can help better reconstructing 3D images.

Regarding network structures, studies showed that convolutional neural network (CNN) models are capable to achieve complex nonlinear regression tasks [11,12]. The network structure of super-resolution convolutional neural network (SRCNN) [1] consists of only three convolutional layers, namely, feature extraction layer, nonlinear mapping layer and reconstructed layer. Even by using a shallow network composed of a few layers, SRCNN achieves superior results over other state-of-the-art SR methods. Therefore, we adopt the commonly used SRCNN structure as the backbone of our reconstruction network.

We propose a SR method called cross-sectional image reconstruction (C-SIR) to construct high-resolution 3D volumes from 2D images based on residual convolutional neural network model using unsupervised learning. The proposed model allows for the use of multiple input data taken from different planes and generates a high resolution 3D image for assisting quantitative analysis. The contributions of this paper are two-folds. First, because lack of high-resolution data for training, we proposed to use complementary information from different views for unsupervised learning. Second, simulated MR images are used for pre-training of the model.

2   Method

2.1   C-SIR Network Structure

The network is implemented based on the SRCNN structure, which aims to generate a 3D volume from a series of cross-sectional images. In order to deepen the network structure and further improve the super resolution effect, Kim et al. [4] proposed a very deep super resolution (VDSR) network. VDSR reconstructs HR images by learning the residual of high-frequency parts between HR images and LR images, which can make it easier to train and optimize the model. VDSR can obtain higher image quality and generate clearer image edges and textures at the same time, and quantitative results show that the performance of VDSR is improved compared with SRCNN. However, one disadvantage of VDSR is that image interpolation and amplification are performed before applying the network, which increases the number of network parameters and computational load. Therefore, we make the network structure deeper and use the deconvolutional layer to upscale the LR image. The mask of the LR image is applied as a weighted attention mechanism to enforce the network focusing only on the brain region and to eliminate the interference of background.

Figure 2 shows the network structure. C-SIR consists of an up-sampling module and three local residual modules. The up-sampling module includes a deconvolutional layer and a convolutional layer, which magnify the LR image to the same size as the HR image. Each local residual module consists of 6 convolutional layers using a kernel of size $3 \times 3 \times 3$, and each convolution is followed by a nonlinear mapping using a LeakyReLU activation function. The residue learned by the local residual module is added to the input data of the module, and the

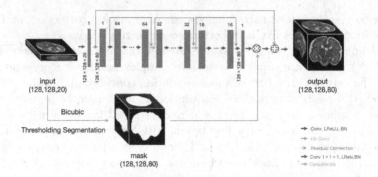

Fig. 2. The structure of SRCNN.

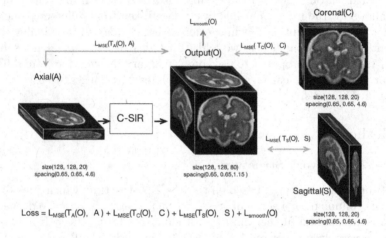

Fig. 3. Unsupervised learning using cross-sectional views.

summation result is used as the input of the next local residuals module. Finally, a global residual block is used to add the high frequency components extracted from the whole network to the output of the deconvolutional layer, and the final output of the network is obtained.

In experiments we fix the reconstruction ratio to 4, so an input image with size $128 \times 128 \times 20$ will generate an output image with size $128 \times 128 \times 80$. The attention mechanism of C-SIR is designed to multiply the brain mask and the residual map learned by the network, so that the residual learning only focuses on the brain region, preventing the interference of irrelevant noise. The global and local residual learning methods are adopted to reduce the complexity of the model and the difficulty of model training. At the same time, local residual learning is used to extract the features of the image as possible, which effectively avoids the degradation and gradient explosion of the model.

2.2   Unsupervised Training Strategy

In order to use the complementary information between multiple sequences, the C-SIR network is trained simultaneously to interpolate and reconstruct 3D high-resolution images from multiple 2D views based on an unsupervised learning strategy shown in Fig. 3.

Specifically, when an axial plane sequence is used as the input, the corresponding coronal and sagittal sequences will be used as the ground truth for training. The same network can be trained by alternating the inputs and the desired outputs using cross-sectional images of the same subject. Because there is no ground-truth 3D high-resolution images, the losses are designed only for the available slices.

There are two types of losses used during training. First, the mean squares error (MSE) loss is used to compute the similarity between input axial images and the output images; between the output and the coronal images; and between the output and the sagittal images. Herein, the similarity loss between input and output are only calculated on the corresponding slices to ensure that the output is similar to the input. For the similarity between different views, the losses are also only computed for voxels available in the coronal and the sagittal images when the inputs are axial images. The second loss is a apatial smoothness loss of the output 3D fetal brain MRI. Equation 1 is the summary of the loss functions used for training.

$$Loss = L_{MSE}(Output, Axial) + L_{smooth}(Output)$$
$$+ L_{MSE}(Output, T_C(Coronal)) \tag{1}$$
$$+ L_{MSE}(Output, T_S(Sagittal)).$$

Because there is possible motion between different sequences, the output image is transformed (registered) to the coronal image and the sagittal image, respectively, before computing the MSE losses with corresponding 2D sequences. Herein, registration among different views can be performed prior to the training. In Fig. 3, T_A, T_C, and T_S represent the transformations between the output space and the axial, coronal, and sagittal spaces respectively. In fact, the output space and the input space are always assumed to be the same, so only two transformations are used in Eq. 1 for each training data. In this study we use rigid registrations. Nonrigid deformable registration networks could also be used for motion correction in the future.

During the training stage, the C-SIR network can be trained by alternating axial, coronal, and sagittal images as the input and the corresponding images as the output, so that one network can be used to reconstruct the 3D images from different 2D sequences. The parameters of the network are updated by back-propagating the above loss. For MSE, we tried both l1 and l2 norms and found that l2 performed better than l1 norm for the image restoration and super-resolution reconstruction problem.

After training the network, for a given fetal subject, we can use the network to predict the 3D images from each 2D sequence separately and then combine the results after applying the registration matrices. Herein an simple average is used to generate the final 3D image.

Pre-training and Testing with Simulated Images. Our network generates a 80-slice high resolution image from a 20-slice low-resolution image. Although multiple sets of sequences can provide complementary information, the slices generated by the middle layer still lack some structural details. This may be caused by relatively small number of samples. So we introduce pre-training to supplement the corresponding information. We use 3D high-resolution isotropic adult brain MRI data to obtain low-resolution two-dimensional sequences in three orthogonal directions for supervised network training, from which we can extract as many common features as possible during pre-training, so as to reduce the learning burden of the model. The pre-training procedure is the same as that described above, and an additional 3D MSE loss is used between the output image and the ground truth, which is computed across the entire image domain.

3   Experiments and Results

Three hundred fetuses, 125 adult 3D (T1-weighted) images and 125 adult 3D (T2-weighted) images are used in our experiments. Each subject has three cross-sectional T2 sequences. Image intensity is normalized to a range between 0 and 200. The use of high-resolution adult images enables us to simulate the down-sampled cross-sectional images for evaluating the performance of the network. Moreover, they are also used for pre-training our network. To simulate the images, first, all the adult images are down-sampled to simulate axial, coronal, and sagittal 2D slices, uisng 1/4 nearest neighbor down-sampling operators in each direction.

During the training stage, 100 (50 T1-weighted, 50 T2-weighted) adult images are mixed and used for supervised pre-training. Then, the pre-trained network is further refined using additional 100 (50 T1-weighted, 50 T2-weighted) adult images (adult-recon) or 100 fetal images to refine the network in unsupervised strategy (fetal-recon), respectively. To demonstrate the usefulness of supervised pre-training, we simultaneously use 100 fetal images for unsupervised training of the model.

After training, the resultant fetal-recon and adult-recon networks are evaluated. Quantitative results are compared to demonstrate the goodness of reconstruction. For adult-recon, 50 (25 T1-weighted, 25 T2-weighted) adult images are used for quantitative comparison between different reconstruction algorithms using structural similarity (SSIM) and peak signal-to-noise ratio (PSNR). The down-sampled, reconstructed, and original 3D images are fed to FSL-FAST segmentation algorithm for tissue segmentation, and Dice coefficients of gray matter(GM) and white matter(WM) are computed.

Figures 4 and 5 show two examples of the reconstruction results for T1 and T2 adult images, respectively. It can be seen that the coronal and sagittal views of the input image is blurry because of thick slice distance (4mm). The results show

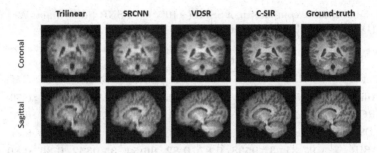

Fig. 4. Sample results for reconstruction of T1 adult images from axial sequences.

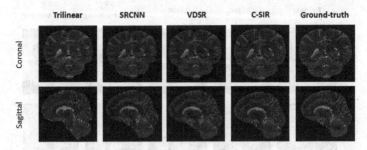

Fig. 5. Sample results for reconstruction of T2 adult images from axial sequences.

that the image reconstructed by trilinear interpolation lacks detailed information and shows an obvious checkerboard shape. Compared with trilinear interpolation, SRCNN has better reconstruction result and can better reflect the boundaries and details of the reconstructed image, but the image is still too smooth in general. The reconstructed images of VDSR are relatively supplemented with more edge details, but the whole image is relatively smooth. In general, the appearance of the reconstructed image of C-SIR is more similar to that of the original 3D image, brain tissues are more clear and can better reflect the details of the brain.

For all the testing images, SSIM and PNSR between the simulated down-sampled images and the original 3D images and those between the reconstructed images and the ground-truths are shown in Table 1. Additionally, Dice of white matter (WM) and gray matter (GM) are computed after segmenting the respective images w.r.t. those of the original adult T1/T2 images. It can be seen that after C-SIR reconstruction, all the metrics have been improved. The PNSR of the T1/T2 images are increased by more than 2.5dB, and the SSIM is increased by about 0.02. The quantitative results of adult MRIs show that our network can effectively improve the image quality of LR to HR mapping and gain for subsequent brain segmentation.

Because we have no 3D HR fetal MR images in hand, only qualitative results are presented for fetal-recon compared with SRCNN and VDSR, including those with and without pre-training (Fig. 6). We can see that SRCNN and VDSR are

Table 1. Quantitative comparison of SSIM ($\times 10^{-2}$) ↑, PSNR ([dB]) ↑ and WM (white matter) (Dice) ↑ and GM (gray matter) (Dice) ↑.

Task	T1 Adult MRI				T2 Adult MRI			
Method	SSIM	PSNR	WM	GM	SSIM	PSNR	WM	GM
Trilinear	95.53	34.9056	0.81	0.73	94.04	32.3014	0.80	0.70
SRCNN [1]	96.44	35.5423	0.84	0.76	95.09	33.7658	0.83	0.76
VDSR [4]	96.82	37.0132	0.85	0.78	95.89	34.5684	0.84	0.77
C-SIR	**97.31**	**37.9538**	**0.87**	**0.82**	**96.54**	**35.0352**	**0.86**	**0.80**

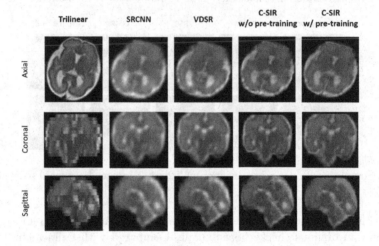

Fig. 6. Sample results for reconstruction of fetal brain images.

lack of details, and MRI reconstructed by C-SIR has been greatly improved with smooth transitions and sharp edges and fine scale details. The image results without pre-training have been able to display more image details, producing clearer image edges, but there are some obvious pixel missing shadows. After refining the pre-trained model, the reconstructed fetal brains are much better.

4   Conclusion

We proposed a cross-sectional image reconstruction (C-SIR) method to generate 3D images from 2D cross-sectional acquisitions based on unsupervised learning. Simulated down-sampled images with known ground-truths can also be used for pre-training the network. Because clinical MR acquisitions are often not using 3D protocols, 3D reconstruction can help compute quantification measures more precisely. For example, fetal brain MRIs acquired using 2D sequences in different views can be fused and mapped to form a 3D volume to automatically measure anatomical structures. In experiments, we used adult images with simulated down-sampled images for supervised pre-training (and evaluation) and then trained separate C-SIR networks for fetus images and additional adult images in unsupervised strategy. The results showed that the proposed approach is promising in terms of image quality and subsequent tissue segmentation accuracy. In the future, more images and modalities should be involved in the study, and deformable fields between different views can also be used to enhance the quality of reconstructed images.

Acknowledgement. This work was partially supported by The Key R and D Program of Guangdong Province, China (grant number 2021B0101420006), National Natural Science Foundation of China (grant number 62131015), Science and Technology Commission of Shanghai Municipality (STCSM) (grant number 21010502600) and the National Key Research and Development Program of China (2022ZD0209000).

References

1. Dong, C., Loy, C.C., He, K., Tang, X.: Image super-resolution using deep convolutional networks. IEEE Trans. Pattern Anal. Mach. Intell. **38**(2), 295–307 (2015)
2. Ebner, M., et al.: An automated framework for localization, segmentation and super-resolution reconstruction of fetal brain MRI. Neuroimage **206**, 116324 (2020)
3. Kainz, B., et al.: Fast volume reconstruction from motion corrupted stacks of 2D slices. IEEE Trans. Med. Imaging **34**(9), 1901–1913 (2015)
4. Kim, J., Lee, J.K., Lee, K.M.: Accurate image super-resolution using very deep convolutional networks. In: Proceedings of The IEEE Conference on Computer Vision and Pattern Recognition, pp. 1646–1654 (2016)
5. Ledig, C., et al.: Photo-realistic single image super-resolution using a generative adversarial network. In: Proceedings of The IEEE Conference on Computer Vision and Pattern Recognition, pp. 4681–4690 (2017)
6. Meshaka, R., Gaunt, T., Shelmerdine, S.C.: Artificial intelligence applied to fetal MRI: a scoping review of current research. Br. J. Radiol. **96**(1147), 20211205 (2023)
7. Nagaraj, U.D., Kline-Fath, B.M.: Clinical applications of fetal MRI in the brain. Diagnostics **12**(3), 764 (2022)
8. Saleem, S.: Fetal MRI: an approach to practice: a review. J. Adv. Res. **5**(5), 507–523 (2014)
9. Sui, Y., Afacan, O., Jaimes, C., Gholipour, A., Warfield, S.K.: Scan-specific generative neural network for MRI super-resolution reconstruction. IEEE Trans. Med. Imaging **41**(6), 1383–1399 (2022)
10. Wu, S., Sun, Y., Zhou, X., Gong, X., Yang, Y.: Magnetic resonance imaging: principles and recent advances. J. Med. Syst. **42**(8), 148 (2018)
11. Zhang, J., Cui, Z., Jiang, C., Guo, S., Gao, F., Shen, D.: Hierarchical organ-aware total-body standard-dose PET reconstruction from low-dose PET and CT images. IEEE Transactions on Neural Networks and Learning Systems (2023)
12. Zhang, J., Cui, Z., Jiang, C., Zhang, J., Gao, F., Shen, D.: Mapping in cycles: dual-domain PET-CT synthesis framework with cycle-consistent constraints. In: Wang, L., Dou, Q., Fletcher, P.T., Speidel, S., Li, S. (eds.) Medical Image Computing and Computer Assisted Intervention – MICCAI 2022. MICCAI 2022. Lecture Notes in Computer Science, vol. 13436, pp. 758–767. Springer, Cham (2022). https://doi.org/10.1007/978-3-031-16446-0_72

Robust Unsupervised Super-Resolution of Infant MRI via Dual-Modal Deep Image Prior

Cheng Che Tsai[1], Xiaoyang Chen[2,3], Sahar Ahmad[2,3],
and Pew-Thian Yap[2,3(✉)]

[1] Department of Computer Science, University of North Carolina, Chapel Hill, NC,
USA
[2] Department of Radiology, University of North Carolina, Chapel Hill, NC, USA
[3] Biomedical Research Imaging Center, University of North Carolina, Chapel Hill,
NC, USA
ptyap@med.unc.edu

Abstract. Magnetic resonance imaging (MRI) is commonly used for studying infant brain development. However, due to the lengthy image acquisition time and limited subject compliance, high-quality infant MRI can be challenging. Without imposing additional burden on image acquisition, image super-resolution (SR) can be used to enhance image quality post-acquisition. Most SR techniques are supervised and trained on multiple aligned low-resolution (LR) and high-resolution (HR) image pairs, which in practice are not usually available. Unlike supervised approaches, Deep Image Prior (DIP) can be employed for unsupervised single-image SR, utilizing solely the input LR image for *de novo* optimization to produce an HR image. However, determining when to stop early in DIP training is non-trivial and presents a challenge to fully automating the SR process. To address this issue, we constrain the low-frequency k-space of the SR image to be similar to that of the LR image. We further improve performance by designing a dual-modal framework that leverages shared anatomical information between T1-weighted and T2-weighted images. We evaluated our model, dual-modal DIP (dmDIP), on infant MRI data acquired from birth to one year of age, demonstrating that enhanced image quality can be obtained with substantially reduced sensitivity to early stopping.

Keywords: Infant MRI · Unsupervised Learning · Super-Resolution · Dual-Modality

1 Introduction

Studies seeking to understand the structural and functional organization of the early developing brain are often hindered by the difficulty in acquiring high-quality infant magnetic resonance (MR) images [5]. The standard protocol for

X. Cao et al. (Eds.): MLMI 2023, LNCS 14348, pp. 42–51, 2024.
https://doi.org/10.1007/978-3-031-45673-2_5

high-resolution (HR) infant MRI might take tens of minutes [1] and is hence challenging for babies who typically cannot lie still in a rumbling scanner for a prolonged period of time. Specialized protocols developed to accelerate infant MRI [10,11] typically involve data undersampling [10] and hence tend to degrade image quality.

Recent advancements in super-resolution (SR) techniques based on deep learning afford new opportunities to overcome challenges in infant brain MRI [2–4,7,12,16]. SR aims to enhance image quality and reconstruct HR images from their low-resolution (LR) counterparts without requiring changes in imaging protocols [6]. Most SR methods are based on supervised learning, which involves training models for generating realistic high-resolution (HR) images by mapping LR images to their paired HR counterparts. However, the applicability of supervised methods is limited when ground truth HR images is unavailable for model training. By contrast, unsupervised learning circumvents this limitation by relying only on the information in LR images to obtain HR images. Therefore, unsupervised methods can be applied to super-resolve LR images to an arbitrary target resolution and are appealing for infant MRI SR since acquiring HR images is typically very challenging.

Deep image prior (DIP) is a relatively recent framework applicable for unsupervised SR [13]. It can be trained with a single input LR image to output the corresponding super-resolved image. This single-image super-resolution (SISR) approach has several benefits when dealing with medical images. First, SISR can be applied to populations with limited training data, such as infant brain MRI. Second, SISR can provide instance-level optimized results and avoid realistic-looking artifacts that are commonly presented in SR models trained with multiple images [17]. However, one of the major challenges of DIP is its sensitivity to the stopping criterion [13]. Stopping too early leads to under-resolved images, whereas stopping too late causes overfitting to noise and artifacts. Moreover, the vanilla DIP [13] disregards rich relationships among different imaging modalities, which can provide complementary information conducive to improving SR [2].

In this paper, we introduce a DIP-inspired model for robust SR in infant brain MRI. Leveraging the fact that the T1-weighted (T1w) and T2-weighted (T2w) images of an identical subject share the same anatomical features, we propose a dual-modal DIP (dmDIP) framework for concurrent SR of T1w and T2w images from their common feature maps. Additionally, we demonstrate that a simple k-space manipulation trick can substantially stabilize SR outcomes and reduce sensitivity to the early stopping criterion. We validate the efficacy of dmDIP, both quantitatively and qualitatively, in comparison with the vanilla DIP.

2   Methods

2.1   Model Architecture

Deep Image Prior (DIP). Let x_0 be a given LR image, the super-resolved HR image $\hat{x}$ can be determined as

$$\hat{x} = f_{\hat{\theta}}(z), \ \hat{\theta} = \arg\min_{\theta} \|x_0 - d(f_\theta(z))\|, \tag{1}$$

where z is the random noise input to the neural network f_θ, which needs to be trained, and $d(\cdot) : \mathbb{R}^{C \times rH \times rW} \to \mathbb{R}^{C \times H \times W}$ is a downsampler with factor r. Downsampling is carried out by retaining only the central k-space, consisting of low-frequency components, of an input image. After removing the high-frequency components, inverse Fourier transform was applied to generate the LR images.

Dual-Modal DIP (dmDIP). Earlier research has demonstrated that T1-weighted (T1w) and T2-weighted (T2w) images capture complementary information that can be beneficial for SR [2,4,9]. We incorporated this concept into the vanilla DIP in two cascading steps (Fig. 1b).

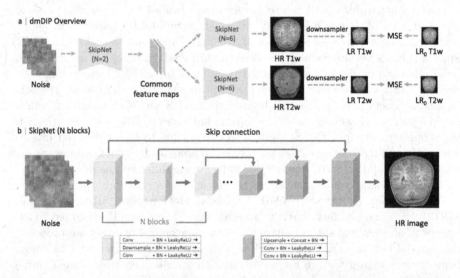

Fig. 1. (a) Overview of dual-modal DIP (dmDIP). (b) SkipNet with N sets of skip-connected downsampling-upsampling blocks.

We employ a SkipNet [15], denoted as f_{θ_1}, to learn feature maps common to both T1w and T2w images with input noise z.

These maps are then fed as inputs to two separate SkipNet arms, f_{θ_2} and f_{θ_3}, for generating in parallel the HR T1w and T2w images, respectively. Our model is trained end-to-end to solve

$$\{\hat{\theta}_1, \hat{\theta}_2, \hat{\theta}_3\} = \arg\min_{\theta_1, \theta_2, \theta_3} \|x_0^{\text{T1w}} - d(f_{\theta_2}(f_{\theta_1}(z)))\| + \|x_0^{\text{T2w}} - d(f_{\theta_3}(f_{\theta_1}(z)))\|. \tag{2}$$

Central K-space Replacement (KR). A significant limitation of DIP is its reliance on early stopping. DIP provides the best SR outcome before it over-fits and starts picking up irrelevant imaging noise [13]. Conversely, stopping the process too early could result in a suboptimal model that is insufficient to fully generate anatomical details. This dilemma necessitates the need for manual inspection of the network output to empirically set a stopping point.

To overcome this limitation, we propose a k-space replacement method to stabilize the outcome of each optimization iteration. The core idea is to ensure that a generated HR image will possess precisely the same low-frequency com-ponents as its corresponding LR image. For each iteration, we substitute the central k-space of the HR image with the k-space of the corresponding LR image. This strategy helps to establish a lower bound on the accuracy of the generated HR image, promoting stability while the optimization proceeds to refine high-frequency components for image details. This stability reduces sensitivity to the point of early stopping and is particularly valuable when an HR image is not available to help manually or numerically determine optimal stopping.

Implementation Details. For both vanilla DIP and dmDIP, we conducted 7,000 training iterations and recorded the SR outcomes every 100 iterations. Both models were optimized using the Adam optimizer with learning rate 0.01 and batch size 1. For fair comparison, we employed the SkipNet used in the vanilla DIP as the foundational architecture for constructing dmDIP. In dmDIP, the common feature SkipNet consisted of two blocks (Fig. 1a).The subsequent two modality-specific SkipNets each consisted of 6 blocks.

All implementation was carried out via PyTorch and models were trained using an NVIDIA RTX 2080 GPU. On average, dmDIP took about 16 min to super-resolve a set of T1w and T2w images, approximately 33% longer than the vanilla DIP (12 min).

3   Results

3.1   Data Processing

For evaluation, we used MRI data collected via the UNC/UMN Baby Connec-tome Project (BCP) [8], which involved children across the first 5 years of life. T1w and T2w images (0.8 mm isotropic resolution) of 5 infants, 0.5, 3, 6, 9, and 12 months of age, to cover a period where MRI tissue contrast exhibits rapid variation. For each subject, we extracted HR images of size 192×192 at 10 ran-dom coronal sections of both T1w and T2w images, giving a total of 100 image slices. These image slices were downsampled to generate the LR images.

3.2   Visual and Numerical Evaluation

We compared dmDIP with bicubic upsampling, sinc interpolation, and vanilla DIP. dmDIP outperforms the other three methods for both undersampling fac-tors $r = 2$ and $r = 4$ in terms of PSNR (Table 1).

Table 1. Comparison of SR methods in terms of PSNR (mean ± S.D.).

Method	r = 2		r = 4	
	T1w	T2w	T1w	T2w
Bicubic	28.01 ± 1.5	26.60 ± 0.88	23.86 ± 1.59	22.90 ± 0.84
Sinc	31.37 ± 1.21	29.17 ± 0.95	26.41 ± 1.6	25.35 ± 0.91
DIP	30.75 ± 1.88	28.85 ± 0.89	26.84 ± 1.51	25.60 ± 0.91
dmDIP	**32.43 ± 1.15**	**29.79 ± 0.89**	**27.69 ± 1.38**	**26.31 ± 1.05**

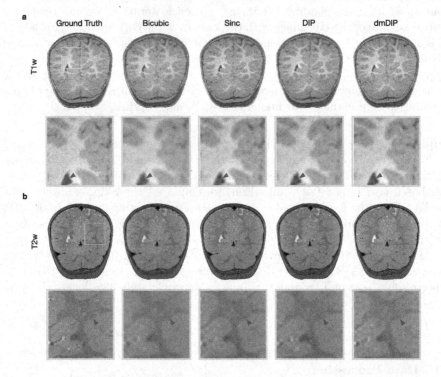

Fig. 2. Coronal views of super-resolved (a) T1w and (b) T2w LR images of a 1-year-old obtained by different methods for factor $r = 2$.

It can be seen from Fig. 2 that dmDIP generates sharp images with distinct boundaries between white matter and lateral ventricles (zoomed-in view in Fig. 2a) and between white matter and cortical gray matter (zoomed-in view in Fig. 2b). While quantitatively better than bicubic interpolation, sinc interpolation suffers from persistent Gibbs ringing artifacts [14] (Fig. 2a).

To ensure the robustness of these findings, we inspected the MR images of a three-month-old baby. At this age, the gray-white matter contrast is inverted when compared to that of a one-year-old infant. Despite this difference, dmDIP

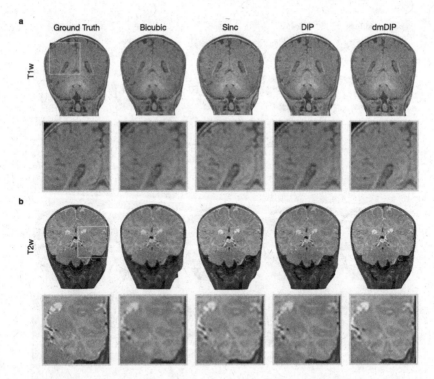

Fig. 3. Coronal views of super-resolved (a) T1w and (b) T2w LR images of a 3-month-old obtained by different method for factor $r = 2$.

consistently yields the best detail-preserving SR outcome compared with the other methods (Fig. 3).

3.3 Beyond the Original Resolution

As an unsupervised method, dmDIP allows SR beyond the original image resolution without requiring images at higher resolution to be acquired. For T1w SR (Fig. 4a), dmDIP outperforms all other methods by generating an image with sharp details and good tissue contrast. For T2w SR (Fig. 4b), DIP and dmDIP yield competitive results; dmDIP tends to preserve more details but also more noise. DIP provides a clean image but with relatively fewer details.

3.4 Ablation Study

To verify the benefits of dual-modal training and KR, we compared the four models below:

– DIP w/o KR: Vanilla DIP.
– DIP w/ KR: Vanilla DIP with KR after each iteration.

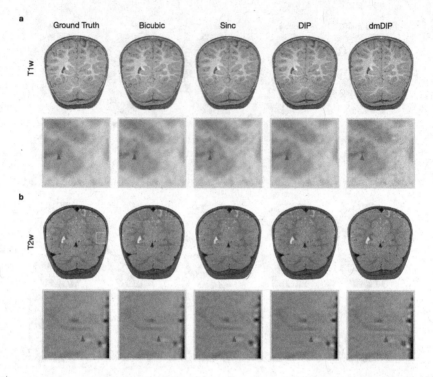

Fig. 4. Coronal views of super-resolved (a) T1w and (b) T2w HR images to beyond the original resolution. The result for a 1-year-old is obtained by different methods for upsampling factor 4.

– dmDIP w/o KR: Dual-modal training without KR.
– dmDIP w/ KR: The proposed model.

The training trajectories of the four models (Fig. 5) suggest that while dmDIP is a slower learner in the first 500 ($r = 2$) or 2000 ($r = 4$) iterations, it performs better than the vanilla DIP thereafter. The performance difference becomes more substantial for the harder SR task ($r = 4$).

Figures 5a–5d show that adding KR improves both DIP and dmDIP by desensitizing them to stopping criteria while maintaining performance. Note that dmDIP with KR consistently achieves the best performance at almost every iteration recorded.

As the contrast between gray and white matter in infant MRI changes rapidly with brain development, we evaluated the performance of models across age (Fig. 6). The SR performance between the four models is consistent across time points. DIP with KR occasionally surpasses dmDIP with KR for SR with $r = 2$. dmDIP with KR outperforms all the other models in every age group when $r = 4$. Lastly, the two models with KR consistently outperform their counterparts without KR.

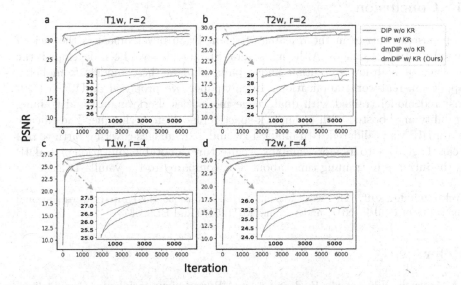

Fig. 5. Utilizing k-space replacement (KR) to boost the performance of DIP and dmDIP. KR desensitizes model training to stopping criterion while maintaining better performance. Differences in training trajectories between DIP and dmDIP become larger for more challenging SR tasks (c, d). All results are presented in terms of average PSNR at each iteration.

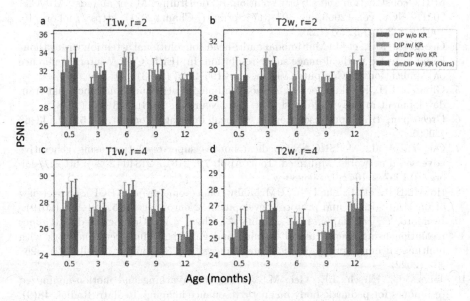

Fig. 6. The benefits of dual-modal training and k-space replacement (KR) are consistent across age.

4   Conclusion

A persistent barrier in the research of infant brain development is the lack of high-quality MR images. Although several SR models have been developed, the absence of high-low resolution image pairs hinders these models from being applied in real-world applications. In this paper, we proposed dmDIP, a DIP-inspired model trained with dual image modalities. Performance and training stability are boosted with a simple k-space replacement technique. The efficacy of dmDIP was validated with infant T1w and T2w images covering birth to one year of age. Due to its more complex architecture, a notable limitation of dmDIP is the increase in training time, about 33% compared to the vanilla DIP.

Acknowledgment. This work was supported in part by the United States National Institutes of Health (NIH) under grants MH125479 and EB008374.

References

1. Antonov, N.K., et al.: Feed and wrap MRI technique in infants. Clin. Pediatr. **56**(12), 1095–1103 (2017)
2. Feng, C.-M., Fu, H., Yuan, S., Xu, Y.: Multi-contrast MRI super-resolution via a multi-stage integration network. In: de Bruijne, M., et al. (eds.) MICCAI 2021. LNCS, vol. 12906, pp. 140–149. Springer, Cham (2021). https://doi.org/10.1007/978-3-030-87231-1_14
3. Feng, C.-M., Yan, Y., Fu, H., Chen, L., Xu, Y.: Task transformer network for joint MRI reconstruction and super-resolution. In: de Bruijne, M., et al. (eds.) MICCAI 2021. LNCS, vol. 12906, pp. 307–317. Springer, Cham (2021). https://doi.org/10.1007/978-3-030-87231-1_30
4. Georgescu, M.I., et al.: Multimodal multi-head convolutional attention with various kernel sizes for medical image super-resolution. In: IEEE/CVF Winter Conference on Applications of Computer Vision, pp. 2195–2205 (2023)
5. Gilmore, J.H., Knickmeyer, R.C., Gao, W.: Imaging structural and functional brain development in early childhood. Nat. Rev. Neurosci. **19**(3), 123–137 (2018)
6. Greenspan, H.: Super-resolution in medical imaging. Comput. J. **52**(1), 43–63 (2009)
7. Gu, Y., et al.: MedSRGAN: medical images super-resolution using generative adversarial networks. Multimed. Tools Appl. **79**, 21815–21840 (2020). https://doi.org/10.1007/s11042-020-08980-w
8. Howell, B.R., et al.: The UNC/UMN baby connectome project (BCP): an overview of the study design and protocol development. Neuroimage **185**, 891–905 (2019)
9. Iwamoto, Y., Takeda, K., Li, Y., Shiino, A., Chen, Y.W.: Unsupervised MRI super resolution using deep external learning and guided residual dense network with multimodal image priors. IEEE Trans. Emerg. Topics Comput. Intell. **7**(2), 426–435 (2022)
10. Jaimes, C., Kirsch, J.E., Gee, M.S.: Fast, free-breathing and motion-minimized techniques for pediatric body magnetic resonance imaging. Pediatr. Radiol. **48**(9), 1197–1208 (2018). https://doi.org/10.1007/s00247-018-4116-x
11. Lindberg, D.M., et al.: Feasibility and accuracy of fast MRI versus CT for traumatic brain injury in young children. Pediatrics **144**(4), e20190419 (2019)

12. Mahapatra, D., Bozorgtabar, B., Garnavi, R.: Image super-resolution using progressive generative adversarial networks for medical image analysis. Comput. Med. Imaging Graph. **71**, 30–39 (2019)

13. Ulyanov, D., Vedaldi, A., Lempitsky, V.: Deep image prior. In: Proceedings of the IEEE Conference on Computer Vision and Pattern Recognition, pp. 9446–9454 (2018)

14. Veraart, J., Fieremans, E., Jelescu, I.O., Knoll, F., Novikov, D.S.: Gibbs ringing in diffusion MRI. Magn. Reson. Med. **76**(1), 301–314 (2016)

15. Wang, X., Yu, F., Dou, Z.-Y., Darrell, T., Gonzalez, J.E.: SkipNet: learning dynamic routing in convolutional networks. In: Ferrari, V., Hebert, M., Sminchisescu, C., Weiss, Y. (eds.) ECCV 2018. LNCS, vol. 11217, pp. 420–436. Springer, Cham (2018). https://doi.org/10.1007/978-3-030-01261-8_25

16. Zhao, X., Zhang, Y., Zhang, T., Zou, X.: Channel splitting network for single MR image super-resolution. IEEE Trans. Image Process. **28**(11), 5649–5662 (2019)

17. Zhu, J., Yang, G., Lio, P.: How can we make GAN perform better in single medical image super-resolution? A lesion focused multi-scale approach. In: IEEE International Symposium on Biomedical Imaging (ISBI), pp. 1669–1673 (2019)

SR4ZCT: Self-supervised Through-Plane Resolution Enhancement for CT Images with Arbitrary Resolution and Overlap

Jiayang Shi[✉], Daniël M. Pelt, and K. Joost Batenburg

Leiden University, Leiden, The Netherlands
{j.shi,d.m.pelt,k.j.batenburg}@liacs.leidenuniv.nl

Abstract. Computed tomography (CT) is a widely used non-invasive medical imaging technique for disease diagnosis. The diagnostic accuracy is often affected by image resolution, which can be insufficient in practice. For medical CT images, the through-plane resolution is often worse than the in-plane resolution and there can be overlap between slices, causing difficulties in diagnoses. Self-supervised methods for through-plane resolution enhancement, which train on in-plane images and infer on through-plane images, have shown promise for both CT and MRI imaging. However, existing self-supervised methods either neglect overlap or can only handle specific cases with fixed combinations of resolution and overlap. To address these limitations, we propose a self-supervised method called SR4ZCT. It employs the same off-axis training approach while being capable of handling arbitrary combinations of resolution and overlap. Our method explicitly models the relationship between resolutions and voxel spacings of different planes to accurately simulate training images that match the original through-plane images. We highlight the significance of accurate modeling in self-supervised off-axis training and demonstrate the effectiveness of SR4ZCT using a real-world dataset.

Keywords: CT · Resolution enhancement · Self-supervised learning

1 Introduction

CT is a valuable tool in disease diagnosis due to its ability to perform non-invasive examinations [5]. The diagnostic accuracy of CT imaging is dependent on the resolution of the images, which can sometimes be insufficient in practice [8]. For medical CT images, the through-plane resolution is often inferior to the in-plane axial resolution [1,17], resulting in anisotropic reconstructed voxels that can create difficulties in identifying lesions and consequently lead to inaccurate diagnoses [1,3,6]. Furthermore, medical CT scans are commonly performed in

Supplementary Information The online version contains supplementary material available at https://doi.org/10.1007/978-3-031-45673-2_6.

a helical trajectory, which can cause overlapping slices along the through-plane axis [2,9]. The overlap caused by certain combinations of through-plane resolution and spacing introduces extra blurriness in sagittal and coronal images and potentially leads to inaccuracies in image interpretation [4,7,15].

To improve the through-plane resolution of CT images, various deep learning-based methods have been proposed. Supervised methods that use pairs of low-(LR) and high-resolution (HR) volumes have demonstrated promising results [11,14,19]. However, HR reference volumes may not always be available. As an alternative, self-supervised methods have been introduced, training on in-plane images and applying the learned model to through-plane images [18,20]. These methods simulate LR training images from HR in-plane images that match the through-plane images. However, they have limitations in handling CT images with overlap or complicated resolution/overlap combinations. For example, [18] can only handle CT images without overlap and integer resolution ratios, while the use of convolutional operations in [20] restricts its usage to certain specific combinations of overlap and resolution.

In this work, we propose a self-supervised method for enhancing the through-plane resolution of CT images by explicitly modeling the relation between the resolutions and spacings of different planes. Our method builds upon the same idea of off-axis training introduced in [18,20], but with the ability to handle CT images with arbitrary resolution/overlap. We use accurate interpolation techniques to simulate LR images from HR images, considering the specific resolution and overlap parameters. Our contributions are: (1) proposing a self-supervised method that can enhance resolution for CT images with arbitrary resolution and overlap, (2) demonstrating the importance of accurate modeling in off-axis training, and (3) applying our method to real-world datasets with complicated resolution/overlap combinations, showcasing its practical applicability.

2 Method

Problem Statement and Method. We present in this part a generalized problem statement for enhancing the resolution of CT images and describe our proposed method SR4ZCT as illustrated in Fig. 1. We denote the reconstructed CT volume by $\mathbf{I}(x,y,z) \in \mathbb{R}^{X \times Y \times Z}$, where X, Y, Z represent the pixel numbers along x, y, z axes, respectively. CT images can be viewed along three different orientations: axial, coronal, and sagittal. We denote the axial image, the coronal image, and the sagittal image at the position z', y', and x' along the z axis, the y axis, and the x axis as $\mathbf{a}_{z'} = \mathbf{I}(:,:,z')$, $\mathbf{c}_{y'} = \mathbf{I}(:,y',:)$, and $\mathbf{s}_{x'} = \mathbf{I}(x',:,:)$, respectively. The voxel size of the CT volume is defined as $r^x \times r^y \times r^z$, where often $r^x = r^y = r^{xy} < r^z$ for medical CT images. The spacing distance d between the centers of neighboring voxels along the z-axis can be smaller than r^z, resulting in the overlap $o^z = r^z - d$ in the through-plane direction. Given the target voxel size $r^x \times r^y \times r^z_{tar}$ and overlap o^z_{tar}, the goal of resolution enhancement for CT through-plane images is to reduce their voxel size and overlap in z direction, hereby increasing the through-plane resolution.

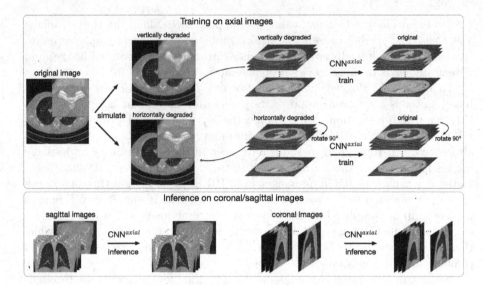

Fig. 1. Overview of our SR4ZCT method.

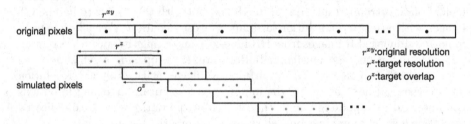

Fig. 2. $\mathcal{F}_{\downarrow}^{ver}$ to simulate arbitrary resolution and overlap. The function linearly interpolates at various points (indicated by different colors) inside the pixels, these interpolated points are then averaged to be the values of the current pixels.

SR4ZCT is based on generating a training dataset that consists of HR axial images as targets and their corresponding virtual through-plane-like axial images as inputs. To achieve this, we define the downscaling function $\mathcal{F}_{\downarrow}^{ver}$ as shown in Fig. 2, which downscales an axial image $\mathbf{a}_i$ with pixel sizes $r^{xy} \times r^{xy}$ by employing linear interpolation in the vertical direction. The output is a virtual through-plane-like image with pixel sizes $r^z \times r^{xy}$ and overlaps o^z between image rows. We similarly define the upscaling function $\mathcal{F}_{\uparrow}^{ver}$ that upscales the virtual through-plane-like image with pixel sizes $r^z \times r^{xy}$ and overlap o^z using linear interpolation to produce a degraded axial image with pixel sizes $r^{xy} \times r^{xy}$ and zero overlap. In addition, we define scaling functions $\mathcal{F}_{\downarrow}^{hor}$ and $\mathcal{F}_{\uparrow}^{hor}$ that apply the scaling in the horizontal direction instead of the vertical direction.

Training data is produced by first creating two network input images for each axial image $\mathbf{a}_i$: $\mathbf{a}_i^{ver} = \mathcal{F}_{\uparrow}^{ver}(\mathcal{F}_{\downarrow}^{ver}(\mathbf{a}_i))$ and $\mathbf{a}_i^{hor} = \mathcal{F}_{\uparrow}^{hor}(\mathcal{F}_{\downarrow}^{hor}(\mathbf{a}_i))$. Next, the simulated images are fed into the neural network f_θ to learn the mapping from degraded images $\mathbf{a}_i^{ver}$ and $\mathbf{a}_i^{hor}$ to their corresponding original axial image $\mathbf{a}_i$. As the resolution enhancement is along the z axis, which is always in the vertical direction for coronal and sagittal images, the horizontally degraded images and their corresponding axial images are rotated 90° using rotation function $\mathcal{R}$. The training is performed on all Z axial images and their corresponding degraded images using the loss function L, resulting in $2 \times Z$ training pairs. The weights θ of the neural network f_θ are determined by minimizing the total loss as

$$\theta^* = \min_\theta \sum_{i=1}^{Z} L(f_\theta(\mathbf{a}_i^{ver}), \mathbf{a}_i) + L(f_\theta(\mathcal{R}(\mathbf{a}_i^{hor})), \mathcal{R}(\mathbf{a}_i)). \tag{1}$$

After training, the network f_θ represents the learned mapping from virtual through-plane-like axial images with resolution $r^z \times r^{xy}$ and overlap o^z between image rows to HR axial image with $r^{xy} \times r^{xy}$ resolution. With the assumption that CT images often share similar features from different orientations, we can extend the learned mapping from axial images to coronal and sagittal images. To enhance the resolution, we first use $\mathcal{F}_\uparrow^{ver}$ to upscale coronal and sagittal images with voxel sizes $r^z \times r^{xy}$ and overlap o^z between image rows to images with resolution $r^{xy} \times r^{xy}$. Subsequently, we apply the trained neural network f_θ directly to the upscaled coronal and sagittal images. The outputs of the neural network correspond to the improved images with enhanced resolution.

In general, any image-to-image neural network could be used as f_θ. In this work, we use 2D MS-D networks [13] with 100 layers, trained for 200 epochs using L2 loss and ADAM optimizer [10]. The code is available on GitHub.[1]

3   Experiments and Results

Comparison with Supervised Learning. We conducted an experiment to evaluate the effectiveness of SR4ZCT in improving the resolution of CT images, comparing it to the supervised learning method. We selected four volumes (nr. 06, 14, 22, 23) from the `Task06 Lung` dataset of the Medical Segmentation Decathlon [16]. These volumes had the same slice thickness ($0.625\,mm$) but varying axial resolutions ranging from $0.74\,mm$ to $0.97\,mm$, resulting in a total of 2170 axial images of size 512×512 pixels. We downscaled the axial images to a resolution of $1\,mm$, creating volumes with voxel sizes of $1\,mm \times 1\,mm \times 0.65\,mm$. The through-plane resolution of the simulated volumes ranged from $2.5\,mm$ with $1.25\,mm$ overlap to $6.25\,mm$ with $3.125\,mm$ overlap. We trained the supervised learning method using three LR and HR volume pairs (numbers 06, 14, 22), and evaluated it on the remaining LR CT volume (nr. 23). We utilized two distinct 2D MS-D networks to train the coronal and sagittal images separately in a supervised manner, thereby achieving the best possible results through supervised learning. We applied SR4ZCT to the LR CT volume (nr. 23) and compared the results with those obtained using the supervised learning method.

---

[1] https://github.com/jiayangshi/SR4ZCT.

Table 1. Comparison of SR4ZCT and supervised learning on simulated CT volumes. The used metrics are PSNR and SSIM calculated from coronal and sagittal images in the central area, after cropping the empty regions.

resolution	overlap	view	original		supervised		ours	
			PSNR	SSIM	PSNR	SSIM	PSNR	SSIM
2.5 mm	1.25 mm	cor	37.71	0.972	**51.94**	**0.998**	46.48	0.995
		sag	39.12	0.977	**52.50**	**0.998**	47.68	0.995
3.75 mm	1.875 mm	cor	33.65	0.934	**42.96**	**0.987**	42.36	0.985
		sag	35.08	0.945	**43.73**	**0.989**	43.39	0.987
5 mm	2.5 mm	cor	31.43	0.895	37.70	0.965	**38.46**	**0.967**
		sag	32.87	0.913	39.36	**0.973**	**39.55**	0.973
6.25 mm	3.125 mm	cor	29.91	0.9859	34.74	0.938	**35.88**	**0.946**
		sag	31.32	0.882	36.02	0.949	**36.96**	**0.955**

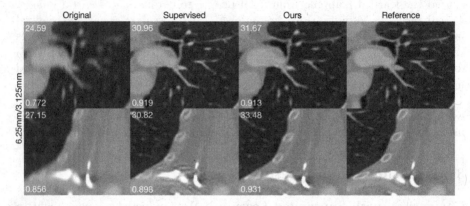

Fig. 3. The comparison of supervised learning and SR4ZCT on 6.25 mm resolution and 3.125 mm overlap dataset. Patches are shown for visualization. The PSNR and SSIM of each patch are shown in the top- and bottom-left corners.

Table 1 shows that SR4ZCT successfully enhanced the resolution of all four simulated CT volumes, as indicated by the improvements in both PSNR and SSIM. While SR4ZCT showed slightly lower PSNR and SSIM values than the supervised learning method for the 2.5 mm/1.25 mm case, its performance approached and surpassed that of the supervised learning method as the resolution decreased. Specifically, for the 3.75 mm/1.875 mm case, SR4ZCT achieved similar results to supervised learning, whereas in the 5 mm/2.5 mm and 6.25 mm/3.125 mm cases, SR4ZCT outperformed supervised learning. Figure 3 presents an example of the 6.25 mm/3.125 mm case, where both the supervised learning method and SR4ZCT performed similarly in the lung area as SR4ZCT, but supervised learning created artifacts on the edge area, highlighting one of its disadvantages as it depends on the quality and quantity of training data. If the training data slightly differs from the testing data or is insufficient in amount, it may perform suboptimally. In contrast, SR4ZCT only makes use of the same data for training and testing, reducing the potential gap between training and testing data and yielding more robust results. Overall, our results demonstrate the effectiveness of SR4ZCT in enhancing the resolution of through-plane images, with similar or even superior performance compared with supervised learning.

Comparison with SMORE. SMORE [20] is a method designed for MRI images and relies on the fact that MRI acquisition is akin to sampling in the Fourier domain. It simulates LR images from HR images by removing the high-frequency components. However, SMORE is not directly applicable to CT imaging. Nonetheless, we implemented a version of SMORE based on its core idea of simulating LR images by convolving HR images. While convolution can simulate certain resolution/overlap combinations by adjusting the filter kernel and stride, it fails to accurately simulate specific resolution/overlap combinations, when the target resolution/overlap is not an integer multiple of the HR resolution/overlap.

To compare our method with SMORE, we used the same volumes of the Medical Segmentation Decathlon dataset [16] and downscaled them in the vertical direction to simulate low through-plane resolution with overlaps of $2\,mm$, $2.5\,mm$, and $3\,mm$ for a through-plane resolution of $5\,mm$. The results in Table 2 demonstrate that SMORE performs better when the simulated LR images using convolution match the through-plane resolution/overlap. For example, applying 1D convolution with a filter width of 5 and a stride of 3 on the $1\,mm \times 1\,mm$ HR axial images is equivalent to simulating a resolution of $5\,mm$ and an overlap of $2\,mm$. It achieved higher PSNR values on the $5\,mm/2\,mm$ volumes than the one using a stride of 2. When the target overlap is $2.5\,mm$, no convolution configuration accurately simulates it by adjusting the stride and resulting in lower PSNRs. The same limitation applies when the ratio of actual and target resolution is non-integer, making it impossible to find an accurate convolution filter with a decimal filter width. These limitations of convolution result in suboptimal performance. In contrast, our method simulates LR images using accurate interpolation, which can handle arbitrary resolution/overlap combinations.

Table 2. Results of SMORE and our method applied on the simulated dataset. The used metric is PSNR, the values are computed based on the average PSNR of the coronal images of all testing volumes. Output images are given in the supplementary file.

	$5\,mm/2\,mm$	$5\,mm/2.5\,mm$	$5\,mm/3\,mm$
SMORE stride 2	37.15	38.39	43.55
SMORE stride 3	41.93	41.86	41.41
ours	**42.93**	**43.30**	**44.75**

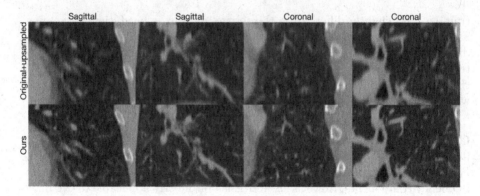

Fig. 4. Results of SR4ZCT applied on L291 from Low Dose CT Grand Challenge [12]. We show two image patches of coronal and sagittal images. The first row contains patches of original coronal and sagittal images, and the second row shows the output of SR4ZCT. Extra results are shown in the supplementary file.

Real-World CT Images Without Reference. We present this experiment designed to evaluate the effectiveness of SR4ZCT on real-world CT images with anisotropic resolution and overlap. We selected a CT volume of patient L291 from the Low Dose CT Challenge [12], which has an in-plane resolution of $0.74\,mm$, through-plane slice thickness of $3\,mm$ and $1\,mm$ overlap. As no reference volume without overlap was available, we present the result only for visual assessment.

Figure 4 illustrates the improvement in resolution of the original coronal and sagittal images using SR4ZCT. The enhanced images exhibit sharper details compared to the original images, as observed in the visual comparison. This experiment provides evidence of SR4ZCT's effectiveness in improving CT image resolution in real-world scenarios where HR reference images are unavailable.

Correct Modeling Is Essential. We present this experiment to demonstrate the importance of correctly modeling the simulated training images from axial images, which are used for training the neural network, to match the resolution and overlap of coronal and sagittal images. To validate this requirement, we intentionally introduced modeling errors into the training images, simulating different combinations of resolution and overlap that deviated by either $0.25\,mm$ or $0.5\,mm$ from the actual resolution of $5\,mm$ and $2.5\,mm$ overlap.

Figure 5 shows SR4ZCT achieved the highest PSNR when the training images accurately reflected a resolution of $5\,mm$ and an overlap of $2.5\,mm$. Even minor deviations, such as $0.25\,mm$ errors, resulted in a decrease in PSNR. Interestingly, we observed that when the ratio of error in resolution and overlap was similar, the performance was better than when the error was present in only one parameter. Figure 6 provides examples where incorrect modeling of resolution or overlap by $0.25\,mm$ led to artifacts, highlighting the impact of inaccurate modeling.

Figure 7 also demonstrates the importance of modeling the training images accurately. We applied supervised methods SAINT [14], RPLHR [19], and self-

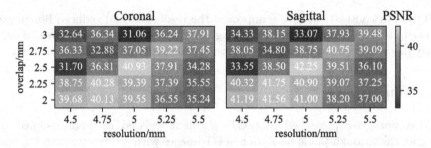

	Coronal					**Sagittal**					**PSNR**
overlap/mm 3	32.64	36.34	31.06	36.24	37.91	34.33	38.15	33.07	37.93	39.48	40
2.75	36.33	32.88	37.05	39.22	37.45	38.05	34.80	38.75	40.75	39.09	
2.5	31.70	36.81	40.93	37.91	34.28	33.55	38.50	42.25	39.51	36.10	
2.25	38.75	40.28	39.39	37.39	35.55	40.32	41.75	40.90	39.07	37.25	35
2	39.68	40.13	39.55	36.55	35.24	41.19	41.56	41.00	38.20	37.00	
	4.5	4.75	5	5.25	5.5	4.5	4.75	5	5.25	5.5	
			resolution/mm					resolution/mm			

Fig. 5. The PSNR of cases where training images were inaccurately modeled. Each block refers to a case where the corresponding resolution/overlap are used to model the training images. The actual resolution/overlap are $5\,mm/2.5\,mm$.

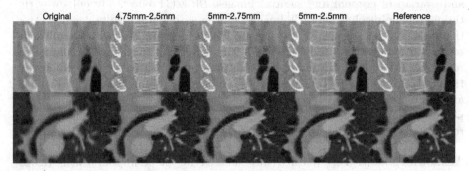

Fig. 6. Results of cases where training images were not correctly modeled. The actual through-plane resolution and overlap are $5\,mm$ and $2.5\,mm$. The second and third column shows cases when modeled training images deviated by $0.25\,mm$.

Fig. 7. Visual comparison of state-of-the-art CT images supervised super-resolution methods SAINT [14], RPLHR [19], self-supervised method SMORE [20] and our method SR4ZCT applied to a real-world sagittal image.

supervised SMORE [20], to Low Dose CT Grand Challenge [12]. We used the provided pre-trained weights for SAINT and trained RPLHR as described on their dataset. SAINT and RPLHR do not consider overlap in their method design. For SMORE, we used convolution with a filter width of 4 and stride 3, which was equivalent to simulating $2.96\,mm$ resolution and $0.74\,mm$ overlap images, the closest possible combination to the actual $3\,mm$ resolution and $1\,mm$ overlap. The blurriness caused by the overlap in the LR image was amplified in the results of SAINT and RPLHR, while artifacts occurred in the output of

SMORE. In contrast, SR4ZCT improved the resolution and reduced blurriness by accurately modeling the training images. This stresses the importance of correctly modeling training images to ensure improvement for resolution effectively.

4   Conclusion

In this work, we presented SR4ZCT, a self-supervised method designed to improve the through-plane resolution of CT images with arbitrary resolution and overlap. The method is based on the same assumption as [18,20] that images of a medical CT volume from different orientations often share similar image features. By accurately simulating axial training images that match the resolution and overlap of coronal and sagittal images, SR4ZCT effectively enhances the through-plane resolution of CT images. Our experimental results demonstrated that SR4ZCT outperformed existing methods for CT images with complicated resolution and overlap combinations. It successfully enhanced the resolution of real-world CT images without the need for reference images. We also emphasized the crucial role of correctly modeling the simulated training images for such off-axis training-based self-supervised method. In the future, our method may have potential applications in a wide range of clinical settings where differences in resolution across volume orientations currently limit the 3D CT resolution.

Acknowledgment. This research was financed by the European Union H2020-MSCA-ITN-2020 under grant agreement no. 956172 (xCTing).

References

1. Angelopoulos, C., Scarfe, W.C., Farman, A.G.: A comparison of maxillofacial CBCT and medical CT. Atlas Oral Maxillofac. Surg. Clin. North Am. **20**(1), 1–17 (2012)
2. Brink, J.A.: Technical aspects of helical (spiral) CT. Radiol. Clin. North Am. **33**(5), 825–841 (1995)
3. Coward, J., et al.: Multi-centre analysis of incidental findings on low-resolution CT attenuation correction images. Br. J. Radiol. **87**(1042), 20130701 (2014)
4. Gavrielides, M.A., Zeng, R., Myers, K.J., Sahiner, B., Petrick, N.: Benefit of overlapping reconstruction for improving the quantitative assessment of CT lung nodule volume. Acad. Radiol. **20**(2), 173–180 (2013)
5. Hansen, P.C., Jørgensen, J., Lionheart, W.R.: Computed Tomography: Algorithms, Insight, and Just Enough Theory. SIAM (2021)
6. He, L., Huang, Y., Ma, Z., Liang, C., Liang, C., Liu, Z.: Effects of contrast-enhancement, reconstruction slice thickness and convolution kernel on the diagnostic performance of radiomics signature in solitary pulmonary nodule. Sci. Rep. **6**(1), 34921 (2016)
7. Honda, O., et al.: Computer-assisted lung nodule volumetry from multi-detector row CT: influence of image reconstruction parameters. Eur. J. Radiol. **62**(1), 106–113 (2007)

8. Iwano, S., et al.: Solitary pulmonary nodules: optimal slice thickness of high-resolution CT in differentiating malignant from benign. Clin. Imaging **28**(5), 322–328 (2004)

9. Kasales, C., et al.: Reconstructed helical CT scans: improvement in z-axis resolution compared with overlapped and nonoverlapped conventional CT scans. AJR Am. J. Roentgenol. **164**(5), 1281–1284 (1995)

10. Kingma, D.P., Ba, J.: Adam: a method for stochastic optimization. In: Bengio, Y., LeCun, Y. (eds.) 3rd International Conference on Learning Representations, ICLR 2015, San Diego, CA, USA, May 7–9, 2015, Conference Track Proceedings (2015). http://arxiv.org/abs/1412.6980

11. Liu, Q., Zhou, Z., Liu, F., Fang, X., Yu, Y., Wang, Y.: Multi-stream progressive upsampling network for dense CT image reconstruction. In: Martel, A.L., et al. (eds.) MICCAI 2020. LNCS, vol. 12266, pp. 518–528. Springer, Cham (2020). https://doi.org/10.1007/978-3-030-59725-2_50

12. McCollough, C.H., et al.: Low-dose CT for the detection and classification of metastatic liver lesions: results of the 2016 low dose CT grand challenge. Med. Phys. **44**(10), e339–e352 (2017)

13. Pelt, D.M., Sethian, J.A.: A mixed-scale dense convolutional neural network for image analysis. Proc. Natl. Acad. Sci. **115**(2), 254–259 (2018)

14. Peng, C., Lin, W.A., Liao, H., Chellappa, R., Zhou, S.K.: Saint: spatially aware interpolation network for medical slice synthesis. In: Proceedings of the IEEE/CVF Conference on Computer Vision and Pattern Recognition, pp. 7750–7759 (2020)

15. Ravenel, J.G., Leue, W.M., Nietert, P.J., Miller, J.V., Taylor, K.K., Silvestri, G.A.: Pulmonary nodule volume: effects of reconstruction parameters on automated measurements-a phantom study. Radiology **247**(2), 400–408 (2008)

16. Simpson, A.L., et al.: A large annotated medical image dataset for the development and evaluation of segmentation algorithms. arXiv preprint arXiv:1902.09063 (2019)

17. Tsukagoshi, S., Ota, T., Fujii, M., Kazama, M., Okumura, M., Johkoh, T.: Improvement of spatial resolution in the longitudinal direction for isotropic imaging in helical CT. Phys. Med. Biol. **52**(3), 791 (2007)

18. Xie, H., et al.: High through-plane resolution CT imaging with self-supervised deep learning. Phys. Med. Biol. **66**(14), 145013 (2021)

19. Yu, P., Zhang, H., Kang, H., Tang, W., Arnold, C.W., Zhang, R.: RPLHR-CT dataset and transformer baseline for volumetric super-resolution from CT scans. In: Wang, L., Dou, Q., Fletcher, P.T., Speidel, S., Li, S. (eds.) Medical Image Computing and Computer Assisted Intervention – MICCAI 2022. MICCAI 2022. Lecture Notes in Computer Science, vol. 13436, pp. 344–353. Springer, Cham (2022). https://doi.org/10.1007/978-3-031-16446-0_33

20. Zhao, C., Dewey, B.E., Pham, D.L., Calabresi, P.A., Reich, D.S., Prince, J.L.: Smore: a self-supervised anti-aliasing and super-resolution algorithm for MRI using deep learning. IEEE Trans. Med. Imaging **40**(3), 805–817 (2020)

unORANIC: Unsupervised Orthogonalization of Anatomy and Image-Characteristic Features

Sebastian Doerrich$^{(\boxtimes)}$, Francesco Di Salvo, and Christian Ledig

xAILab, University of Bamberg, Bamberg, Germany
sebastian.doerrich@uni-bamberg.de

Abstract. We introduce unORANIC, an unsupervised approach that uses an adapted loss function to drive the orthogonalization of anatomy and image-characteristic features. The method is versatile for diverse modalities and tasks, as it does not require domain knowledge, paired data samples, or labels. During test time unORANIC is applied to potentially corrupted images, orthogonalizing their anatomy and characteristic components, to subsequently reconstruct corruption-free images, showing their domain-invariant anatomy only. This feature orthogonalization further improves generalization and robustness against corruptions. We confirm this qualitatively and quantitatively on 5 distinct datasets by assessing unORANIC's classification accuracy, corruption detection and revision capabilities. Our approach shows promise for enhancing the generalizability and robustness of practical applications in medical image analysis. The source code is available at github.com/sdoerrich97/unORANIC.

Keywords: Feature Orthogonalization · Robustness · Corruption Revision · Unsupervised learning · Generalization

1 Introduction

In recent years, deep learning algorithms have shown promise in medical image analysis, including segmentation [23,24], classification [8,17], and anomaly detection [10,19]. However, their generalizability across diverse imaging domains remains a challenge [7] especially for their adoption in clinical practice [25] due to domain shifts caused by variations in scanner models [12], imaging parameters [13], corruption artifacts [21], or patient motion [20]. A schematic representation of this issue is presented in Fig. 1a. To address this, two research areas have emerged: domain adaptation (DA) and domain generalization (DG). DA aligns feature distributions between source and target domains [14], while DG trains models on diverse source domains to learn domain-invariant features [15]. Within this framework, some methods aim to disentangle anatomy features from modality factors to improve generalization. Chartsias et al. introduce SDNet, which uses segmentation labels to factorize 2D medical images into spatial anatomical and non-spatial modality factors for robust cardiac image segmentation [4].

© The Author(s), under exclusive license to Springer Nature Switzerland AG 2024
X. Cao et al. (Eds.): MLMI 2023, LNCS 14348, pp. 62–71, 2024.
https://doi.org/10.1007/978-3-031-45673-2_7

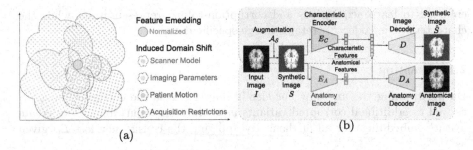

Fig. 1. a) Schematic visualization of a domain disparity caused by variations in the imaging process. **b)** Visualization of unORANIC's fundamental idea.

Robert et al. present HybridNet to learn invariant class-related representations via a supervised training concept [22]. Dewey et al. propose a deep learning-based harmonization technique for MR images under limited supervision to standardize across scanners and sites [6]. In contrast, Zuo et al. propose unsupervised MR image harmonization to address contrast variations in multi-site MR imaging [29]. However, all of these approaches are constrained by either requiring a certain type of supervision, precise knowledge of the target domain, or specific inter-/intra-site paired data samples.

To address these limitations, we present unORANIC, an approach to orthogonalize anatomy and image-characteristic features in an unsupervised manner, without requiring domain knowledge, paired data, or labels of any kind. The method enables bias-free anatomical reconstruction and works for a diverse set of modalities. For that scope, we jointly train two encoder-decoder branches. One branch is used to extract true anatomy features and the other to model the characteristic image information discarded by the first branch to reconstruct the input image in an autoencoder objective. A high-level overview of the approach is provided in Fig. 1b. Results on a diverse dataset demonstrate its feasibility and potential for improving generalization and robustness in medical image analysis.

2   Methodology

For the subsequent sections, we consider the input images as bias-free and uncorrupted (I). We further define $\mathcal{A}_S$ as a random augmentation that distorts an input image I for the purpose to generate a synthetic, corrupted version S of that image. As presented in Fig. 1b, such a synthetic image S is obtained via the augmentation $\mathcal{A}_S$ applied to I and subsequently fed to both the anatomy encoder E_A and the characteristic encoder E_C simultaneously. The resulting embeddings are concatenated ($\oplus$) and forwarded to a convolutional decoder D to create the reconstruction $\hat{S}$ with its specific characteristics such as contrast level or brightness. By removing these characteristic features in the encoded embeddings of E_A, we can reconstruct a distortion-free version ($\hat{I}_A$) of the original input image I. To allow this behavior, the anatomy encoder, E_A, is actively enforced to learn acquisition- and corruption-robust representations while the characteristic encoder E_C retains image-specific details.

2.1   Training

During training the anatomy encoder is shared across the corrupted image S as well as additional corrupted variants v_i of the same input image I to create feature embeddings for all of them. By applying the consistency loss $\mathcal{L}_C$ given as:

$$\mathcal{L}_C = \frac{1}{Z} \left(\sum_{\forall \{v_i, v_j\} \in V, \ v_i \neq v_j} ||E^A(v_i) - E^A(v_j)||_2 \right) \quad \text{with } Z = \binom{V}{2} \quad (1)$$

on the resulting feature maps, the anatomy encoder is forced to learn distortion-invariant features. Here, V is the set of all variants including S, v_i and v_j are two distinct variants as well as $E^A(v_i)$ and $E^A(v_j)$ are the corresponding anatomy feature embeddings. To further guide the anatomy encoder to learn representative anatomy features, the consistency loss is assisted by a combination of the reconstruction loss of the synthetic image $\mathcal{L}_{R_S}$ and the reconstruction loss of the original, distortion-free image $\mathcal{L}_{R_I}$ given as:

$$\mathcal{L}_{R_S} = \frac{1}{NM} \left|\left| S - \hat{S} \right|\right|_2 = \frac{1}{NM} ||S - D\left(E_A(S) \oplus E_C(S)\right)||_2 \quad (2)$$

$$\mathcal{L}_{R_I} = \frac{1}{NM} \left|\left| I - \hat{I}_A \right|\right|_2 = \frac{1}{NM} ||I - D_A(E_A(I))||_2 \quad (3)$$

with N and M being the image height and width respectively. Using $\mathcal{L}_{R_I}$ to update the encoder E_A, as well as the decoder D_A, allows the joint optimization of both. This yields more robust results than updating only the decoder with that loss term. The complete loss for the anatomy branch is therefore given as:

$$\mathcal{L}_{\text{total}} = \lambda_{\text{reconstruction}} \left(\mathcal{L}_{R_I} + \mathcal{L}_{R_S} \right) + \lambda_{\text{consistency}} \mathcal{L}_C \quad (4)$$

where $\lambda_{\text{reconstruction}}$ and $\lambda_{\text{consistency}}$ are two non-negative numbers that control the trade-off between the reconstruction and consistency loss. In contrast, the characteristic encoder E_C and decoder D are only optimized via the reconstruction loss of the synthetic image $\mathcal{L}_{R_S}$.

2.2   Implementation

The entire training pipeline is depicted in Fig. 2. Both encoders consist of four identical blocks in total, where each block comprises a residual block followed by a downsampling block. Each residual block itself consists of twice a convolution followed by batch normalization and the Leaky ReLU activation function. The downsampling blocks use a set of a strided convolution, a batch normalization, and a Leaky ReLU activation to half the image dimension while doubling the channel dimension. In contrast, the decoders mirror the encoder architecture by replacing the downsampling blocks with corresponding upsampling blocks for which the convolutions are swapped with transposed convolutional layers. During each iteration, the input image I is distorted using augmentation $\mathcal{A}_S$ to generate S and subsequently passed through the shared anatomy encoder E_A in combination with two different distorted variants V_1 and V_2. These variants comprise the same anatomical information but different distortions than S. The consistency loss $\mathcal{L}_C$ is computed using the encoded feature embeddings, and the reconstruction loss of the original, distortion-free image $\mathcal{L}_{R_I}$ is calculated by passing the feature embedding of S through the anatomy decoder D_A to generate the anatomical reconstruction $\hat{I}_A$. Furthermore, the anatomy and characteristic feature embeddings are concatenated and used by the image decoder D to reconstruct $\hat{S}$ for the calculation of $\mathcal{L}_{R_S}$. $\mathcal{A}_S$, $\mathcal{A}_{v_1}$ and $\mathcal{A}_{v_2}$ are independent of each other and can be chosen randomly from a set of augmentations such as the Albumentations library [3]. The network is trained until convergence using a batch size of 64 as well as the Adam optimizer with a cyclic learning rate. Experiments were performed on 28×28 pixel images with a latent dimension of 256, but the approach is adaptable to images of any size. The number of variants used for the anatomy encoder training is flexible, and in our experiments, three variants were utilized.

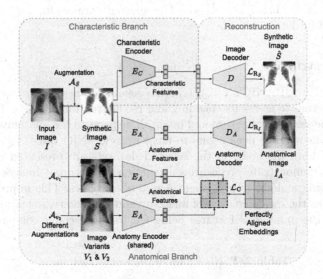

Fig. 2. Training pipeline of unORANIC for CT images.

3   Experiments and Results

We evaluate the model's performance in terms of its reconstruction ability as well as its capability to revise existing corruptions within an image. Further,

Table 1. Details of the selected set of datasets

MedMNIST Dataset	Modality	Task (# Classes)	# Train / Val / Test
Blood [1]	Blood Cell Microscope	Multi-Class (8)	$11,959 / 1,712 / 3,421$
Breast [2]	Breast Ultrasound	Binary-Class (2)	$546 / 78 / 156$
Derma [5,26]	Dermatoscope	Multi-Class (7)	$7,007 / 1,003 / 2,005$
Pneumonia [11]	Chest X-Ray	Binary-Class (2)	$4,708 / 524 / 624$
Retina [16]	Fundus Camera	Ordinal Regression (5)	$1,080 / 120 / 400$

we assess its applicability for the execution of downstream tasks by using the anatomy feature embedding, as well as its capacity to detect corruptions by using the characteristic feature embedding. Last, we rate the robustness of our model against different severity and types of corruption.

3.1 Dataset

To demonstrate the versatility of our method, all experiments were conducted on a diverse selection of datasets of the publicly available MedMNIST v2 benchmark[1] [27,28]. The MedMNIST v2 benchmark is a comprehensive collection of standardized, 28×28 dimensional biomedical datasets, with corresponding classification labels and baseline methods. The selected datasets for our experiments are described in detail in Table 1.

3.2 Reconstruction

We compared unORANIC's reconstruction results with a vanilla autoencoder (AE) architecture to assess its encoding and reconstruction abilities. The vanilla AE shares the same architecture and latent dimension as the anatomy branch of our model. The average peak signal-to-noise ratio (PSNR) values for the reconstructions of both methods on the selected datasets are presented in Table 2. Both models demonstrate precise reconstruction of the input images, with our model achieving a slight improvement across all datasets. This improvement is attributable to the concatenation of anatomy and characteristic features, resulting in twice the number of features being fed to the decoder compared to the

Table 2. PSNR values of the reconstructions

Methods	Blood	Breast	Derma	Pneumonia	Retina
AE	30.837	28.725	38.125	34.787	35.872
unORANIC	**31.700**	**29.390**	**38.569**	**36.040**	**36.309**

[1] Yang, J., et al. *MedMNIST v2-A large-scale lightweight benchmark for 2D and 3D biomedical image classification.* Scientific Data. 2023. License: CC BY 4.0. Zenodo. https://zenodo.org/record/6496656.

AE model. The reconstruction quality of both models is additionally depicted visually in Fig. 3 using selected examples.

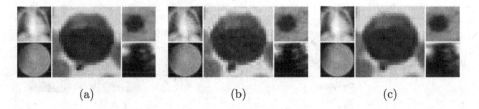

(a)   (b)   (c)

Fig. 3. a) Reconstruction unORANIC **b)** Original input **c)** Reconstructions AE.

3.3 Corruption Revision

We now evaluate the capability of unORANIC's anatomical branch to revise existing corruptions in an input image. For this, all input images in the test set $\{I_1, I_2, ..., I_n\}$ are intentionally corrupted using a set of corruptions from the Albumentations library [3] to generate synthetic distorted versions $\{S_1, S_2, ..., S_n\}$ first, before passing those through the unORANIC network afterward. Each corruption is hereby uniformly applied to all test images. Despite those distortions, the anatomy branch of the model should still be able to reconstruct the original, uncorrupted input images $\hat{I}_i$. To assess this, we compute the average PSNR value between the original images I_i and their corrupted versions S_i. Afterward, we compare this value to the average PSNR between the original uncorrupted input images I_i and unORANIC's anatomical reconstructions $\hat{I}_{A_i}$. Both of these values are plotted against each other in Fig. 4a, for the BloodM-NIST dataset. For reference purposes, the figure contains the PSNR value for uncorrupted input images as well. It can be seen that the anatomical reconstruction quality is close to the uncorrupted one and overall consistent across all applied corruptions which proves unORANIC's corruption revision capability. As it can be seen in Fig. 4b, this holds true even for severe corruptions such as solarization.

3.4 Classification

We proceed to assess the representativeness of the encoded feature embeddings, comprising both anatomical and characteristic information, and their suitability for downstream tasks or applications. To this end, we compare our model with the ResNet-18 baseline provided by MedMNIST v2 on two distinct tasks. The first task involves the classification of each dataset, determining the type of disease (e.g., cancer or non-cancer). For the second task, the models are assigned to detect whether a corruption (the same ones used in the previous revision experiment) has or has not been applied to the input image in the form of a

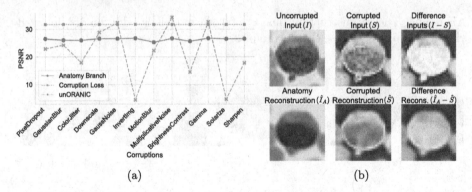

Fig. 4. a) Corruption revision results of unORANIC for a set of corruptions. **b)** Visualization of unORANIC's corruption revision for a solarization corruption. The method reconstructs I via $\hat{I}_A$ while seeing only the corrupted input S.

binary classification. To do so, we freeze our trained network architecture and train a single linear layer on top of each embedding for its respective task. The same procedure is applied to the AE architecture, while the MedMNIST baseline architecture is retrained specifically for the corruption classification task. Results are summarized in Table 3 and Table 4, with AUC denoting the Area under the ROC Curve and ACC representing the Accuracy of the predictions. Overall, our model's classification ability is not as strong as that of the baseline model. However, it is important to note that our model was trained entirely in an unsupervised manner, except for the additional linear layer, compared to the supervised baseline approach. Regarding the detection of corruptions within the input images, our model outperforms the baseline method. Furthermore, it should be emphasized that our model was trained simultaneously for both tasks, while the reference models were trained separately for each individual task.

Table 3. Classification results on each selected dataset

Methods	Blood		Breast		Derma		Pneumonia		Retina	
	AUC	ACC	AUC	ACC	AUC	ACC	AUC	ACC	AUC	ACC
Baseline	**0.998**	**0.958**	**0.901**	**0.863**	**0.917**	**0.735**	0.944	0.854	**0.717**	0.524
AE	0.966	0.806	0.819	0.801	0.787	0.694	0.904	0.840	0.678	0.470
unORANIC	0.977	0.848	0.812	0.808	0.776	0.699	**0.961**	**0.862**	0.691	**0.530**

3.5 Corruption Robustness

Lastly, we assess unORANIC's robustness against unseen corruptions during training. To accomplish this, we adopt the image corruptions introduced by [9,18], following a similar methodology as outlined in Sect. 3.3. The used corruptions are presented in Fig. 5a for an example of the PneumoniaMNIST dataset.

Table 4. Corruption detection on each selected dataset

Methods	Blood		Breast		Derma		Pneumonia		Retina	
	AUC	ACC	AUC	ACC	AUC	ACC	AUC	ACC	AUC	ACC
Baseline	0.548	0.184	0.565	0.571	0.517	0.046	0.614	0.524	0.500	0.283
AE	0.746	0.736	0.576	0.545	**0.655**	**0.640**	0.648	0.657	0.827	0.785
unORANIC	**0.755**	**0.746**	**0.612**	**0.590**	0.643	0.620	**0.667**	**0.660**	**0.847**	**0.823**

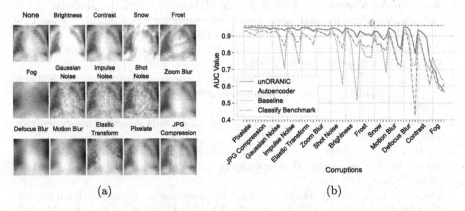

(a)   (b)

Fig. 5. a) Visualization of the applied corruptions. **b)** Classification results of unO-RANIC compared to the vanilla AE and the baseline when exposed to the applied corruptions with increasing severity.

Sequentially, we apply each corruption to all test images first and subsequently pass these corrupted images through the trained unORANIC, vanilla AE and baseline model provided by the MedMNIST v2 benchmark for the datasets associated classification task. This process is repeated for each individual corruption. Additionally, we vary the severity of each corruption on a scale of 1 to 5, with 1 representing minor corruption and 5 indicating severe corruption. By collecting the AUC values for each combination of corruption and severity on the PneumoniaMNIST dataset and plotting them for each method individually, we obtain the results presented in Fig. 5b. Notably, our model demonstrates greater overall robustness to corruption, particularly for noise, compared to both the baseline and the simple autoencoder architecture.

4   Conclusion

The objective of this study was to explore the feasibility of orthogonalizing anatomy and image-characteristic features in an unsupervised manner, without requiring specific dataset configurations or prior knowledge about the data. The conducted experiments affirm that unORANIC's simplistic network architecture can explicitly separate these two feature categories. This capability enables the training of corruption-robust architectures and the reconstruction of unbiased anatomical images. The findings from our experiments motivate further investigations into extending this approach to more advanced tasks, architectures as well as additional datasets, and explore its potential for practical applications.

References

1. Acevedo, A., Merino, A., Alférez, S., Ángel Molina, Boldú, L., Rodellar, J.: A dataset of microscopic peripheral blood cell images for development of automatic recognition systems. Data Brief **30**, 105474 (2020)
2. Al-Dhabyani, W., Gomaa, M., Khaled, H., Fahmy, A.: Dataset of breast ultrasound images. Data Brief **28**, 104863 (2020)
3. Buslaev, A., Iglovikov, V.I., Khvedchenya, E., Parinov, A., Druzhinin, M., Kalinin, A.A.: Albumentations: fast and flexible image augmentations. Information **11**, 125 (2020)
4. Chartsias, A., et al.: Disentangled representation learning in cardiac image analysis. Med. Image Anal. **58**, 101535 (2019)
5. Codella, N., et al.: Skin lesion analysis toward melanoma detection 2018: a challenge hosted by the international skin imaging collaboration (isic). arXiv:1902.03368 (2019)
6. Dewey, B.E., et al.: A disentangled latent space for cross-site MRI harmonization. In: Martel, A.L., et al. (eds.) MICCAI 2020. LNCS, vol. 12267, pp. 720–729. Springer, Cham (2020). https://doi.org/10.1007/978-3-030-59728-3_70
7. Eche, T., Schwartz, L.H., Mokrane, F.Z., Dercle, L.: Toward generalizability in the deployment of artificial intelligence in radiology: Role of computation stress testing to overcome underspecification. Radiol. Artificial Intell. **3**, e210097 (2021)
8. He, K., Zhang, X., Ren, S., Sun, J.: Deep residual learning for image recognition. In: Proceedings of the IEEE Computer Society Conference on Computer Vision and Pattern Recognition 2016-December, pp. 770–778 (2016)
9. Hendrycks, D., Dietterich, T.: Benchmarking neural network robustness to common corruptions and perturbations. In: Proceedings of the International Conference on Learning Representations (2019)
10. Jeong, J., Zou, Y., Kim, T., Zhang, D., Ravichandran, A., Dabeer, O.: Winclip: zero-/few-shot anomaly classification and segmentation. In: Proceedings of the IEEE/CVF Conference on Computer Vision and Pattern Recognition (CVPR), pp. 19606–19616 (2023)
11. Kermany, D.S., et al.: Identifying medical diagnoses and treatable diseases by image-based deep learning. Cell **172**, 1122-1131.e9 (2018)
12. Khan, A., et al.: Impact of scanner variability on lymph node segmentation in computational pathology. J. Pathol. Inf. **13**, 100127 (2022)
13. Lafarge, M.W., Pluim, J.P.W., Eppenhof, K.A.J., Moeskops, P., Veta, M.: Domain-Adversarial Neural Networks to Address the Appearance Variability of Histopathology Images. In: Cardoso, M.J., et al. (eds.) DLMIA/ML-CDS -2017. LNCS, vol. 10553, pp. 83–91. Springer, Cham (2017). https://doi.org/10.1007/978-3-319-67558-9_10
14. Li, B., Wang, Y., Zhang, S., Li, D., Keutzer, K., Darrell, T., Zhao, H.: Learning invariant representations and risks for semi-supervised domain adaptation. In: Proceedings of the IEEE Computer Society Conference on Computer Vision and Pattern Recognition, pp. 1104–1113 (2021)

15. Li, D., Yang, Y., Song, Y.Z., Hospedales, T.M.: Learning to generalize: Meta-learning for domain generalization. In: Proceedings of the AAAI Conference on Artificial Intelligence, vol. 32, pp. 3490–3497 (2018)

16. Liu, R., et al.: Deepdrid: diabetic retinopathy-grading and image quality estimation challenge. Patterns **3**, 100512 (2022)

17. Manzari, O.N., Ahmadabadi, H., Kashiani, H., Shokouhi, S.B., Ayatollahi, A.: MedVit: a robust vision transformer for generalized medical image classification. Comput. Biol. Med. **157**, 106791 (2023)

18. Michaelis, C., et al.: Benchmarking robustness in object detection: Autonomous driving when winter is coming. arXiv preprint arXiv:1907.07484 (2019)

19. Ngo, P.C., Winarto, A.A., Kou, C.K.L., Park, S., Akram, F., Lee, H.K.: Fence GAN: towards better anomaly detection. In: Proceedings - International Conference on Tools with Artificial Intelligence, ICTAI 2019-November, pp. 141–148 (2019)

20. Oksuz, I., et al.: Deep learning-based detection and correction of cardiac MR motion artefacts during reconstruction for high-quality segmentation. IEEE Trans. Med. Imaging **39**, 4001–4010 (2020)

21. Priyanka, Kumar, D.: Feature extraction and selection of kidney ultrasound images using GLCM and PCA. Procedia Comput. Sci. **167**, 1722–1731 (2020)

22. Robert, T., Thome, N., Cord, M.: HybridNet: classification and reconstruction cooperation for semi-supervised learning. In: Ferrari, V., Hebert, M., Sminchisescu, C., Weiss, Y. (eds.) ECCV 2018. LNCS, vol. 11211, pp. 158–175. Springer, Cham (2018). https://doi.org/10.1007/978-3-030-01234-2_10

23. Rondinella, A., et al.: Boosting multiple sclerosis lesion segmentation through attention mechanism. Comput. Biol. Med. **161**, 107021 (2023)

24. Ronneberger, O., Fischer, P., Brox, T.: U-Net: convolutional networks for biomedical image segmentation. In: Navab, N., Hornegger, J., Wells, W.M., Frangi, A.F. (eds.) MICCAI 2015. LNCS, vol. 9351, pp. 234–241. Springer, Cham (2015). https://doi.org/10.1007/978-3-319-24574-4_28

25. Stacke, K., Eilertsen, G., Unger, J., Lundstrom, C.: Measuring domain shift for deep learning in histopathology. IEEE J. Biomed. Health Inform. **25**, 325–336 (2021)

26. Tschandl, P., Rosendahl, C., Kittler, H.: The ham10000 dataset, a large collection of multi-source dermatoscopic images of common pigmented skin lesions. Sci. Data **5**(1), 1–9 (2018)

27. Yang, J., Shi, R., Ni, B.: Medmnist classification decathlon: A lightweight automl benchmark for medical image analysis. In: Proceedings - International Symposium on Biomedical Imaging 2021-April, pp. 191–195 (2020)

28. Yang, J., et al.: Medmnist v2 - a large-scale lightweight benchmark for 2d and 3d biomedical image classification. Sci. Data **10**(1), 1–10 (2023)

29. Zuo, L., et al.: Unsupervised MR harmonization by learning disentangled representations using information bottleneck theory. Neuroimage **243**, 118569 (2021)

An Investigation of Different Deep Learning Pipelines for GABA-Edited MRS Reconstruction

Rodrigo Berto[1,4]([✉]), Hanna Bugler[1,4], Roberto Souza[1,3,5], and Ashley Harris[2,4]

[1] Department of Biomedical Engineering, University of Calgary, Calgary, Canada
[2] Department of Radiology, University of Calgary, Calgary, Canada
[3] Hotchkiss Brain Institute, University of Calgary, Calgary, Canada
[4] Alberta Children's Hospital Research Institute, University of Calgary, Calgary, Canada
rodrigo.pommotber1@ucalgary.ca
[5] Department of Electrical and Software Engineering, University of Calgary, Calgary, Canada

Abstract. Edited magnetic resonance spectroscopy (MRS) can provide localized information on gamma-aminobutyric acid (GABA) concentration *in vivo*. However, edited-MRS scans are long due to the fact that many acquisitions, known as transients, need to be collected and averaged to obtain a high-quality spectrum for reliable GABA quantification. In this work, we investigate Deep Learning (DL) pipelines for the reconstruction of GABA-edited MRS spectra using only a quarter of the transients typically acquired. We compared two neural network architectures: a 1D U-NET and a proposed dimension-reducing 2D U-NET (Rdc-UNET2D) that we proposed. We also compared the impact of training the DL pipelines using solely *in vivo* data or pre-training the models on simulated followed by fine-tuning on *in vivo* data. Results for this study showed the proposed Rdc-UNET2D model pre-trained on simulated data and fine-tuned on *in vivo* data had the best performance among the different DL pipelines compared. This model obtained a higher SNR and a lower fit error than a conventional reconstruction pipeline using the full amount of transients typically acquired. This indicates that through DL it is possible to reduce GABA-edited MRS scan times by four times while maintaining or improving data quality. In the spirit of open science, the code and data to reproduce our pipeline are publicly available.

Keywords: GABA-edited MRS · MRS Denoising · Deep Learning

1 Introduction

Gamma-Aminobutyric Acid (GABA) is a neurochemical of widespread interest in neuroscience. For example, GABA concentration has been investigated and

Supplementary Information The online version contains supplementary material available at https://doi.org/10.1007/978-3-031-45673-2_8.

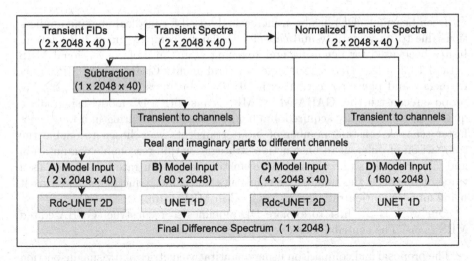

Fig. 1. Flowchart of the different pipelines investigated. Pipelines differ in their usage of edit-ON and edit-OFF transients or difference transients and their model architecture (1D U-NET or Rdc-UNET2D), which requires different dimensionalities. The output is the same for all pipelines: a GABA-difference spectrum

found to be associated with depression [23,24], developmental disorders [7], as well as other disorders [2,19,25,31].

Edited magnetic resonance spectroscopy (MRS) techniques, such as Merscher-Garwood point resolved spectroscopy (MEGA-PRESS) [12], are used to measure GABA concentration *in vivo*. During half of the measurements (edit-ON subspectra), an editing pulse is applied at 1.9 ppm to modulate the GABA peak at 3 ppm while not affecting the overlapping creatine (Cr) peak at 3 ppm. During the other half of the measurements (edit-OFF subspectra), there is no editing pulse at 1.9 ppm, leaving GABA and Cr intact. By subtracting the edit-OFF subspectra from the edit-ON, non-modulated peaks, specifically Cr, are removed, and the GABA peak is isolated for quantification [9,21,29]. However, this subtraction decreases the signal magnitude while increasing the noise, resulting in a noisy GABA difference spectrum, which cannot be reliably quantified. To improve the quality of the final spectrum, many transients (i.e., measurements) are averaged to increase the signal-to-noise ratio (SNR). Typical GABA-edited MRS scans acquire 320 transients (160 edit-ON and 160 edit-OFF) and last over 10 min [14], leading to discomfort and risks of motion artefacts or other scanner instabilities [8]. In addition to averaging the transients, other denoising techniques are typically performed post-acquisition to improve signal quality, such as removing corrupted transients and frequency and phase correction (FPC) [18]. These methods, along with the subspectra subtraction and the time domain to frequency domain conversion are collectively named GABA-edited MRS preprocessing or reconstruction.

Deep Learning (DL) has been increasingly used for medical imaging applications due to its versatility and capacity to handle complex image information. It has been used for denoising and reducing scan times of conventional MRS data [3,10,30]. However, for the lower spectral quality GABA-edited MRS, only frequency and phase correction with DL has been investigated [11,26,28]. We propose reconstructing GABA-edited MRS scans with a DL model using only a quarter of the typically acquired number of transients, resulting in a four times faster scan. A constant number of input transients was chosen to enable the comparison of different reconstruction strategies within reasonable computation and time restraints. The four times scan time reduction rate was chosen as a trade-off between the number of transients and time reduction, as well as it being an acceleration rate commonly used in MRI [1,16].

Our work is the first to explore DL pipelines to reconstruct GABA-edited MRS scans. The contributions of our paper are:

- The proposal and comparison using quantitative metrics and visual inspection of different DL pipelines to reconstruct GABA-edited MRS with four times less data than traditional scans.
- The analysis of the impact of different training procedures using either *in vivo* data only or pre-training the models on simulated data followed by fine-tuning on *in vivo* data.
- A public benchmark to evaluate different GABA-edited MRS reconstruction methods.

The code and data to reproduce our pipeline are publicly available in the following repository: https://github.com/rodrigopberto/Edited-MRS-DL-Reconstruction.

2 Materials and Methods

2.1 Data

Given the scarcity of publicly available GABA-edited MRS raw data and the lack of ground truths for *in vivo* scans, both simulated and *in vivo* datasets were used for training.

The simulated dataset consisted of simulated ground-truth pairs (edit-ON and edit-OFF) generated using the MEGA-PRESS variant *Mega-PressShaped_fast* function [32] from the FID-A MATLAB toolbox [27]. The simulation function used the following parameters: 3 T magnetic field strength, 2 s repetition time (TR), 68 ms echo time (TE), 2 kHz spectral width and 2048 spectral points. These are the most present parameters in the *in vivo* scans. Twenty-two neurochemicals, as well as common lipids and proteins, were used for the simulation. Concentrations were randomly distributed ($\pm 10\%$) about their theoretical means found in the literature [4,6,17]. Detailed parameters for the neurochemical concentrations are available in the supplementary material. Five hundred ground-truth pairs (edit-OFF and edit-ON) were generated for training. Random Gaussian amplitude noise, frequency shifts, and phase shifts were added to the ground truths to simulate the realistic distribution of data quality of *in vivo* data.

The *in vivo* dataset was extracted from the big GABA repository [13,15,20] and consists of 144 scans from GE, Phillips and Siemens scanners across twelve different sites and with twelve scans per site. The scans were split into 84 for training, 24 for validation and 36 for testing, and each subset had the same number of scans per site. Gannet [5], a GABA-edited MRS processing tool, was used to extract the data.

2.2 Proposed Pipelines

The pipelines used to reconstruct GABA-edited difference spectra consist of two steps: 1) processing the raw input data (time-domain transients) into the input expected by the model, which is either the normalized difference transients or the normalized edit-OFF and edit-ON transients. 2) generating a single difference spectrum from the model input using the DL model.

In step 1, the raw MRS data, consisting of 80 time-domain transients (40 edit-ON and 40 edit-OFF), was converted into four slightly different formats depending on the model architecture and input format, as shown in Fig. 1. For all pipelines, the transients were converted to the frequency domain via an inverse Fourier transform and normalized by dividing all transients by the mean of the real-valued maximum of the N-acetylaspartate (NAA) peaks of the edit-OFF subspectra. This normalization was selected since the NAA peak is

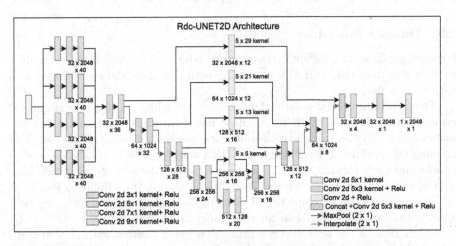

Fig. 2. Diagram of the proposed model Rdc-UNET2D. The first section consists of four parallel convolutional sequences with different 1-dimensional kernels. The next section is a 2D U-NET without padding along the y dimension, so this dimension decreases along the network. Convolutional layers are added to the skip connections to ensure y-dimension integrity in the decoder arm. The U-NET section ends with two final convolutional layers to convert the data to the correct shape.

the highest within the region of interest for GABA-edited MRS (2.5 - 4 ppm) and it avoids offsetting the spectra. Subsequently, the data processing changed depending on the model configuration. For half of the pipelines, edit-OFF and edit-ON transients are kept in separate channels, and for the other half, the edit-OFF transients are subtracted from the edit-ON transients to obtain difference transients. For the pipelines using the 1D U-NET, the dimensionality is reduced by merging the transient dimension into the channel dimension. Finally, for all pipelines, the real and imaginary values are separated into different channels. These are the four processing paths to prepare the input for the four different model configurations.

The first model used was a 1D U-NET [22] with 64 filters for the first convolution, 4 pools, a kernel of size 5, and ReLu activation functions. The model has approximately 17 million parameters. A detailed diagram of the model is provided in the supplementary material.

The second model is our proposed dimension-reducing 2D U-NET (Rdc-UNET2D). Its first section consists of 4 parallel 2D convolutional networks of 3 layers, each network with a different x-axis kernel size and a y-axis kernel size of 1. This section extracts features from each individual transient using the same kernels for all transients. The output of each network is concatenated and passed into a 2D U-NET with padding only along the x-axis and a kernel of 5×3, resulting in a decreased size along the transient dimension on each convolution. This results in slowly combining the features from different transients throughout the network. To ensure continuity in the transient dimension, the skip connections have an added convolutional layer. Figure 2 illustrates the model and the different kernels involved. The network has approximately 16 million parameters.

2.3   Training Procedure

For each pipeline, two different training procedures were compared: 1) training with *in vivo* data only and 2) pre-training with simulated data and fine-tuning with *in vivo* data.

The pre-training on simulated data consisted of adding noise to the simulated ground-truths at each batch to simulate noisy transients and comparing the model output to the ground-truth difference spectra to calculate the loss. The training was performed for 200 epochs with a learning rate that halved every 10 epochs and early stopping after 10 epochs without a decrease in validation loss. The *in vivo* validation dataset was used for calculating the validation loss.

The *in vivo* training consisted of randomly sampling 40 transients per sub-spectra from each *in vivo* scan for the model input and comparing the model reconstruction to the reference reconstruction using Gannet with all 320 transients of the scan to calculate the loss. The training was performed for 500 epochs with a learning rate that halved every 50 epochs and early stopping after 50 epochs without a decrease in the validation loss.

For the *in vivo* training procedure, only the *in vivo* training was used. For the fine-tuning procedure, the simulated pre-training was used followed by the *in vivo* training. Figure 3 illustrates the simulated and *in vivo* data training procedures, highlighting the difference in defining both the input to the pipeline and the reference for calculating the loss.

For all training procedures, a composition of mean absolute error (MAE) losses of different regions of the spectra was used. This ensured that the two most influential features in the GABA-difference spectrum, the 3 ppm GABA-peak and the 3.75 ppm Glutamate + Glutamine (Glx) peak were most relevant in loss calculation. The loss function used was the following and the weights of each spectral region were adjusted empirically:

$$Loss = \frac{(MAE_{global} + 3 \times MAE_{3.55-3.95ppm} + 6 \times MAE_{2.8-3.2ppm})}{10}$$

2.4   Testing and Quality Assessment Metrics

To assess the quality of the reconstructions, the pipelines were tested on the *in vivo* testing dataset. The results were compared to the reconstructions by Gannet, when using all the transients (*Gannet full*) and when using only a quarter of the transients (*Gannet quarter*). The metrics used to evaluate the quality of the reconstructions were chosen to provide complementary analysis of their quality with traditional MRS metrics and a typical DL metric, as follows:

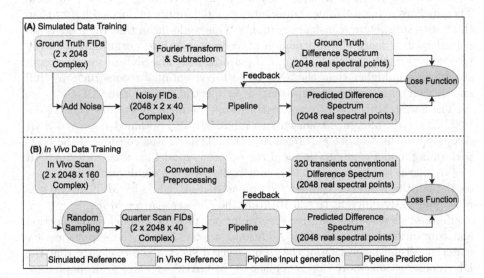

Fig. 3. The training procedure for A) simulated data and B) *in vivo* data. Preparing the pipeline input consists of generating an accelerated scan, either by A) adding noise to simulated ground-truths or B) selecting a reduced number of *in vivo* transients. The pipeline then outputs a difference spectrum which is compared to either A) the ground-truth difference spectrum or B) the conventional reconstruction with 320 transients.

- **Visual Inspection**: The spectra were qualitatively inspected regarding the shapes of the GABA (3 ppm) and Glx (3.75 ppm) peaks and any substantial subtraction artifacts or other significant discrepancies present.
- **Mean Squared Error (MSE)**: The MSE was measured between 2.5 ppm and 4 ppm. The reference for this metric was the *Gannet full* reconstruction.
- **Signal-to-Noise Ratio**: The SNR of each spectrum was calculated as the GABA peak height over twice the standard deviation of the 10 ppm to 12 ppm region.
- **GABA Fit Error**: The reconstructed spectra were fitted using Gannet and the GABA fitting error was compared among the different spectra reconstructions.

A Wilcoxon signed rank test with $\alpha = 0.05$ was used to determine statistical significance. This significance test was chosen because it compares paired data and makes no assumption about the underlying distribution.

3 Results and Discussion

Table 1 summarizes the MSE, SNR, and fit error metrics for the *in vivo* testing dataset for all four pipelines and two training procedures. The fine-tuned models performed the best for each pipeline, indicating that the weights pre-trained on simulations provided a better starting point for training with *in vivo* data. The pipelines using difference transients as model inputs performed better than those with subspectra transients as inputs. This could mean that the models did not adequately learn the subtraction procedure and thus, performing it before passing it through the model improved performance. The SNR of the DL reconstructions was higher than the SNR of both conventional reconstructions (full and quarter), indicating an improvement in spectral quality using DL.

The lowest MSE and fit error metrics were obtained by the fine-tuned Rdc-UNET2D with difference transients as inputs. It also obtained improved SNR compared to the *Gannet Full* reconstruction. Thus, it was considered the best-performing model configuration and the proposed model for this work. Its SNR was significantly higher ($p < 0.01$), and its fit error was significantly lower ($p = 0.04$) than the reference *Gannet full* reconstruction, while its MSE metric was not significantly different ($p = 0.10$) from the *Gannet quarter* MSE.

Figure 4 shows representative spectra for our proposed model reconstruction and conventional preprocessing. Additional spectra are in the supplementary material. Visual inspection of the reconstructed spectra shows clearly that the DL reconstructions more closely replicated the GABA and Glx peak shape compared to the *Gannet quarter* reconstructions. However, for the non-peak regions within the 2.5 - 4 ppm window used to calculate MSE, the *Gannet quarter* reconstruction was closer to the reference. These regions between the peaks are not relevant for the end goal of quantification, and it is believed that they are responsible for the lack of statistical significance between the MSE of the model reconstructions and the *Gannet quarter* reconstructions. The DL reconstructions

Table 1. Metrics (*mean ± std*) for different model and training configurations for the testing dataset. Conventional reconstructions using Gannet were also provided as reference, the *Gannet Full*, which uses all the available transients and is the target, and the *Gannet Quarter*, which uses only a quarter of the available transients.

Pipeline	MSE (2.5-4 ppm)	SNR	Fit Error (%)
Gannet Full	*N/A*	19.6 ± 5.5	5.0 ± 2.4
Gannet Quarter	0.033 ± 0.080	10.0 ± 3.1	7.8 ± 5.1
In Vivo Training			
U-NET 1D Subspectra	0.044 ± 0.036	22.5 ± 8.5	4.6 ± 1.8
U-NET 1D Difference	0.035 ± 0.023	23.3 ± 9.0	5.8 ± 3.0
Rdc-UNET2D Subspectra	0.041 ± 0.030	32.1 ± 15.8	6.3 ± 3.3
Rdc-UNET2D Difference	0.035 ± 0.018	25.8 ± 8.9	5.1 ± 3.9
Fine-Tune Training			
U-NET 1D Subspectra	0.036 ± 0.030	23.7 ± 9.6	6.8 ± 4.6
U-NET 1D Difference	0.029 ± 0.022	30.7 ± 13.1	4.7 ± 2.5
Rdc-UNET2D Subspectra	0.041 ± 0.032	39.8 ± 20.0	5.0 ± 6.2
Rdc-UNET2D Difference	$\mathbf{0.023 \pm 0.017}$	$\mathbf{45.3 \pm 21.5}$	$\mathbf{4.4 \pm 2.4}$

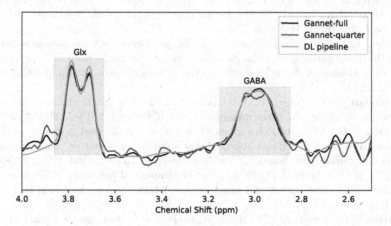

Fig. 4. Representative spectra reconstructions of: 1) Conventional pipeline with all the transients (*Gannet Full* - black), this is the reference reconstruction. 2) Conventional pipeline with a quarter of the transients (*Gannet Quarter* - green). 3) DL pipeline reconstruction (Model - Orange). The proposed DL reconstruction had high similarity to the reference for both the GABA and Glx peaks, and had less variation in the non-peak regions, demonstrating a high-quality reconstruction.

are smoother in these non-peak regions, indicating a smaller impact of noise in the reconstructions, which is reflected in the smaller fit error.

Given the improvement in fit error and SNR compared to the target, we can conclude that the proposed model obtained equivalent spectra to the full reconstruction using four times less data, reducing acquisition time by 75%

4   Conclusion

Among the compared pipelines, the proposed model, Rdc-UNET2D, obtained the best results. It obtained improved GABA fit error and GABA SNR metrics on reconstructions with 80 transients compared to conventional reconstructions with 320 transients, demonstrating the possibility of using DL to reduce scan times by 4 times while improving spectral quality.

References

1. Beauferris, Y., Teuwen, J., Karkalousos, D., et al.: Multi-coil MRI reconstruction challenge-assessing brain MRI reconstruction models and their generalizability to varying coil configurations. Front. Neurosci. **16**, 919186 (2022)
2. Brady, R.O., McCarthy, J.M., Prescot, A.P., et al.: Brain gamma-aminobutyric acid (GABA) abnormalities in bipolar disorder. Bipolar Disord. **15**(4), 434–439 (2013)
3. Chen, D., Hu, W., Liu, H., et al.: Magnetic resonance spectroscopy deep learning denoising using few in vivo data. IEEE Trans. Comput. Imaging **9**, 448–458 (2023)
4. De Graaf, R.A.: In vivo NMR spectroscopy: Principles and Techniques. NJ, third edition edn, Wiley, Hoboken (2019)
5. Edden, R.A., Puts, N.A., Harris, A.D., et al.: Gannet: A batch-processing tool for the quantitative analysis of gamma-aminobutyric acid-edited MR spectroscopy spectra: Gannet: GABA Analysis Toolkit. J. Magn. Reson. Imaging **40**(6), 1445–1452 (2014)
6. Govindaraju, V., Young, K., Maudsley, A.A.: Proton NMR chemical shifts and coupling constants for brain metabolites. NMR Biomed. **13**(3), 129–153 (2000)
7. Harris, A.D., Gilbert, D.L., Horn, P.S., et al.: Relationship between GABA levels and task-dependent cortical excitability in children with attention-deficit/hyperactivity disorder. Clin. Neurophysiol. **132**(5), 1163–1172 (2021)
8. Harris, A.D., Glaubitz, B., Near, J., et al.: Impact of frequency drift on gamma-aminobutyric acid-edited MR spectroscopy. Magn. Reson. Med. **72**(4), 941–948 (2014)
9. Harris, A.D., Saleh, M.G., Edden, R.A.: Edited [1] H magnetic resonance spectroscopy in vivo: Methods and metabolites: Edited [1] H MRS. Magn. Reson. Med. **77**(4), 1377–1389 (2017)
10. Lee, H.H., Kim, H.: Intact metabolite spectrum mining by deep learning in proton magnetic resonance spectroscopy of the brain. Magn. Reson. Med. **82**(1), 33–48 (2019)
11. Ma, D.J., Le, H.A.M., Ye, Y., et al.: MR spectroscopy frequency and phase correction using convolutional neural networks. Magn. Reson. Med. **87**(4), 1700–1710 (2022)

12. Mescher, M., Merkle, H., Kirsch, J., et al.: Simultaneous in vivo spectral editing and water suppression. NMR Biomed. **11**(6), 266–272 (1998)
13. Mikkelsen, M., Barker, P.B., Bhattacharyya, P.K., et al.: Big GABA: edited MR spectroscopy at 24 research sites. Neuroimage **159**, 32–45 (2017)
14. Mikkelsen, M., Loo, R.S., Puts, N.A., et al.: Designing GABA-edited magnetic resonance spectroscopy studies: considerations of scan duration, signal-to-noise ratio and sample size. J. Neurosci. Methods **303**, 86–94 (2018)
15. Mikkelsen, M., Rimbault, D.L., Barker, P.B., et al.: Big GABA II: water-referenced edited MR spectroscopy at 25 research sites. Neuroimage **191**, 537–548 (2019)
16. Muckley, M.J., Riemenschneider, B., Radmanesh, A., et al.: Results of the 2020 fastMRI Challenge for Machine Learning MR Image Reconstruction. IEEE Trans. Med. Imaging **40**(9), 2306–2317 (2021)
17. Near, J., Andersson, J., Maron, E., et al.: Unedited in vivo detection and quantification of γ-aminobutyric acid in the occipital cortex using short-TE MRS at 3 T. NMR Biomed. **26**(11), 1353–1362 (2013)
18. Near, J., Harris, A.D., Juchem, C., et al.: Preprocessing, analysis and quantification in single-voxel magnetic resonance spectroscopy: experts' consensus recommendations. NMR Biomed. **34**(5) (2021)
19. Öngür, D., Prescot, A.P., McCarthy, J., et al.: Elevated gamma-aminobutyric acid levels in chronic schizophrenia. Biol. Psychiat. **68**(7), 667–670 (2010)
20. Považan, M., Mikkelsen, M., Berrington, A., et al.: Comparison of. Radiologymultivendor single-voxel MR spectroscopy data acquired in healthy brain at 26 sites **295**(1), 171–180 (2020)
21. Puts, N.A., Edden, R.A.: In vivo magnetic resonance spectroscopy of GABA: a methodological review. Prog. Nucl. Magn. Reson. Spectrosc. **60**, 29–41 (2012)
22. Ronneberger, O., Fischer, P., Brox, T.: U-Net: convolutional networks for biomedical image segmentation. In: Navab, N., Hornegger, J., Wells, W.M., Frangi, A.F. (eds.) MICCAI 2015. LNCS, vol. 9351, pp. 234–241. Springer, Cham (2015). https://doi.org/10.1007/978-3-319-24574-4_28
23. Sanacora, G., Mason, G.F., Rothman, D.L., Krystal, J.H.: Increased occipital cortex GABA concentrations in depressed patients after therapy with selective serotonin reuptake inhibitors. Am. J. Psychiatry **159**(4), 663–665 (2002)
24. Sanacora, G., Mason, G.F., Rothman, D.L., et al.: Increased cortical GABA concentrations in depressed patients receiving ECT. Am. J. Psychiatry **160**(3), 577–579 (2003)
25. Schür, R.R., Draisma, L.W., Wijnen, J.P., et al.: Brain GABA levels across psychiatric disorders: a systematic literature review and meta-analysis of [1] H-MRS studies: Brain GABA Levels Across Psychiatric Disorders. Hum. Brain Mapp. **37**(9), 3337–3352 (2016)
26. Shamaei, A., Starcukova, J., Pavlova, I., Starcuk, Z.: Model-informed unsupervised deep learning approaches to frequency and phase correction of MRS signals. Magn. Reson. Med. **89**(3), 1221–1236 (2023)
27. Simpson, R., Devenyi, G.A., Jezzard, P., et al.: Advanced processing and simulation of MRS data using the FID appliance (FID-A)- an open source. MATLAB-based Toolkit Magn. Reson. Med. **77**(1), 23–33 (2017)
28. Tapper, S., Mikkelsen, M., Dewey, B.E., et al.: Frequency and phase correction of J-difference edited MR spectra using deep learning. Magn. Reson. Med. **85**(4), 1755–1765 (2021)
29. Waddell, K.W., Avison, M.J., Joers, J.M., Gore, J.C.: A practical guide to robust detection of GABA in human brain by J-difference spectroscopy at 3 T using a standard volume coil. Magn. Reson. Imaging **25**(7), 1032–1038 (2007)

30. Wang, J., Ji, B., Lei, Y., et al.: Denoising Magnetic Resonance Spectroscopy (MRS) Data Using Stacked Autoencoder for Improving Signal-to-Noise Ratio and Speed of MRS. arXiv:2303.16503v1 (2023)

31. Yoon, J.H., Maddock, R.J., Rokem, A., et al.: GABA concentration is reduced in visual cortex in schizophrenia and correlates with orientation-specific surround suppression. J. Neurosci. **30**(10), 3777–3781 (2010)

32. Zhang, Y., An, L., Shen, J.: Fast computation of full density matrix of multispin systems for spatially localized in vivo magnetic resonance spectroscopy. Med. Phys. **44**(8), 4169–4178 (2017)

Towards Abdominal 3-D Scene Rendering from Laparoscopy Surgical Videos Using NeRFs

Khoa Tuan Nguyen[1,2(✉)], Francesca Tozzi[4,6], Nikdokht Rashidian[5,6], Wouter Willaert[4,6], Joris Vankerschaver[2,3], and Wesley De Neve[1,2]

[1] IDLab, ELIS, Ghent University, Ghent, Belgium
{khoatuan.nguyen,wesley.deneve}@ghent.ac.kr
[2] Center for Biosystems and Biotech Data Science, Ghent University Global Campus, Ghent University Global Campus, Incheon, Korea
joris.vankerschaver@ghent.ac.kr
[3] Department of Applied Mathematics, Informatics and Statistics, Ghent University, Ghent, Belgium
[4] Department of GI Surgery, Ghent University Hospital, Ghent, Belgium
{francesca.tozzi,wouter.willaert}@ugent.be
[5] Department of HPB Surgery and Liver Transplantation, Ghent University Hospital, Ghent, Belgium
nikdokht.rashidian@ugent.be
[6] Department of Human Structure and Repair, Ghent University, Ghent, Belgium

Abstract. Given that a conventional laparoscope only provides a two-dimensional (2-D) view, the detection and diagnosis of medical ailments can be challenging. To overcome the visual constraints associated with laparoscopy, the use of laparoscopic images and videos to reconstruct the three-dimensional (3-D) anatomical structure of the abdomen has proven to be a promising approach. Neural Radiance Fields (NeRFs) have recently gained attention thanks to their ability to generate photo-realistic images from a 3-D static scene, thus facilitating a more comprehensive exploration of the abdomen through the synthesis of new views. This distinguishes NeRFs from alternative methods such as Simultaneous Localization and Mapping (SLAM) and depth estimation. In this paper, we present a comprehensive examination of NeRFs in the context of laparoscopy surgical videos, with the goal of rendering abdominal scenes in 3-D. Although our experimental results are promising, the proposed approach encounters substantial challenges, which require further exploration in future research.

Keywords: 3-D reconstruction · Laparoscopy · Neural Rendering · View Synthesis

1 Introduction

Laparoscopy, also known as keyhole surgery or minimally invasive surgery (MIS), is a surgical technique that enables a surgeon to access the inside of the abdomen

X. Cao et al. (Eds.): MLMI 2023, LNCS 14348, pp. 83–93, 2024.
https://doi.org/10.1007/978-3-031-45673-2_9

Fig. 1. Illustration of NeRFs and 3-D reconstruction using estimated depth information. From left to right: a video frame extracted from a laparoscopic surgical video, an image rendered by NeRFs, depth estimation for the given video frame, and reconstructions in both mesh and voxel formats based on the estimated depth information.

without the need for making large incisions in the skin. The surgeon inserts a slender tool, with a light and a camera attached, through small skin incisions, which makes it then possible for the surgeon to see inside the abdomen, negating the need for large skin cuts and a long recovery period. However, the view of the surgeon during MIS is limited to the perspective of the camera. To address this limitation, various approaches have been developed to provide surgeons with additional information such as depth, segmentation, 3-D reconstruction, and augmented surgery techniques [1, 10, 24]. Despite substantial advancements, current methods still face challenges in achieving high-quality and photorealistic 3-D reconstruction results, as shown in Fig. 1.

Neural Radiance Fields (NeRFs) [13, 15] have recently emerged as a popular method for view synthesis using natural images. However, their application in the medical field has thus far remained limited [4, 8, 26]. In this paper, we focus on the task of understanding the underlying scene in MIS through the utilization of NeRFs. Specifically, we investigate how well NeRFs can be applied to laparoscopic surgical videos, particularly during the preparatory phase of a surgical procedure where surgical tools, surgical actions, scenes with blood, and moving organs are absent. In summary, our study makes the following contributions:

- We present a workflow that analyzes laparoscopic surgical videos by employing state-of-the-art NeRFs-based methods to render a complete abdominal environment.
- We adopt mask-based ray sampling to better mitigate the presence of artefacts that are associated with the black outer areas commonly found in laparoscopic surgical videos.
- We investigate the performance of each adopted NeRFs-based method, identifying and discussing the challenges associated with their usage.

2   Related Work

One approach to achieve 3-D scene reconstruction is through the utilization of Simultaneous Localization And Mapping (SLAM), which enables the mapping of 3-D coordinates and the direct reconstruction of 3-D surfaces from a given camera view [10]. Guodong *et al.* proposed an enhanced SLAM method called MKCF-SLAM [27, 28], incorporating multi-scale feature patch tracking to reconstruct laparoscopic scenes characterized by texture deficiency, high levels of reflection, and changes in target scale. In contrast, Haoyin *et al.* presented a 2-D non-rigid SLAM system known as EMDQ-SLAM [31,32], capable of compensating for pixel deformation and performing real-time image mosaicking. Another approach for 3-D scene reconstruction involves depth estimation. Recasen *et al.* proposed Endo-Depth-and-Motion [17], a pipeline for 3-D scene reconstruction from monocular videos that combines self-supervised depth estimation, photometric odometry, and volumetric fusion. Additionally, Baoru *et al.* [7] leveraged stereo image pairs to generate corresponding point cloud scenes and minimize the point distance between the two scenes. Shuwei *et al.* [22] proposed a fusion of depth prediction, appearance flow prediction, and image reconstruction to achieve robust training.

Unlike the methods discussed previously, NeRFs are based on deep neural networks that focus on learning to represent 3-D scenes, as proposed by Mildenhall *et al.* [13]. To the best of our knowledge, the application of NeRFs to surgical videos, and endoscopic videos in particular, was first discussed in the work of Wang *et al.*, resulting in an approach known as EndoNeRF [26]. EndoNeRF specifically operates on prostatectomy data and leverages the D-NeRF method [16] to render a short dynamic scene and inpaint the surgical tools. However, whereas the focus of EndoNeRF is on rendering a specific area, the focus of our research effort is on rendering an entire abdominal environment. Moreover, our research effort primarily focuses on rendering static scenes, with a minimal presence of surgical tools, surgical actions, and moving organs. In particular, we generate static scenes through the use of two NeRFs-based methods that offer fast convergence: NerfAcc [9,25] and Instant-NGP [15].

3   Method

As depicted in Fig. 2, we present a workflow to apply NeRFs to laparoscopic surgical videos. We describe each step of this workflow in the following sections.

3.1   Video Frame Extraction

In this section, we describe the process of extracting video frames from a selected laparoscopic surgical video to create a dataset. In particular, we extract a total of n_{frame} equidistant frames from the given video. These extracted frames are then stored for subsequent analysis. By varying the value of n_{frame}, we generate multiple datasets, each containing a different number of video frames. This allows us to investigate the impact of the variable n_{frame} in our experiments, which encompass a broad spectrum of sparsity levels, ranging from sparsely sampled views (with only a few frames) to densely sampled views (with numerous frames).

3.2   Camera Pose Estimation

Upon obtaining a dataset consisting of n_{frame} extracted video frames, it becomes necessary to determine the camera pose for each individual frame. In order to

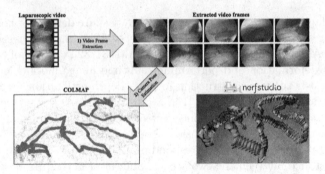

Fig. 2. NeRFs-based workflow for rendering of the abdomen from a laparoscopic surgical video. The workflow consists of three major steps, including two pre-processing steps and a training step. First, a dataset is created by extracting a total of n_{frame} video frames from the laparoscopic surgical video. Second, we estimate the camera pose for each extracted video frame, resulting in n_{pose} camera poses. The figure displays n_{pose} camera poses obtained using COLMAP after the second step. Finally, NeRFs-based models are trained using the extracted images and their corresponding camera poses, utilizing the Nerfstudio library.

accomplish this task, we employ Structure-from-Motion (SfM) algorithms that make use of COLMAP[1] [20,21] with Scale-Invariant Feature Transform (SIFT) features, as well as the hierarchical localization toolbox (HLOC)[2] [18]. Specifically, the HLOC toolbox utilizes COLMAP with SuperPoint [2] for feature extraction and SuperGlue [19] for image matching. In addition, we also make use of LocalRF [12] as an alternative camera pose estimation method. This method is capable of handling long-range camera trajectories and large unbounded scenes. By leveraging these tools, from the initial n_{frame} frames, we can estimate camera poses for a total of n_{pose} frames. Ideally, by using LocalRF [12], we would have $n_{frame} = n_{pose}$.

3.3 Training and Mask-Based Ray Sampling

For training the NeRFs-based models, we utilize the Nerfstudio [25] library, which includes the NerfAcc [9] and Instant-NGP [15] models.

As depicted in Fig. 3, our approach involves sampling rays from the input image and its corresponding camera pose, with each ray passing through a pixel on the image. It is important to note that we exclude the red lines in order to prevent the occurrence of artifacts at the intersections between these red lines and the blue lines from overlapping images. To that end, we propose sampling rays with masks. Instead of creating a mask for each individual image, we opted to use a single mask that applies to all images. This one-for-all mask takes the form of a circle located at the center of the image, with the radius calculated

---

[1] https://colmap.github.io/.
[2] https://github.com/cvg/Hierarchical-Localization.

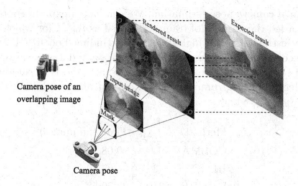

Fig. 3. Ray sampling for NeRFs training. Each line represents a ray. Green lines indicate the required sampled rays. Red lines represent rays in the black area that should not be sampled. Artifacts occur in the rendered result when an image overlaps with the black area of other images. This is demonstrated by the existence of artifacts within the blue circle, which illustrates the point where the blue line intersects with the red lines from other images (represented by the thick dashed red line). To obtain the desired result, we create a mask that only samples rays from pixels with color values. (Color figure online)

based on the circular view of the laparoscope. This mask allows us to selectively sample rays only from pixels that contain color values. By employing this mask, the model is able to infer the content of the black areas solely based on the color information from the overlapping regions in other images.

4   Experiments

To experiment with the use of NeRFs for rendering an abdominal scene, we chose a laparoscopic surgical video from a collection of videos recorded at Ghent University Hospital. This chosen video captures the preparatory phase of a laparoscopic surgery and does not involve the use of surgical tools. It provides a laparoscopic view recorded during an examination of the abdomen, captured at a resolution of 3840×2160 pixels. To optimize the video for our purposes, we cropped out 6.6% from the left and 9.1% from the right, removing redundant black areas. This resulted in video frames having a resolution of 3237×2160 pixels. The recorded footage has a duration of 93 seconds, and the frame rate is approximately 29.935 fps (frames per second), leading to a total of 2784 available video frames. All execution times were obtained on a machine equipped with a single RTX A6000 GPU and 128 GB RAM.

4.1   Camera Pose Estimation

We extract video frames from a selected laparoscopic surgical video, with n_{frame} having the following values: 310, 1392, and 2784 frames. This results in three

Table 1. Results of camera pose estimation. We conducted experiments with different settings and recorded the number of frames, denoted as n_{pose}, for which camera poses were successfully estimated. The highest n_{pose} (excluding LocalRF [12]) is shown in bold, and the second best is underscored for each n_{frame} dataset. LocalRF is capable of estimating all camera poses, albeit with a trade-off in terms of running time. Similarly, HLOC suffers from long running times. Consequently, we did not use HLOC and LocalRF for $n_{frame} = 2784$.

n_{frame}	Method	n_{pose}	Time (hh:mm:ss)
310	COLMAP	62	00:08:26
	HLOC	43	01:08:39
	LocalRF	310	23:20:52
1392	COLMAP	376	00:49:29
	HLOC	647	21:51:34
	LocalRF	1392	75:10:54
2784	COLMAP	2667	04:32:53

datasets, each having a different sampling interval. By doing so, we can assess the effectiveness of SfM methods on datasets with varying levels of sparsity. The best results of COLMAP and HLOC can be found in Table 1. Furthermore, we employ LocalRF [12] to ensure we can estimate all camera poses. We have made modifications to the code provided in the LocalRF paper to restrict the sampling of rays solely to valid color pixels. Table 1 indicates that COLMAP exhibits a substantially higher speed than HLOC, with a speedup factor of 8 and 26.5 for $n_{frame} = \{310, 1392\}$, respectively. However, COLMAP only performs well in the case of a dense view with $n_{frame} = 2784$, whereas HLOC demonstrates better effectiveness in other scenarios with $n_{frame} = \{310, 1392\}$. Although LocalRF [12] is capable of estimating all camera poses, its extensive running time presents a drawback. Consequently, LocalRF is only suitable for application in sparse views, where the number of images that needs to be determined is limited.

4.2 Qualitative and Quantitative Rendering Results

Based on the results of the previous experiment, we created individual image sets for five possible n_{pose} values: $\{62, 310, 647, 1392, 2667\}$. We then trained NeRFs-based models on each of these sets. For training, we took 90% of n_{pose}, comprising equally spaced images, as the training set. The remaining images were allocated to the evaluation set. To avoid memory limitations when working with a large number of 4K images, we down-scaled the input images by a factor of four. We trained the NerfAcc [9] and Instant-NGP [15] models using the default hyperparameter settings provided by the Nerfstudio library [25]. We observed that training with a sampling mask increased the processing time because of the underlying Nerfstudio implementation. Therefore, we adjusted the number

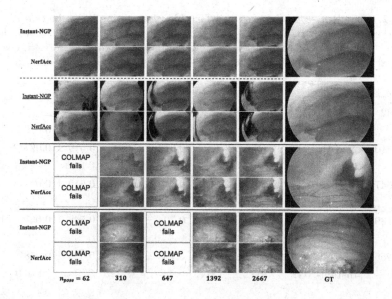

Fig. 4. Novel view synthesis for a select number of unseen video frames in the evaluation set. Training with a mask is shown in bold (Color figure online), whereas training without a mask is shown as underscored. Black patterns occur when training without a mask. The green box indicates a failure of COLMAP to estimate the camera pose of the video frame in the n_{pose} set (rendering a video frame is not possible without the correct camera perspective). The frames generated by Instant-NGP [15] are better in subjective quality and less blurry than those produced by NerfAcc [9].

of training iterations to be $\{3000, 15000, 30000, 50000, 100000\}$, corresponding to the number of images in each n_{pose} set, respectively.

For qualitative evaluation, Fig. 3 shows the render results obtained for a select number of images in the evaluation set. As explained in Sect. 3.3, training without a mask resulted in black patterns, while training with a mask improved the image quality and expanded the field of view. However, it is important to note that training with a mask currently presents a trade-off between image quality and processing time. Overall, the rendered frames from Instant-NGP [15] exhibited higher quality and less blurriness compared to NerfAcc [9]. Furthermore, among the different values of n_{pose}, the set with $n_{pose} = 2667$ (dense views) yielded the highest quality rendered frames compared to the other sets. This finding is consistent with the notion that a NeRFs-based method generally performs better when this method is applied to dense and overlapping views.

To evaluate the objective quality of the synthesized novel views, we measure the PSNR, SSIM, and LPIPS [30] scores, which are conventional image quality metrics [13,26], between the synthesized views and the corresponding ground truth video frames. The results obtained for training with a mask are presented in Table 2. We can observe that these results are in line with our qualitative findings, suggesting that Instant-NGP [15] outperforms NerfAcc [9].

Table 2. Quantitative results obtained for video frames in the evaluation set. We report the average PSNR/SSIM scores (higher is better) and LPIPS [30] scores (lower is better). The highest score for each n_{pose} evaluation set is indicated in bold. Overall, the Instant-NGP [15] model outperforms the NerfAcc [9] model across all n_{pose} sets.

n_{pose}	Method	PSNR ↑	SSIM ↑	LPIPS ↓
62	NerfAcc	12.760 ± 0.424	$\mathbf{0.764 \pm 0.018}$	$\mathbf{0.448 \pm 0.020}$
	Instant-NGP	$\mathbf{12.769 \pm 0.231}$	0.763 ± 0.008	0.470 ± 0.028
310	NerfAcc	12.520 ± 2.439	0.674 ± 0.084	0.619 ± 0.097
	Instant-NGP	$\mathbf{13.558 \pm 1.620}$	$\mathbf{0.714 \pm 0.051}$	$\mathbf{0.609 \pm 0.099}$
647	NerfAcc	13.817 ± 1.447	0.730 ± 0.044	0.510 ± 0.084
	Instant-NGP	$\mathbf{14.136 \pm 1.350}$	$\mathbf{0.744 \pm 0.034}$	$\mathbf{0.472 \pm 0.053}$
1392	NerfAcc	13.602 ± 1.598	0.704 ± 0.044	0.548 ± 0.055
	Instant-NGP	$\mathbf{14.694 \pm 1.244}$	$\mathbf{0.746 \pm 0.025}$	$\mathbf{0.538 \pm 0.051}$
2667	NerfAcc	14.226 ± 1.526	0.713 ± 0.039	0.529 ± 0.058
	Instant-NGP	$\mathbf{14.512 \pm 1.549}$	$\mathbf{0.757 \pm 0.025}$	$\mathbf{0.450 \pm 0.050}$

5   Conclusions

In this paper, we introduced a workflow for applying state-of-the-art NeRFs-based methods to laparoscopic surgical videos in order to render a complete abdominal environment, identifying and overcoming several challenges.

A first challenge we encountered is the presence of artifacts in recorded video frames (e.g., blurriness caused by laparoscope movement), leading to failures in camera pose estimation using COLMAP. To mitigate these failures, we utilized LocalRF [12], which enabled us to estimate all camera poses but required a substantial amount of computational time (e.g., 75 h for 1392 images).

A second challenge we faced was determining the appropriate frame resolution and the number of extracted frames (n_{frame}). Opting for frames with a higher resolution, such as 4K or even 8K [29], resulted in rendered images of a higher quality but also increased training time. This is due to the need to randomly sample rays (pixels) in each training iteration to ensure full pixel coverage. Also, larger n_{pose} values substantially increased the training time. Consequently, a trade-off exists between training time and rendered image quality.

A third challenge we came across is the limited view provided by a laparoscope, resulting in recorded videos that typically contain black outer areas. To address this challenge, we employed sampling with a mask as a workaround. As shown by our experimental results, the proposed one-for-all mask approach enhanced the quality of the rendered outcomes and broadened the perspective.

A fourth challenge is related to the camera trajectory followed, which is — in the context of laparoscopy surgery — usually free in nature, rather than being a regular forward-facing trajectory that captures extensive overlapping views. Such an elongated and unconstrained trajectory poses difficulties for NeRFs-based models to fully render an abdominal scene.

In future work, we will keep focussing on overcoming the primary challenge of the limited surgical view provided by a laparoscope. To that end, we aim at combining NeRFs with diffusion models [3,6,23], such as DiffRF [14], NerfDiff [5], and RealFusion [11]. Additionally, we plan to employ NeRFs to predict intraoperative changes during laparoscopic surgery, for instance addressing the issue of organ deformation between preoperative and intraoperative states.

References

1. Ali, S.: Where do we stand in AI for endoscopic image analysis? deciphering gaps and future directions. NPJ Digital Med. 5(1), 184 (2022)
2. DeTone, D., Malisiewicz, T., Rabinovich, A.: Superpoint: self-supervised interest point detection and description. In: Proceedings of the IEEE Conference on Computer Vision and Pattern Recognition Workshops, pp. 224–236 (2018)
3. Dhariwal, P., Nichol, A.: Diffusion models beat GANs on image synthesis. Adv. Neural. Inf. Process. Syst. 34, 8780–8794 (2021)
4. Gerats, B.G., Wolterink, J.M., Broeders, I.A.: Depth-supervise NeRF for multi-view RGB-D operating room images. arXiv preprint arXiv:2211.12436 (2022)
5. Gu, J., Trevithick, A., Lin, K.E., Susskind, J., Theobalt, C., Liu, L., Ramamoorthi, R.: NerfDiff: single-image View Synthesis with NeRF-guided Distillation from 3D-aware Diffusion. In: International Conference on Machine Learning (2023)
6. Ho, J., Jain, A., Abbeel, P.: Denoising diffusion probabilistic models. Adv. Neural. Inf. Process. Syst. 33, 6840–6851 (2020)
7. Huang, B., et al.: Self-supervised depth estimation in laparoscopic image using 3D geometric consistency. In: International Conference on Medical Image Computing and Computer-Assisted Intervention, pp. 13–22. Springer (2022). https://doi.org/10.1007/978-3-031-16449-1_2
8. Huy, P.N., Quan, T.M.: Neural radiance projection. In: 2022 IEEE 19th International Symposium on Biomedical Imaging (ISBI), pp. 1–5 (2022). https://doi.org/10.1109/ISBI52829.2022.9761457
9. Li, R., Gao, H., Tancik, M., Kanazawa, A.: NerfAcc: efficient sampling accelerates NeRFs. arXiv preprint arXiv:2305.04966 (2023)
10. Lin, B., Sun, Y., Qian, X., Goldgof, D., Gitlin, R., You, Y.: Video-based 3D reconstruction, laparoscope localization and deformation recovery for abdominal minimally invasive surgery: a survey. Int. J. Med. Robot. Comput. Assist. Surg. 12(2), 158–178 (2016)
11. Melas-Kyriazi, L., Rupprecht, C., Laina, I., Vedaldi, A.: RealFusion: 360° reconstruction of any object from a single image. arXiv:2302.10663v2 (2023)
12. Meuleman, A., Liu, Y.L., Gao, C., Huang, J.B., Kim, C., Kim, M.H., Kopf, J.: Progressively optimized local radiance fields for robust view synthesis. In: CVPR (2023)
13. Mildenhall, B., Srinivasan, P.P., Tancik, M., Barron, J.T., Ramamoorthi, R., Ng, R.: NeRF: representing scenes as neural radiance fields for view synthesis. In: Vedaldi, A., Bischof, H., Brox, T., Frahm, J.-M. (eds.) ECCV 2020. LNCS, vol. 12346, pp. 405–421. Springer, Cham (2020). https://doi.org/10.1007/978-3-030-58452-8_24
14. Müller, N., Siddiqui, Y., Porzi, L., Bulo, S.R., Kontschieder, P., Nießner, M.: Diffrf: rendering-guided 3d radiance field diffusion. In: Proceedings of the IEEE/CVF Conference on Computer Vision and Pattern Recognition, pp. 4328–4338 (2023)

15. Müller, T., Evans, A., Schied, C., Keller, A.: Instant Neural Graphics Primitives with a Multiresolution Hash Encoding. ACM Trans. Graph. **41**(4), 102:1–102:15 (2022). https://doi.org/10.1145/3528223.3530127

16. Pumarola, A., Corona, E., Pons-Moll, G., Moreno-Noguer, F.: D-nerf: neural radiance fields for dynamic scenes. In: Proceedings of the IEEE/CVF Conference on Computer Vision and Pattern Recognition, pp. 10318–10327 (2021)

17. Recasens, D., Lamarca, J., Fácil, J.M., Montiel, J., Civera, J.: Endo-depth-and-motion: reconstruction and tracking in endoscopic videos using depth networks and photometric constraints. IEEE Robot. Autom. Lett. **6**(4), 7225–7232 (2021)

18. Sarlin, P.E., Cadena, C., Siegwart, R., Dymczyk, M.: From Coarse to Fine: robust Hierarchical Localization at Large Scale. In: CVPR (2019)

19. Sarlin, P.E., DeTone, D., Malisiewicz, T., Rabinovich, A.: SuperGlue: learning feature Matching with graph neural networks. In: CVPR (2020)

20. Schönberger, J.L., Frahm, J.M.: Structure-from-motion revisited. In: Conference on Computer Vision and Pattern Recognition (CVPR) (2016)

21. Schönberger, J.L., Zheng, E., Frahm, J.-M., Pollefeys, M.: Pixelwise view selection for unstructured multi-view stereo. In: Leibe, B., Matas, J., Sebe, N., Welling, M. (eds.) ECCV 2016. LNCS, vol. 9907, pp. 501–518. Springer, Cham (2016). https://doi.org/10.1007/978-3-319-46487-9_31

22. Shao, S., et al.: Self-supervised monocular depth and ego-motion estimation in endoscopy: appearance flow to the rescue. Med. Image Anal. **77**, 102338 (2022)

23. Sohl-Dickstein, J., Weiss, E., Maheswaranathan, N., Ganguli, S.: Deep unsupervised learning using nonequilibrium thermodynamics. In: International Conference on Machine Learning, pp. 2256–2265. PMLR (2015)

24. Soler, L., Hostettler, A., Pessaux, P., Mutter, D., Marescaux, J.: Augmented surgery: an inevitable step in the progress of minimally invasive surgery. In: Gharagozloo, F., Patel, V.R., Giulianotti, P.C., Poston, R., Gruessner, R., Meyer, M. (eds.) Robotic Surgery, pp. 217–226. Springer, Cham (2021). https://doi.org/10.1007/978-3-030-53594-0_21

25. Tancik, M., et al.: Nerfstudio: a modular framework for neural radiance field development. In: ACM SIGGRAPH 2023 Conference Proceedings. SIGGRAPH '23 (2023)

26. Wang, Y., Long, Y., Fan, S.H., Dou, Q.: Neural rendering for stereo 3D reconstruction of deformable tissues in robotic surgery. In: International Conference on Medical Image Computing and Computer-Assisted Intervention, pp. 431–441. Springer (2022). https://doi.org/10.1007/978-3-031-16449-1_41

27. Wei, G., Feng, G., Li, H., Chen, T., Shi, W., Jiang, Z.: A novel SLAM method for laparoscopic scene reconstruction with feature patch tracking. In: 2020 International Conference on Virtual Reality and Visualization (ICVRV), pp. 287–291. IEEE (2020)

28. Wei, G., Yang, H., Shi, W., Jiang, Z., Chen, T., Wang, Y.: Laparoscopic scene reconstruction based on multiscale feature patch tracking method. In: 2021 International Conference on Electronic Information Engineering and Computer Science (EIECS), pp. 588–592 (2021). https://doi.org/10.1109/EIECS53707.2021.9588016

29. Yamashita, H., Aoki, H., Tanioka, K., Mori, T., Chiba, T.: Ultra-high definition (8K UHD) endoscope: our first clinical success. Springerplus **5**(1), 1–5 (2016)

30. Zhang, R., Isola, P., Efros, A.A., Shechtman, E., Wang, O.: The unreasonable effectiveness of deep features as a perceptual metric. In: Proceedings of the IEEE Conference on Computer Vision and Pattern Recognition, pp. 586–595 (2018)

31. Zhou, H., Jayender, J.: EMDQ-SLAM: Real-Time High-Resolution Reconstruction of Soft Tissue Surface from Stereo Laparoscopy Videos. In: de Bruijne, M., et al. (eds.) MICCAI 2021. LNCS, vol. 12904, pp. 331–340. Springer, Cham (2021). https://doi.org/10.1007/978-3-030-87202-1_32

32. Zhou, H., Jayender, J.: Real-time nonrigid mosaicking of laparoscopy images. IEEE Trans. Med. Imaging **40**(6), 1726–1736 (2021). https://doi.org/10.1109/TMI.2021.3065030

Brain MRI to PET Synthesis and Amyloid Estimation in Alzheimer's Disease via 3D Multimodal Contrastive GAN

Yan Jin[1]([✉]), Jonathan DuBois[2], Chongyue Zhao[3], Liang Zhan[3], Audrey Gabelle[1], Neda Jahanshad[4], Paul M. Thompson[4], Arie Gafson[1], and Shibeshih Belachew[1]

[1] Biogen Digital Health, Biogen Inc., Cambridge, MA, USA
yjinz@ucla.edu
[2] Research Biomarkers, Biogen Inc., Cambridge, MA, USA
[3] Dept. of Electrical and Computer Engineering, University of Pittsburgh, Pittsburgh, PA, USA
[4] Mark and Mary Stevens Neuroimaging and Informatics Institute, University of Southern California, Los Angeles, CA, USA

Abstract. Positron emission tomography (PET) can detect brain amyloid-β (Aβ) deposits, a diagnostic hallmark of Alzheimer's disease and a target for disease modifying treatment. However, PET-Aβ is expensive, not widely available, and, unlike magnetic resonance imaging (MRI), exposes the patient to ionizing radiation. Here we propose a novel 3D multimodal generative adversarial network with contrastive learning to synthesize PET-Aβ images from cheaper, more accessible, and less invasive MRI scans (T1-weighted and fluid attenuated inversion recovery [FLAIR] images). In tests on independent samples of paired MRI/PET-Aβ data, our synthetic PET-Aβ images were of high quality with a structural similarity index measure of 0.94, which outperformed previously published methods. We also evaluated synthetic PET-Aβ images by extracting standardized uptake value ratio measurements. The synthetic images could identify amyloid positive patients with a balanced accuracy of 79%, holding promise for potential future use in a diagnostic clinical setting.

1 Introduction

Alzheimer's disease (AD) is a chronic neurodegenerative disease that causes cognitive and behavioral symptoms and leads to progressive loss of autonomy. In order to optimize the care pathway of the AD continuum, early diagnosis is critical [1, 2]. Among all the diagnostic imaging modalities, magnetic resonance imaging (MRI) and positron emission tomography (PET) amyloid-β (Aβ) (PET-Aβ) are the most common imaging modalities able to detect the key hallmarks of AD at an early phase. The key imaging hallmarks include volumetric MRI atrophy in brain regions where neurodegeneration is associated with formation of neurofibrillary tangles and an abnormal level of aggregated Aβ deposits measured by the PET standardized uptake value ratio (SUVR).

As the Aβ plaques are typically the first biomarker abnormalities across the AD continuum and have recently become a clinically-validated target for disease modifying therapies [3, 4], PET-Aβ has been widely used in research studies and clinical trials. However, the scalability of PET-Aβ remains complex, as PET-Aβ is an expensive modality, the access to Aβ tracers is limited, and there is some risk to patients associated with radiation exposure. In contrast, MRI scanners are widely available, do not involve exposure to ionizing radiation, and the cost remains feasible. As such, a solution utilizing brain MRI to reconstruct PET-Aβ could be an important advance for the field of AD diagnosis.

Major progress has been made in medical image synthesis including deep learning architectures. In the field of brain MRI to PET synthesis, Sikka et al. [5] proposed using a deep convolutional autoencoder based on U-Net architecture to synthesize fluorodeoxyglucose (FDG)-PET from T1-weighted (T1w) MRI. Recently, Generative Adversarial Networks (GANs) have shown superior image synthesis performance relative to other methods. Hu et al. [6] formulated a bidirectional GAN to synthesize FDG-PET from T1w MRI. Zhang et al. [7] also established a BPGAN for a similar task. However, these methods and results have limitations: (1) The datasets used to train such models usually combine different stages of the disease. For example, in Alzheimer's Disease Neuroimaging Initiative (ADNI), AD, mild cognitive impairment (MCI), and normal control (NC) have differing amyloid burdens and a model blindly trained on the entire dataset without considering individual differences will diminish its differentiation capability; (2) The performance of the model is assessed solely based on the criteria of image reconstruction quality such as mean absolute error (MAE), peak signal-to-noise ratio (PSNR), and structural similarity index measure (SSIM), but the real clinical utilities of PET images for diagnostic purpose have not been evaluated; (3) In contrast to FDG-PET, which is more closely linked to T1w MRI due to structural atrophy, Aβ aggregation generally occurs prior to structural changes and may require additional MRI contrasts to accurately evaluate. Therefore, using single modality (e.g., only the T1w MRI) in the models may not be sufficient.

To address these issues, we propose a novel multimodal contrastive GAN architecture to use both T1w and fluid attenuated inversion recovery (FLAIR) images to synthesize PET-Aβ. The novel contributions of our methodology are threefold: First, by leveraging the concept of supervised contrastive learning [8], we synthesize more discriminative images that depend on clinical labels such as AD, MCI, and NC. Second, using both T1w and FLAIR provides complementary information for synthesis. Third, we evaluate our synthetic PET images by extracting SUVR measures for standardized regions of interest (ROIs) to show how to use synthetic images for real clinical applications. This study allows us for the first time to design models based on intrinsic data-clinical information relations between brain MRI and PET-Aβ imaging and we demonstrate the utility of synthetic images in a clinical diagnostic setting, beyond simple reconstruction evaluation.

2 Methods

2.1 Brain MRI and PET Image Processing

Only T1w and FLAIR brain MR images and static PET images were used in the study. The original T1w and FLAIR images of the same subject were first processed through N4 field bias correction and then skull stripped with SynthStrip [9]. The FLAIR image was linearly registered to the T1w image with Advanced Normalization Tools (ANTs) [10]. The T1w image was segmented into anatomically meaningful ROIs with FreeSurfer [11].

The static PET image was linearly registered to the T1w image with ANTs and skull stripped with the brain mask of the T1w image. One means of assessing amyloid load is using SUVR. SUVR measurement requires normalization of PET images to a reference region to account for non-displaceable radiotracer binding. For PET-Aβ, cerebellar gray matter is often considered as a gold standard reference region [12]. The cerebellar gray matter was retrieved from the segmented T1w image, and the mean intensity of the region was calculated. Then the voxelwise intensity values of the entire PET image were divided by the mean intensity of the cerebellar gray matter to obtain the PET SUVR image.

The set of the T1w, FLAIR, and PET SUVR images were aligned to the common template used in the OASIS project [13]. The PET SUVR image was also smoothed with a full width at half maximum (FWHM) of 6 mm. The three images were finally normalized to have an intensity level between 0 and 1.

2.2 Multimodal Contrastive GAN

Framework Overview. Our Multimodal Contrastive GAN (MC-GAN) consists of two generators G_1 and G_2 and one discriminator D for multimodal multi-label domain translation. Specifically, we trained our G_1 to map multimodal inputs X_{T1w} and X_{FLAIR} into an output Y_{PET} under the clinical label L, which is one of the diagnostic groups: AD, MCI, and NC, i.e., $Y_{PET} = G_1(X_{T1w}, X_{FLAIR}, L)$. Here, we used a three-channel binary volume with one-hot encoding for labeling. It can be written as $L = [l_1, l_2, l_3]$, where [] is the channelwise concatenation operator, and l_n is a one-channel binary volume for domain labeling (the same size as X_{T1w} and X_{FLAIR}), i.e., $L = [1, 0, 0]$ represents AD; $[0, 1, 0]$ represents NC; and $[0, 0, 1]$ represents MCI. The three inputs were concatenated together and fed to G_1. Similarly, G_2 mapped inputs Y_{PET} back to $[Y_{T1w}, Y_{FLAIR}]$, i.e., $[Y_{T1w}, Y_{FLAIR}] = G_2(Y_{PET}, L)$, to form a cycleGAN [14], which increases network stability and speeds up convergence. Our discriminator D performed two functions simultaneously, that is, distinguishing synthetic from real PET images and controlling the distances between embedded image features depending on clinical labels. The framework of our MC-GAN is illustrated in Fig. 1.

Loss Function. To ensure that the synthetic PET image is as close as possible to the original PET image, we proposed the following loss components:

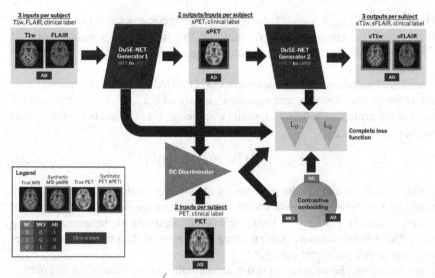

Fig. 1. The architecture of the proposed MC-GAN that demonstrates a representative training cycle for an individual with AD. DuSE-NET: 3D Dual Squeeze-and-Excitation Network; DC: Deep Convolutional

(1) *Pair Loss:* Given the T1w and FLAIR MR images, we compared pairs of real and synthetic PET images for our pair training. The loss can be represented as follows:

$$L_{pair} = E\left[\|G_1(X_{T1w}, X_{FLAIR}, L) - Y_{PET}^{GT}\|_1\right] \quad (1)$$

where G_1 generates a PET volume $G_1(X_{T1w}, X_{FLAIR}, L) \rightarrow Y_{PET}$ with inputs of X_{T1w} and X_{FLAIR} under the clinical label L. Here, L_{pair} aims to minimize the ℓ_1 difference, which is better than ℓ_2 in image restoration [15], between the ground truth PET volume Y_{PET}^{GT} and the predicted PET volume Y_{PET}.

(2) *Adversarial Loss:* To ensure that the synthetic PET is indistinguishable from the real one, we used the Wasserstein loss as:

$$L_{adv} = min_{G_1} max_D E\left[D\left(Y_{PET}^{GT}\right) - D(G_1(X_{T1w}, X_{FLAIR}, L))\right] \quad (2)$$

where D tries to distinguish the synthetic PET image $G_1(X_{T1w}, X_{FLAIR}, L)$ from the real PET image Y_{PET}^{GT}, so D tends to maximize the loss. On the other hand, G_1 tries to confuse D, so G_1 tends to minimize the loss. Since the Wasserstein loss is not bounded, the discriminator (or critic) D is allowed to improve without degrading its feedback to the generator G_1 [16].

(3) *Cyclic Reconstruction Loss:* Adversarial loss alone cannot guarantee that the learned function can map an input to its desired output [14]. The cycleGAN uses an additional extension to the architecture, termed cycle consistency, in which the output of the first generator is used to the input of the second generator and the output of the second generator should match the original image. In our case, it can be represented as:

$$L_{rec} = E\left[\|[X_{T1w}, X_{FLAIR}] - G_2(G_1(X_{T1w}, X_{FLAIR}, L), L)\|_1\right] \quad (3)$$

where $G_2(G_1(X_{T1w}, X_{FLAIR}, L), L)$ is the cyclic reconstructed T1w and FLAIR images using the composition of G_1 and G_2. This loss acts as a regularization of generator models and guides the generation process to converge more stably and quickly.

(4) *Gradient Penalty Loss:* The Wasserstein GAN (WGAN) requires 1-Lipschitz continuity for the loss function, and a gradient penalty adds a constraint on the gradient norm of the critic D to enforce Lipschitz continuity [17]. The gradient penalty loss can be represented as:

$$L_{gp} = E_{\hat{x}}\left[\left(\|\nabla_{\hat{x}} D(\hat{x})\|_2 - 1\right)^2\right] \tag{4}$$

where $\hat{x} \in P_{\hat{x}}$, and $P_{\hat{x}}$ is the distribution obtained by uniformly sampling along a straight line between the real and synthetic PET image domains. This gradient penalty forces the gradients of the critic's output regarding the inputs to have unit norm. This allows even more stable training of the network than WGAN and requires very little hyper-parameter tuning.

(5) *Contrastive Loss:* Since the training dataset is usually heterogeneous, which contains patients with different clinical labels, we need to consider both data-to-label and data-to-data relations during training. Here, we adopted the concept of contrastive learning to propose a contrastive loss [18]. Let $X = [X_1, \ldots, X_n]$, where $X_i = [X_{PET}]_i \in R^{LXWXH}$ are randomly selected batch samples, and $Y = [Y_1, \ldots, Y_n]$, where $Y_i \in R$ are corresponding clinical labels. Then we used the convolutional layers in D as an encoder $E(X) \in R^t$ and a projection layer P to embed $E(X)$ onto a unit hypersphere, i.e., $P : R^t \to S^d$. We mapped the data space to the hypersphere with the composition $R(X) = P(E(X))$. We also employed a label embedding function $C(Y) : R \to R^d$. Then the contrastive loss can be formulated as:

$$L_{contra}(X_i, Y_i; t) = -\log\left(\frac{\exp\left(\frac{R(X_i)^T C(Y_i)}{t}\right) + \sum_{k=1}^n I_{Y_k = Y_i} \exp\left(\frac{R(X_i)^T R(X_k)}{t}\right)}{\exp\left(\frac{R(X_i)^T C(Y_i)}{t}\right) + \sum_{k=1}^n I_{k \neq i} \exp\left(\frac{R(X_i)^T R(X_k)}{t}\right)}\right) \tag{5}$$

where t is a temperature to control push and pull force. We considered those samples Y_i's are positive samples, and those whose $k \neq i$ are negative samples. Equation (5) pulls a positive embedded sample $R(X_i)$ closer to the embedding label $C(Y_i)$ and pushes all negative samples away with the exception of those negative samples that have the same label, i.e., $Y_k = Y_i$.

The complete loss function is comprised of these five components and can be written as:

$$L_G = L_{adv} + \lambda_{rec} L_{rec} + \lambda_{pair} L_{pair} + \lambda_{contra} L_{contra} \tag{6}$$

$$L_D = L_{adv} + \lambda_{gp} L_{gp} + \lambda_{contra} L_{contra} \tag{7}$$

where λ_{rec}, λ_{pair}, λ_{gp}, and λ_{contra} are the weights for L_{rec}, L_{pair}, L_{gp} and L_{contra}, respectively. During training, we optimized D using L_D with G_1 and G_2 fixed, and then optimized G_1 and G_2 using L_G with D fixed. We chose Adam as our optimization algorithm.

To fully optimize the loss function, all the components in the function need to have low values so that the synthetic PET image shares high similarity with the real one.

Network Design. Our discriminator was a deep convolutional (DC) discriminator, which consisted of 5 strided convolutional layers without max pooling and fully connected layers. After the last convolutional layer, we added two embedded layers for both images and labels to compute L_{contra}. Our generator is a 3D U-Net based autoencoder with squeeze-and-excitation (SE) enhancement [19]. We added a dual SE block at the output of each layer to fuse channelwise and spatial features. At the latent space layer, we also added 3 sequential residual convolutional layers to further strengthen feature representations.

2.3 SUVR ROI Calculation

For each synthetic PET image, we extracted ROIs including the frontal cortex, lateral temporal cortex, medial temporal cortex, parietal cortex, anterior cingulate cortex, and posterior cingulate cortex, based on the FreeSurfer segmentation. These cortical regions are expected to be significantly altered in AD and commonly used to measure aggregated Aβ deposition in PET SUVR images [20]. We combined all these regions and created a composite cortex to calculate the mean composite SUVR value.

3 Experiments and Results

3.1 Experiment Setup

Data Preparation. We collected 408 sets (212 females and 196 males) of T1w, FLAIR, and static Florbetaben-PET images from ADNI3 (77 AD: mean age 75.4, mean Mini-Mental State Examination [MMSE] 22; 158 MCI: mean age 71.9, mean MMSE 27.5; 173 NC: mean age 70, mean MMSE 29.3) [21]. The image size of T1w was $208 \times 240 \times 256$ with 1 mm isotropic voxel size. FLAIR was $160 \times 256 \times 256$ with $1.2 \times 1 \times 1$ mm voxel size. PET was $224 \times 224 \times 81$ with $1 \times 1 \times 2$ mm voxel size. For each set, the MRI acquisition time was within 3 months of the PET acquisition time to ensure they represented the subject's brain in the same state of atrophy. All images were processed with the pipeline described in Sect. 2.1. All images were aligned to the OASIS template and cropped to a size of $192 \times 192 \times 144$ with 1-mm isotropic voxel size before feeding into MC-GAN. The cropping was necessary due to the MC-GAN architecture and for computation efficiency.

Implementation Details. Following processing, the imaging dataset was randomly divided into 80% training and 20% testing, proportional to the three clinical labels. During the training, we randomly sampled 3D patches of size $128 \times 128 \times 128$ from each pair of T1w and FLAIR images and conducted image augmentation by rotating images by 90° and flipping along axes randomly. We empirically set hyperparameters $\lambda_{rec} = 5$, $\lambda_{pair} = 15$, $\lambda_{gp} = 10$, $\lambda_{contra} = 1$, and $t = 0.08$. We used 4 32 GB NVIDIA Tesla V100-SXM2 GPUs to train the MC-GAN for up to 500 epochs. At test time, the full-size $192 \times 192 \times 144$ T1w and FLAIR images, along with the clinical label, were input to the trained generator G_1 to reconstruct the full-size PET-Aβ image.

3.2 Synthesis Quality Evaluation

Ablation Study on Multimodality and Contrastivity. We were able to generate synthetic PET-Aβ images with the MR images from ADNI3. The quality of the synthetic PET-Aβ images was assessed by the MAE, the PSNR and the SSIM. We conducted an ablation study to show the significance of those essential components of MC-GAN, that is, multimodal inputs and contrastive learning. Using our developed MC-GAN, we, obtained 0.06, 26.64 and 0.94 for the MAE, PSNR and SSIM, respectively. The MC-GAN outperforms all compared methods that were missing one or both components (Table 1). The differences are statistically significant. Furthermore, although it is an indirect comparison, the mean SSIM of our MC-GAN (0.94) outperforms those of Hu et al. [6] (0.89) and Zhang et al. [7] (0.73). Figure 2 shows a representative set of T1w, FLAIR, real and synthetic PET-Aβ images for three clinical labels.

Table 1. Quantitative comparison of the performance of brain MRI to PET-Aβ synthesis methods. SS-T1w – Single-modal (T1w) Single Class, SS-FLAIR – Single-modal (FLAIR) Single Class, MS – Multimodal Single Class, SC-T1w – Single-modal (T1w) Contrastive, SC-FLAIR – Single-modal (FLAIR) Contrastive, MC – Multimodal Contrastive.

Methods	MAE	PSNR	SSIM
SS-T1w-GAN	0.10 ± 0.02	21.29 ± 1.33	0.82 ± 0.04
SS-FLAIR-GAN	0.09 ± 0.03	22.02 ± 2.26	0.78 ± 0.08
MS-GAN	0.09 ± 0.03	20.67 ± 2.73	0.86 ± 0.06
SC-T1w-GAN	0.08 ± 0.03	24.78 ± 2.80	0.91 ± 0.03
SC-FLAIR-GAN	0.09 ± 0.03	22.95 ± 2.88	0.85 ± 0.12
MC-GAN	$\mathbf{0.06 \pm 0.02}$	$\mathbf{26.64 \pm 2.81}$	$\mathbf{0.94 \pm 0.01}$

3.3 Normalized SUVR Comparison

Figure 3(a) compares the normalized SUVRs (between 0 and 1) of the composite region for real and synthetic PET-Aβ images for each group. The mean normalized SUVR for the AD, MCI, and NC were 0.57, 0.56, 0.45 for real PET-Aβ images, and 0.60, 0.51, 0.45 using synthetic PET-Aβ images, respectively. We conducted t-tests and the differences were statistically significant on the SUVRs of synthetic PET-Aβ images between AD vs NC ($p < 0.0001$) and MCI vs NC ($p < 0.0001$), whose differences were also significant in real PET-Aβ images.

Figure 3(b) illustrates the mean percentage errors between real and synthetic PET-Aβ images for each group. All the mean percentage errors are within 15%, which is considered as a clinically acceptable error range. Due to the small number of AD patients in both the training set (n = 61) and the test set (n = 16), the standard deviation of the percentage errors in this group is much larger than those of the other two groups. The heterogeneity of the MCI group also made it challenging to obtain more precise results.

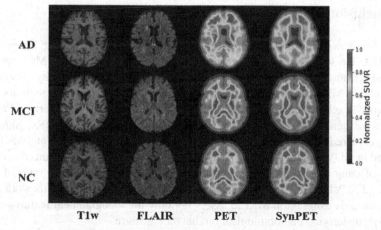

Fig. 2. A set of T1w, FLAIR, real and synthetic PET-Aβ normalized SUVR images for an AD, MCI, and NC patient from ADNI3, respectively. The regions in warmer color in PET-Aβ normalized SUVR images demonstrate greater Aβ deposition in the brain.

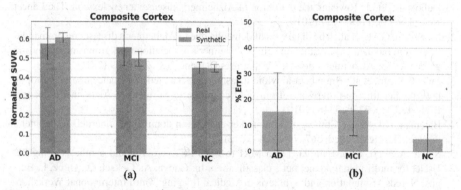

Fig. 3. (a) Comparison of the normalized SUVRs of the composite cortex for real and synthetic PET-Aβ for AD, MCI, and NC; (b) The mean percentage errors between the composite normalized SUVRs for real and synthetic PET-Aβ for AD, MCI, and NC, respectively.

3.4 Aβ Positivity Classification

Aβ deposition is a pathological hallmark of AD and its detection (by PET-Aβ or cerebrospinal fluid) is part of the diagnostic workup of individuals suspected to have the disease [22]. Clinically, a patient whose SUVR is above 1.4 is considered PET-Aβ positive [23]. Based on our dataset, it is approximately equal to the normalized SUVR 0.5. The number of participants with positive PET-Aβ in the test set was 31 using real data. Among them, 24 positives were identified in the synthetic data. Thus, the sensitivity was 77%. On the other hand, 51 participants were PET-Aβ negative using real data, 41 negatives were observed with our synthetic data demonstrating 80% specificity. Therefore, the balanced accuracy was 79%.

4 Conclusions

We developed an end-to-end deep learning framework generating PET-Aβ images from brain MRI. We tested our model on a representative sample of AD, MCI, and NC participants from the ADNI3 database. Our MC-GAN used complementary information from multiple MRI modalities and fused it with contrastive clinical labels to guide the network to achieve high-quality and state-of-the-art results. The generated PET-Aβ images were then compared to the real PET-Aβ images in terms of SUVRs, which has not been explored in prior work. Our synthetic images discriminated AD and MCI from NC based on SUVRs and identified positive PET-Aβ patients with a balanced accuracy of 79%, offering a benchmark for future methods. The percentage error rates of mean normalized SUVRs were acceptable across the three clinical groups. This work paves the way towards using brain MRI to assess not only the topographical features of AD but also its underlying pathobiological biomarker signature.

References

1. Dubois, B., et al.: Revising the definition of Alzheimer's disease: a new lexicon. The Lancet Neurol. **9**(11), 1118–1127 (2010)
2. McKhann, G.M., et al.: The diagnosis of dementia due to Alzheimer's disease: recommendations from the national institute on aging-Alzheimer's association workgroups on diagnostic guidelines for Alzheimer's disease. Alzheimers Dement. **7**(3), 263–269 (2011)
3. Jack, C.R., Jr., et al.: Brain beta-amyloid measures and magnetic resonance imaging atrophy both predict time-to-progression from mild cognitive impairment to Alzheimer's disease. Brain **133**(11), 3336–3348 (2010)
4. Bateman, R.J., et al.: Clinical and biomarker changes in dominantly inherited Alzheimer's disease. N. Engl. J. Med. **367**, 795–804 (2012)
5. Sikka, A., Peri, S.V., Bathula, D.R.: MRI to FDG-PET: cross-modal synthesis using 3D U-Net for multi-modal Alzheimer's classification. In: Gooya, A., Goksel, O., Oguz, I., Burgos, N. (eds.) Simulation and Synthesis in Medical Imaging: Third International Workshop, SASHIMI 2018, Held in Conjunction with MICCAI 2018, Granada, Spain, September 16, 2018, Proceedings 3: Springer, Cham, vol. 11037, pp. 80–89 (2018). https://doi.org/10.1007/978-3-030-00536-8_9
6. Hu, S., Lei, B., Wang, S., Wang, Y., Feng, Z., Shen, Y.: Bidirectional mapping generative adversarial networks for brain MR to PET synthesis. IEEE Trans. Med. Imaging **41**(1), 145–157 (2022)
7. Zhang, J., He, X., Qing, L., Gao, F., Wang, B.: BPGAN: brain PET synthesis from MRI using generative adversarial network for multi-modal Alzheimer's disease diagnosis. Comput. Methods Programs Biomed. **217**, 106676 (2022)
8. Khosla, P., et al.: Supervised contrastive learning. Adv. Neural. Inf. Process. Syst. **33**, 18661–18673 (2020)
9. Hoopes, A., Mora, J.S., Dalca, A.V., Fischl, B., Hoffmann, M.: SynthStrip: skull-stripping for any brain image. Neuroimage **260**, 119474 (2022)
10. Avants, B.B., Tustison, N.J., Stauffer, M., Song, G., Wu, B., Gee, J.C.: The Insight ToolKit image registration framework. Front. Neuroinform. **8**, 44 (2014)
11. Desikan, R.S., et al.: An automated labeling system for subdividing the human cerebral cortex on MRI scans into gyral based regions of interest. Neuroimage **31**(3), 968–980 (2006)

12. Bullich, S., et al.: Optimal reference region to measure longitudinal amyloid-beta change with (18)F-Florbetaben PET. J. Nucl. Med. **58**(8), 1300–1306 (2017)

13. Marcus, D.S., Wang, T.H., Parker, J., Csernansky, J.G., Morris, J.C., Buckner, R.L.: Open Access Series of Imaging Studies (OASIS): cross-sectional MRI data in young, middle aged, nondemented, and demented older adults. J. Cogn. Neurosci. **19**(9), 1498–1507 (2007)

14. Zhu, J.-Y., Park T., Isola P., Efros, A.A.: Unpaired image-to-image translation using cycle-consistent adversarial networks. In: Proceedings of the IEEE International Conference on Computer Vision, pp. 2223–2232 (2017)

15. Zhao, H., Gallo, O., Frosio, I., Kautz, J.: Loss functions for image restoration with neural networks. IEEE Trans. Comput. Imaging. **3**(1), 47–57 (2016)

16. Arjovsky, M., Chintala, S., Bottou, L.: Wasserstein generative adversarial networks. In: Proceedings of the 34th International Conference on Machine Learning, PMLR, pp. 214–223 (2017)

17. Gulrajani, I., Ahmed, F., Arjovsky, M., Dumoulin, V., Courville, A.: Improved training of Wasserstein GANs. In: Proceedings of the 31st International Conference on Neural Information Processing Systems. Long Beach, California, USA, pp. 5769–5779. Curran Associates Inc. (2017)

18. Kang, M., Park, J.: ContraGAN: contrastive learning for conditional image generation. Adv. Neural. Inf. Process. Syst. **33**, 21357–21369 (2020)

19. Hu, J., Shen, L., Sun, G.: Squeeze-and-excitation networks. In: Proceedings of the IEEE Conference on Computer Vision and Pattern Recognition, pp. 7132–7141 (2018)

20. Chiao, P., et al.: Impact of reference and target region selection on amyloid PET SUV ratios in the phase 1b PRIME study of Aducanumab. J. Nucl. Med. **60**(1), 100–106 (2019)

21. Veitch, D.P., et al.: Using the Alzheimer's disease neuroimaging initiative to improve early detection, diagnosis, and treatment of Alzheimer's disease. Alzheimer's Dement. **18**(4), 824–857 (2022)

22. Ma, C., Hong, F., Yang, S.: Amyloidosis in Alzheimer's disease: pathogeny, etiology, and related therapeutic directions. Molecules **27**(4), 1210 (2022)

23. Jack, C.R., et al.: Amyloid-first and neurodegeneration-first profiles characterize incident amyloid PET positivity. Neurology **81**(20), 1732–1740 (2013)

Accelerated MRI Reconstruction via Dynamic Deformable Alignment Based Transformer

Wafa Alghallabi[1]([envelope]) [iD], Akshay Dudhane[1] [iD], Waqas Zamir[2] [iD],
Salman Khan[1,3] [iD], and Fahad Shahbaz Khan[1,4] [iD]

[1] Mohamed bin Zayed University of AI, Abu Dhabi, United Arab Emirates
wafa.alghallabi@mbzuai.ac.ae
[2] Inception Institute of AI, Abu Dhabi, United Arab Emirates
[3] Australian National University, Canberra, Australia
[4] Linköping University, Linköping, Sweden

Abstract. Magnetic resonance imaging (MRI) is a slow diagnostic technique due to its time-consuming acquisition speed. To address this, parallel imaging and compressed sensing methods were developed. Parallel imaging acquires multiple anatomy views simultaneously, while compressed sensing acquires fewer samples than traditional methods. However, reconstructing images from undersampled multi-coil data remains challenging. Existing methods concatenate input slices and adjacent slices along the channel dimension to gather more information for MRI reconstruction. Implicit feature alignment within adjacent slices is crucial for optimal reconstruction performance. Hence, we propose MFormer: an accelerated MRI reconstruction transformer with cascading MFormer blocks containing multi-scale **D**ynamic **D**eformable **S**win **T**ransformer (DST) modules. Unlike other methods, our DST modules implicitly align adjacent slice features using dynamic deformable convolution and extract local non-local features before merging information. We adapt input variations by aggregating deformable convolution kernel weights and biases through a dynamic weight predictor. Extensive experiments on Stanford2D, Stanford3D, and large-scale FastMRI datasets show the merits of our contributions, achieving state-of-the-art MRI reconstruction performance. Our code and models are available at https://github.com/wafaAlghallabi/MFomer.

Keywords: MRI reconstruction · Alignment · Dynamic convolution

1 Introduction

Magnetic resonance imaging (MRI) employs strong magnetic fields for detailed images of a patient's anatomy and physiology. Obtaining fully-sampled data is

Supplementary Information The online version contains supplementary material available at https://doi.org/10.1007/978-3-031-45673-2_11.

time-consuming; hence, accelerated MRI scans use undersampled k-space [21]. However, it leads to undesired aliasing artifices, making MRI reconstruction unfeasible due to insufficient s-space points. Compressed Sensing [6] has become a classical framework for accelerated MRI reconstruction. Recently, significant progress has been made in developing deep learning algorithms for MRI reconstruction [2,16]. Furthermore, encoder-decoder architectures such as the U-Net [24] have been widely used in the medical imaging field. The hierarchical structure of U-Net has demonstrated a strong ability in MRI reconstructions [31] [30]. Another approach for tackling this task is the unrolled network [9,11,32]. It involves unrolling the procedure of an iterative optimization algorithm, such as gradient descent, into a deep network that allows the network to refine the image progressively and enhance the reconstruction quality. Sriram et al. [13] proposes VarNet, a deep learning method that combines compressed sensing and variational regularization to improve the quality of reconstructed images from undersampled MRI measurements. Then, E2E-VarNet [27] was introduced to extend VarNet [13] by applying end-to-end learning. However, these methods rely on the use of convolutional neural networks and require large-scale datasets for training to obtain high-quality MRI reconstructions [20].

Transformer-based methods have emerged as promising alternatives in addressing various medical imaging issues [1,17,18,26,28,33]. Yet, there have been limited efforts to apply transformers in MRI reconstruction. In [10], an end-to-end transformer-based network has been proposed for joint MRI reconstruction and super-resolution. Another transformer-based method, Recon-Former [12], utilizes recurrent pyramid transformer layers to reconstruct high-quality MRI images from under-sampled k-space data. Huang et al. [15] propose using a swin transformer for MRI acceleration. Mao et al. [9] present HUMUS-net, swin transformer-based network that achieves benchmark performance on the fastMRI [34] dataset. HUMUS-net is an unrolled and multi-scale network that combines the benefits of convolutions and transformer blocks.

HUMUS-Net [9] processes a given MRI slice (current slice) along with its adjacent slices and obtains a denoised output. However, the adjacent slices may have a misalignment with the current slice, which is likely to cause a distorted reconstruction. Misalignment with adjacent slices can occur from scanner imperfections, subject variability, slice thickness/gap, etc. To tackle this, we propose MForemer: MRI reconstruction transformer, which utilizes a cascade of MFormer blocks, each having multiple **D**ynamic **D**eformable **S**win **T**ransformer (DST) modules. The proposed MFormer block implicitly aligns the adjacent slices with the current slice at multiple scales through our DST module before merging their information towards the MRI reconstruction. The proposed DST module uses new dynamic deformable convolution for alignment and extracts local non-local features through the residual swin transformer sub-module. Our DST module is adaptive to the input data. It dynamically aggregates the weights and biases of the given deformable convolution, used in feature alignment, through the dynamic weight predictor block. With such input adaptability, the proposed

MFormer possesses better representational power compared to existing methods for MRI reconstruction in the literature. Our key contributions are as follows:

- We propose an approach, named MFormer, that utilizes a dynamic deformable swin transformer module to implicitly align the features of adjacent slices with the current slice at multiple scales, followed by their integration. Within the DST module, to perform feature alignment, we introduce a dynamic deformable convolution technique.
- In order to accommodate input variations, the proposed dynamic weight predictor aggregates the weights and biases of deformable convolution kernels. It learns the aggregation coefficient based on the variances observed in the center and adjacent slices.
- We conduct extensive experiments to validate the effectiveness of our MFormer on three datasets. Our MFormer demonstrated superior performance for MRI reconstruction and outperformed the recent HUMUS-Net [9].

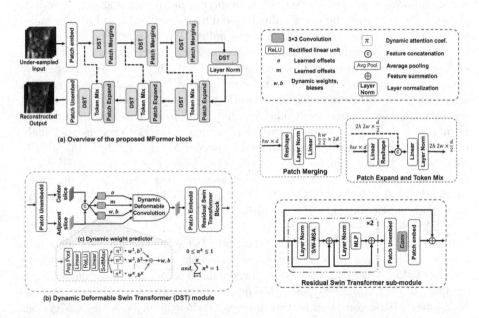

Fig. 1. The proposed MFormer uses a sequence of sub-networks or cascades of unrolled MFormer blocks. **(a)** Here, we show one such sub-network, the MFormer block, that is used in MRI reconstruction. The MFormer block comprises multiple **D**ynamic **D**eformable **S**win **T**ransformer (DST) modules. **(b)**, Here, we present an overview of a single DST module that implicitly aligns adjacent slice features with the current slice at multiple scales. Within the DST module, we introduce a dynamic deformable convolution for implicit feature alignment and a residual swin transformer sub-module to extract local and non-local features. **(c)** To adapt the input variations, weights and biases of deformable kernels are aggregated through our dynamic weight predictor. As per variations in the center and adjacent slices, it learns the aggregation coefficient. Patch merging and patch expand are used to down-scale and upscale the features.

2 Proposed Approach

Undersampling k-space has become a widely adopted technique to accelerate the acquisition speed of MRI scans. However, the drawback of this approach is the resulting undersampled data, which presents a significant challenge in achieving high-quality images. In an attempt to address this issue, previous methods like HUMUS-Net attempt to concatenate current and adjacent slices along the channel dimension and process through the network for MRI reconstruction. Nonetheless, this approach fails to consider the misalignment between the adjacent slices and the center slice, leading to suboptimal reconstruction outcomes. To overcome this limitation and enhance the quality of MRI reconstruction, we propose a novel approach called MFormer: an accelerated MRI reconstruction transformer. Our method implicitly aligns the features of adjacent slices with the current slice before integrating their information. The alignment step maximizes the performance of MRI reconstruction and generate high-quality images that accurately represent the underlying anatomy and physiology.

The proposed MFormer comprises a sequence of sub-networks or cascades (eight in total), each representing an unrolled iteration of an underlying optimization algorithm. Figure 1(a) shows such a sub-network, called MFormer block. Our MFormer block is a multi-scale hierarchical architecture design that aligns adjacent slices with the current slice and extracts local non-local features at multiple scales using the proposed Dynamic Deformable Swin Transformer (DST) module. The DST module introduces dynamic deformable convolution for aligning features at multiple scales and a residual shifted window transformer (RSwin) sub-module for local non-local feature extraction. The detailed explanation of the MFormer block and our DST module is as follows.

2.1 MFormer Block

The proposed MFormer block takes an intermediate reconstruction from the previous cascade and carries out a single round of de-noising to generate a better reconstruction for the following cascade. Let $I^t \in \mathbb{R}^{H \times W \times c_{in}}$ be the tth noisy, complex-valued input MRI slice derived from under-sampled k-space data, where the real and imaginary parts of the image are concatenated along the channel dimension and $D^t \in \mathbb{R}^{H \times W \times c_{in}}$ be the reconstructed output. Here, H, W, c_{in} represents input height, width, and channels, respectively. Our goal is to reconstruct the response from current slice I^t, and the $2N$ neighboring slices $\{I^{t-N}, ... I^{t-1}, I^{t+1}, ... I^{t+N}\}$. Therefore, our MFormer block takes $2N+1$ consecutive MR slices $\{I^i\}_{i-N}^{t+N}$ as an input and processes it at multiple scales through the DST modules to reconstruct the response for the tth MR image. The overall pipeline of the proposed Mformer block and DST module are shown in Fig. 1(a) and (b), respectively.

DST Module: has dynamic deformable convolution for feature alignment and residual swin transformer sub-module for local non-local feature extraction.

(a) *Dynamic Deformable Convolution:* In the existing literature [7,22,23, 29], deformable convolution [3,5,8] has been the most preferred choice to align the adjacent image features with the reference/base image features. However, its weights and biases are fixed and are not dynamic with respect to the input features. Therefore, we propose dynamic deformable convolution as shown in Fig. 1(b). Let F^t, $F^i \in \mathbb{R}^{h \times w \times f}$ be the initial latent representation of the center (I^t) and adjacent slice (I^i) respectively. Where, $h = \frac{H}{s}$, and $w = \frac{W}{s}$; here, s is a downscaling factor (To process HR input, we firstly downscale it through the simple strided convolution with sride $s = 2$). As shown in Fig. 1(b), we process (F^t, F^i) through simple convolutions and predicts sampling parameters (offsets, modulation scalar) for F^i and dynamic attention coefficients to aggregate the weights and biases of modulated deformable convolution.

$$A^i = g(w^d_{(F^i)}(F^i, o, m) + b_{(F^i)}), \quad \{o, m\} = w^o \left\{ F^t, F^i \right\} \tag{1}$$

where o, m represents offsets and modulated scalars respectively obtained through the offset convolution (w^o). $w^d_{(F^i)}$ and biases $b_{(F^i)}$ are functions of F^i and represent the aggregated weights, biases of dynamic deformable convolution as:

$$w^d_{(F^i)} = \sum_{k=1}^{K} \pi^k_{(F^i)} \cdot w^k \quad \text{and} \quad b_{(F^i)} = \sum_{k=1}^{K} \pi^k_{(F^i)} \cdot b^k$$

Here, w^k, b^k represents k^{th} convolution weights and biases, where $k = 1 : K$. In our experiments, we consider $K = 4$. The attention coefficients $\pi^k_{(F^i)}$ are dynamic and vary for each input F^i. They represent the optimal aggregation of weights and biases of deformable convolution for a given input. We compute these attention coefficients through the *dynamic weight predictor* as shown in Fig. 1(c). Input features F^i is processed through a global average pooling followed by a series of linear and ReLU layers as shown to obtain the attention coefficients. As the aggregated weights and biases are dynamic with respect to the input features, the proposed dynamic deformable convolution has more representational capabilities than the conventional deformable convolution.

(b) *Residual Swin Transformer Sub-module:* The obtained F^t and A^i are further passed through the patch embedding and processed through series of the residual swin transformer (RSwin) sub-modules [9] to extract local non-local features. The schematic of the RSwin is shown in Fig. 1.

Finally, the aligned and improved features of the adjacent and current slice are merged along the feature dimension, followed by a series of simple convolution operations which integrates the information of the adjacent and current slice. As a result, the proposed MFormer block aligns and integrates the 2N adjacent slices with the center slice to reduce the noise content in the current slice.

2.2 Iterative Unrolling

We adopt the same setting as in [9] for iterative unrolling to address the inverse problem. This approach decomposes the problem into smaller sub-problems, treating the architecture as a cascade of simpler denoisers (our MFormer blocks). Each block progressively refines the estimate obtained from the previous unrolled iteration, making the problem easier to solve and enhancing solution robustness. The block diagram of this operation is in the supplementary. After applying the inverse Fourier transform to obtain the image I_i, we employ the following steps: first, we utilize the reduce operation R to merge coil images using their corresponding sensitivity maps. Then, the expand operation ε transforms the reconstruction obtained from the MFormer block back into individual coil images.

Table 1. Comparison of the performance of various methods on the Stanford2D, Stanford3D, and FastMRI (validation set) datasets on 8× acceleration factor. For Stanford2D, Stanford3D, mean and standard error of the runs are shown with multiple train-validation splits.

Dataset	Method	SSIM(↑)	PSNR(↑)	NMSE(↓)
Stanford2D	E2E-VarNet [27]	0.8928 ± 0.0168	33.9 ± 0.7	0.0339 ± 0.0037
	HUMUS-Net [9]	0.8954 ± 0.0136	33.7 ± 0.6	0.0337 ± 0.0024
	MFormer (Ours)	**0.9100 ± 0.0094**	**34.9 ± 0.6**	**0.0299 ± 0.0035**
Stanford3D	E2E-VarNet [27]	0.9432 ± 0.0063	40.0 ± 0.6	0.0203 ± 0.0006
	HUMUS-Net [9]	0.9459 ± 0.0065	40.4 ± 0.7	0.0184 ± 0.0008
	MFormer (Ours)	**0.9483 ± 0.0040**	**40.6 ± 0.8**	**0.0173 ± 0.0007**
FastMRI	E2E-VarNet [27]	0.8900	36.89	0.0090
	HUMUS-Net [9]	0.8936	37.03	0.0090
	MFormer (Ours)	**0.8970**	**37.45**	**0.0080**

Additionally, we incorporate the data consistency module, which computes a correction map to ensure the reconstructed image accurately represents the measured data. Our MFormer utilizes eight cascades of the MFormer block. Among these cascades, only the first MFormer block aligns the adjacent slices with respect to the current slice, while the rest of the MFormer blocks are without alignment. During experiments, we observed that including the alignment strategy in the subsequent cascades burdens the MFormer regarding parameters and processing time.

3 Experiments

We validate the proposed MFormer on 3 datasets for accelerated MRI reconstruction. To compare with the SOTA methods, we followed the fastMRI multi-coil knee track setup with 8× acceleration [34]. First, we undersample the fully-sampled k-space by preserving 4% of the lowest frequency neighboring k-space lines and randomly subsampling 12.5% of all k-space lines in the phase encoding direction. The resolution of the input image is adjusted to 384 × 384 by center-cropping and padding for reflections. In all our experiments, we use 2 adjacent slices along with the current slice. Our MFormer block consists of 7 DST modules, each consists one dynamic deformable convolution followed by one RSwin sub-module with an embedding dimension of 66 (which doubled at every patch merging and halved at every patch expand). With this configuration, our MFormer has $111M$ parameters and 777.10 GFlops. We train our MFormer using Adam [19] for 50 epochs with an initial learning rate is 0.0001, which is dropped by a factor of 10 at epoch 40.

3.1 Benchmark Datasets

-**FastMRI** [34] is the biggest public dataset for MRI scans developed by a collaborative research project of Facebook AI Research (FAIR) and NYU Langone Health. In our work, we conducted our experiments on the FastMRI multi-coil knee dataset, which contains more than 46k slices in 1290 volumes. We trained MFormer on the training set that consists of more than 35k slices in 973 volumes only, with random mask generation.

-**Stanford2D** [4] **dataset** comprises fully-sampled volumes from diverse anatomical regions, totaling 89 volumes. In order to train our model, 80% of the volumes were randomly selected for training, and the remaining 20% were used for validation [9]. We randomly generated three train-validation splits for the dataset to ensure unbiased results. -**Stanford Fullysampled 3D FSE Knees (Stanford3D)** [25] dataset includes 19 volumes of knee MRI scans, each containing 256 slices. We split the data into train-validation sets using a method similar to Stanford 2D, running three training sessions. Our MFormer is trained for 25 epochs.

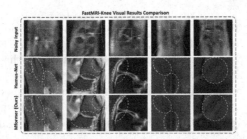

Fig. 2. Visual results of our MFormer and HUMUS-Net on FastMRI [34] for accelerated MRI reconstruction. Our approach achieves sharper boundaries with better-reconstruction results (see dotted white regions), compared to the HUMUS-Net [9].

3.2 Experimental Results

The superiority of the proposed MFormer over existing methods is demonstrated in Table 1, where it outperforms the competitive methods on the validation set of the three datasets. In FastMRI, the proposed MFormer achieves 0.897 SSIM, which is higher than that of HUMUS-Net 0.893 and 0.89 of E2E-VarNet. Similar trends can be observed in terms of average peak signal-to-noise ratio

(PSNR) and normalized mean square error (NMSE) values, further highlighting performance improvements with the proposed MFormer. As in Stanford2D, we achieve a PSNR gain of 1.2 dB over the second-best HUMUS-Net. Similarly, our MFormer achieves a notable improvement over the existing HUMUS-Net in the Stanford3D dataset. Furthermore, in Table 2, our MFormer shows considerable improvement over other methods on the FastMRI test set, highlighting the efficacy of our approach. More results on different acceleration factors (×4, ×6) can be found in the supplementary. In Fig. 2 and Fig. 3, we present visual comparisons of our MFormer method and the existing HUMUS-Net for accelerated MRI reconstruction (with 8x acceleration factor) on a few samples of the three datasets. Our approach demonstrates improved boundary sharpness and better reconstruction results when compared to HUMUS-Net.

Ablation Studies: To investigate the impact of aligning adjacent slices with the center slice on MRI reconstruction performance, we conducted a series of experiments using the Stanford3D dataset and evaluated the results using the SSIM. We compared the effectiveness of existing deformable convolution techniques [5] with our proposed dynamic deformable convolution method (Sect. 2.1(a)) for feature alignment. Table 3 presents the results of our experiments. Initially, we observed that using three adjacent slices as input yielded better results compared to using a single slice. We then explored a single-scale strategy using deformable convolution, where the initial latent representation of adjacent slices was aligned with the center slice. Additionally, we employed a multi-scale strategy, aligning the features of adjacent slices with those of the center slice, which resulted in further improvements. Subsequently, we replaced the deformable convolution with our Deformable Swin Transformer module, which outperformed the existing deformable convolution technique. We also attempted to apply dilation to our alignment module in order to enlarge the receptive field, but this adversely affected the performance. Finally, we replaced the Deformable Swin Transformer module with our Dynamic Deformable Swin Transformer, and this led to a significant improvement in performance. Our proposed Dynamic Deformable Swin Transformer proved to be an effective technique for enhancing reconstruction performance in comparison to existing methods.

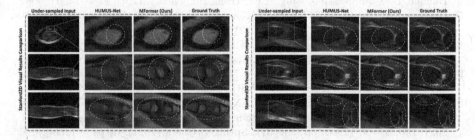

Fig. 3. Visual results of our MFormer and HUMUS-Net on Stanford2D [4] and Stanford3D [25] for accelerated MRI reconstruction. Our approach achieves sharper boundaries with better reconstruction results (denoted by dotted white regions) compared to the HUMUS-Net [9]. Best viewed zoomed in. More results are in the supplementary.

Table 2. Quantitative evaluation on FastMRI knee multi-coil ×8 test set. All the methods are trained only on the training set of FastMRI dataset.

Method	SSIM(↑)	PSNR(↑)	NMSE(↓)
E2E-VarNet [27]	0.8900	36.9	0.0089
XPDNet [32]	0.8893	37.2	0.0083
∑-Net [14]	0.8877	36.7	0.0091
U-Net [34]	0.8640	34.7	0.0132
HUMUS-Net [9]	0.8936	37.0	0.0086
MFormer (Ours)	0.8976	37.3	0.0081

Table 3. Our ablation studies which involves different alignment strategies on Stanford3D dataset (single run).

Method	Additional input	SSIM
HUMUS-Net [9]	No	0.9445
HUMUS-Net [9]	2 adjacent slices	0.9455
Deform-Conv (single scale)	2 adjacent slices	0.9457
Deform-Conv (multi-scale)	2 adjacent slices	0.9463
Deformable Swin module	2 adjacent slices	0.9478
Dilated Deformable Swin module	2 adjacent slices	0.9461
Dynamic Deformable Swin module	2 adjacent slices	**0.9483**

4   Conclusion

We show that the proper alignment of undersampled MRI slices is crucial to enhance the MRI reconstruction quality. Introducing MFormer, that enhances MRI reconstruction performance using a cascade of MFormer blocks. Each block contains Dynamic Deformable Swin Transformer (DST) modules, which utilizes our dynamic deformable convolution to align adjacent slice features with the current slice. Extracting local non-local features via residual Swin Transformer sub-modules, we then merge the information to achieve SOTA results for MRI reconstruction across three datasets.

References

1. Adjei-Mensah, I., et al.: Investigating vision transformer models for low-resolution medical image recognition. In: ICCWAMTIP (2021)
2. Ahishakiye, E., Bastiaan Van Gijzen, M., Tumwiine, J., Wario, R., Obungoloch, J.: A survey on deep learning in medical image reconstruction. Intell. Med. 1, 118–127 (2021)
3. Bhat, G., et al.: NTIRE 2022 burst super-resolution challenge. In: CVPR (2022)
4. Cheng, J.Y.: Stanford 2D FSE (2018). http://mridata.org/list?project=Stanford2DFSE. Accessed 11 June 2023
5. Dai, J., et al.: Deformable convolutional networks. In: ICCV (2017)
6. Donoho, D.: Compressed sensing. IEEE Trans. Inf. Theory 52(4), 1289–1306 (2006)
7. Dudhane, A., Zamir, S.W., Khan, S., Khan, F.S., Yang, M.H.: Burst image restoration and enhancement. In: CVPR (2022)
8. Dudhane, A., Zamir, S.W., Khan, S., Khan, F.S., Yang, M.H.: Burstormer: burst image restoration and enhancement transformer. In: CVPR (2023)
9. Fabian, Z., Tinaz, B., Soltanolkotabi, M.: HUMUS-Net: hybrid unrolled multi-scale network architecture for accelerated MRI reconstruction. In: NeurIPS (2022)
10. Feng, C.-M., Yan, Y., Fu, H., Chen, L., Xu, Y.: Task transformer network for joint MRI reconstruction and super-resolution. In: de Bruijne, M., et al. (eds.) MICCAI 2021. LNCS, vol. 12906, pp. 307–317. Springer, Cham (2021). https://doi.org/10.1007/978-3-030-87231-1_30
11. Gilton, D., Ongie, G., Willett, R.: Deep equilibrium architectures for inverse problems in imaging. IEEE Trans. Comput. Imaging 7, 1123–1133 (2021)

12. Guo, P., Mei, Y., Zhou, J., Jiang, S., Patel, V.M.: Reconformer: accelerated MRI reconstruction using recurrent transformer. arXiv preprint arXiv:2201.09376 (2022)

13. Hammernik, K., et al.: Learning a variational network for reconstruction of accelerated MRI data. Magn. Reson. Med. **79**(6), 3055–3071 (2018)

14. Hammernik, K., Schlemper, J., Qin, C., Duan, J., Summers, R.M., Rueckert, D.: Sigma-net: systematic evaluation of iterative deep neural networks for fast parallel MR image reconstruction. arXiv preprint arXiv:1912.09278 (2019)

15. Huang, J., et al.: Swin transformer for fast MRI (2022)

16. Hyun, C.M., Kim, H.P., Lee, S.M., Lee, S., Seo, J.K.: Deep learning for undersampled MRI reconstruction. Phys. Med. Biol. (2018). https://doi.org/10.1088/1361-6560/aac71a

17. Karimi, D., Dou, H., Gholipour, A.: Medical image segmentation using transformer networks. IEEE Access **10**, 29322–29332 (2022)

18. Khan, S., Naseer, M., Hayat, M., Zamir, S.W., Khan, F.S., Shah, M.: Transformers in vision: a survey. ACM Comput. Surv. **54**, 1–41 (2022)

19. Kingma, D.P., Ba, J.: Adam: a method for stochastic optimization. arXiv preprint arXiv:1412.6980 (2014)

20. Knoll, F., et al.: Deep learning methods for parallel magnetic resonance image reconstruction. arXiv preprint (2019)

21. Lustig, M., Donoho, D.L., Santos, J.M., Pauly, J.M.: Compressed sensing MRI. IEEE Signal Process. Mag. **25**(2), 72–82 (2008)

22. Mehta, N., Dudhane, A., Murala, S., Zamir, S.W., Khan, S., Khan, F.S.: Adaptive feature consolidation network for burst super-resolution. In: CVPR (2022)

23. Mehta, N., Dudhane, A., Murala, S., Zamir, S.W., Khan, S., Khan, F.S.: Gated multi-resolution transfer network for burst restoration and enhancement. In: CVPR (2023)

24. Ronneberger, O., Fischer, P., Brox, T.: U-Net: convolutional networks for biomedical image segmentation. In: Navab, N., Hornegger, J., Wells, W.M., Frangi, A.F. (eds.) MICCAI 2015. LNCS, vol. 9351, pp. 234–241. Springer, Cham (2015). https://doi.org/10.1007/978-3-319-24574-4_28

25. Sawyer, A., et al.: Creation of fully sampled MR data repository for compressed sensing of the knee. Ge Healthcare (2013)

26. Shamshad, F., et al.: Transformers in medical imaging: a survey. Med. Image Anal. (2023)

27. Sriram, A., et al.: End-to-end variational networks for accelerated MRI reconstruction. In: Martel, A.L., et al. (eds.) MICCAI 2020. LNCS, vol. 12262, pp. 64–73. Springer, Cham (2020). https://doi.org/10.1007/978-3-030-59713-9_7

28. Wang, B., Wang, F., Dong, P., Li, C.: Multiscale transunet++: dense hybrid U-net with transformer for medical image segmentation. Signal Image Video Process. **16**(6), 1607–1614 (2022)

29. Wang, L., et al.: NTIRE 2023 challenge on stereo image super-resolution: methods and results. In: CVPR (2023)

30. Yang, G., et al.: DAGAN: deep de-aliasing generative adversarial networks for fast compressed sensing MRI reconstruction. IEEE Trans. Med. Imaging **37**(6), 1310–1321 (2018)

31. Zabihi, S., Rahimian, E., Asif, A., Mohammadi, A.: Sepunet: depthwise separable convolution integrated U-net for MRI reconstruction. In: ICIP (2021)

32. Zaccharie Ramzi, Philippe Ciuciu, J.L.S.: Xpdnet for MRI reconstruction: an application to the 2020 fastmri challenge. arXiv preprint arXiv:2010.07290v2 (2020)

33. Zamir, S.W., Arora, A., Khan, S., Hayat, M., Khan, F.S., Yang, M.H.: Restormer: efficient transformer for high-resolution image restoration. In: CVPR (2022)
34. Zbontar, J., et al.: fastMRI: an open dataset and benchmarks for accelerated MRI. arXiv preprint arXiv:1811.08839 (2018)

Deformable Cross-Attention Transformer for Medical Image Registration

Junyu Chen[1(✉)], Yihao Liu[2], Yufan He[3], and Yong Du[1]

[1] Russell H. Morgan Department of Radiology and Radiological Science, Johns Hopkins Medical Institutes, Baltimore, MD, USA
{jchen245,duyong}@jhmi.edu
[2] Department of Electrical and Computer Engineering, Johns Hopkins University, Baltimore, MD, USA
yliu236@jhmi.edu
[3] NVIDIA Corporation, Bethesda, MD, USA
yufanh@nvidia.com

Abstract. Transformers have recently shown promise for medical image applications, leading to an increasing interest in developing such models for medical image registration. Recent advancements in designing registration Transformers have focused on using cross-attention (CA) to enable a more precise understanding of spatial correspondences between moving and fixed images. Here, we propose a novel CA mechanism that computes windowed attention using deformable windows. In contrast to existing CA mechanisms that require intensive computational complexity by either computing CA globally or locally with a fixed and expanded search window, the proposed deformable CA can selectively sample a diverse set of features over a large search window while maintaining low computational complexity. The proposed model was extensively evaluated on multi-modal, mono-modal, and atlas-to-patient registration tasks, demonstrating promising performance against state-of-the-art methods and indicating its effectiveness for medical image registration. The source code for this work is available at http://bit.ly/47HcEex.

Keywords: Image Registration · Transformer · Cross-attention

1 Introduction

Deep learning-based registration methods have emerged as a faster alternative to optimization-based methods, with promising registration accuracy across a range of registration tasks [1,10]. These methods often adopt convolutional neural networks (ConvNets), particularly U-Net-like networks [21], as the backbone

Supplementary Information The online version contains supplementary material available at https://doi.org/10.1007/978-3-031-45673-2_12.

architecture [1,10]. Yet, due to the locality of the convolution operations, the effective receptive fields (ERFs) of ConvNets are only a fraction of their theoretical receptive fields [12,16]. This limits the performance of ConvNets in image registration, which often requires registration models to establish long-range spatial correspondences between images.

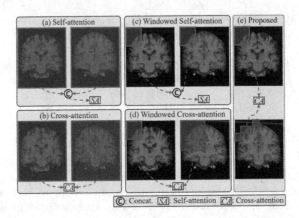

Fig. 1. Graphical illustrations of different attention mechanisms. (a) The conventional self-attention [6,26] used in `ViT-V-Net` [5] and `DTN` [29], which computes attention for the concatenated tokens of moving and fixed images. (b) Cross-attention used in `Attention-Reg` [25], which computes attention between the tokens of moving and fixed images. (c) Windowed self-attention [15] used in `TransMorph` [4], which computes attention for the concatenated tokens of moving and fixed images within a local window. (d) Windowed cross-attention proposed in `XMorpher` [23], which computes attention between the tokens of fixed and moving images, specifically between two local windows of different sizes. (e) The proposed deformable cross-attention mechanism, which computes attention between tokens within a rectangular window and a deformed window with an arbitrary shape but the same size as the rectangular window.

Transformers, which originated from natural language processing tasks [26], have shown promise in a variety of medical imaging applications [12], including registration [4,5,29]. Transformers employ the self-attention (SA) mechanism, which can either be a global operation [6] or computed locally within large windows [15]. Consequently, Transformers have been shown to capture long-range spatial correspondences for registration more effectively than ConvNets [4].

Several recent advancements in Transformer-based registration models have focused on developing cross-attention (CA) mechanisms, such as `XMorpher` [23] and `Attention-Reg` [25]. CA improves upon SA by facilitating the efficient fusion of high-level features between images to improve the comprehension of spatial correspondences. However, the existing CA mechanisms still have drawbacks; either they compute CA globally [25], which prevents hierarchical feature extraction and applies only to low-resolution features, or they compute CA within a fixed but expanded window [23], which significantly increases computational complexity.

In this paper, we present a hybrid Transformer-ConvNet model based on a novel deformable CA mechanism for image registration. As shown in Fig. 1, the proposed deformable CA module differs from existing SA and CA modules in that it employs the windowed attention mechanism [15] with a learnable offset. This allows the sampling windows of the reference image to take on any shapes based on the offsets, offering several advantages over existing methods: **1)** In contrast to the CA proposed in [23], which calculates attention between windows of varying sizes, the proposed deformable CA module samples tokens from a larger search region, which can even encompass the entire image. Meanwhile, the attention computation is confined within a uniform window size, thereby keeping the computational complexity low. **2)** The deformable CA enables the proposed model to focus more on the regions where the disparity between the moving and fixed images is significant, in comparison to the baseline ConvNets and SA-based Transformers, leading to improved registration performance. Comprehensive evaluations were conducted on mono- and multi-modal registration tasks using publicly available datasets. The proposed model competed favorably against existing state-of-the-art methods, showcasing its promising potential for a wide range of image registration applications.

2 Background and Related Works

Self-attention. SA [6,26] is typically applied to a set of tokens (*i.e.*, embeddings that represent patches of the input image). Let $x \in \mathbb{R}^{N \times D}$ be a set of N tokens with D-dimensional embeddings. The tokens are first encoded by a fully connected layer $U_{q,k,v} \in \mathbb{R}^{D \times D_{q,k,v}}$ to obtain three matrix representations, Queries Q, Keys K, and Values V: $[Q, K, V] = x U_{q,k,v}$. Subsequently, the scaled dot-product attention is calculated using $SA(x) = \text{softmax}(\frac{QK^{\top}}{\sqrt{D_k}})V$. In general, SA computes a normalized score for each token based on the dot product of Q and K. This score is then used to decide which Value token to attend to.

Cross-Attention. CA is a frequently used variant of SA for inter- and intra-modal tasks in computer vision [2,11,28] and has been investigated for its potential in image registration [14,23,25]. CA differs from SA in terms of how the matrix representations are computed. As CA is typically used between two modalities or images (*i.e.*, a base image and a reference image), the matrices Q, K, and V are generated using different inputs:

$$[K_b, V_b] = x_b U_{k,v}, \quad Q_r = x_r U_q, \quad CA(x) = \text{softmax}(\frac{Q_r K_b^{\top}}{\sqrt{D_k}})V_b, \quad (1)$$

where x_b and x_r denote, respectively, the tokens of the base and the reference. In [25], Song *et al.* introduced `Attention-Reg`, which employs Eq. 1 to compute CA between a moving and a fixed image. To ensure low computational complexity, CA is computed globally between the downsampled features extracted

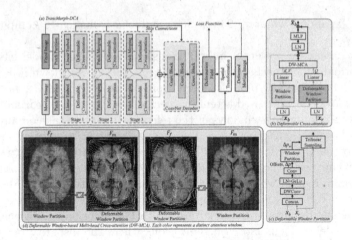

Fig. 2. The overall framework of the proposed method. (a) The proposed network architecture, which is composed of parallel Transformer encoders and a ConvNet decoder to generate a deformation field. (b) The deformable CA, which fuses features between encoders. (c) The schematic of the deformable window partitioning strategy. (d) An example of deformable CA computation in the DW-MCA.

by ConvNets. However, because CA is only applied to a single resolution, it does not provide hierarchical feature fusion across different resolutions, a factor that is deemed important for several successful registration models [19,20]. More recently, Shi *et al.* introduced XMorpher [23], which is based on the Swin Transformer [15]. In XMorpher, CA is computed between the local windows of the tokens of different resolutions, enabling hierarchical feature fusion. As shown in Fig. 1(d), the local windows are of different sizes, with a base window of size $N_b = h \times w \times d$ and a larger search window of size $N_s = \alpha h \times \beta w \times \gamma d$, where α, β, and γ are set equally to 3. Using a larger search window facilitates the effective establishment of spatial correspondence, but it also increases the computational complexity of each CA module. Specifically, if the same window size of N_b is used, the complexity of CA is approximately $O(N_b^2 D_k)$. However, if windows of different sizes, N_b and N_s, are used, the complexity becomes $O(N_b N_s D_k) = O(\alpha\beta\gamma N_b^2 D_k)$, where $\alpha\beta\gamma = 3 \times 3 \times 3 = 27$. This means that using a larger search window would increase the computational complexity dramatically (by 27 times) and quickly become computationally infeasible.

Enlarging the search space while keeping computational costs low is challenging for 3D medical image registration. In this paper, we try to solve it with a deformable CA module that operates on equal-sized windows. This module not only provides hierarchical feature fusion, but also allows more efficient token sampling over a larger region than previously mentioned CA modules. Additionally, the proposed CA maintains a low computational complexity.

3   Proposed Method

The proposed model is depicted in Fig. 2(a), which has dual Transformer encoders with deformable CA modules that enable effective communication between them. Each encoder is similar to the Swin [15] used in `TransMorph` [4], but the SA modules are replaced with the deformable CA modules. To integrate the features between each stage of the two encoders, we followed [25] by adding the features and passing them to the decoder via skip connections. In contrast to `XMorpher` [23], which uses a Transformer for the decoder, we opted for the ConvNet decoder introduced in [3,4]. This choice was motivated by the inductive bias that ConvNets bring in, which Transformers typically lack [12]. ConvNets are also better at refining features for subsequent deformation generation, owing to the locality of convolution operations. Moreover, ConvNets have fewer parameters, making them efficient and hence speeding up the training process.

Cross-Attention Transformer. Our model employs parallel deformable CA encoders to extract hierarchical features from the moving and fixed images in the encoding stage. At each resolution of the encoder, k successive deformable CA modules are applied to vertically fuse features between the two encoders. The deformable cross-attention module takes in a base (*i.e.*, x_b) and a reference (*i.e.*, x_r), and computes the attention between them, with the reference guiding the network on where to focus within the base. As shown in Fig. 2(a), one encoding path uses the moving and fixed images as the base and reference, respectively, whereas the other encoding path switches the roles of the base and reference, using the moving image as the reference and the fixed image as the base.

Deformable Cross-Attention. Figure 2(b) depicts the core element of the proposed model, the deformable CA. The module first applies *LayerNorm* (LN) to x_b and x_r, then partitions x_b into non-overlapping rectangular equal-sized windows, following [15]. Next, the x_b is projected into K_b and V_b embeddings through a linear layer. This process is expressed as $[K_b, V_b] = \text{WP}(\text{LN}(x_b))U_{k,v}$, where WP($\cdot$) denotes the window partition operation. On the other hand, the window partitioning for x_r is based on the offsets, Δp, learned by a lightweight offset network. As shown in Fig. 2(c), this network comprises two consecutive convolutional layers (depth-wise and regular convolutional layers) and takes the added x_b and x_r as input. The offsets, Δp, shift the sampling positions of the rectangular windows beyond their origins, allowing tokens to be sampled outside these windows. Specifically, Δp are first divided into equal-sized windows, Δp_w, and tokens in x_r are subsequently sampled based on Δp_w using trilinear interpolation. Note that this sampling process is analogous to first resampling the tokens based on the offsets and then partitioning them into windows. We generated a different set of Δp_w for each head in the multi-head attention, thereby enabling diverse sampling of the tokens across heads. The proposed deformable window-based multi-head CA (DW-MCA) is then expressed as:

$$[K_b, V_b] = \text{WP}(\text{LN}(x_b))U_{k,v},$$
$$\Delta p = \theta_{\Delta p}(x_b, x_r), \quad \Delta p_w = \text{WP}(\Delta p), \quad Q_r = \psi(x_r; p + \Delta p_w)U_k, \qquad (2)$$
$$\text{DW-MCA}(x) = \text{softmax}(\frac{Q_r K_b^\top}{\sqrt{D_k}})V_b,$$

where $\theta_{\Delta p}$ denotes the offset network and $\psi(\cdot;\cdot)$ is the interpolation function. To introduce cross-window connections, the shifted window partitioning strategy [15] was implemented in successive Transformer blocks.

The attention computation of the deformable CA is nearly identical to the conventional window-based SA employed in Swin [15], with the addition of a lightweight offset network whose complexity is approximately $O(m^3 N_b D_k)$ (m is the convolution kernel size and $m^3 \approx N_b$). As a result, the overall complexity of the proposed CA module is $O(2N_b^2 D_k)$, which comprises the complexity of the offset network and the CA computation. In comparison, the CA used in XMorpher [23] has a complexity of $O(27N_b^2 D_k)$, as outlined in Sect. 2. This highlights the three main advantages of the deformable CA module: 1) it enables token sampling beyond a pre-defined window, theoretically encompassing the entire image size, 2) it allows sampling windows to overlap, improving communication between windows, and 3) it maintains fixed-size windows for the CA computation, thereby retaining low computational complexity.

The deformable CA, the deformable attention (DA) [27], and the Swin DA (SDA) [9] share some similarities, but there are fundamental differences. Firstly, DA computes attention globally within a single modality or image, whereas the deformable CA utilizes windowed attention and a hierarchical architecture to fuse features of different resolutions across images or modalities. Secondly, the offset network in DA and SDA is applied solely to the Query embeddings of input tokens, and SDA generates offsets based on window-partitioned tokens, leading to square-shaped "windowing" artifacts in the sampling grid, as observed in [9]. In contrast, in the deformable CA, the offset network is applied to all tokens of both the reference and the base to take advantage of their spatial correspondences, resulting in a smoother and more meaningful sampling grid, as demonstrated in Fig. 2(d). Lastly, while DA and SDA use a limited number of reference points to interpolate tokens during sampling, deformable CA employs a dense set of reference points with the same resolution as the input tokens, allowing deformable CA to sample tokens more diversely.

4   Experiments

Dataset and Pre-processing. The proposed method was tested on three publicly available datasets to evaluate its performance on three registration tasks: 1) inter-patient multi-modal registration, 2) inter-patient mono-modal registration, and 3) atlas-to-patient registration. The dataset used for the first task is the ALBERTs dataset [7], which consists of T1- and T2-weighted brain MRIs of 20 infants. Manual segmentation of the neonatal brain was provided, each consisting of 50 ROIs. The patients were randomly split into three sets with a ratio of 10:4:6. We performed inter-patient T1-to-T2 registration, which resulted in 90, 12, and 30 image pairs for training, validation, and testing, respectively. For the second and third registration tasks, we used the OASIS dataset [17] from the Learn2Reg challenge [8] and the IXI dataset[1] from [4], respectively. The former includes 413 T1 brain MRI images, of which 394 were assigned for training and 19 for testing. The latter consists of 576 T1 brain MRI images, which were distributed as 403 for training, 58 for validation, and 115 for testing. For the third task, we used a moving image, which was a brain atlas image obtained from [10]. All images from the three datasets were cropped to the dimensions of $160 \times 192 \times 224$.

Evaluation Metrics. To assess the registration performance, the Dice coefficient was used to measure the overlap of the anatomical label maps. Moreover, for the OASIS dataset, we additionally used Hausdorff distance (HdD95) to evaluate performance and the standard deviation of the Jacobian determinant (SDlogJ) to assess deformation invertibility, in accordance with Learn2Reg. For the ALBERTs and IXI datasets, we used two metrics, the percentage of all non-positive Jacobian determinant ($\%|J| \leq 0$) and the non-diffeomorphic volume (%NDV), both proposed in [13], to evaluate deformation invertibility since they are more accurate measures under the finite-difference approximation.

Table 1. Quantitative results for mono-modal (OASIS) and multi-modal inter-patient (ALBERTs) registration tasks, as well as atlas-to-patient (IXI) registration tasks. Note that part of the OASIS results was obtained from Learn2Reg leaderboard [8].

OASIS (Mono-modality)				IXI (Atlas-to-patient)			
Method	Dice↑	HdD95↓	SDlogJ↓	Method	Dice↑	$\%\|J\|\leq 0$ ↓	%NDV↓
ConvexAdam [24]	0.846±0.016	1.500±0.304	0.067±0.005	VoxelMorph [1]	0.732±0.123	6.26%	1.04%
LapIRN [18]	0.861±0.015	1.514±0.337	0.072±0.007	CycleMorph [10]	0.737±0.123	6.38%	1.15%
TransMorph [4]	0.862±0.014	1.431±0.282	0.128±0.021	TM-bspl [4]	0.761±0.128	0%	0%
TM-TVF [3]	*0.869±0.014*	**1.396±0.295**	0.094±0.018	TM-TVF [3]	0.756±0.122	2.05%	0.36%
XMorpher* [23]	0.854±0.012	1.647±0.346	0.100±0.016	XMorpher* [23]	0.751±0.123	0%	0%
TM-DCA	**0.873±0.015**	*1.400±0.368*	0.105±0.028	TM-DCA	**0.763±0.128**	0%	0%

ALBERTs (Multi-modality)			
Method	Dice↑	$\%\|J\|\leq 0$ ↓	%NDV↓
VoxelMorph [24]	0.651±0.159	0.04%	0.02%
TransMorph [4]	0.672±0.159	0.15%	0.04%
TM-TVF [3]	*0.722±0.132*	0.13%	0.03%
XMorpher* [23]	0.710±0.135	0.11%	0.03%
TM-DCA	**0.724±0.131**	0.24%	0.07%

Results and Discussion. The proposed model, TM-DCA, was evaluated against several state-of-the-art models on the three registration tasks, and the corresponding quantitative outcomes are presented in Table 1. Note that our GPU

---

[1] https://brain-development.org/ixi-dataset/.

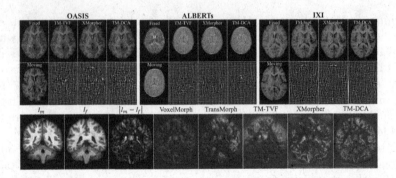

Fig. 3. Qualitative results and a visualization of Grad-CAM [22] heat maps of the comparative registration models generated for a pair of images.

was unable to accommodate the original XMorpher [23] (>48 GB) for the image size used in this study. This is likely due to the large window CA computation and the full Transformer architecture used by the model. To address this, we used the encoder of XMorpher in combination with the decoder of TM-DCA (denoted as XMorpher*) to reduce GPU burden and facilitate a more precise comparison between the deformable CA in TM-DCA and the CA used in XMorpher. On the OASIS dataset, TM-DCA achieved the highest mean Dice score of 0.873, which was significantly better than the second-best performing method, TM-TVF [3], as confirmed by a paired t-test with $p < 0.01$. On the IXI dataset, TM-DCA achieved the highest mean Dice score of 0.763, which was significantly better than TM-bspl with $p < 0.01$. Remarkably, TM-DCA produced diffeomorphic registration with almost no folded voxels, using the same decoder as TM-bspl. Finally, on the ALBERTs dataset, TM-DCA again achieved the highest mean Dice score of 0.713, which was significantly better than TM-TVF with $p = 0.04 < 0.05$, thus demonstrating its superior performance in multi-modal registration. It is important to note that the proposed TM-DCA model and its CA (*i.e.*, XMorpher) and SA (*i.e.*, TM-TVF and TM-bspl) counterparts differed only in their encoders, while the decoder used was identical for all models. TM-DCA consistently outperformed the baselines across the three applications, supporting the effectiveness of the proposed CA module.

Qualitative comparison results between the registration models are presented in Fig. 3. In addition, we conducted a comparison of the *Grad-CAM* [22] heat map for various learning-based registration models. The heat maps were generated by computing the NCC between the deformed moving image and the fixed image, and were then averaged across the convolutional layers at the end of the decoder, just prior to the final layer that predicts the deformation field. Notably, VoxelMorph exhibited inadequate focus on the differences between the image pair, which may be attributed to ConvNets' limited ability to explicitly comprehend contextual information in the image. The SA-based models (TransMorph and TM-TVF) showed similar trends, wherein they focused reasonably well on regions with significant differences but relatively less attention was given to areas with minor differences. In contrast, the attention of CA-based models was more uniformly distributed, with the proposed `TM-DCA` method more effectively capturing differences than `XMorpher`. The presented heat maps highlight the superior performance of the proposed CA mechanism in effectively interpreting contextual information and accurately capturing spatial correspondences between images. In combination with the observed improvements in performance across various registration tasks, these results suggest that `TM-DCA` has significant potential as the preferred attention mechanism for image registration applications.

5   Conclusion

In this study, we introduced a Transformer-based network for unsupervised image registration. The proposed model incorporates a novel CA module that computes attention between the features of the moving and fixed images. Unlike the SA and CA mechanisms used in existing methods, the proposed CA module computes attention between tokens sampled from a square window and a learned window of arbitrary shape. This enables the efficient computation of attention while allowing the extraction of useful features from a large window to accurately capture spatial correspondences between images. The proposed method was evaluated against several state-of-the-art methods on multiple registration tasks and demonstrated significant performance improvements compared to the baselines, highlighting the effectiveness of the proposed CA module.

References

1. Balakrishnan, G., Zhao, A., Sabuncu, M.R., Guttag, J., Dalca, A.V.: Voxelmorph: a learning framework for deformable medical image registration. IEEE Trans. Med. Imaging **38**(8), 1788–1800 (2019)
2. Chen, C.F.R., Fan, Q., Panda, R.: Crossvit: cross-attention multi-scale vision transformer for image classification. In: Proceedings of the IEEE/CVF International Conference on Computer Vision, pp. 357–366 (2021)
3. Chen, J., Frey, E.C., Du, Y.: Unsupervised learning of diffeomorphic image registration via transmorph. In: Hering, A., Schnabel, J., Zhang, M., Ferrante, E., Heinrich, M., Rueckert, D. (eds.) WBIR 2022. LNCS, vol. 13386, pp. 96–102. Springer, Cham (2022). https://doi.org/10.1007/978-3-031-11203-4_11
4. Chen, J., Frey, E.C., He, Y., Segars, W.P., Li, Y., Du, Y.: Transmorph: transformer for unsupervised medical image registration. Med. Image Anal. **82**, 102615 (2022)
5. Chen, J., He, Y., Frey, E., Li, Y., Du, Y.: ViT-V-Net: vision transformer for unsupervised volumetric medical image registration. In: Medical Imaging with Deep Learning (2021)
6. Dosovitskiy, A., et al.: An image is worth 16x16 words: transformers for image recognition at scale. arXiv preprint arXiv:2010.11929 (2020)
7. Gousias, I.S., et al.: Magnetic resonance imaging of the newborn brain: manual segmentation of labelled atlases in term-born and preterm infants. Neuroimage **62**(3), 1499–1509 (2012)

8. Hering, A., et al.: Learn2Reg: comprehensive multi-task medical image registration challenge, dataset and evaluation in the era of deep learning. IEEE Trans. Med. Imaging **42**(3), 697–712 (2022)

9. Huang, J., Xing, X., Gao, Z., Yang, G.: Swin deformable attention U-net transformer (SDAUT) for explainable fast MRI. In: Wang, L., Dou, Q., Fletcher, P.T., Speidel, S., Li, S. (eds.) MICCAI 2022. LNCS, vol. 13436, pp. 538–548. Springer, Cham (2022). https://doi.org/10.1007/978-3-031-16446-0_51

10. Kim, B., Kim, D.H., Park, S.H., Kim, J., Lee, J.G., Ye, J.C.: Cyclemorph: cycle consistent unsupervised deformable image registration. Med. Image Anal. **71**, 102036 (2021)

11. Kim, H.H., Yu, S., Yuan, S., Tomasi, C.: Cross-attention transformer for video interpolation. In: Proceedings of the Asian Conference on Computer Vision, pp. 320–337 (2022)

12. Li, J., Chen, J., Tang, Y., Landman, B.A., Zhou, S.K.: Transforming medical imaging with transformers? A comparative review of key properties, current progresses, and future perspectives. arXiv preprint arXiv:2206.01136 (2022)

13. Liu, Y., Chen, J., Wei, S., Carass, A., Prince, J.: On finite difference jacobian computation in deformable image registration. arXiv preprint arXiv:2212.06060 (2022)

14. Liu, Y., Zuo, L., Han, S., Xue, Y., Prince, J.L., Carass, A.: Coordinate translator for learning deformable medical image registration. In: Li, X., Lv, J., Huo, Y., Dong, B., Leahy, R.M., Li, Q. (eds.) MMMI 2022. LNCS, vol. 13594, pp. 98–109. Springer, Cham (2022). https://doi.org/10.1007/978-3-031-18814-5_10

15. Liu, Z., et al.: Swin transformer: hierarchical vision transformer using shifted windows. In: Proceedings of the IEEE/CVF International Conference on Computer Vision, pp. 10012–10022 (2021)

16. Luo, W., Li, Y., Urtasun, R., Zemel, R.: Understanding the effective receptive field in deep convolutional neural networks. In: Advances in Neural Information Processing Systems, vol. 29 (2016)

17. Marcus, D.S., Wang, T.H., Parker, J., Csernansky, J.G., Morris, J.C., Buckner, R.L.: Open access series of imaging studies (OASIS): cross-sectional MRI data in young, middle aged, nondemented, and demented older adults. J. Cogn. Neurosci. **19**(9), 1498–1507 (2007)

18. Mok, T.C.W., Chung, A.C.S.: Conditional deformable image registration with convolutional neural network. In: de Bruijne, M., Cattin, P.C., Cotin, S., Padoy, N., Speidel, S., Zheng, Y., Essert, C. (eds.) MICCAI 2021. LNCS, vol. 12904, pp. 35–45. Springer, Cham (2021). https://doi.org/10.1007/978-3-030-87202-1_4

19. Mok, T.C., Chung, A.: Affine medical image registration with coarse-to-fine vision transformer. In: Proceedings of the IEEE/CVF Conference on Computer Vision and Pattern Recognition, pp. 20835–20844 (2022)

20. Mok, T.C.W., Chung, A.C.S.: Large deformation diffeomorphic image registration with laplacian pyramid networks. In: Martel, A.L., et al. (eds.) MICCAI 2020. LNCS, vol. 12263, pp. 211–221. Springer, Cham (2020). https://doi.org/10.1007/978-3-030-59716-0_21

21. Ronneberger, O., Fischer, P., Brox, T.: U-Net: convolutional networks for biomedical image segmentation. In: Navab, N., Hornegger, J., Wells, W.M., Frangi, A.F. (eds.) MICCAI 2015. LNCS, vol. 9351, pp. 234–241. Springer, Cham (2015). https://doi.org/10.1007/978-3-319-24574-4_28

22. Selvaraju, R.R., Cogswell, M., Das, A., Vedantam, R., Parikh, D., Batra, D.: Gradcam: visual explanations from deep networks via gradient-based localization. In:

Proceedings of the IEEE International Conference on Computer Vision, pp. 618–626 (2017)

23. Shi, J., et al.: XMorpher: full transformer for deformable medical image registration via cross attention. In: Wang, L., Dou, Q., Fletcher, P.T., Speidel, S., Li, S. (eds.) MICCAI 2022. LNCS, vol. 13436, pp. 217–226. Springer, Cham (2022). https://doi.org/10.1007/978-3-031-16446-0_21

24. Siebert, H., Hansen, L., Heinrich, M.P.: Fast 3D registration with accurate optimisation and little learning for Learn2Reg 2021. In: Aubreville, M., Zimmerer, D., Heinrich, M. (eds.) MICCAI 2021. LNCS, vol. 13166, pp. 174–179. Springer, Cham (2022). https://doi.org/10.1007/978-3-030-97281-3_25

25. Song, X., et al.: Cross-modal attention for multi-modal image registration. Med. Image Anal. **82**, 102612 (2022)

26. Vaswani, A., et al.: Attention is all you need. In: Advances in Neural Information Processing Systems, vol. 30 (2017)

27. Xia, Z., Pan, X., Song, S., Li, L.E., Huang, G.: Vision transformer with deformable attention. In: Proceedings of the IEEE/CVF Conference on Computer Vision and Pattern Recognition, pp. 4794–4803 (2022)

28. Xu, X., Wang, T., Yang, Y., Zuo, L., Shen, F., Shen, H.T.: Cross-modal attention with semantic consistence for image-text matching. IEEE Trans. Neural Netw. Learn. Syst. **31**(12), 5412–5425 (2020)

29. Zhang, Y., Pei, Y., Zha, H.: Learning dual transformer network for diffeomorphic registration. In: de Bruijne, M., et al. (eds.) MICCAI 2021. LNCS, vol. 12904, pp. 129–138. Springer, Cham (2021). https://doi.org/10.1007/978-3-030-87202-1_13

Deformable Medical Image Registration Under Distribution Shifts with Neural Instance Optimization

Tony C. W. Mok[1,2(✉)], Zi Li[1,2], Yingda Xia[1], Jiawen Yao[1,2], Ling Zhang[1], Jingren Zhou[1,2], and Le Lu[1]

[1] DAMO Academy, Alibaba Group, Hangzhou, China
`mokchi-wing.mcw@alibaba-inc.com`
[2] Hupan Lab, Hangzhou 310023, China

Abstract. Deep-learning deformable image registration methods often struggle if test-image characteristic shifts from the training domain, such as the large variations in anatomy and contrast changes with different imaging protocols. Gradient descent-based instance optimization is often introduced to refine the solution of deep-learning methods, but the performance gain is minimal due to the high degree of freedom in the solution and the absence of robust initial deformation. In this paper, we propose a new instance optimization method, Neural Instance Optimization (NIO), to correct the bias in the deformation field caused by the distribution shifts for deep-learning methods. Our method naturally leverages the inductive bias of the convolutional neural network, the prior knowledge learned from the training domain and the multi-resolution optimization strategy to fully adapt a learning-based method to individual image pairs, avoiding registration failure during the inference phase. We evaluate our method with gold standard, human cortical and subcortical segmentation, and manually identified anatomical landmarks to contrast NIO's performance with conventional and deep-learning approaches. Our method compares favourably with both approaches and significantly improves the performance of deep-learning methods under distribution shifts with 1.5% to 3.0% and 2.3% to 6.2% gains in registration accuracy and robustness, respectively.

1 Introduction

Deformable medical image registration aims to establish non-linear correspondence of anatomy between image scans, which is essential in a comprehensive medical image processing and analysis pipeline. Deep learning-based image registration (DLIR) methods can achieve remarkable results on training and testing images from the same distribution, as evidenced by tremendous medical image registration benchmarks [6,10]. However, DLIR method remains notoriously vulnerable to distribution shifts. Distribution shift refers to the existence of significant divergence between the distributions of the training and the

X. Cao et al. (Eds.): MLMI 2023, LNCS 14348, pp. 126–136, 2024.
https://doi.org/10.1007/978-3-031-45673-2_13

test data [29]. Registration accuracy and robustness of DLIR method could be degraded substantially when the test-image characteristic shifts from the training domain. Different from the learning-based methods, conventional image registration methods [1,22] often register images with an iterative optimization strategy, which circumvents the need for learning. Although conventional image registration methods excel in robustness and diffeomorphic properties under distribution shifts, the resulting solutions are often suboptimal and can be time-consuming with high-resolution 3D image volumes, as indicated by recent studies [3,19].

Combining an initial deformation field predicted by DLIR with instance optimization [3,9,10,27] or further learning fine-tuning steps [28] has drawn growing attention in the community. Instance optimization refers to the process of iteratively optimizing the deformation field along with the optimizer, which is similar to classic optical flow estimation [21]. Instance optimization often serves as a post-processing step to refine the solution from DLIR method during the inference phase. Nevertheless, there is only minimal performance gain using instance optimization under distribution shifts. First, the effectiveness of instance optimization relies on the assumption that the model in DLIR is capable of estimating a robust initial deformation prediction for instance optimization. This assumption cannot be held when test-image characteristic shifts from the training domain, such as large variations in brain anatomy and contrast changes from diverse imaging protocols. Second, taking into account that non-linear image registration is an ill-posed problem [23], the searching space of instance optimization with a high degree of freedom is intractable, resulting in a minimal performance gain. Third, while the multilevel optimization technique has been proven very efficient in conventional variational registration approaches to avoid local minima [2], this technique is not applicable to instance-specific optimization when initialized with the dense and fine-resolution solution from the DLIR method as the fine details of the deformation field from the DLIR method will be distorted during downsampling in the coarse-to-fine optimization pipeline.

While there are vast research studies [7,13,14,17,20,26] on deformable image registration, the robustness and generalizability of DLIR method against distribution shifts are less explored in the context of image registration. Zhu et al. [31] use a cascaded U-Net structure to minimize the distribution shifts between training and target domain by performing test-time training. To against the diverse contrast change in brain MR registration, Hoffmann et al. [11] propose to train networks with synthetic images generated with synthetic anatomical label maps, which requires access to the anatomical delineation of the training set. Different from the above studies, we focus on maximizing the registration accuracy and robustness of instance (individual) image pair under distribution shifts without access to the delineation of the anatomical structures in the training or test set.

To this end, we present neural instance optimization (NIO), a new instance optimization method that is capable of significantly improving the registration performance of deep-learning methods during the inference phase, leading to robust generalizability under domain shift.

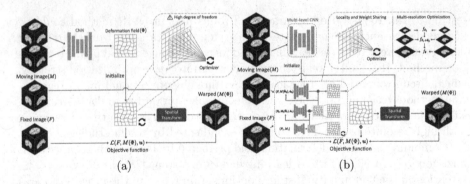

Fig. 1. Overview of the (a) deep-learning method followed by instance optimization approach and (b) the proposed neural instance optimization method. Instead of directly manipulating the deformation field with instance-specific optimization, our proposed method implicitly updates the deformation field by adapting the CNN to each target image pair with the multiresolution optimization strategy.

2   Methods

Deformable image registration aims to establish a dense non-linear correspondence between a fixed image F and a moving image M, subject to a smoothness regularization to penalize implausible solutions. Existing DLIR methods often formulate deformable image registration as a learning problem $\Phi = f_\theta(F, M)$, where f_θ is parameterized with a convolutional neural network (CNN) and θ denotes the set of learning parameters in CNN. Inherit from the learning nature, DLIR methods are prone to distribution shifts and the registration performance could be degraded. Our goal is to refine the deformation field in the inference phase to avoid failure in registration due to distribution shifts and maintain the sub-pixel accuracy of the registration simultaneously.

2.1   Neural Instance Optimization

Formulation of Instance Optimization. Recall that the instance optimization method can be adopted as a post-processing step to refine the predicted deformation field predicted from the DLIR method in the inference phase. Formally, the formulation of instance optimization is defined as follows:

$$\Phi^* = \arg\min_{\Phi_\theta} \mathcal{L}_{sim}(F, M(\Phi_\theta)) + \lambda\mathcal{L}_{reg}(\boldsymbol{u}_\theta), \qquad (1)$$

where $\mathcal{L}_{sim}$ denotes the dissimilarity measure, $\mathcal{L}_{reg}$ is the smoothness regularization and λ is a hyperparameter. The deformation field is defined as $\Phi_\theta = \boldsymbol{u}_\theta + \boldsymbol{Id}$, where $\boldsymbol{Id}$ denotes the identity transform in mutual space Ω. Specifically, Φ_θ in instance optimization is first initialized with the displacement from a feed-forward prediction of DLIR network, followed by an update on $\boldsymbol{u}_\theta$ with gradient descent to maximize the similarity between F and $M(\Phi_\theta)$ in an iterative manner, as shown in Fig. 1(a). While instance optimization could be used to refine DLIR's solution, the performance gain to registration is minimal under moderate or severe distribution shifts due to the high degree of freedom in Φ_θ and a robust initial deformation prediction cannot be guaranteed by DLIR under distribution shifts, violating the underlying assumption of instance optimization approach.

Neural Instance Optimization (NIO). To circumvent the pitfalls in the instance optimization approach, we implicitly refine the resulting deformation field by updating the parameters in the CNN instead of directly manipulating the deformation field as in the instance optimization approach. To exemplify the idea, we first parametrize an example of the function f_θ with the deep Laplacian pyramid image registration network (LapIRN) [18], i.e., $\Phi = f_\theta(F, M)$. The overview of our proposed method is illustrated in Fig. 1(b). Mathematically, our method reformulates the problem from Eq. 1 as follows:

$$\theta^* = \arg\min_\theta \mathcal{L}_{sim}(F_t, M_t \circ f_\theta(F_t, M_t)) + \lambda\mathcal{L}_{reg}(u), \qquad (2)$$

where F_t and M_t belong to the fixed and moving image pair in the target dataset $\mathcal{D}_{test}$, and $\circ$ denotes the composition operator, i.e., $M_t \circ f_\theta(F_t, M_t) = M_t(\Phi)$. During the inference phase of each image pair, the parameters in the network θ are first initialized with the pre-trained θ_0 on the training dataset. Then, we use gradient descent for η iterations searching for the optimal set of θ^* that adapt the model f_θ for individual (F_t, M_t) pair in $\mathcal{D}_{test}$. Compared with the instance optimization approach, our proposed method is two-fold: first, the intrinsic inductive bias, i.e., weight sharing and locality, embedded in the CNN implicitly reduces the degree of freedom of the solution, reducing the searching space in the optimization problem; second, the prior knowledge of registering the images in the training set can be transferred to the individual test image pair during the optimization, avoiding the sub-optimal solution in the instance optimization.

2.2   Multi-resolution Optimization Strategy with NIO

Optimizing the CNN f_θ during the inference phase could be computationally intensive, especially for DLIR methods with multilevel network architecture. This restricts the maximum number of iterations of NIO in the inference phase. To alleviate this issue, we propose a multi-resolution optimization strategy for NIO. Specifically, given a L-level LapIRN framework f_θ with $\theta \in \{\theta_1, \ldots, \theta_L\}$ such that θ_i represents the parameters of the network in level i. We first create the image pyramid for input images by downsampling the input images with trilinear interpolation to obtain $F_i \in \{F_1, \ldots, F_L\}$ (and $M_i \in \{M_1, \ldots, M_L\}$), where F_i denotes the downsampled F with a scale factor $0.5^{(L-i)}$ and $F_L = F$. The network f_θ is then optimized in a coarse-to-fine manner, starting with the coarsest input pair (F_1 and M_1) and network f_{θ_1}. Then, we progressively add the network in the next level $f_{\theta_{i+1}}$ into optimization until the optimization of the final level f_θ is completed, as shown in Fig. 1 (b). The output displacement field $\boldsymbol{u}$ of f_θ is formed by aggregating the upsampled displacement fields $\hat{\boldsymbol{u}}_i$ from each level via element-wise addition. We set $L = 3$ throughout this paper. By adjusting the number of iterations η_i in each level of the optimization, most of the computation is deployed with images at a lower resolution and the optimization naturally inherits the advantage of the conventional multi-resolution strategy, enabling a better trade-off between registration accuracy and runtime.

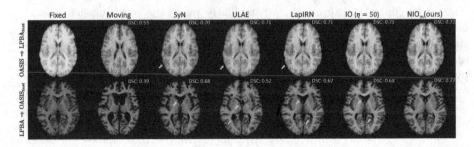

Fig. 2. Example axial MR slices from the moving, atlas and resulting warped images from SyN, ULAE, LapIRN, LapIRN followed by IO and NIO_m for the cross-dataset brain MR registration. Major artifacts are highlighted with yellow arrows. (Color figure online)

2.3 Loss Function

To further accelerate the runtime in the inference phase, we discard the similarity pyramid in the vanilla LapIRN [19] and adopt the local normalized cross-correlation (LNCC) with window size 7^3 as similarity function $\mathcal{L}_{sim}$. We adopt a diffusion regularizer as $\mathcal{L}_{reg}$ to encourage smooth solutions and penalise implausible solutions. We follow [13] to further normalize the cost function with λ, i.e., $\lambda \in [0, 1]$. The loss function for level i in the proposed multi-resolution optimization scheme is defined as:

$$\mathcal{L}_i = (1 - \lambda)\text{LNCC}(F, M(\hat{\boldsymbol{\Phi}}_i)) + \lambda||\nabla\hat{\boldsymbol{u}}_i|| \quad i \in \{1, 2, \ldots, L\}, \tag{3}$$

where $\hat{(\cdot)}$ is the upsampling operator with trilinear interpolation, which upsample $\boldsymbol{u}_i$ and $\boldsymbol{\Phi}_i$ to match the size of F and M, maintaining the consistency of the objective function among all the levels.

3   Experiments

Data and Pre-processing. We evaluate our method on cross-dataset brain atlas registration and intra-patient inspiration-expiration Chest CT registration. For brain atlas registration, we use 414 T1-weighted brain MR scans from the

Table 1. Quantitative results of brain MR atlas registration. $X^n_{train} \Rightarrow Y_{test}$ represents the experiment of training on n scans from the X dataset and registering images in the test set of Y dataset to pre-defined atlases. The subscript of DSC indicates the number of anatomical structures involved. Initial: Affine spatial normalization. ↑: higher is better, and ↓: lower is better. †: $p < 0.001$, in comparison to NIO_m. *: $p < 0.05$, in comparison to NIO_m.

Method	$\text{OASIS}^{n=250}_{train} \Rightarrow \text{LPBA}_{test}$			$\text{LPBA}^{n=20}_{train} \Rightarrow \text{OASIS}_{test}$			$\text{LPBA}^{n=20}_{train} \Rightarrow \text{LPBA}_{test}$			T_{test} (sec) ↓
	DSC_{54} ↑	DSC30_{54} ↑	$std(|J_\phi|)$ ↓	DSC_{35} ↑	DSC30_{35} ↑	$std(|J_\phi|)$ ↓	DSC_{54} ↑	DSC30_{54} ↑	$std(|J_\phi|)$ ↓	
Initial	$0.535 \pm 0.05^\dagger$	$0.476 \pm 0.02^\dagger$	-	$0.598 \pm 0.08^\dagger$	$0.498 \pm 0.05^\dagger$	-	$0.535 \pm 0.05^\dagger$	$0.476 \pm 0.02^\dagger$	-	-	-
(Iterative) NiftyReg [22]	$0.690 \pm 0.02^\dagger$	$0.664 \pm 0.02^*$	0.345 ± 0.04	$0.763 \pm 0.04^\dagger$	$0.714 \pm 0.03^\dagger$	0.655 ± 0.11	$0.690 \pm 0.02^\dagger$	$0.664 \pm 0.02^\dagger$	0.345 ± 0.04	126.7 ± 23.0
(Iterative) SyN [1]	$0.692 \pm 0.02^\dagger$	$0.666 \pm 0.01^*$	0.255 ± 0.02	$0.772 \pm 0.03^\dagger$	$0.729 \pm 0.03^\dagger$	0.342 ± 0.04	$0.692 \pm 0.02^\dagger$	$0.666 \pm 0.01^\dagger$	0.255 ± 0.02	892.4 ± 56.0
MPR [14]	$0.639 \pm 0.03^\dagger$	$0.598 \pm 0.02^\dagger$	$\mathbf{0.224 \pm 0.02}$	$0.662 \pm 0.08^\dagger$	$0.561 \pm 0.05^\dagger$	$\mathbf{0.325 \pm 0.01}$	$0.656 \pm 0.03^\dagger$	$0.620 \pm 0.02^\dagger$	$\mathbf{0.206 \pm 0.07}$	$\mathbf{0.08 \pm 0.01}$
ULAE [26]	$0.665 \pm 0.03^\dagger$	$0.632 \pm 0.02^\dagger$	0.471 ± 0.04	$0.607 \pm 0.05^\dagger$	$0.543 \pm 0.03^\dagger$	0.675 ± 0.04	$0.700 \pm 0.02^\dagger$	$0.676 \pm 0.01^\dagger$	0.382 ± 0.04	0.20 ± 0.00
LapIRN [18]	$0.694 \pm 0.02^\dagger$	$0.665 \pm 0.02^*$	0.435 ± 0.05	$0.793 \pm 0.04^\dagger$	$0.743 \pm 0.04^\dagger$	0.654 ± 0.06	$0.710 \pm 0.02^\dagger$	$0.688 \pm 0.01^*$	0.410 ± 0.04	0.14 ± 0.01
LapIRN + IO ($\eta = 50$)	$0.696 \pm 0.02^\dagger$	$0.667 \pm 0.02^*$	0.440 ± 0.05	$0.801 \pm 0.04^\dagger$	$0.752 \pm 0.04^\dagger$	0.684 ± 0.05	$0.712 \pm 0.02^*$	$0.690 \pm 0.01^*$	0.426 ± 0.04	33.37 ± 0.01
LapIRN + IO ($\eta = 200$)	$0.697 \pm 0.02^\dagger$	$0.668 \pm 0.02^*$	0.450 ± 0.04	$0.805 \pm 0.04^\dagger$	$0.759 \pm 0.04^\dagger$	0.714 ± 0.06	$0.713 \pm 0.02^*$	$0.692 \pm 0.01^*$	0.441 ± 0.04	60.80 ± 0.11
NIO ($\eta = 50$)	0.708 ± 0.03	0.678 ± 0.02	0.443 ± 0.04	0.815 ± 0.02	0.785 ± 0.02	0.605 ± 0.04	0.719 ± 0.01	0.701 ± 0.01	0.432 ± 0.03	47.22 ± 0.91
NIO_m ($\eta = [100, 60, 30]$)	$\mathbf{0.711 \pm 0.02}$	$\mathbf{0.685 \pm 0.02}$	0.434 ± 0.03	$\mathbf{0.817 \pm 0.02}$	$\mathbf{0.789 \pm 0.02}$	0.600 ± 0.04	$\mathbf{0.721 \pm 0.01}$	$\mathbf{0.704 \pm 0.01}$	0.425 ± 0.03	52.71 ± 0.57

OASIS dataset [15,16] and 40 brain MR scans from the LPBA dataset [24,25]. Three experiments are conducted to assess the registration accuracy, robustness and plausibility of our method with dataset shifts and insufficient training data on brain atlas registration: 1) training on OASIS dataset and testing on LPBA dataset; 2) training on LPBA and testing on OASIS; and 3) train on LPBA and testing on LPBA datasets. We follow the pre-processing pipeline in [13] to preprocess the OASIS dataset. We divide the OASIS dataset into 250, 5, and 159 scans, and divide the LPBA dataset into 20, 3, and 17 scans for training, validation, and test sets, respectively. 35 and 54 anatomical structures are included in the evaluation for OASIS and LPBA datasets, respectively. We resample all MR scans to isotropic voxel sizes of 1 mm and center-cropped all scans to $144 \times 160 \times 192$. We select Case 285 and (S24, S25) from the test set of OASIS and LPBA as atlases, respectively.

For Chest CT registration, we use publicly available 4D chest CT from the DIR-Lab dataset [4] for the inter-patient chest CT registration task. The dataset consists of ten 4D chest CT scans, which encompass a full breathing cycle in ten timepoints. The axial isotropic resolutions of each scan range from 256×256 to 512×512 (0.97 mm to 1.16 mm per voxel, with a slice thickness and increment of 2.5 mm. We follow [30] to perform leave-one-out cross-validation during evaluation. We use U-net [12] to delineate the lung lobe of each scan and use the segmentation to define the region of interest in this task. The learning-based method was trained with intra-patient registration by taking random timepoints per patient as fixed and moving images. We leverage the 300 manually identified anatomical landmarks annotated at maximum inspiration and maximum expiration for each case to quantify the registration accuracy.

Implementation. Our proposed method and the other baseline methods are implemented with PyTorch 1.9 and deployed on the same machine equipped with an Nvidia RTX4090 GPU and an Intel Core (i7-13700) CPU. We build our method on top of conditional LapIRN [18] framework. For the multi-resolution optimization strategy, we set the iteration of each level η to $[100, 60, 30]$ and

Table 2. Mean ± standard deviation of the target registration error in millimetre (mm) determined on DIR-Lab 4D-CT dataset. *: method trained on external data.

Scan	Initial	DLIR [30]	B-splines [5]	LungRegNet* [8]	LapIRN [19]	IO ($\eta = 200$)	NIO_m (ours)
Case 1	3.89 ± 2.78	1.27 ± 1.16	1.2 ± 0.6	**0.98 ± 0.54**	1.05 ± 0.47	1.00 ± 0.46	0.99 ± 0.46
Case 2	4.34 ± 3.90	1.20 ± 1.12	1.1 ± 0.6	0.98 ± 0.52	1.04 ± 0.52	0.97 ± 0.47	**0.97 ± 0.47**
Case 3	6.94 ± 4.05	1.48 ± 1.26	1.6 ± 0.9	1.14 ± 0.64	1.18 ± 0.62	1.11 ± 0.61	**1.10 ± 0.61**
Case 4	9.83 ± 4.85	2.09 ± 1.93	1.6 ± 1.1	1.39 ± 0.99	1.40 ± 0.99	1.35 ± 0.98	**1.33 ± 0.95**
Case 5	7.48 ± 5.50	1.95 ± 2.10	2.0 ± 1.6	1.43 ± 1.31	1.41 ± 1.21	**1.34 ± 1.20**	**1.34 ± 1.21**
Case 6	10.89 ± 6.96	5.16 ± 7.09	1.7 ± 1.0	2.26 ± 2.93	1.36 ± 0.77	1.20 ± 0.67	**1.16 ± 0.66**
Case 7	11.03 ± 7.42	3.05 ± 3.01	1.9 ± 1.2	1.42 ± 1.16	1.31 ± 0.68	1.20 ± 0.64	**1.16 ± 0.62**
Case 8	14.99 ± 9.00	6.48 ± 5.37	2.2 ± 2.3	3.13 ± 3.77	1.73 ± 2.31	1.59 ± 2.20	**1.19 ± 0.96**
Case 9	7.92 ± 3.97	2.10 ± 1.66	1.6 ± 0.9	1.27 ± 0.94	1.34 ± 0.73	1.20 ± 0.69	**1.16 ± 0.65**
Case 10	7.30 ± 6.34	2.09 ± 2.24	1.7 ± 1.2	1.93 ± 3.06	1.25 ± 0.77	1.14 ± 0.74	**1.08 ± 0.58**
Mean	8.46 ± 6.58	2.64 ± 4.32	1.66 ± 1.14	1.59 ± 1.58	1.31 ± 0.19	1.21 ± 0.87	**1.15 ± 0.71**

$[100, 60, 60]$ for brain atlas and chest 4DCT registration, respectively. λ is set to 0.4 for both the training and inference phases. We adopt Adam optimizer with a fixed learning rate of $1e^{-4}$ in both training and inference phases. We train all the deep-learning methods from scratch for 150,000 iterations and select the model with the highest DSC (lowest TRE) in the validation set.

Measurement. In brain atlas registration, we register each scan in the test set to an atlas, propagate the anatomical segmentation map of the moving scan using the resulting deformation field, and measure the overlap of the segmentation maps using the Dice similarity coefficient (DSC). We also quantify the robustness of each method by measuring the 30% lowest DSC (DSC30) of all cases. In chest 4DCT registration, we register scans of the maximum expiration phase to that of the maximum inspiration phase and measure the target registration error (TRE) using the manually identified anatomical landmarks. The standard deviation of Jacobian determinant on the deformation field $(\text{std}(|J_\Phi|))$ is measured, which quantifies the smoothness and local orientation consistency of the deformation field. Furthermore, we measure the average registration time per case (T_{test}).

Baseline Methods. We compare our method with two conventional methods (denoted as NiftyReg [22] and SyN [1]), and three state-of-the-art learning-based methods (denoted as MPR [14], ULAE [26] and LapIRN [18,19]) for brain atlas registration. NiftyReg and SyN use the multilevel iterative optimization strategy and do not suffer from the distribution shifts issue. For MPR and ULAE, we adopt the official implementation maintained by the authors and use the best hyper-parameters reported in the paper [14,26]. Besides, we compare the baseline LapIRN with the Adam-based instance optimization approach [27](denoted as IO). Empirically, the performance gains of IO saturated within 80-150 iterations. We set the number of iterations for IO to [50, 200] and 200 for brain atlas registration and chest 4DCT registration tasks, respectively. NIO with the proposed multiresolution strategy is denoted as NIO_m. We follow [27] to induce additional smoothness of the displacement field by adding a B-spline deformation model and affine augmentation to LapIRN, IO and NIO$_m$ during the inference phase for chest 4DCT registration.

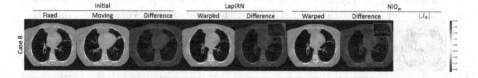

Fig. 3. Example axial slices from the inspiration-expiration registration results. From left to right: the fixed image, moving image, and the results from LapIRN and NIO$_m$. The difference map overlays the fixed (in red) and moving/warped images (in green). (Color figure online)

Results and Discussions. Table 1 presents comprehensive results of the brain atlas registration. Fig. 2 illustrates the qualitative results of the cross-dataset brain atlas registration. Two out of three learning-based methods (MPR and ULAE) fail spectacularly in the cross-dataset brain atlas registration tasks and achieve consistently inferior registration accuracy and robustness to the conventional methods (NiftyReg and SyN) among three tasks in brain atlas registration, suggesting the distribution shift can significantly degrade the solution's quality of learning-based methods. Interestingly, the registration performance of LapIRN diverged from the other learning-based methods and achieved comparable results to the conventional methods under distribution shifts. Comparing NIO$_m$ to LapIRN, our method significantly improves the registration accuracy (+1.5% to +3.0% gain) and robustness (+2.3% to +6.2% gain) of LapIRN in brain atlas registration, reaching state-of-the-art results among three tasks with severe distribution shifts at the cost of registration time. As shown in Fig. 2, our method can drastically improve the registration result for extreme cases with large inter-variation in anatomical structures presented in fixed and moving images. Comparing NIO to the IO, NIO achieves consistently superior registration results with less number of iterations, suggesting the effectiveness of NIO and capable of avoiding sub-optimal solutions in IO. The results of each case on the DIR-Lab 4DCT dataset are shown in Table 2. The results demonstrate that our method not only significantly improves the registration accuracy (an average of 12% drop in TRE compared to the LapIRN), but also corrects the outlier with large initial registration errors. For instance, the average TRE in case 8 of the baseline decreased from 1.73 mm to 1.19 mm using our method. Our method can also correct subtle misalignment inside the lung lobe, as highlighted in Fig. 3. The fraction of landmark pairs with <1.2 mm error of LapIRN improves from 55% to 64% with NIO. The results suggested that our method enables learning-based methods to achieve state-of-the-art registration results even with insufficient training data (20 and 90 scans in LPBA and DIR-Lab 4DCT dataset, respectively) and distribution shifts.

4   Conclusion

We have presented a novel neural instance optimization to improve the registration accuracy and robustness of learning-based deformable image registration methods under distribution shifts. Our method naturally leverages the inductive bias of the convolutional neural network, the prior knowledge learned from the training domain and the multi-resolution optimization strategy to optimize the solution during the inference phase. Extensive experiments on brain atlas and chest 4DCT registration have been carried out, demonstrating that our proposed method achieves state-of-the-art results even with severe distribution shifts or limited training data.

References

1. Avants, B.B., Epstein, C.L., Grossman, M., Gee, J.C.: Symmetric diffeomorphic image registration with cross-correlation: evaluating automated labeling of elderly and neurodegenerative brain. Med. Image Anal. **12**(1), 26–41 (2008)
2. Bajcsy, R., Kovačič, S.: Multiresolution elastic matching. Comput. Vis. Graph. Image Process. **46**(1), 1–21 (1989)
3. Balakrishnan, G., Zhao, A., Sabuncu, M.R., Guttag, J., Dalca, A.V.: Voxelmorph: a learning framework for deformable medical image registration. IEEE Trans. Med. Imaging **38**(8), 1788–1800 (2019)
4. Castillo, R., et al.: A framework for evaluation of deformable image registration spatial accuracy using large landmark point sets. Phys. Med. Biol. **54**(7), 1849 (2009)
5. Delmon, V., et al.: Registration of sliding objects using direction dependent b-splines decomposition. Phys. Med. Biol. **58**(5), 1303 (2013)
6. Eisenmann, M., et al.: Biomedical image analysis competitions: the state of current participation practice. arXiv preprint arXiv:2212.08568 (2022)
7. Falta, F., Hansen, L., Heinrich, M.P.: Learning iterative optimisation for deformable image registration of lung CT with recurrent convolutional networks. In: Wang, L., Dou, Q., Fletcher, P.T., Speidel, S., Li, S. (eds.) MICCAI 2022. LNCS, vol. 13436, pp. 301–309. Springer, Cham (2022). https://doi.org/10.1007/978-3-031-16446-0_29
8. Fu, Y., et al.: Lungregnet: an unsupervised deformable image registration method for 4D-CT lung. Med. Phys. **47**(4), 1763–1774 (2020)
9. Heinrich, M.P., Hansen, L.: Voxelmorph++ going beyond the cranial vault with keypoint supervision and multi-channel instance optimisation. In: Hering, A., Schnabel, J., Zhang, M., Ferrante, E., Heinrich, M., Rueckert, D. (eds.) WBIR 2022. LNCS, vol. 13386, pp. 85–95. Springer, Cham (2022). https://doi.org/10.1007/978-3-031-11203-4_10
10. Hering, A., Hansen, L., Mok, T.C., et al.: Learn2Reg: comprehensive multi-task medical image registration challenge, dataset and evaluation in the era of deep learning. IEEE Trans. Med. Imaging **42**(3), 697–712 (2022)
11. Hoffmann, M., Billot, B., Greve, D.N., Iglesias, J.E., Fischl, B., Dalca, A.V.: Synthmorph: learning contrast-invariant registration without acquired images. IEEE Trans. Med. Imaging **41**(3), 543–558 (2021)

12. Hofmanninger, J., Prayer, F., Pan, J., Röhrich, S., Prosch, H., Langs, G.: Automatic lung segmentation in routine imaging is primarily a data diversity problem, not a methodology problem. Eur. Radiol. Exp. **4**(1), 1–13 (2020)

13. Hoopes, A., Hoffmann, M., Fischl, B., Guttag, J., Dalca, A.V.: HyperMorph: amortized hyperparameter learning for image registration. In: Feragen, A., Sommer, S., Schnabel, J., Nielsen, M. (eds.) IPMI 2021. LNCS, vol. 12729, pp. 3–17. Springer, Cham (2021). https://doi.org/10.1007/978-3-030-78191-0_1

14. Liu, R., Li, Z., et al.: Learning deformable image registration from optimization: perspective, modules, bilevel training and beyond. IEEE Trans. Pattern Anal. Mach. Intell. **44**(11), 7688–7704 (2022)

15. Marcus, D.S., Wang, T.H., Parker, J., Csernansky, J.G., Morris, J.C., Buckner, R.L.: Open access series of imaging studies (OASIS): cross-sectional MRI data in young, middle aged, nondemented, and demented older adults. J. Cogn. Neurosci. **19**(9), 1498–1507 (2007)

16. Marcus, D.S., Wang, T.H., et al.: Oasis brains - open access series of imaging studies. https://www.oasis-brains.org/. Accessed 01 Mar 2021

17. Mok, T.C., Chung, A.: Fast symmetric diffeomorphic image registration with convolutional neural networks. In: Proceedings of the IEEE/CVF Conference on Computer Vision and Pattern Recognition, pp. 4644–4653 (2020)

18. Mok, T.C.W., Chung, A.C.S.: Conditional deformable image registration with convolutional neural network. In: de Bruijne, M., et al. (eds.) MICCAI 2021. LNCS, vol. 12904, pp. 35–45. Springer, Cham (2021). https://doi.org/10.1007/978-3-030-87202-1_4

19. Mok, T.C.W., Chung, A.C.S.: Large deformation diffeomorphic image registration with laplacian pyramid networks. In: Martel, A.L., et al. (eds.) MICCAI 2020. LNCS, vol. 12263, pp. 211–221. Springer, Cham (2020). https://doi.org/10.1007/978-3-030-59716-0_21

20. Mok, T.C., Chung, A.C.: Unsupervised deformable image registration with absent correspondences in pre-operative and post-recurrence brain tumor mri scans. In: Wang, L., Dou, Q., Fletcher, P.T., Speidel, S., Li, S. (eds.) MICCAI 2022. LNCS, vol. 13436, pp. 25–35. Springer, Cham (2022). https://doi.org/10.1007/978-3-031-16446-0_3

21. Papenberg, N., Bruhn, A., Brox, T., Didas, S., Weickert, J.: Highly accurate optic flow computation with theoretically justified warping. Int. J. Comput. Vision **67**, 141–158 (2006)

22. Rueckert, D., Sonoda, L.I., Hayes, C., Hill, D.L., Leach, M.O., Hawkes, D.J.: Non-rigid registration using free-form deformations: application to breast MR images. IEEE Trans. Med. Imaging **18**(8), 712–721 (1999)

23. Ruthotto, L., Modersitzki, J.: Non-linear image registration. In: Handbook of Mathematical Methods in Imaging: Volume 1, 2nd edn, pp. 2005–2051 (2015)

24. Shattuck, D.W., et al.: Construction of a 3D probabilistic atlas of human cortical structures. Neuroimage **39**(3), 1064–1080 (2008)

25. Shattuck, D.W., Mirza, M., et al.: LPBA40 atlases download. https://resource.loni.usc.edu/resources/atlases-downloads/. Accessed 01 Mar 2021

26. Shu, Y., Wang, H., Xiao, B., Bi, X., Li, W.: Medical image registration based on uncoupled learning and accumulative enhancement. In: de Bruijne, M., et al. (eds.) MICCAI 2021. LNCS, vol. 12904, pp. 3–13. Springer, Cham (2021). https://doi.org/10.1007/978-3-030-87202-1_1

27. Siebert, H., Hansen, L., Heinrich, M.P.: Fast 3D registration with accurate optimisation and little learning for Learn2Reg 2021. In: Aubreville, M., Zimmerer, D.,

Heinrich, M. (eds.) MICCAI 2021. LNCS, vol. 13166, pp. 174–179. Springer, Cham (2022). https://doi.org/10.1007/978-3-030-97281-3_25

28. Teed, Z., Deng, J.: RAFT: recurrent all-pairs field transforms for optical flow. In: Vedaldi, A., Bischof, H., Brox, T., Frahm, J.-M. (eds.) ECCV 2020. LNCS, vol. 12347, pp. 402–419. Springer, Cham (2020). https://doi.org/10.1007/978-3-030-58536-5_24

29. Torralba, A., Efros, A.A.: Unbiased look at dataset bias. In: CVPR 2011, pp. 1521–1528. IEEE (2011)

30. de Vos, B.D., Berendsen, F.F., Viergever, M.A., Sokooti, H., Staring, M., Išgum, I.: A deep learning framework for unsupervised affine and deformable image registration. Med. Image Anal. **52**, 128–143 (2019)

31. Zhu, W., Huang, Y., Xu, D., Qian, Z., Fan, W., Xie, X.: Test-time training for deformable multi-scale image registration. In: 2021 IEEE International Conference on Robotics and Automation (ICRA), pp. 13618–13625. IEEE (2021)

Implicitly Solved Regularization for Learning-Based Image Registration

Jan Ehrhardt[1,2](✉) and Heinz Handels[1,2]

[1] Institute of Medical Informatics, University of Lübeck, Lübeck, Germany
jan.ehrhardt@uni-luebeck.de
[2] German Research Center for Artificial Intelligence, Lübeck, Germany

Abstract. Deformable image registration is a fundamental step in many medical image analysis tasks and has attracted a large amount of research to develop efficient and accurate unsupervised machine learning approaches. Although much attention has been paid to finding suitable similarity measures, network architectures and training methods, much less research has been devoted to suitable regularization techniques to ensure the plausibility of the learned deformations.

In this paper, we propose implicitly solved regularizers for unsupervised and weakly supervised learning of deformable image registration. In place of pure gradient descent with automatic differentiation, we combine efficient implicit solvers for the regularization term with the established gradient-based optimization regarding the network parameters. As a result, our approach is broadly applicable and can be combined with a range of similarity measures and network architectures. Our experiments with state-of-the-art network architectures show that the proposed approach has the potential to increase the smoothness, i.e. the plausibility, of the learned deformations and the registration accuracy measured as dice overlaps. Furthermore, we show that due to efficient GPU implementations of the implicit solvers, this increase in plausibility and accuracy comes at almost no additional cost in terms of computational time.

Keywords: image registration · regularization · implicit solvers

1 Introduction

Image registration is a fundamental step in many medical image analysis tasks, and aims to find a plausible spatial transformation to align two images, i.e. to compensate for morphological or positional differences between the imaged subjects. In recent years, much attention has been paid to the development of efficient and accurate machine learning methods for deformable medical image registration [12], especially unsupervised and weakly supervised approaches [17]. Research has focused on the development of similarity measures [7,10,14], network architectures [1,5,6,26], transformation models [21,27], and optimization

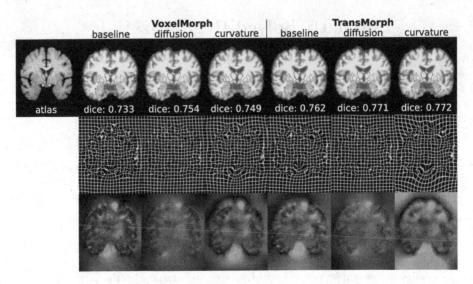

Fig. 1. Qualitative registration results for atlas-to-image registration of two deep learning models trained with diffusion regularization by automatic differentiation (baseline) and using implicit solvers for diffusion and curvature regularization as examples. The deformations are different for a comparable level of registration accuracy, with our approach producing smoother deformations.

approaches [30]. Less attention has been paid to the development of regularization techniques related to the plausibility of the estimated transformation. Few approaches use regularizers such as curvature or bending energy [5,26] or developed task-specific approaches by incorporating population models [2] or biomechanical priors [16,24]. However, the vast majority of unsupervised learning approaches use a diffusion regularization, e.g. 10 out of 11 in the Learn2Reg challenge 2021 [15]. This is surprising since for traditional iterative registration methods a whole range of regularization approaches with specific properties have been developed [3,8,11,20,22,25] and studies show their impact on the properties and hence the plausibility of the generated deformations [29].

In this work, we propose to combine efficient implicit solvers for regularization with the established explicit gradient-descent optimization of the network parameters by using a decoupled energy term as loss function. Figure 1 highlights that both the chosen regularizer and the optimization strategy have an impact on the learned deformation. To investigate the impact of this training strategy, we evaluate three implicitly solved regularizers known from iterative registration (diffusion, curvature, and linear elasticity) for two deep learning approaches (voxelmorph [1] and transmorph [5]) in an unsupervised and a weakly supervised setting on 3D brain MRI and abdominal CT data. The experiments show that the proposed approach can increase both the smoothness of the learned deformations and the registration accuracy measured as Dice score. Furthermore, we show that due to efficient GPU implementations, this increase in plausibility and accuracy is possible at almost no additional cost in terms of computational time.

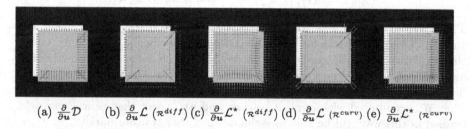

(a) $\frac{\partial}{\partial u}\mathcal{D}$ (b) $\frac{\partial}{\partial u}\mathcal{L}$ $(\mathcal{R}^{diff})$ (c) $\frac{\partial}{\partial u}\mathcal{L}^*$ $(\mathcal{R}^{diff})$ (d) $\frac{\partial}{\partial u}\mathcal{L}$ $(\mathcal{R}^{curv})$ (e) $\frac{\partial}{\partial u}\mathcal{L}^*$ $(\mathcal{R}^{curv})$

Fig. 2. An illustrative example showing a fixed image (white square in background) and a moving image (shifted gray square in foreground) overlaid by the gradients w.r.t u computed for similarity measure $\|I_f - I_m \circ \varphi\|^2$ (a), and loss functions $\mathcal{L}$ (Eq. 1: (b)+(d)) and $\mathcal{L}^*$ (Eq. 2: (c)+(e)) with diffusion (b)+(c) and curvature regularization (d)+(e). (computed with Pytorch's autograd function [23], colors show magnitude, arrows are scaled for better visibility.) (Color figure online)

2   Methods

Medical image registration aims to find a spatial transformation $\varphi : \Omega \to \Omega, \Omega \subset \mathbb{R}^d$, to align a moving image I_m with the fixed image I_f. Often φ is represented by a dense displacement vector field u as $\varphi(x) = x + u(x)$. In most unsupervised learning-based settings φ_θ (i.e. u_θ) is estimated using a neural network with parameters θ that is trained to minimize a loss function that can be written as:

$$\mathcal{L}(\theta) = \mathcal{D}(I_f, I_m \circ \varphi_\theta) + \alpha\mathcal{R}(\varphi_\theta), \tag{1}$$

where $\mathcal{D}$ measures the similarity between two images and $\mathcal{R}$ imposes regularization to enforce plausibility of the deformation. Network parameters are usually learned by backpropagating the loss in Eq. 1 using automatic differentiation methods and gradient descent. This represents an explicit finite difference scheme. In many iterative registration approaches, the partial differential equations arising from Eq. 1 are solved by semi-implicit optimization schemes that use efficient implicit solvers for the regularization term [20,29]. Semi-implicit schemes exhibit improved numerical stability, allow large step sizes, and thus faster convergence compared to explicit schemes. Since the similarity term $\mathcal{D}$ is the driving force of the registration method and $\frac{\partial}{\partial u}\mathcal{D}$ is non-zero only at the edges of the (deformed) moving image, implicit regularizers have the advantage that they propagate the gradient information over the entire image domain in each iteration, whereas in explicit methods this information is propagated only step-wise. This is illustrated by a toy example in Fig. 2.

Our work aims to exploit and evaluate efficient implicit regularizers for the training of neural networks. Following a history of work on quadratic relaxation in the context of iterative methods [4,9,13], we use an auxiliary vector field $\overline{u}$ to decouple data term and regularizer:

$$\mathcal{L}^\star(\theta; \overline{u}) = \mathcal{D}(I_f, I_m \circ \varphi_\theta) + \frac{\sigma}{2} \|u_\theta - \overline{u}\|^2 + \sigma\gamma\mathcal{R}(\overline{u}). \tag{2}$$

The advantage of Eq. 2 is that for a given estimate of the displacement u_θ, globally optimal solutions for $\overline{u}$ can be computed by efficient algorithms. Given the regularized field $\overline{u}$, gradients with respect to the network parameters are independent of $\mathcal{R}$ and no longer contain spatial derivatives of u. Figure 2(c+e) illustrates that Eq. 2 leads to smooth gradients w.r.t. the displacement field u when $\overline{u}$ is computed with implicit solvers, and we expect that neural networks trained in this way will be able to estimate smoother transformations.

The loss in Eq. 2 contains two parameters σ and γ, where γ directly influences the smoothness of the deformation $\overline{u}$ and σ balances intensity similarities and smoothness and is thus related to the noise level of the images [4]. To employ Eq. (2) for network training, a two-step procedure is derived by rewriting as

$$\mathcal{L}^\diamond(\theta) = \mathcal{D}(I_f, I_m \circ \varphi_\theta) + \frac{\sigma}{2} \|u_\theta - \overline{u}\|^2 \tag{3}$$

$$\text{s.t.} \quad \overline{u} = \arg\min_u \frac{1}{2} \|u_\theta - u\|^2 + \gamma\mathcal{R}(u). \tag{4}$$

In each training iteration, Eq. 4 is used to compute $\overline{u}$ from the current prediction u_θ, followed by a backpropagation of $\mathcal{L}^\diamond$ to update the network parameters. For many regularizers Eq. 4 can be solved efficiently by a linear equation system

$$\text{vec}(\overline{u}) = (\mathbf{I} - \gamma\mathbf{A})^{-1} \text{vec}(u_\theta), \tag{5}$$

where $\text{vec}(\cdot)$ is a vectorized representation, $\mathbf{I}$ is the identity matrix and $\mathbf{A}$ results from a differential linear operator depending on the chosen regularizer. We investigate three regularizers with associated differential linear operators $\boldsymbol{A}^{diff}[\boldsymbol{u}] = \Delta\boldsymbol{u}$ (diffusion), $\boldsymbol{A}^{curv}[\boldsymbol{u}] = \Delta^2\boldsymbol{u}$ (curvature), and $\boldsymbol{A}^{elas}[\boldsymbol{u}] = \mu\Delta\boldsymbol{u} + (\mu + \lambda)\nabla\text{div}\boldsymbol{u}$ (linear elastic) and μ, λ the Lamé constants.

For diffusion regularization the linear system in Eq. 5 is solved by additive operator splitting [28], resulting in an efficient $\mathcal{O}(n)$ algorithm. For curvature and linear elastic regularization Eq. 5 is solved by convolution

$$\overline{u} = \mathcal{F}^{-1}\left(\mathcal{F}(u_\theta) / (\mathbf{I} - \gamma\mathcal{F}(\boldsymbol{A}))\right), \tag{6}$$

where $\mathcal{F}$ is the discrete Fourier transform (DFT) and $\mathcal{F}(\boldsymbol{A})$ an eigendecomposition of the linear operator $\boldsymbol{A}$ that can be pre-computed (see [20] for derivations of $\mathcal{F}(\boldsymbol{A}^{curv})$ and $\mathcal{F}(\boldsymbol{A}^{elas})$). Note that Eq. 6 represents a point-wise operation resulting in an $\mathcal{O}(n \log n)$ algorithm due to the necessary DFT and its inverse.

The presented solution methods for Eq. (4) allow efficient parallelized GPU implementations for 2D and 3D image data using modern deep learning frameworks. Our approach is model-agnostic in the sense that we make no assumptions about the underlying network architecture and can be combined with different similarity measures and optimizers. Thus, a variety of existing unsupervised or weakly supervised deep learning approaches can be used by plugging in the loss given in Eq. (2).

3   Experiments

To demonstrate the versatility and advantages of the proposed approach, we perform experiments on two conceptual different model architectures in two different registration scenarios. The widely used CNN-based voxelmorph model [1] and a recent transformer-based approach [5], which is among the currently best performing deep learning approaches, are selected as model architectures. As baselines, both models were trained with explicit diffusion regularization according to Eq. 1. The baseline models were compared to three regularizers using the proposed implicit solving strategy (Eq. 2). The implicit regularizers were implemented in PyTorch v1.13[1], and all experiments were performed on an NVIDIA RTX A5000 GPU.

Atlas-to-Patient Brain MRI Registration. The first scenarios performs an atlas-to-patient registration on 576 T1-weighted brain MRI images of the publicly available IXI database[2] and an atlas brain MRI obtained from [19]. Freesurfer segmentations of 30 anatomical structures created by [5] are available for all image data, but are used for evaluation purposes only. To maximize the comparability, reproducibility, and transparency, this experiment is based on the repository given in [5], using the same data, data splits (403/58/115 for train/validation/test), similarity measures (local normalized cross-correlation), and optimizer (ADAM optimization with learning rate 0.0001). For the baseline models, the regularization hyperparameter adapted to this dataset ($\alpha = 1$) were used for the explicit diffusion regularization, as given in [5].

Weakly Supervised Inter-patient Abdominal CT Registration. The second scenario performs a weakly supervised inter-patient registration on 200 abdominal CT images with 15 labeled organ categories [18]. The images are resampled to $2 \times 2 \times 2 mm^3$, affinely pre-aligned and splitted into 160/20/20 images for train/validation/test. Weakly supervision [17] is performed by additionally using organ labels for loss evaluation during the training but not as network input. Network architectures and optimizer settings are identical to the first experiment and a combination of mean squared intensity difference and dice loss was used as similarity measure. Grid search was performed to select regularization hyperparameters.

Evaluation Metrics. Registration accuracy is quantified using the Dice score between segmentation masks of the warped and fixed image. Smoothness and plausibility of the deformations are measured by the standard deviation of the logarithmic Jacobian determinant $std(\log(|\boldsymbol{J}_\varphi|))$, with $|\boldsymbol{J}_\varphi| = \det \nabla \cdot \varphi$ and the absolute average of negative Jacobian determinant values ($\frac{1}{n}\||\boldsymbol{J}_\varphi| < 0\|_1$), where n is the number of voxels.

---

[1] git.opendfki.de/jan.ehrhardt/implicitly-solved-regularization.

[2] www.braindevelopment.org.

Table 1. Quantitative evaluation results for 115 brain MRI test images (differences of diffusion, elastic, curvature are significant with respect to the baseline ($p < 0.001$)).

Model		Dice score	$\text{std}(\log(|\boldsymbol{J_\varphi}|))$	$\frac{1}{n}\||\boldsymbol{J_\varphi}| < 0\|_1$
	affine	0.386 ± 0.189	–	–
	SyN	0.645 ± 0.140	–	–
Voxel-morph	baseline	0.731 ± 0.115	0.456 ± 0.023	0.00412
	diffusive	0.744 ± 0.123	$\mathbf{0.257 \pm 0.012}$	**0.00019**
	elastic	$\mathbf{0.747 \pm 0.124}$	0.265 ± 0.012	0.00055
	curvature	0.735 ± 0.121	0.361 ± 0.029	0.00085
Trans-morph	baseline	0.750 ± 0.120	0.458 ± 0.025	0.00377
	diffusive	$\mathbf{0.763 \pm 0.119}$	0.316 ± 0.012	0.00080
	elastic	0.763 ± 0.123	$\mathbf{0.286 \pm 0.010}$	**0.00075**
	curvature	0.755 ± 0.122	0.417 ± 0.025	0.00162

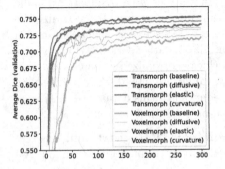 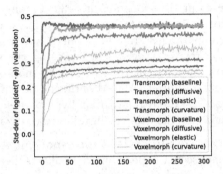

Fig. 3. Validation Dice scores (left) and smoothness $\text{std}(\log(|\boldsymbol{J_\varphi}|)$ (right) during training of VoxelMorph and TransMorph using the baseline method and proposed implicit regularization for diffusion, elastic and curvature.

4   Results and Discussion

4.1   Brain MRI Atlas-to-Patient Registration

Table 1 shows the quantitative evaluation of the atlas-to-patient registration on the brain MRI test data. A grid search was applied to determine regularization hyperparameters $\sigma = 1, \gamma^{diff} = \gamma^{elas} = 2, \gamma^{curv} = 5$, setting $\mu = 1$ and $\lambda = 0$ for the elastic regularization as suggested in [20]. All approaches using the implicit regularization given in Eq. 2 outperform their baseline method with respect to smoothness and accuracy. With elastic and diffusive regularization smoother and more accurate results are obtained compared to curvature regularization. Figure 3 shows dice scores and Jacobian measures of the validation set over the course of the network training. For dice scores, the learning curves of the implicit methods are steeper, suggesting that implicit regularization better

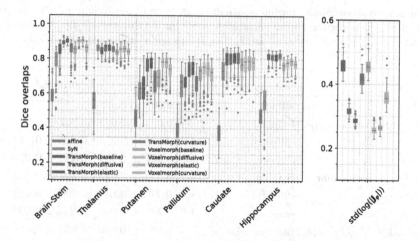

Fig. 4. Boxplots for quantitative comparison on the MRI atlas-to-image registration task for VoxelMorph and TransMorph models using standard optimization (baseline) and proposed optimization with implicit regularization for diffusion, elastic and curvature regularizer. Left: Dice scores for different brain structures, Right: smoothness of the deformations measured as $std(\log(|J_\varphi|))$.

guides the registration towards an optimum. Figure 4 shows quantitative results of 115 test MR images as boxplots. The implicit elastic and diffusive regularization show a higher mean and lower standard deviation for the Dice score for many brain structures compared to the baseline method. At the same time, the estimated deformations are consistently smooth for all test data. The curvature regularization generates less smooth deformation fields measured by J_φ and generally requires higher regularization weights. This behavior is consistent, since first-order gradients are not penalized.

Regarding the run-time, we found no significant differences between the baseline method and all three implicit regularizers ($917 \pm 64\,ms$ per training iteration, $160 \times 192 \times 224$ voxels, batch size 1, data load and pre-processing excluded), despite the increased complexity for solving Eq. 5. We conclude that due to the efficient GPU-based DFT, there is no runtime penalty from our approach.

4.2   Weakly Supervised Inter-patient Registration of Abdominal CT

In general, abdominal data show large differences due to the individual anatomical variations and respiratory influences, and registration of these data is difficult. In our experiment, we investigated explicit diffusive (baseline) and implicit diffusive and elastic regularization. Dice loss and MSE were equally weighted and a grid search strategy was used to find appropriate regularization hyperparameters. Training was performed for 500 epochs and then the parameters with the best dice results on the validation data and the best results for a pre-specified smoothness of $std(\log(|J_\varphi|)) \leq 0.2$ were selected.

Table 2. Evaluation results for inter-patient registration of the abdominal CT test set, "best dice" selects regularization parameters according to the average dice score on the validation set, "smooth" use $std(\log(|J_\varphi|)) \leq 0.2$ as additional constraint (significance to the baseline is shown in bold ($p < 0.005$)).

Model			α	σ	γ	Dice score	$std(\log(|J_\varphi|))$
		affine	–	–	–	0.401 ± 0.189	–
best dice	Voxel-morph	baseline	1.0	–	–	0.603 ± 0.083	0.301 ± 0.022
		diffusive	–	1.0	1.0	0.601 ± 0.078	$\mathbf{0.287 \pm 0.018}$
		elastic	–	2.0	2.0	0.597 ± 0.082	$\mathbf{0.189 \pm 0.018}$
	Trans-morph	baseline	1.0	–	–	0.699 ± 0.092	0.299 ± 0.022
		diffusive	–	1.0	1.0	$\mathbf{0.709 \pm 0.088}$	0.296 ± 0.023
		elastic	–	2.0	1.0	0.701 ± 0.080	$\mathbf{0.216 \pm 0.015}$
smooth	Voxel-morph	baseline	4.0	–	–	0.586 ± 0.086	0.193 ± 0.018
		diffusive	–	2.0	2.0	0.579 ± 0.074	0.196 ± 0.017
		elastic	–	2.0	2.0	0.597 ± 0.082	0.189 ± 0.018
	Trans-morph	baseline	4.0	–	–	0.683 ± 0.068	0.194 ± 0.022
		diffusive	–	2.0	2.0	$\mathbf{0.701 \pm 0.090}$	0.203 ± 0.015
		elastic	–	2.0	2.0	$\mathbf{0.699 \pm 0.080}$	0.200 ± 0.015

The results on image pairs build from 20 test data can be seen in Table 2. The same levels of regularization strength were selected for explicit and implicit diffusive regularization in each task ($\alpha = \sigma\gamma$). No significant differences in the obtained Dice scores were found for the voxelmorph model, but the transmorph model shows a slight but significant improvement for implicit diffusive regularization, which increases with higher smoothness. This may indicate that high-capacity models benefit more from implicit regularization for this task. Regarding elastic regularization, and in line with previous studies [29], we observed that this operator performs better with higher regularization strength, suggesting that it is more suitable for scenarios where moderate or high regularity of the deformation is required.

5   Conclusion

In this work, we combine efficient implicit solvers for the regularization term with the established explicit gradient-based optimization of the network parameters in unsupervised or weakly supervised learning of image registration. This allows to efficiently utilize a range of known regularizers from traditional iterative registration methods in the training of neural networks. The experiments show that the excellent numerical properties of implicit solvers may lead to smoother and also more accurate deformations estimated by the network. Our approach can be used with different network architectures and image similarity measures.

Three regularizers and their implicite solving methods were investigated: diffusion, curvature and linear elastic. To our knowledge, this is the first application of linear elasticity in deep learning-based registration. The comparison of explicit and implicit diffusion regularization reveals the positive impact of the proposed solving strategy. Furthermore, our results show that elastic and diffusion regularization behave similarly in terms of accuracy and smoothness, while curvature regularization seems to be less suitable for the considered task and evaluation measures. Finally, due to efficient GPU implementations, the increase in plausibility and accuracy comes at almost no additional cost in terms of computational time. Overall, our experiments suggest that further investigation of alternative regularization techniques might be useful for a range of registration tasks.

References

1. Balakrishnan, G., Zhao, A., Sabuncu, M.R., Guttag, J., Dalca, A.V.: VoxelMorph: a learning framework for deformable medical image registration. IEEE Trans. Med. Imaging **38**(8), 1788–1800 (2019)
2. Bhalodia, R., Elhabian, S.Y., Kavan, L., Whitaker, R.T.: A cooperative autoencoder for population-based regularization of CNN image registration. In: Shen, D., et al. (eds.) MICCAI 2019. LNCS, vol. 11765, pp. 391–400. Springer, Cham (2019). https://doi.org/10.1007/978-3-030-32245-8_44
3. Burger, M., Modersitzki, J., Ruthotto, L.: A hyperelastic regularization energy for image registration. SIAM J. Sci. Comput. **35**(1), B132–B148 (2013)
4. Cachier, P., Bardinet, E., Dormont, D., Pennec, X., Ayache, N.: Iconic feature based nonrigid registration: the PASHA algorithm. Comput. Vis. Image Underst. **89**(2), 272–298 (2003)
5. Chen, J., Frey, E.C., He, Y., Segars, W.P., Li, Y., Du, Y.: TransMorph: transformer for unsupervised medical image registration. Med. Image Anal. **82**, 102615 (2022)
6. Chen, J., He, Y., Frey, E., Li, Y., Du, Y.: ViT-V-Net: vision transformer for unsupervised volumetric medical image registration. arXiv preprint arXiv:2104.06468 (2022)
7. Cheng, X., Zhang, L., Zheng, Y.: Deep similarity learning for multimodal medical images. Comput. Methods Biomech. Biomed. Eng. Imag. Visualiz. **6**(3), 248–252 (2018)
8. Christensen, G., Rabbitt, R., Miller, M.: Deformable templates using large deformation kinematics. IEEE Trans. Image Process. **5**(10), 1435–1447 (1996)
9. Cohen, L.D.: Auxiliary variables and two-step iterative algorithms in computer vision problems. J. Math. Imaging Vision **6**(1), 59–83 (1996)
10. Ferrante, E., Dokania, P.K., Silva, R.M., Paragios, N.: Weakly supervised learning of metric aggregations for deformable image registration. IEEE J. Biomed. Health Inform. **23**(4), 1374–1384 (2019)
11. Fischer, B., Modersitzki, J.: A unified approach to fast image registration and a new curvature based registration technique. Linear Algebra Appl. **380**, 107–124 (2004)
12. Fu, Y., Lei, Y., Wang, T., Curran, W.J., Liu, T., Yang, X.: Deep learning in medical image registration: a review. Phys. Med. Biol. **65**(20), 20TR01 (2020)
13. Geman, D., Yang, C.: Nonlinear image recovery with half-quadratic regularization. IEEE Trans. Image Process. **4**(7), 932–946 (1995)

14. Haskins, G., et al.: Learning deep similarity metric for 3D MR-TRUS image registration. Int. J. Comput. Assist. Radiol. Surg. **14**(3), 417–425 (2019)

15. Hering, A., et al.: Learn2Reg: comprehensive multi-task medical image registration challenge, dataset and evaluation in the era of deep learning. IEEE Trans. Med. Imaging 1 (2022)

16. Hu, Y., et al.: Adversarial deformation regularization for training image registration neural networks. In: Frangi, A.F., et al. (eds.) MICCAI 2018. LNCS, vol. 11070, pp. 774–782. Springer, Cham (2018). https://doi.org/10.1007/978-3-030-00928-1_87

17. Hu, Y., et al.: Weakly-supervised convolutional neural networks for multimodal image registration. Med. Image Anal. **49**, 1–13 (2018)

18. Ji, Y., et al.: Amos: a large-scale abdominal multi-organ benchmark for versatile medical image segmentation. Adv. Neural. Inf. Process. Syst. **35**, 36722–36732 (2022)

19. Kim, B., Kim, D.H., Park, S.H., Kim, J., Lee, J.G., Ye, J.C.: CycleMorph: cycle consistent unsupervised deformable image registration. Med. Image Anal. **71**, 102036 (2021)

20. Modersitzki, J.: Numerical methods for image registration. In: Numerical Mathematics and Scientific Computation. Oxford University Press, Oxford (2003)

21. Mok, T.C.W., Chung, A.C.S.: Fast symmetric diffeomorphic image registration with convolutional neural networks. In: Proceedings of the IEEE/CVF Conference on Computer Vision and Pattern Recognition, pp. 4644–4653 (2020)

22. Pace, D.F., Aylward, S.R., Niethammer, M.: A locally adaptive regularization based on anisotropic diffusion for deformable image registration of sliding organs. IEEE Trans. Med. Imaging **32**(11), 2114–2126 (2013)

23. Paszke, A., et al.: Automatic differentiation in PyTorch. In: NIPS 2017 Workshop Autodiff (2017). https://openreview.net/forum?id=BJJsrmfCZ

24. Qin, C., Wang, S., Chen, C., Qiu, H., Bai, W., Rueckert, D.: Biomechanics-informed neural networks for myocardial motion tracking in MRI. In: Martel, A.L., et al. (eds.) MICCAI 2020. LNCS, vol. 12263, pp. 296–306. Springer, Cham (2020). https://doi.org/10.1007/978-3-030-59716-0_29

25. Schmidt-Richberg, A., Werner, R., Handels, H., Ehrhardt, J.: Estimation of slipping organ motion by registration with direction-dependent regularization. Med. Image Anal. **16**, 150–159 (2012)

26. de Vos, B.D., Berendsen, F.F., Viergever, M.A., Sokooti, H., Staring, M., Išgum, I.: A deep learning framework for unsupervised affine and deformable image registration. Med. Image Anal. **52**, 128–143 (2019)

27. Wang, J., Zhang, M.: DeepFLASH: an efficient network for learning-based medical image registration. In: Proceedings of the IEEE/CVF Conference on Computer Vision and Pattern Recognition, pp. 4444–4452 (2020)

28. Weickert, J., Romeny, B., Viergever, M.: Efficient and reliable schemes for nonlinear diffusion filtering. IEEE Trans. Image Process. **7**(3), 398–410 (1998)

29. Werner, R., Schmidt-Richberg, A., Handels, H., Ehrhardt, J.: Estimation of lung motion fields in 4D CT data by variational non-linear intensity-based registration: a comparison and evaluation study. Phys. Med. Biol. **59**(15), 4247–4260 (2014)

30. Xu, Z., et al.: Double-uncertainty guided spatial and temporal consistency regularization weighting for learning-based abdominal registration. In: Wang, L., Dou, Q., Fletcher, P.T., Speidel, S., Li, S. (eds.) Medical Image Computing and Computer Assisted Intervention, MICCAI 2022, LNCS, pp. 14–24. Springer, Cham (2022). https://doi.org/10.1007/978-3-031-16446-0_2

BHSD: A 3D Multi-class Brain Hemorrhage Segmentation Dataset

Biao Wu[1], Yutong Xie[1], Zeyu Zhang[1,2], Jinchao Ge[1], Kaspar Yaxley[3],
Suzan Bahadir[3], Qi Wu[1], Yifan Liu[1,4], and Minh-Son To[1,3(✉)]

[1] University of Adelaide, Adelaide, Australia
[2] Australian National University, Canberra, Australia
[3] Flinders University, Adelaide, Australia
minhson.to@flinders.edu.au
[4] ETH Zürich, Zürich, Switzerland

Abstract. Intracranial hemorrhage (ICH) is a pathological condition characterized by bleeding inside the skull or brain, which can be attributed to various factors. Identifying, localizing and quantifying ICH has important clinical implications, in a bleed-dependent manner. While deep learning techniques are widely used in medical image segmentation and have been applied to the ICH segmentation task, existing public ICH datasets do not support the multi-class segmentation problem. To address this, we develop the Brain Hemorrhage Segmentation Dataset (BHSD), which provides a 3D multi-class ICH dataset containing 192 volumes with pixel-level annotations and 2200 volumes with slice-level annotations across five categories of ICH. To demonstrate the utility of the dataset, we formulate a series of supervised and semi-supervised ICH segmentation tasks. We provide experimental results with state-of-the-art models as reference benchmarks for further model developments and evaluations on this dataset. The dataset and checkpoint is available at https://github.com/White65534/BHSD.

Keywords: Intracranial hemorrhage · Segmentation · Multi-class

1   Introduction

Intracranial hemorrhage (ICH) refers to bleeding that occurs inside the skull or brain. There are various types, based on the anatomical relation of the bleeding with the brain and its surrounding membranes. These types include extradural hemorrhage (EDH), subdural hemorrhage (SDH), subarachnoid hemorrhage (SAH), intraparenchymal hemorrhage (IPH), and intraventricular hemorrhage (IVH). The causes of ICH are diverse [7,11,18], encompassing factors such as trauma, vascular malformations, tumors, hypertension, and venous thrombosis.

Supplementary Information The online version contains supplementary material available at https://doi.org/10.1007/978-3-031-45673-2_15.

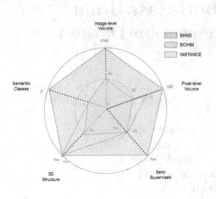 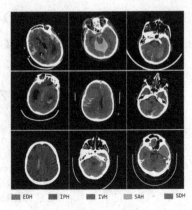

(a) Comparison with other datasets.   (b) Example annotations.

Fig. 1. Overview of the BHSD. (a) Comparison of the BCIHM dataset, INSTANCE dataset and BHSD in terms of data characteristics. The BHSD provides more data, and more annotation information at different levels. (b) Representative examples of the diverse hemorrhage annotations provided in the BHSD. The different colors correspond to different hemorrhage classes, as indicated by the legend. (Color figure online)

The detection and characterization of ICH are typically done using non-contrast CT scans, which can reveal the type and distribution of bleeding. Accurate localization, quantification, and classification of hemorrhages are crucial as they have significant clinical implications, allowing clinicians to estimate severity, predict outcomes, and monitor progress [1,6,10,20]. Treatment options may also vary depending on the quantity and type of bleed.

The rapid development of deep learning and large-scale labeled dataset has accelerated the automation of medical image segmentation [15,25]. Despite the importance of delineating different classes of hemorrhage, many existing public ICH datasets focus either on hemorrhage classification [12] or single-class segmentation [16], specifically foreground versus background. Thus, it is essential to develop tools that can accurately segment different classes of hemorrhage rather than solely perform foreground versus background segmentation. Unfortunately, the lack of public datasets with multi-class, pixel-level annotations hinders the development of class-specific segmentation techniques.

The purpose of this work is to augment a large, public ICH dataset [5] to produce a 3D, multi-class ICH dataset with pixel-level hemorrhage annotations, hereafter referred to as the brain hemorrhage segmentation dataset (BHSD). Our approach leverages the existing high-quality slice-level annotations performed by neuroradiologists and subsequently relabels a subset of CT scans with multi-class pixel-level annotations. We demonstrate the utility of this dataset by performing a series of experiments and providing benchmarks on supervised and semi-supervised segmentation tasks.

2 Multi-class Brain Hemorrhage Segmentation Dataset

2.1 Brain Hemorrhage Datasets

In this section, we describe existing, public brain hemorrhage datasets. The BCIHM dataset consists of 82 non-contrast CT scans of patients with traumatic brain injury [12]. The dataset is provided in NIfTI format. Each slice of the scans was reviewed by two radiologists who recorded hemorrhage types if hemorrhage occurred or if a fracture occurred. Hemorrhage regions in each slice were also segmented, however, the annotations only support a foreground versus background segmentation. In contrast, the INSTANCE dataset, introduced as part of a MICCAI 2022 Challenge, consists of 200 volumes, of which 130 have foreground and background segmentation labels [16]. The volumes in this dataset are also provided in NIfTI format, but again all bleed types are combined into a single foreground class.

The CQ500 dataset consists of 491 CT scans with 193,317 slices in DICOM format [3]. The scans have been read by three radiologists, and the annotations provided indicate, at the scan level, the presence, type and location of hemorrhage. While this dataset does not support segmentation tasks, an augmentation of this dataset, BHX, provides bounding box annotations for five types of hemorrhage [19]. Specifically, BHX contains 39,668 bounding boxes in 23,409 images. Another key brain hemorrhage dataset was published by the Radiological Society of North America (RSNA) [5]. This dataset is a public collection of 874,035 CT head images in DICOM format from a mixed patient cohort with and without ICH. The dataset is multi-institutional and multi-national and includes slice-level expert annotations from neuroradiologists about the presence and type of bleed. This dataset was used for the RSNA 2019 Machine Learning Challenge for detecting brain hemorrhages, *i.e.*, a classification, not segmentation problem.

2.2 Reconstruction and Annotation Pipeline of the BHSD

The BHSD is a high-quality medical imaging dataset comprising 2192 high-resolution 3D CT scans of the brain, each containing between 24 to 40 slices of 512×512 pixels in size (Fig. 1a). The original RSNA dataset was provided as a collection of randomly sorted slices in DICOM format with slice-level annotations. Important contextual information in adjacent slices may be lost in single slices, hence the first step was to reconstitute 3D head scans. Since the anonymized patient identifiers were provided and the DICOM files retained geometric/positional data, the original 3D head scans could be reconstructed and converted to NIfTI format [14]. The slice-level hemorrhage labels provided with the original RSNA dataset were mapped to the corresponding slices in the reconstructed head scans. A subset of 192 scans with one or more of five categories of ICH, namely EDH, IPH, IVH, SAH, and/or SDH, was subsequently selected for further annotation.

Pixel-level annotations were performed by three medical imaging experts in two stages. Hemorrhages on individual head scans were independently segmented

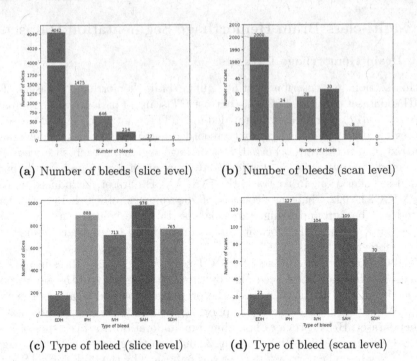

(a) Number of bleeds (slice level)   (b) Number of bleeds (scan level)

(c) Type of bleed (slice level)   (d) Type of bleed (scan level)

Fig. 2. Summary composition of the BHSD, at the slice and scan levels, by number and type of bleed.

using ITK-SNAP [27] by two trained medical imaging experts and radiology residents, both with over one year of experience reading CT head scans, using the original image-level hemorrhage annotations as a guide. These annotations were then reviewed by a board-certified radiologist with over 5 years post-fellowship experience, ensuring the quality of the annotations. To supplement the 192 volumes with pixel-level annotations, we also collected corresponding image-level annotations from the RSNA dataset and provide an additional 2000 3D CT scans with slice-level annotations, including scans with no bleed. The composition of the BHSD is shown in Fig. 2.

By covering both image-level and pixel-level annotations, the BHSD allows a more comprehensive interrogation of brain hemorrhage imaging, and as we show, enables the development of more varied deep learning methods for ICH segmentation.

2.3   Segmentation Applications Using the BHSD

In this section, we describe two segmentation applications designed to leverage the proposed BHSD.

Supervised Segmentation. Supervised multi-class segmentation refers to the classification of all individual pixels in an image into distinct classes, using segmentation mask annotations for supervision. For ICH, multi-class segmentation is clinically more significant while technically more challenging compared with foreground and background segmentation. Multi-class segmentation can more accurately identify and segment different types of ICH, which is important for diagnosis, treatment planning and prognostication.

However, multi-class segmentation also has many challenges since ICH come in different shapes, sizes, and densities, and multiple different types of bleeding may occur simultaneously. This task may therefore require more data than the foreground/background problem. We evaluate the performance of supervised segmentation methods under different conditions using the BHSD.

Semi-supervised with Pixel-Labeled and Unlabeled Data. Acquiring labeled medical imaging data can be a costly and challenging process, whereas unlabeled data is usually more obtainable. Annotating medical images also requires specialized domain expertise, which can pose a significant barrier to the widespread development of deep learning methods for clinical practice. Semi-supervised learning (SSL) addresses this challenge by using a small amount of labeled data and a large amount of unlabeled data for model training. Using the BHSD, we simulate the SSL scenario by discarding the image-level annotations to build an unlabeled dataset. Our approach combines data with pixel-level annotations and unlabeled data to evaluate the performance of SSL methods.

3   Experiments and Benchmarking Methods

To demonstrate the utility of the BHSD, we perform a series of segmentation experiments under different conditions and provide benchmarks for future model evaluations using this dataset.

Evaluation Metrics. Segmentation performance was evaluated using the Dice similarity coefficient (DSC) [4], which compares the similarity between the predicted and true segmentation.

Implementation Details. All experiments were performed on a single A6000 GPU. Following nnUnet [13], we first truncated the HU values of each scan using the range of [-40,120], and then normalized the truncated voxel values by subtracting 40 and dividing by 80. We randomly cropped sub-volumes of size [32,128,128] from CT scans as the input. Other parameters in different models retain the official default settings. In the BHSD, the 192 volumes were divided evenly into training and testing sets. The sets were balanced in terms of the number and types of bleeds, each containing 96 volumes. Furthermore, by verifying the original patient identifiers, no patient was contained in both sets.

3.1   Supervised 3D Segmentation

In Experiment 1, we conducted a comprehensive evaluation of state-of-the-art 3D semantic segmentation models using the BHSD dataset. We evaluate five 3D models with SOTA backbones designed for supervised semantic segmentation, namely UNETR [9], Swin UNETR [8,17], CoTr [26], nnFormer [22,28], and nnUNet [13] (Table 1).

Table 1. Benchmark 3D supervised segmentation performance using the BHSD.

	EDH	IPH	IVH	SAH	SDH	Mean
UNETR	1.64	28.28	22.08	4.36	3.63	11.99
Swin UNETR	2.53	34.18	29.28	10.07	8.43	16.89
CoTr	1.63	48.62	53.55	17.88	15.44	27.43
nnFormer	0.00	69.75	25.78	25.94	10.31	29.19
nnUnet3D	4.81	54.12	51.48	21.57	15.23	29.44

The results indicated that nnUnet achieved superior performance compared to other models with an average Dice of 29.44. However, it is important to acknowledge the limitation of the dataset, specifically the low occurrence of epidural hematoma *i.e.,* EDH class. Consequently, the segmentation performance for this class was considerably inferior to other classes. Further refinement and adaptation are necessary to enhance the segmentation performance for the EDH class, considering its limited representation in the dataset. Confusion matrix analysis may also allow identification of imbalances and biases in model predictions (Supplementary Fig. 1).

3.2   Incorporation of Scans with No Hemorrhage

In Experiment 2, we sought to enhance the model's performance through a gradual augmentation of the training set with negative samples within a supervised experimental setup. This also allowed us to address the issue of false positives. The model achieved optimal performance with 200 negative samples (Table 2), both in terms of hemorrhage segmentation performance and suppression of false positives.

This outcome suggests the presence of a balance point, where an insufficient number of negative samples hampers the model's ability to effectively learn the discriminating features, leading to inadequate suppression of false positives, while excessive negative samples may cause the model to overly focus on them, resulting in difficulties distinguishing between target and non-target categories (Supplementary Fig. 2). It is worth noting that selection of the optimal number of negative samples requires a comprehensive consideration of the model's recognition and generalization abilities, as well as the characteristics and requirements of the dataset. Future research can explore combinations of varying sample

Table 2. Incorporation of scans with (B) and without bleeding (NB). The false positive (FP) rate is also indicated.

Heading level	EDH	IPH	IVH	SAH	SDH	Mean	FP
96B	4.81	54.12	51.48	21.57	15.23	29.44	1.41
96B + 200NB	9.77	56.90	58.53	29.98	21.73	35.38	0.84
96B + 400NB	4.46	39.45	39.90	15.43	9.30	21.71	1.30
96B + 600NB	3.98	36.24	26.25	6.41	6.59	15.90	1.85
96B + 800NB	2.14	28.21	29.57	4.47	2.78	13.43	2.41

quantities to gain a deeper understanding and achieve more precise performance optimization.

3.3 Single Class, Multiple Models Versus Multiple Class, Single Model

The utilization of existing brain hemorrhage data for multi-class semantic segmentation necessitates training a single-class detection model and merging the prediction outcomes. To further investigate the advantages of the BHSD dataset compared to existing datasets, in Experiment 3 we performed separate training iterations on BHSD, focusing on one category at a time. We repeated this process five times, resulting in distinct models capable of recognizing individual categories. Subsequently, we combined the inference results from these five models. Through comparative analysis, we observed that the multi-class semantic method surpassed its predecessor in the fusion result (Table 3). However, it is important to note that the application of model fusion for single-category segmentation yielded considerable challenges, such as extensive overlapping and conflicting prediction outcomes (Supplementary Fig. 3). This finding underscores the ineffectiveness of the fusion approach for multi-class semantic segmentation.

Table 3. Multiple single class models versus single multi-class model.

Heading level	EDH	IPH	IVH	SAH	SDH	Mean
1class-EDH	5.15	-	-	-	-	-
1class-IPH	-	37.55	-	-	-	-
1class-IVH	-	-	23.60	-	-	-
1class-SAH	-	-	-	14.50	-	-
1class-SDH	-	-	-	-	17.59	
5class-Merge	1.25	3.42	2.57	2.43	2.16	19.68
5class-Single	4.81	54.12	51.48	21.57	15.23	29.44

3.4   Semi-supervised Segmentation

Experiment 4 was conducted with the aim of investigating the utilization of unlabelled data from clinical settings to enhance the performance of the dataset. Through the implementation of semi-supervised experiments, we sought to assess the potential for improvement. The unlabelled data introduced in this experiment were randomly sampled, with no available information regarding the health status or presence of hemorrhage. To evaluate this setting, four methods were applied to the BHSD, namely mean teacher [21], cross pseudo supervision (CPS) [2], entropy minimization [24], and interpolation consistency [23]. In these experiments, the training set of 96 volumes was supplemented by 500 unlabeled data. The same test set of 96 volumes was retained. For this task, we also merged hemorrhage labels to a single foreground mask.

The experimental findings revealed that the model's performance could indeed be further enhanced by employing an appropriate semi-supervised approach (Table 4). This observation highlights the efficacy of incorporating unlabelled data in clinical settings. Moreover, the merging of all categories served to accentuate the extent of performance improvement achieved through this approach. While there was mixed performance, we find that CPS improves on supervised learning and achieves 49.50% DSC in the binary segmentation task. The CPS method [2] uses pseudo-labels generated by one model to train another model, and then using this new model to generate new pseudo-labels and so forth, iterating this process to improve the accuracy of the pseudo-labelled data and improve the performance of the model segmentation.

Table 4. Semi-supervised performance based on nnUNet. To highlight the advantages of semi-supervision, we merged all the hemorrhage categories and report foreground and background semantic segmentation results.

Method	unlabeled samples	Dice
SupOnly	0	45.10 ± 0.21
Entropy Minimization	500	36.91 ± 0.16
Mean Teacher	500	44.63 ± 0.18
Interpolation Consistency	500	45.38 ± 0.33
Cross Pseudo Supervision	500	49.50 ± 0.19

4   Conclusion

We describe a 3D CT head dataset, the BHSD, for intracranial hemorrhage segmentation. This dataset includes a diverse mix of head scans with pixel-level and slice-level annotations, as well as scans with and without hemorrhage. To qualitatively and quantitatively scrutinize the characteristics of the BHSD, we compare popular SOTA models and diverse training techniques and draw three key insights from our benchmarking experiments. Firstly, the BHSD can significantly enhance the performance of SOTA models for multi-class segmentation of ICH. Multi-class segmentation significantly outperforms the fusion of single-class segmentation models. Secondly, incorporation of scans without hemorrhage can enhance segmentation performance. However, the right balance needs to be found. Third, the BHSD improves model performance using a semi-supervised approach even when the volumes are not annotated at the pixel-level. Hence, the BHSD is a valuable dataset for ICH segmentation models, providing an opportunity to study how segmentation tasks can make better use of unlabeled and weakly labeled data. This will in turn facilitate the development and validation of computer-aided diagnostic tools for clinical practice.

References

1. Auer, L.M., et al.: Endoscopic surgery versus medical treatment for spontaneous intracerebral hematoma: a randomized study. J. Neurosurg. **70**(4), 530–535 (1989)
2. Chen, X., Yuan, Y., Zeng, G., Wang, J.: Semi-supervised semantic segmentation with cross pseudo supervision. In: Proceedings of the IEEE/CVF Conference on Computer Vision and Pattern Recognition, pp. 2613–2622 (2021)
3. Chilamkurthy, S., et al.: Development and validation of deep learning algorithms for detection of critical findings in head CT scans (2018)
4. Dice, L.R.: Measures of the amount of ecologic association between species. Ecology **26**(3), 297–302 (1945)
5. Flanders, A.E., et al.: Construction of a machine learning dataset through collaboration: the RSNA 2019 brain CT hemorrhage challenge. Radiol. Artif. Intell. **2**(3), e190211 (2020)
6. Frontera, J.A., et al.: Prediction of symptomatic vasospasm after subarachnoid hemorrhage: the modified fisher scale. Neurosurgery **59**(1), 21–27 (2006)
7. Grønbæk, H., et al.: Liver cirrhosis, other liver diseases, and risk of hospitalisation for intracerebral haemorrhage: a danish population-based case-control study. BMC Gastroenterol. **8**, 1–6 (2008)
8. Hatamizadeh, A., Nath, V., Tang, Y., Yang, D., Roth, H.R., Xu, D.: Swin UNETR: swin transformers for semantic segmentation of brain tumors in MRI images. In: Crimi, A., Bakas, S. (eds.) MICCAI 2021. LNCS, vol. 12962, pp. 272–284. Springer, Cham (2022). https://doi.org/10.1007/978-3-031-08999-2_22
9. Hatamizadeh, A., et al.: UNETR: transformers for 3D medical image segmentation. In: Proceedings of the IEEE/CVF Winter Conference on Applications of Computer Vision, pp. 574–584 (2022)
10. Hemphill, J.C., III., et al.: Guidelines for the management of spontaneous intracerebral hemorrhage: a guideline for healthcare professionals from the american heart association/american stroke association. Stroke **46**(7), 2032–2060 (2015)
11. Howard, G., et al.: Risk factors for intracerebral hemorrhage: the reasons for geographic and racial differences in stroke (regards) study. Stroke **44**(5), 1282–1287 (2013)
12. Hssayeni, M., Croock, M., Salman, A., Al-khafaji, H., Yahya, Z., Ghoraani, B.: Computed tomography images for intracranial hemorrhage detection and segmentation. Intracranial hemorrhage segmentation using a deep convolutional model. Data **5**(1), 14 (2020)

13. Isensee, F., Jaeger, P.F., Kohl, S.A., Petersen, J., Maier-Hein, K.H.: nnU-Net: a self-configuring method for deep learning-based biomedical image segmentation. Nat. Methods **18**(2), 203–211 (2021)

14. Larobina, M., Murino, L.: Medical image file formats. J. Digit. Imaging **27**, 200–206 (2014)

15. Lee, H., Kim, M., Do, S.: Practical window setting optimization for medical image deep learning. arXiv preprint arXiv:1812.00572 (2018)

16. Li, X., et al.: The state-of-the-art 3D anisotropic intracranial hemorrhage segmentation on non-contrast head CT: the instance challenge. arXiv preprint arXiv:2301.03281 (2023)

17. Liu, Z., et al.: Swin transformer: hierarchical vision transformer using shifted windows. In: Proceedings of the IEEE/CVF International Conference on Computer Vision, pp. 10012–10022 (2021)

18. McCarron, M.O., Nicoll, J.A., Ironside, J.W., Love, S., Alberts, M.J., Bone, I.: Cerebral amyloid angiopathy-related hemorrhage: interaction of apoe $\varepsilon 2$ with putative clinical risk factors. Stroke **30**(8), 1643–1646 (1999)

19. Reis, E.P., et al.: Brain hemorrhage extended (BHX): bounding box extrapolation from thick to thin slice CT images. PhysioNe **101**(23), e215-20 (2020)

20. Steiner, T., et al.: European stroke organisation (ESO) guidelines for the management of spontaneous intracerebral hemorrhage. Int. J. Stroke **9**(7), 840–855 (2014)

21. Tarvainen, A., Valpola, H.: Mean teachers are better role models: weight-averaged consistency targets improve semi-supervised deep learning results. In: Advances in Neural Information Processing Systems, vol. 30 (2017)

22. Vaswani, A., et al.: Attention is all you need. In: Advances in Neural Information Processing Systems, vol. 30 (2017)

23. Verma, V., et al.: Interpolation consistency training for semi-supervised learning. Neural Netw. **145**, 90–106 (2022)

24. Vu, T.H., Jain, H., Bucher, M., Cord, M., Pérez, P.: Advent: adversarial entropy minimization for domain adaptation in semantic segmentation. In: Proceedings of the IEEE/CVF Conference on Computer Vision and Pattern Recognition, pp. 2517–2526 (2019)

25. Wang, X., et al.: A deep learning algorithm for automatic detection and classification of acute intracranial hemorrhages in head CT scans. NeuroImage Clin. **32**, 102785 (2021)

26. Xie, Y., Zhang, J., Shen, C., Xia, Y.: CoTr: efficiently bridging CNN and transformer for 3D medical image segmentation. In: de Bruijne, M., et al. (eds.) MICCAI 2021. LNCS, vol. 12903, pp. 171–180. Springer, Cham (2021). https://doi.org/10.1007/978-3-030-87199-4_16

27. Yushkevich, P.A., Gerig, G.: ITK-SNAP: an intractive medical image segmentation tool to meet the need for expert-guided segmentation of complex medical images. IEEE Pulse **8**(4), 54–57 (2017)

28. Zhou, H.Y., Guo, J., Zhang, Y., Yu, L., Wang, L., Yu, Y.: nnFormer: interleaved transformer for volumetric segmentation. arXiv preprint arXiv:2109.03201 (2021)

Contrastive Learning-Based Breast Tumor Segmentation in DCE-MRI

Shanshan Guo[1,2,3], Jiadong Zhang[1], Dongdong Gu[2], Fei Gao[3], Yiqiang Zhan[2], Zhong Xue[2(✉)], and Dinggang Shen[1,2,4(✉)]

[1] School of Biomedical Engineering, ShanghaiTech University,
Shanghai 201210, China
[2] Shanghai United Imaging Intelligence Co., Ltd., Shanghai 200230, China
zhong.xue@uii-ai.com, dgshen@shanghaitech.edu.cn
[3] School of Computer Science and Technology, ShanghaiTech University,
Shanghai 201210, China
[4] Shanghai Clinical Research and Trial Center, Shanghai 201210, China

Abstract. Precise and automated segmentation of tumors from breast dynamic contrast-enhanced magnetic resonance images (DCE-MRI) is crucial for obtaining quantitative morphological and functional information, thereby assisting subsequent diagnosis and treatment. However, many existing methods mainly focus on features within tumor regions and neglect enhanced background tissues, leading to the potential over-segmentation problem. To better distinguish tumor tissues from complex background structures (e.g., enhanced vessels), we propose a novel approach based on contrastive feature learning. Our method involves pre-training a highly sensitive encoder using contrastive learning, where tumor and background patches are utilized as paired positive-negative samples, to emphasize tumor tissues and to enhance their discriminative features. Furthermore, the well-trained encoder is employed for accurate tumor segmentation by using a feature fusion module in a global-to-local manner. Through extensive validations using a large dataset of breast DCE-MRI scans, our proposed model demonstrates superior segmentation performance, effectively reducing over-segmentation on enhanced tissue regions as expected.

Keywords: Breast tumor · Contrastive learning · Segmentation

1 Introduction

Breast cancer is the most common cancer among women and remains a leading cause of cancer-related fatalities [11]. Recent studies have shown that diagnosis and treatment of breast cancers at early stages can greatly improve the survival rate [10]. Clinicians commonly use breast X-ray (mammography) [8], ultrasound [5], and MRI [7] as diagnostic imaging modalities. Dynamic contrast-enhanced magnetic resonance imaging (DCE-MRI) is a highly sensitive modality for soft

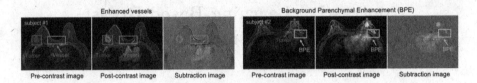

Fig. 1. Breast DCE-MRI of two typical cases: each case includes pre-contrast, post-contrast, and their subtraction images. The enhanced tumor is highlighted in red, vessels in green, and background parenchymal enhancement (BPE) in orange. (Color figure online)

tissue that provides comprehensive information of the breast and is increasingly used in clinical [14]. The standard breast DCE-MRI collects a series of T1-weighted MRIs before and after contrast agent injection [16]. As illustrated in Fig. 1, the pre-contrast images typically exhibit similar intensity contrasts between tumors and other tissues, while post-contrast images show higher signal contrasts for tumor regions relative to the background. Additionally, the subtraction images (i.e., the difference between post-contrast and pre-contrast images after proper registration) provide better tumor visualization. However, physicians still cannot accurately distinguish tumors from other enhanced tissues, such as vessels and glands, by merely detecting intensity changes. Consequently, accurate and automated segmentation of tumors from DCE-MRI remains a challenge.

Based on different feature extraction modes existing breast tumor segmentation methods can be categorized into two classes: hand-crafted and deep learning methods. For the first category, Chang et al. [1] extracts focal tumor breast lesions using kinetic and morphological features from DCE-MRIs. Gubern et al. [4] employs blob and relative enhancement voxel features to localize lesions. However, these methods usually conduct feature extraction and perform training separately, and could lead to poor segmentation performance when lesion boundaries appear blurry or with similar intensities compared to surrounding tissues. The second category involves deep learning, which has gained popularity in lesion segmentation due to its ability to leverage end-to-end learning and automatic feature extraction mechanisms. Zhang et al. [17] applies a multi-stage fully convolutional network (FCN) to perform breast tumor segmentation in a coarse-to-fine framework. Wang et al. [12] uses a combination of 2D and 3D networks with a multi-scale feature extraction module for breast DCE-MRI segmentation. However, most of the existing works primarily focus on tumor-enhanced region extraction and ignore the fact that other tissues in the background may also be enhanced, which potentially leads to over-segmentation. To address this issue and to improve the specificity of tumor segmentation while maintaining high sensitivity, we propose a tumor-background contrastive feature learning framework for breast tumor DCE-MRI segmentation.

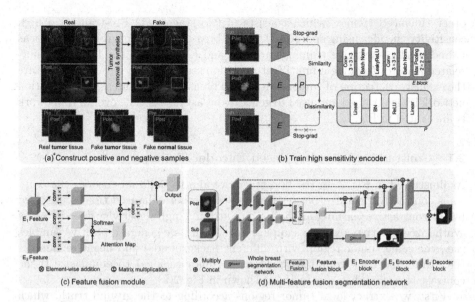

Fig. 2. The proposed breast tumor segmentation framework. (a) Constructing positive and negative samples; (b) pre-training a high sensitivity encoder by learning from contrast features; (c) details of the feature fusion block; (d) overall segmentation network for breast tumor segmentation.

The main idea is to bring tumor-related features closer and to push the features of enhanced tissues in the background further away. Specifically, we adopt the SimSiam [2] framework of contrastive learning and build paired positive-negative samples of tumor-background to pre-train a high sensitivity encoder. Different from traditional augmentation methods such as rotation or cropping operations, we generate fake breast DCE-MRIs by removing and synthesizing tumor-background tissues, including (1) the same tumor region with different backgrounds and (2) the same background with different tumor regions, as paired positive-negative samples. In this way, the pre-trained encoder can be highly sensitive to distinguish features from tumors with other enhanced tissues. We incorporate this encoder into the downstream segmentation network using the feature fusion mechanism to guide feature extraction for tumor segmentation. In experiments, we evaluate the performance of our framework on 385 DCE-MRIs and obtain better quantitative results compared with other breast tumor segmentation methods. Particularly, the segmentation metrics like Dice are significantly improved. Over-segmentation on the background has been effectively eliminated as expected.

2   Method

The tumor-background contrastive feature extraction framework aims to improve segmentation performance by effectively distinguishing tumors from other enhanced tissues, which consists of two stages: (1) Pre-training a high sensitivity encoder using specially designed breast tumor-background patches as paired positive-negative training samples, and (2) utilizing the learned tumor-related features to guide a multi-feature fusion segmentation network. To ensure the exclusive detection of tumors within the breast, a whole breast segmentation network is also trained and used to generate breast masks. The overall framework is illustrated in Fig. 2.

2.1   Contrastive Learning-Based Encoder Training

As illustrated in Fig. 2. In the first stage, we aim to build an encoder that captures contrastive features between tumors and other enhanced tissues. Inspired by the contrastive learning method, we pre-train a high sensitivity encoder using synthesized positive-negative samples. Specifically, we generate positive samples (i.e., the same tumor region with different backgrounds) and negative samples (i.e., the same background with different tumor regions) to emphasize tissue contrast to accomplish our tasks, as shown in Fig. 2(a).

First, we extract local tumor regions according to the ground truth, which are then mirrored to the opposite normal breast so that they are well embedded in the normal background with a Gaussian filter. Similarly, normal tissues in the healthy breast are removed and patched onto the tumor position in the opposite breast to build fake normal tissues. Therefore, for each case we obtain three patches: 1) real tumor tissue with real background (red), 2) the same tumor tissue with a different background (pink), and 3) the same background with fake normal tissue (green). Considering bilateral breast cancer is uncommon, we exclude those rare cases to build the training samples.

Inspired by the contrastive learning framework [2], three patches are inputted into the same encoder E consisting of a backbone (e.g., RU-Net [15]) and a projection MLP to learn the tumor-related features, as illustrated in Fig. 2(b). Compared with real tumor tissues, fake tumor tissues have the same tumor region but a totally different background with other enhanced tissues. To emphasize the tumor regions, features from two patches should be closer. On the other hand, fake normal tissues have the same background but without corresponding tumor tissue. Thus, to reduce the influence caused by the background (especially the enhanced non-tumor region), the features from real tumor patches and fake normal patches should be pushed away from each other. The predictive MLP layer P optimizes the representation of learned features, and a stop-gradient strategy is added on both sides to prevent model collapse. The loss function Eq. (1) is a combination of the similarity measures between real and fake tumor patches (L_{pos}) and those between real tumor and fake normal patches (L_{neg}):

$$\mathcal{L} = \mathcal{L}_{pos} + \mathcal{L}_{neg}$$
$$= \frac{1}{2}((D(P_2, stopgrad(E_1)) + D(P_1, stopgrad(E_2)))$$
$$- \frac{1}{2}((D(P_2, stopgrad(E_3)) + D(P_3, stopgrad(E_2))), \tag{1}$$

where D is the negative cosine similarity, ranging from -1 to 1. The encoder extracts features from fake tumor patches, real tumor patches, and fake normal patches, denoted as E_1, E_2 and E_3, which are then projected using P, resulting in corresponding projection vectors P_1, P_2 and P_3.

2.2 Multi-feature Fusion Segmentation Network

After well-training the encoder, we incorporate it into the downstream tumor segmentation task shown in Fig. 2(d). Rather than directly fine-tuning or fixing the parameters of the encoder, we employ a compromise approach to guide tumor-related feature learning. Specifically, we replicate the well-trained encoder, one with fixed parameters (E_1) and the other with trainable parameters (E_2). With the feature fusion module (detailed in Fig. 2(c)), the network can better capture features in a global-to-local manner by using tumor-background contrastive feature extraction.

The feature fusion module is described as follows. Given the features from two encoders, the $(1 \times 1 \times 1)$ convolutions and multiplication operations are used to calculate the similarity matrices among their feature maps, which further sever as the attention maps focusing on tumor-related features. In this way, global features from E_1 can guide local feature extraction and improve segmentation performance. We adopt the Dice loss (L_{DICE}) to solve the imbalance problem. In addition, the binary cross-entropy loss (L_{BCE}) is used to verify the segmentation results at the voxel level. The segmentation network is trained end-to-end using the overall loss function Eq. (2) as follows:

$$\mathcal{L} = \lambda \mathcal{L}_{DICE} + \mathcal{L}_{BCE} = \lambda(1 - \frac{2|F(pos, sub) \cap G|}{|F(pos, sub)| \cup |G|})$$
$$- G \cdot log(F(pos, sub)) - (1 - G) \cdot log(1 - F(pos, sub)), \tag{2}$$

where λ is a weighting coefficient to trade-off between the two terms and is set to 5. pos and sub denote the post-contrast image and the corresponding subtraction image, respectively. G is the binary ground truth label mask, and F represents the probability map predicted by the segmentation network. Finally, it is worth noting that we also localize the whole breast with a segmentation network S^{Breast} to improve tumor segmentation accuracy by removing over-segmentation on non-breast regions.

3 Experiment and Result

Dataset and Implementation Details. We collected 385 subjects with breast tumors (including Bi-rads 4 and 5), who underwent DCE-MRI scans using 3T Philips devices, and each case has one pre-contrast MRI and multiple post-contrast MRIs. The sampling interval of DCE-MRI is 1 min. Two experienced radiologists annotated the breast tumors based on the second-phase MRI as the ground truth, considering this phase has the best tissue enhancement. The slice

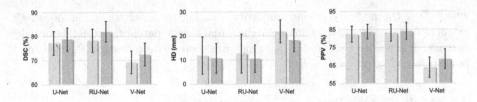

Fig. 3. Comparison of DSC (%), PPV (%) and HD (mm) using different encoder backbones (i.e., U-Net [3], RU-Net [15], and V-Net [6]). Yellow: without pre-training, blue: with pre-training. (Color figure online)

thickness varies from 0.8 mm to 1.2 mm, and the inter-slice resolution varies from 0.5 mm to 1.0 mm. To maintain consistency for all experiments, we partitioned the data by randomly selecting 300 cases for training and 85 cases for testing.

We choose RU-Net [15] as the backbone for tumor segmentation as well as breast segmentation. 3D patches with size of $128 \times 128 \times 32$ mm^3 are sampled for network training, and for testing the whole 3D images are processed with sliding patches. The initial learning rates of the whole breast segmentation model, high sensitivity encoder, and breast tumor segmentation model are set to 0.005, 0.001, and 0.005 respectively, which are decayed to half every 50 epochs.

Pre-train Encoder Performance. We conducted rigorous ablation experiments to evaluate the efficacy of pre-trained encoders for segmentation networks. We pre-trained encoders using various baselines (e.g., U-Net [3], RU-Net [15], and V-Net [6]) and compared them with and without pre-training for tumor segmentation. The results of Dice similarity coefficient (DSC), Positive predictive value (PPV), and Hausdorff distance (HD) metrics are shown in Fig. 3. It can be seen that overall RU-Net outperforms other networks, and the results with pre-training also perform better.

Furthermore, we compared our pre-training strategy with the RU-Net [15] backbone, and a traditional contrastive learning method SimSiam [2] with the same backbone. In SimSiam each subject·is augmented with multiple methods like blurring, horizontal and vertical flipping, Gaussian noising, and scaling at different probabilities. SimSiam contains only the positive sample branches and randomly selects two positive samples as inputs. The trained encoder was then applied to segment breast tumors. As illustrated in Table 1, compared to direct tumor segmentation without pre-training, SimSiam can slightly improve segmentation performance. Additionally, benefiting from the pre-trained encoder with specifically constructed positive and negative tumor samples that highlight the contrast features between tumor and background regions, our method demonstrated more notable improvement such as DSC by 3.7%.

Tumor Segmentation Performance. In order to improve the accuracy of breast tumor segmentation, we propose a specialized training strategy that incorporates the pre-trained encoder and the feature fusion module to enhance the

Table 1. Quantitative comparison of our pre-trained method with traditional contrastive learning method (e.g., SimSiam [2]) and RU-Net [15] without pre-training, in terms of DSC (%), TPR (%), PPV and HD (mm).

Models	DSC (%) ↑	TPR (%) ↑	PPV (%) ↑	HD (mm) ↓
RU-Net	78.23 ± 4.86	80.11 ± 4.55	83.25 ± 4.62	12.71 ± 8.06
RU-Net + SimSiam	79.44 ± 5.07	78.14 ± 5.59	83.94 ± 4.82	11.62 ± 7.23
RU-Net + Our Pre-trained	$\mathbf{81.93 \pm 4.31}$	$\mathbf{83.41 \pm 3.78}$	$\mathbf{84.12 \pm 4.75}$	$\mathbf{10.53 \pm 5.64}$

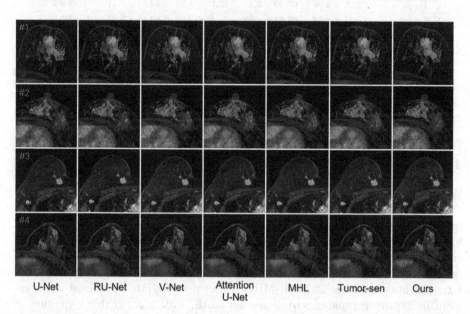

| U-Net | RU-Net | V-Net | Attention U-Net | MHL | Tumor-sen | Ours |

Fig. 4. Visual comparison of the segmentation results by using different methods on four typical samples. Each row shows a segmentation results for a subject using different methods. Green illustrates the boundaries of segmentation results, red means the ground truth. (Color figure online)

guidance provided by the pre-trained encoder. To evaluate the effectiveness of our proposed framework, we compared its performance with classic methods (e.g., U-Net [3], RU-Net [15], V-Net [6] and Attention U-Net (Att U-Net) [9]), as well as with other state-of-the-art breast tumor segmentation models (e.g., Tumor-sen [13], MHL [17]). Five typical metrics are used to evaluate the performance of segmentation, including DSC, Intersection over union (IOU), True positive rate (TPR), PPV, and HD.

Quantitative results are illustrated in Table 2, and it can be seen that our method yielded significantly better results than others for DSC (i.e., increases of 6.6% compared with RU-Net, and p-value = 0.02). Furthermore, smaller HD metrics indicate that the predicted boundaries and surfaces generated using the pre-trained high sensitivity encoder and feature fusion module are closer to the ground truth. Figure 4 illustrates four typical results, demonstrating a noticeable reduction in over-segmentation, and that our results are more consistent with the ground truth.

Table 2. Comparison of tumor segmentation results using different methods (Bold denotes highest)

Models	DSC (%) ↑	IOU (%) ↑	TPR (%) ↑	PPV (%) ↑	HD (*mm*) ↓
U-Net [3]	77.26 ± 4.96	67.39 ± 5.15	75.79 ± 4.70	82.51 ± 4.47	11.92 ± 7.76
V-Net [6]	69.11 ± 4.67	60.64 ± 4.93	74.43 ± 3.97	63.86 ± 5.67	21.87 ± 4.75
RU-Net [15]	78.23 ± 4.86	68.68 ± 5.17	80.11 ± 4.55	83.25 ± 4.62	12.71 ± 8.06
Att U-Net [9]	74.09 ± 5.22	66.71 ± 5.43	80.25 ± 4.97	79.60 ± 5.51	11.86 ± 4.28
Tumor-sen [13]	75.40 ± 5.61	66.03 ± 5.75	77.32 ± 5.61	83.96 ± 4.89	12.28 ± 7.42
MHL [17]	77.80 ± 4.93	67.99 ± 5.05	79.23 ± 4.68	82.95 ± 4.99	10.69 ± 5.21
Ours	**84.79 ± 3.17**	**75.85 ± 3.82**	**84.01 ± 3.02**	**89.28 ± 3.33**	**6.39 ± 3.98**

4   Conclusion

We have proposed a novel pre-training strategy for breast tumor segmentation based on tumor-background contrastive feature learning. Specifically, tissue patches are generated as paired positive-negative samples and used to train the high sensitivity encoder, which can easily distinguish tumor regions from other enhanced tissues. The encoder is further applied to the tumor segmentation network by incorporating the feature fusion module for accurate segmentation. Experiments on the breast DCE-MRI dataset demonstrate significant improvement in metrics compared to the ground truth, with a noticeable reduction in over-segmentation of tumor regions, indicating the advantages of the proposed training strategy.

Acknowledgements. This work was supported in part by The Key R&D Program of Guangdong Province, China (grant number 2021B0101420006), National Natural Science Foundation of China (grant number 62131015), and Science and Technology Commission of Shanghai Municipality (STCSM) (grant number 21010502600).

References

1. Chang, Y., Huang, Y., Huang, C., et al.: Computerized breast lesions detection using kinetic and morphologic analysis for dynamic contrast-enhanced MRI. Magn. Reson. Imaging **32**(5), 514–522 (2014)
2. Chen, X., He, K.: Exploring simple siamese representation learning. In: CVPR, pp. 15750–15758 (2021)
3. Çiçek, Ö., Abdulkadir, A., Lienkamp, S.S., Brox, T., Ronneberger, O.: 3D U-Net: learning dense volumetric segmentation from sparse annotation. In: Ourselin, S., Joskowicz, L., Sabuncu, M.R., Unal, G., Wells, W. (eds.) MICCAI 2016. LNCS, vol. 9901, pp. 424–432. Springer, Cham (2016). https://doi.org/10.1007/978-3-319-46723-8_49

4. Gubern-Mérida, A., Martí, R., Melendez, J., et al.: Automated localization of breast cancer in DCE-MRI. Med. Image Anal. **20**(1), 265–274 (2015)
5. Guo, R., Lu, G., Qin, B., Fei, B.: Ultrasound imaging technologies for breast cancer detection and management: a review. Ultrasound Med. Biol. **44**(1), 37–70 (2018)
6. Milletari, F., Navab, N., Ahmadi, S.: V-Net: fully convolutional neural networks for volumetric medical image segmentation. In: 3DV, pp. 565–571. IEEE (2016)
7. Morrow, M., Waters, J., Morris, E.: MRI for breast cancer screening, diagnosis, and treatment. Lancet **378**(9805), 1804–1811 (2011)
8. Nazari, S.S., Mukherjee, P.: An overview of mammographic density and its association with breast cancer. Breast Cancer **25**, 259–267 (2018)
9. Oktay, O., et al.: Attention U-net: learning where to look for the pancreas. arXiv preprint arXiv:1804.03999 (2018)
10. Scully, O., Bay, B., Yip, G., et al.: Breast cancer metastasis. Cancer Genomics Proteomics **9**, 311–320 (2012)
11. Sharma, G., Dave, R., Sanadya, J., et al.: Various types and management of breast cancer: an overview. J. Adv. Pharm. Technol. Res. **1**(2), 109 (2010)
12. Wang, H., Cao, J., Feng, J., Xie, Y., Yang, D., Chen, B.: Mixed 2D and 3D convolutional network with multi-scale context for lesion segmentation in breast DCE-MRI. Biomed. Signal Process. Control **68**, 102607 (2021)
13. Wang, S., Sun, K., Wang, L., et al.: Breast tumor segmentation in DCE-MRI with tumor sensitive synthesis. IEEE Trans. Neural Netw. Learn Syst. **7** (2021)
14. Yankeelov, T.E., et al.: Integration of quantitative DCE-MRI and ADC mapping to monitor treatment response in human breast cancer: initial results. Magn. Reson. Imaging **25**(1), 1–13 (2007)
15. Yu, L., Yang, X., Chen, H., et al.: Volumetric ConvNets with mixed residual connections for automated prostate segmentation from 3D MR images. In: AAAI AI (2017)
16. Zhang, J., et al.: A robust and efficient AI assistant for breast tumor segmentation from DCE-MRI via a spatial-temporal framework. Patterns (2023)
17. Zhang, J., Saha, A., Zhu, Z., Mazurowski, M.A.: Hierarchical convolutional neural networks for segmentation of breast tumors in MRI with application to radiogenomics. IEEE Trans. Med. Imaging **38**(2), 435–447 (2018)

FFPN: Fourier Feature Pyramid Network for Ultrasound Image Segmentation

Chaoyu Chen[1,2,3], Xin Yang[1,2,3], Rusi Chen[1,2,3], Junxuan Yu[1,2,3],
Liwei Du[1,2,3], Jian Wang[4], Xindi Hu[5], Yan Cao[5], Yingying Liu[6],
and Dong Ni[1,2,3]([✉])

[1] National -Regional Key Technology Engineering Laboratory for Medical
Ultrasound, School of Biomedical Engineering, Health Science Center,
Shenzhen University, Shenzhen, China
nidong@szu.edu.cn
[2] Medical Ultrasound Image Computing (MUSIC) Lab, Shenzhen University,
Shenzhen, China
[3] Marshall Laboratory of Biomedical Engineering, Shenzhen University, Shenzhen,
China
[4] School of Biomedical Engineering and Informatics, Nanjing Medical University,
Nanjing, China
[5] Shenzhen RayShape Medical Technology Co., Ltd., Shenzhen, China
[6] Shenzhen People's Hospital, Second Clinical Medical College of Jinan University,
Shenzhen, China

Abstract. Ultrasound (US) image segmentation is an active research
area that requires real-time and highly accurate analysis in many scenarios. The detect-to-segment (DTS) frameworks have been recently proposed to balance accuracy and efficiency. However, existing approaches
may suffer from inadequate contour encoding or fail to effectively leverage
the encoded results. In this paper, we introduce a novel Fourier-anchor-
based DTS framework called Fourier Feature Pyramid Network (FFPN)
to address the aforementioned issues. The contributions of this paper
are two fold. First, the FFPN utilizes Fourier Descriptors to adequately
encode contours. Specifically, it maps Fourier series with similar ampli-
tudes and frequencies into the same layer of the feature map, thereby
effectively utilizing the encoded Fourier information. Second, we propose
a Contour Sampling Refinement (CSR) module based on the contour proposals and refined features produced by the FFPN. This module extracts
rich features around the predicted contours to further capture detailed
information and refine the contours. Extensive experimental results on
three large and challenging datasets demonstrate that our method outperforms other DTS methods in terms of accuracy and efficiency. Furthermore, our framework can generalize well to other detection or segmentation tasks.

C. Chen and X. Yang—Contribute equally to this work.

X. Cao et al. (Eds.): MLMI 2023, LNCS 14348, pp. 166–175, 2024.
https://doi.org/10.1007/978-3-031-45673-2_17

1 Introduction

Recently, real-time, accurate and low-resource image segmentation methods have gained wide attention in the field of ultrasound (US) image analysis. These methods provide the basis for many clinical tasks, e.g. structure recognition [11], bio-metric measurement [13] and surgical navigation [2]. Figure 1 illustrates the segmentation tasks we have accomplished in this paper, including the apical two-chambers heart(2CH) dataset, Camus dataset [8] and Fetal Head (FH) dataset.

Fig. 1. Segmentation tasks in this paper. (a)-(b): 2CH data. (c)-(d): FH images. (e): Camus data.

Numerous segmentation methods based on deep learning have been proposed, most of which mainly rely on U-shaped networks, such as U-Net [12], nnU-Net [6], and SwinU-Net [1]. The excessive skip-connections and upsampling operations in these methods make the model sacrifices efficiency and resources to ensure accuracy. DeeplabV3 [3] is another commonly used segmentation framework, and it also sacrifices efficiency due to the design of multiscale embedding. In addition, these methods all face the issue of false positive segmentation due to the blurred boundaries in US images. Thus, Mask R-CNN [5] uses the bounding box (b-box) as a constraint on the segmentation region, which reduces the false positive. However, the serial scheme limits its efficiency and excessive reliance on the b-box's output also affects its segmentation performance.

In order to balance resource consumption, efficiency and performance, detect-to-segment (DTS) framework has received significant attention in recent years. The core idea of DTS is to transform the pixel classification problem of an image into the regression and classification problem for each point on the feature map. By this way, each point on the feature map can predict multiple contour proposals. PolarMask [16] used a polar coordinate system for image segmentation, and it is difficult to process the situation that a ray intersects the object multiple times at a special direction. Point-Set [15] uniformly sampled several ordered points to represent the contour. PolySnake [4] designed a multiscale contour refinement module to refine the initial coarse contour. While these sample-based encoding methods are capable of representing contours, many of the points within them do not contain valid information and there is a lack of correlation between points. Consequently, these encoding methods may not be suitable for segmentation tasks. To enhance the encoding results of complex contours, Ellipse Fourier Description [7] (EFD) scheme (Fig. 2 (a)) entered the vision of many researchers [14,17,18]. FCENet [18] and FANet [17] represent the text instance in the Fourier domain, allowing for fast and accurate representation of complex contours. CPN [14] revisited the EFD scheme to represent cell instance segmentation and achieved promising results. Although these EFD-based methods achieves high performance comparable to Pixel-based methods, they only consider the scale characteristics but ignore the frequency characteristics of the Fourier expansion, as shown in Fig. 2 (b), resulting in sub-optimal performance.

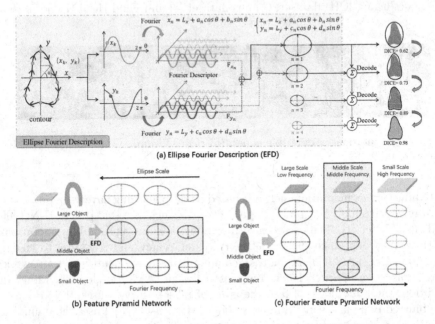

Fig. 2. (a) Detailed illustration of EFD scheme. (b) Previous methods only consider the scale and ignore the frequency characteristics. (c) Our FFPN focus on both the scale and frequency between different levels of Fourier series.

In this study, we revisited the EFD from another perspective, as shown in Fig. 2 (b) and (c). When the closed contour is expanded with Fourier, we focus on both the scale and frequency among different levels of Fourier series, and ingeniously incorporating it into the FPN [10] to devise a novel Fourier-anchor-based framework, named Fourier Feature Pyramid Network (FFPN). Our contributions are two fold. *First*, we design FFPN to assign Fourier series with similar information to the same feature map for collective learning (Fig. 2 (c)). This approach enhances the consistency of feature representation and improves the model's ability to predict encoded results with better accuracy. *Second*, considering the complexity and blurring of object contours in US images, we propose a Contour Sampling Refinement (CSR) module to further improve the model's ability to fit them. Specifically, we aggregate features at different scales in FFPN and extracted the relevant features around the contours on the feature map to rectify the original contour. Experimental results demonstrate that FFPN can stably outperform DTS competitors. Furthermore, our proposed FFPN is promising to generalize to more detection or segmentation tasks.

2   Methodology

Figure 3 is the overview of our FFPN framework. Given an image, we first use the backbone with FPN to extract the pyramidal features. Then these features are unsampled to obtain Fourier pyramidal features. Next, we feed Fourier pyramidal features into different detector heads to generate different levels of Fourier series offsets, location offsets and classification scores. EFD decodes these predictions to the contour proposals. In CSR, the Fourier pyramid features are concatenated to generate the refinement features. Finally, the contour proposals and refinement features together are used to produce the final results.

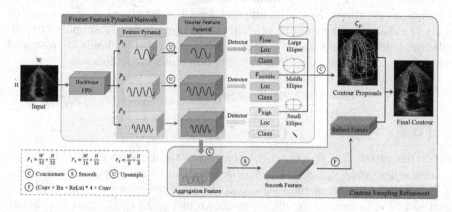

Fig. 3. The overall pipeline of our proposed FFPN. In yellow blocks, F denotes Fourier serie offset, Loc denotes location offset, $Class$ denotes classification score. (Color figure online)

2.1   Fourier Feature Pyramid Network(FFPN)

Inspired by EFD (Fig. 2 (a)), the contour can be described as the Fourier series:

$$x_N(t) = L_x + \sum_{n=1}^{N}(a_n sin(\tfrac{2n\pi t}{T}) + b_n cos(\tfrac{2n\pi t}{T})),$$

$$y_N(t) = L_y + \sum_{n=1}^{N}(c_n sin(\tfrac{2n\pi t}{T}) + d_n cos(\tfrac{2n\pi t}{T})),$$

$$(1)$$

where $(x_N(t), y_N(t))$ is the t-th sampled point on the contour, and the number of sampled points are set to T, e.g. $t \in [0, T]$. N is the number of Fourier series expansions. (L_x, L_y) indicates the coordinates of the center point of the contour. a_n, b_n denote the parameters obtained by Fourier coding of the x-coordinates of all contour points, and c_n, d_n denote the corresponding parameters in the y-coordinates. Thus, the goal of FFPN is to accurately predict the $4N + 2$ parameters to represent the contour ($N = 7$ by default).

According to EFD (Fig. 2 (a)), the low-level Fourier series represent the contour's low-frequency information and main scales, while the high-level Fourier series capture the contour's high-frequency information and shape details. This is consistent with the extracted pyramidal features, where the low-level features (P_3) contain more detailed information, while the high-level features (P_1) capture semantics. Therefore, the proposed FFPN effectively aggregates Fourier series of similar scales and frequencies into the same feature map (as illustrated in Fig. 2 (c)). Then, different level features of feature pyramid are fed into different detector heads with the same architecture (three sibling 3×3 Conv-BN-ReLUs followed by a 3×3 Conv) to generate Fourier offsets, location offsets and classification scores (F, Loc and $Class$ with yellow blocks in Fig. 3.). The up-sampling operations on the P_1 and P_2 features are only to align the scale and simplify subsequent operations. Next, the different level Fourier offsets are concatenated along the channel dimension to predict contour proposals. To simplify the calculation, location offsets and classification scores at different levels are averaged along the channel dimension. The learning objectives are as follows:

$$\begin{cases} \Delta F_{ai} = \frac{G_{ai} - A_{ai}}{E_{xi}} \\ \Delta F_{bi} = \frac{G_{bi} - A_{bi}}{E_{xi}} \end{cases} \quad \begin{cases} \Delta F_{ci} = \frac{G_{ci} - A_{ci}}{E_{yi}} \\ \Delta F_{di} = \frac{G_{di} - A_{di}}{E_{yi}} \end{cases} \quad \begin{cases} \Delta L_x = \frac{G_{Lx} - A_{Lx}}{E_{x1}} \\ \Delta L_y = \frac{G_{Ly} - A_{Ly}}{E_{y1}}, \end{cases} \quad (2)$$

where the tuple $(G_{ai}, G_{bi}, G_{ci}, G_{di})$ represents the Ground truth(GT) of the Fourier series, and $(A_{ai}, A_{bi}, A_{ci}, A_{di})$ denotes the Fourier series of anchor. $i \in [1, N]$ denotes the i-th of the Fourier expansion. The Fourier expansion of each level can be expressed as an ellipse, as shown in Fig. 2 (a). E_{xi} and E_{yi} denote the width and height of the ellipse at level i, respectively. (G_{Lx}, G_{Ly}) is the center point coordinates of the GT contour, and (A_{Lx}, A_{Ly}) is the center point coordinates of the anchor. $(\Delta F_{ai}, \Delta F_{bi}, \Delta F_{ci}, \Delta F_{di}, \Delta L_x, \Delta L_y)$ are the learning objectives. Through the encoding operation of Eq. 2, we normalize the amplitude of Fourier series and center point coordinates to the same scale, simplifying the learning difficulty. The loss function is defined as follows:

$$\mathcal{L} = \mathcal{L}_{Loc} + \mathcal{L}_{Fou} + \mathcal{L}_{Con} + \mathcal{L}_{Cls} \quad (3)$$

The loss function about Location and Fourier are Smooth L1 Loss, and the classification loss is $\mathcal{L}_{Cls} = \alpha BCELoss + \beta FocalLoss$, where $\alpha = 0.25$ and $\beta = 0.75$. The contour loss is defined as $\mathcal{L}_{Con} = 1 - PolarIoU * BoxIoU$, where the $PolarIoU$ follows [16] and the $BoxIoU$ is defined as the IoU between outline bounding boxes. The method to calculate IoU in this paper is defined as $IoU = PolarIoU * BoxIoU$, which is a simple but effective way.

2.2   Contour Sampling Refinement (CSR) Module

Considering the complexity of contour representation, we adopt the two-stage strategy inspired from the detection frameworks [5,10]. Unlike CPN [14] where the contour is directly optimized to destroy its smoothness or FANet [17] where the Fourier series of the contour needs to be refined iteratively, we refine once the Fourier series of the aggregation averaged contour. This strategy improves the model accuracy while ensures the boundary smoothness. Specifically, as shown in Fig. 3, we fuse pyramidal features as the refined-feature. And the top-n ($n = 20$) outputs of FFPN are the contour proposals (C_p) to represent the same object. Thus, the refined-feature and the C_p are the input of CSR module.

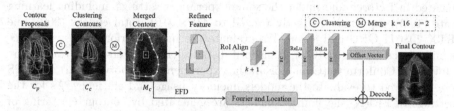

Fig. 4. Illustration of our proposed Contour Sampling Refinement (CSR) module.

Figure 4 demonstrates the workflow of CSR. *First*, we extract the closely clustered contours (C_c) in C_p. The definition of closely clustered is that: at least one contour exists, and the IoU of this contour with all the remaining contours is greater than a threshold ($t = 0.7$). *Second*, we obtain the merged contour (M_c) of the object by averaging the C_c directly. We think that, as the network learns iteratively, M_c has a high confidence in fitting the object. Thus, we extract the feature around M_c to further refine M_c. Concretely, we use the uniform point sampling approach to sample out ($k + 1$) sampling points, indicating the k boundary points and one center point of the contour($k = 16$). To extract the features around the contour, we use the sampled points as center points to generate ($k + 1$) boxes to obtain rich information on refined-feature. RoI-Align module is used in this stage. Finally, these extracted features are passed through a three-layer Multi-Perceptron module to generate the Fourier and center point offsets. The losses used in CSR are same as FFPN, including L_{Fou}, L_{Loc} and L_{Con}. It is worth noting that the anchor information used at this stage is generated by M_c, which draws on the common configuration of a two-stage detection framework.

3 Experimental Results

To validate the performance of our FFPN, we conducted comprehensive comparisons among our method and other segmentation methods, including DTS methods and Pixel-based methods. Additionally, we also compare FFPN with FFPN+CSR (FFPN-R) to validate the effectiveness of CSR.

Datasets. We assess FFPN's performance across three datasets (Fig. 1). Approved by local IRB, the 2CH dataset, comprising 1731 US images, is utilized for left ventricle (LV) segmentation, while the FH dataset, containing 2679

US images, is employed for fetal head segmentation. The Camus [8] dataset contains 700 US images and requires the segmentation of three structures, including the LV, left atrium (LA) and myocardium (MC). Both the 2CH dataset and the FH dataset have been manually annotated by experts using the Pair [9] annotation software package. Each dataset undergoes a random split into training (70%), validation (10%), and testing (20%) subsets.

Experimental Settings. To conduct a fair comparison, all methods implemented in Pytorch and under the same experiment settings, including learning rate (1e-3), input size (directly resized to 416×416), total epochs (200), one RTX 2080Ti GPU and so on. We evaluate the model performance using dice similarity coefficient (DICE), Hausdorff distance (HD), Intersection over Union (IoU) and Conformity (Conf) for all the experiments. Memory (Mem) and FPS are employed to evaluate the models' memory usage and efficiency. As for the settings of FFPN, the Fourier-anchors are generated by contour clustering of the training set, each dataset has 9 base anchors. The IoU threshold for positive and negative samples are 0.25 and 0.10 respectively, and the others are ignored samples.

Table 1. Quantitative evaluation of mean(std) results on 2CH and FH datasets.

DATASET	2CH				FH			Model	
Methods	Metrics								
	DICE(%)↑	IoU(%)↑	HD(pixel)↓	Conf(%)↑	DICE(%)↑	HD(pixel)↓	Conf(%)↑	Mem(G)↓	FPS↑
U-NET	88.95(4.95)	80.45(7.55)	22.41(18.80)	74.40(13.95)	94.92(3.27)	18.1(16.43)	90.78(6.87)	1.33	19.71
DeepLabV3	88.92(5.06)	80.40(7.69)	19.98(12.14)	74.26(14.55)	95.89(2.56)	16.63(22.59)	91.27(6.33)	0.49	15.83
Swin U-NET	88.83(4.63)	80.21(7.21)	20.25(10.28)	74.20(12.67)	95.81(1.38)	15.30(5.67)	91.21(3.10)	2.27	9.81
Mask RCNN	79.50(6.71)	66.48(9.04)	47.53(13.87)	46.51(22.90)	82.60(2.33)	58.24(6.73)	57.64(7.10)	0.43	20.33
PolarMask	84.54(14.65)	77.42(13.15)	34.13(25.81)	58.96(78.22)	94.25(5.31)	25.96(15.12)	64.28(489.90)	0.52	15.43
PolySnake	86.50(6.05)	76.67(8.73)	22.75(13.05)	67.48(18.82)	92.11(4.39)	22.95(11.46)	82.32(11.43)	0.17	14.73
CPN	87.62(6.23)	78.47(9.02)	23.81(15.81)	72.61(15.30)	94.00(5.93)	22.73(22.89)	86.17(16.83)	0.20	27.20
FFPN	88.16(5.97)	79.30(8.88)	21.10(14.34)	71.93(18.15)	95.56(2.40)	13.90(8.84)	90.55(6.56)	0.20	41.52
FFPN-R	89.08(5.24)	80.70(8.10)	19.76(12.52)	74.64(14.55)	96.73(1.11)	10.24(4.13)	93.21(2.43)	0.23	33.52

Quantitative and Qualitative Analysis. As demonstrated in Table 1, FFPN exhibits superior performance compared to feasible DTS methods and achieves comparable performance when compared to effective Pixel-based methods in both the 2CH dataset and FH dataset. Specifically, comparing to the recently proposed method PolySnake, FFPN increases DICE by 1.66% and 3.45%, while simultaneously reducing HD by 1.65 pixels and 9.05 pixels on the 2CH dataset and FH dataset, respectively. This demonstrates the effectiveness of our encoding approach based on EFD. Moreover, in comparison to CPN, which follows the same encoding approach, FFPN improves DICE by 0.54% and 1.56%, and decreases HD by 2.71 pixels and 8.83 pixels on the two datasets, respectively. This fully proves that FFPN effectively utilizes Fourier information by mapping Fourier series with similar scales and frequencies into the same layer of the feature map. Furthermore, FFPN-R is capable of outperforming all methods in all metrics on the two datasets. This demonstrates the effectiveness of CSR, particularly in its ability to capture detailed information and refine contours.

Experimental results on computing resource consumption and efficiency are shown in the *Model* column of Table 1. FFPN achieves an impressive inference speed of 41.52 FPS with a memory consumption of only 0.2 GB, making it the fastest among all DTS and Pixel-based methods. Compared with PolySnake which has the smallest memory footprint, FFPN increases memory usage by only 17.6% while improving its speed by 182.0%. Furthermore, by incorporating CSR, FFPN-R achieves the best accuracy performance at a slight cost of increasing memory consumption by 15.0% and reducing inference speed by 19.5% FPS. Despite this minor trade-off, FFPN-R still exhibits superior resource consumption and efficiency compared to all Pixel-based and most DTS methods.

Table 2. Quantitative evaluation of mean(std) results on Camus dataset.

Methods	LA		MC		LV		Mean	
Metrics	DICE(%)↑	HD(pixel)↓	DICE(%)↑	HD(pixel)↓	DICE(%)↑	HD(pixel)↓	DICE(%)↑	HD(pixel)↓
U-NET	87.74(9.81)	28.88(32.99)	86.03(4.45)	22.43(21.84)	92.06(4.01)	22.43(28.94)	88.61(6.09)	24.58(27.78)
DeepLabV3	88.93(6.34)	22.84(21.91)	87.84(4.04)	17.49(13.05)	92.79(4.02)	16.51(9.36)	89.85(4.80)	18.69(18.80)
Swin U-NET	87.28(9.47)	26.39(24.97)	86.38(4.32)	19.31(8.60)	91.73(4.71)	18.68(11.69)	88.46(6.17)	21.46(15.10)
Mask RCNN	76.83(18.61)	43.38(40.57)	55.99(26.54)	66.78(84.70)	75.16(16.99)	53.95(32.10)	69.33(20.71)	54.70(52.46)
PolarMask	81.56(20.44)	40.91(16.25)	27.28(21.16)	59.92(12.98)	81.56(20.44)	42.67(37.64)	65.75(16.23)	47.83(22.29)
PolySnake	83.77(17.23)	22.99(19.92)	43.65(32.27)	21.82(11.43)	89.58(9.80)	19.72(11.30)	72.33(19.77)	21.51(14.21)
CPN	87.11(13.68)	27.35(37.09)	69.17(25.75)	41.58(58.22)	92.08(5.06)	19.26(13.62)	82.79(14.83)	29.40(36.31)
FFPN	87.37(8.39)	23.81(17.18)	82.52(6.59)	23.63(9.83)	91.35(4.36)	20.02(11.12)	87.08(7.57)	22.48(13.23)
FFPN-R	88.76(7.85)	21.10(17.65)	85.03(4.54)	20.25(7.43)	92.39(4.08)	16.97(9.36)	88.72(6.48)	19.44(12.43)

In addition, we have also validated the effectiveness of our framework in the multi-class segmentation task. As shown in Table 2, FFPN-R achieves a more significant improvement on the Camus dataset compared to other DTS approaches, as compared to the single-class segmentation tasks. This further illustrates the generalizability and scalability of our framework. On the Camus dataset, FFPN-R outperforms most Pixel-based methods. However, the wrapping of the myocardium around the left ventricle poses a challenge in accurately assigning positive and negative samples within the detection framework, resulting in our results being slightly inferior to DeeplabV3. Figure 5 shows the segmentation results of U-NET, DeepLab V3, CPN and FFPN-R. It demonstrates the superior performance of FFPN-R on the segmentation task in US images.

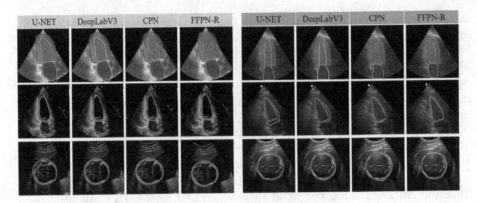

Fig. 5. Segmentation results of different methods on three datasets. For the first row (Camus), the masks are ground truth, and the contours are predictions. For the second (2CH) and third row (FH), the green contours denote ground truth, and the red contours represent predictions. (Color figure online)

4   Conclusion

In this study, we utilize the effectiveness of Fourier Descriptors to represent contours and propose Fourier Feature Pyramid Network (FFPN), a Fourier-anchor-based framework, to describe the segmentation region. It is found that FFPN achieves the best performance against other DTS methods. Moreover, we design a Contour Sampling Refinement (CSR) module to more accurately fit complex contours, which makes our method achieve further improvement and exhibit powerful capabilities on segmentation tasks. These experiments further demonstrate the well-balanced among performance, resource consumption and efficiency of our framework.

Acknowledge. This work was supported by the grant from National Natural Science Foundation of China (Nos. 62171290, 62101343), Shenzhen-Hong Kong Joint Research Program (No. SGDX20201103095613036), and Shenzhen Science and Technology Innovations Committee (No. 20200812143441001).

References

1. Cao, H., et al.: Swin-Unet: Unet-like pure transformer for medical image segmentation. In: Karlinsky, L., Michaeli, T., Nishino, K. (eds.) Computer Vision – ECCV 2022 Workshops: Tel Aviv, Israel, October 23–27, 2022, Proceedings, Part III, pp. 205–218. Springer, Cham (2023). https://doi.org/10.1007/978-3-031-25066-8_9
2. Chen, J., et al.: Transunet: Transformers make strong encoders for medical image segmentation. arXiv preprint arXiv:2102.04306 (2021)
3. Chen, L.C., Papandreou, G., Schroff, F., Adam, H.: Rethinking atrous convolution for semantic image segmentation. arXiv preprint arXiv:1706.05587 (2017)

4. Feng, H., Zhou, W., Yin, Y., Deng, J., Sun, Q., Li, H.: Recurrent contour-based instance segmentation with progressive learning. arXiv preprint arXiv:2301.08898 (2023)
5. He, K., Gkioxari, G., Dollár, P., Girshick, R.: Mask r-CNN. In: Proceedings of the IEEE International Conference on Computer Vision, pp. 2961–2969 (2017)
6. Isensee, F., Jaeger, P.F., Kohl, S.A., Petersen, J., Maier-Hein, K.H.: NNU-net: a self-configuring method for deep learning-based biomedical image segmentation. Nat. Methods 18(2), 203–211 (2021)
7. Kuhl, F.P., Giardina, C.R.: Elliptic Fourier features of a closed contour. Comput. Graph. Image Process. 18(3), 236–258 (1982)
8. Leclerc, S., et al.: Deep learning for segmentation using an open large-scale dataset in 2D echocardiography. IEEE Trans. Med. Imaging 38(9), 2198–2210 (2019)
9. Liang, J., et al.: Sketch guided and progressive growing gan for realistic and editable ultrasound image synthesis. Med. Image Anal. 79, 102461 (2022)
10. Lin, T.Y., Dollár, P., Girshick, R., He, K., Hariharan, B., Belongie, S.: Feature pyramid networks for object detection. In: Proceedings of the IEEE Conference on Computer Vision and Pattern Recognition, pp. 2117–2125 (2017)
11. Painchaud, N., Duchateau, N., Bernard, O., Jodoin, P.M.: Echocardiography segmentation with enforced temporal consistency. IEEE Trans. Med. Imaging 41(10), 2867–2878 (2022)
12. Ronneberger, O., Fischer, P., Brox, T.: U-net: Convolutional networks for biomedical image segmentation. In: Medical Image Computing and Computer-Assisted Intervention-MICCAI 2015: 18th International Conference, Munich, Germany, October 5–9, 2015, Proceedings, Part III 18, pp. 234–241. Springer (2015)
13. Sobhaninia, Z., et al.: Fetal ultrasound image segmentation for measuring biometric parameters using multi-task deep learning. In: 2019 41st Annual International Conference of the IEEE Engineering in Medicine and Biology Society (EMBC), pp. 6545–6548. IEEE (2019)
14. Upschulte, E., Harmeling, S., Amunts, K., Dickscheid, T.: Contour proposal networks for biomedical instance segmentation. Med. Image Anal. 77, 102371 (2022)
15. Wei, F., Sun, X., Li, H., Wang, J., Lin, S.: Point-set anchors for object detection, instance segmentation and pose estimation. In: Computer Vision-ECCV 2020: 16th European Conference, Glasgow, UK, August 23–28, 2020, Proceedings, Part X 16, pp. 527–544. Springer (2020)
16. Xie, E., et al.: Polarmask: Single shot instance segmentation with polar representation. In: Proceedings of the IEEE/CVF conference on Computer Vision and Pattern Recognition, pp. 12193–12202 (2020)
17. Zhao, Y., Cai, Y., Wu, W., Wang, W.: Explore faster localization learning for scene text detection. arXiv preprint arXiv:2207.01342 (2022)
18. Zhu, Y., Chen, J., Liang, L., Kuang, Z., Jin, L., Zhang, W.: Fourier contour embedding for arbitrary-shaped text detection. In: Proceedings of the IEEE/CVF Conference on Computer Vision and Pattern Recognition, pp. 3123–3131 (2021)

Mammo-SAM: Adapting Foundation Segment Anything Model for Automatic Breast Mass Segmentation in Whole Mammograms

Xinyu Xiong[1], Churan Wang[2(✉)], Wenxue Li[3], and Guanbin Li[1(✉)]

[1] School of Computer Science and Engineering, Sun Yat-sen University, Guangzhou, China
liguanbin@mail.sysu.edu.cn
[2] School of Computer Science, Peking University, Beijing, China
churanwang@pku.edu.cn
[3] School of Future Technology, Tianjin University, Tianjin, China

Abstract. Automated breast mass segmentation from mammograms is crucial for assisting radiologists in timely and accurate breast cancer diagnosis. Segment Anything Model (SAM) has recently demonstrated remarkable success in natural image segmentation, suggesting its potential for enhancing artificial intelligence-based automated diagnostic systems. Unfortunately, we observe that the zero-shot performance of SAM in mass segmentation falls short of usability. Therefore, fine-tuning SAM for transfer learning is necessary. However, full-tuning is cost-intensive for foundation models, making it unacceptable in clinical practice. To tackle this problem, in this paper, we propose a parameter-efficient fine-tuning framework named Mammo-SAM, which significantly improves the performance of SAM on the challenging task of mass segmentation. Our key insight includes a tailored adapter to explore multi-scale features and a re-designed CNN-style decoder for precise segmentation. Extensive experiments on the public datasets CBIS-DDSM and INbreast demonstrate that our proposed Mammo-SAM surpasses existing mass segmentation methods and other tuning paradigms designed for SAM, achieving new state-of-the-art performance.

1 Introduction

Breast cancer is one of the leading causes of death from cancer in women [22]. Thankfully, screening mammography can identify breast cancer in its early stage, allowing for more effective treatment so as to improve survival rates [23]. As manual screening is error-prone and relies on experienced radiologists [12], various methods for mammogram mass segmentation [10,15,21] have been proposed to assist human experts in diagnosing breast cancer [25,26]. However, accurate automated identification of masses is always challenging for two reasons.

X. Xiong—Work done when interning at Deepwise AI Lab.

X. Cao et al. (Eds.): MLMI 2023, LNCS 14348, pp. 176–185, 2024.
https://doi.org/10.1007/978-3-031-45673-2_18

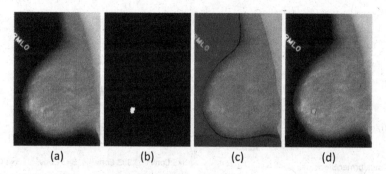

(a)   (b)   (c)   (d)

Fig. 1. (a) A whole mammogram. (b) Ground truth mask of the corresponding breast mass. (c) Zero-shot segmentation result of the segment anything model (SAM) [8]. (d) Segmentation result by our proposed Mammo-SAM, where prediction is marked in red and ground truth is marked in blue. (Color figure online)

Firstly, mammograms are high-resolution in nature and contain complex contexts. Secondly, many breast masses are camouflaged in normal glands, making it difficult to distinguish. Existing mass segmentation models are usually small in size. Therefore, their ability to efficiently process global-local information in mammograms to accurately identify masses is still limited.

Foundation models, such as DALL·E [18], CLIP [17], and GPT-4 [14], have gained much attention due to their impressive performance in vision and language tasks. Recently, the Segment Anything Model (SAM) [8] has garnered significant recognition as a remarkable milestone in computer vision with its exceptional versatility. One compelling factor contributing to the cutting-edge performance of SAM is its utilization of the large-scale segmentation dataset named SA-1B, which consists of more than 1 billion masks and 11 million images. Unfortunately, medical data is not fully considered in the SA-1B. As reported in recent research [4,20,30], due to the serious domain gap between medical and natural images, zero-shot segmentation results of SAM on medical images fall short of usability. We observe similar results on mass segmentation: SAM can only coarsely divide the mammogram image into several regions, failing to recognize the hidden masses, as shown in Fig. 1.

Fortunately, fine-tuning SAM can effectively explore its powerful representation capabilities and boost performance on new downstream tasks. Since foundation models are huge in size, full-tuning all their parameters is inefficient and entails exorbitant computational costs. Therefore, parameter-efficient fine-tuning is necessary. Chen *et al.* [2] inserted adapters [5] to SAM, significantly improving its colon polyp segmentation performance. Zhang *et al.* [29] applied the low-rank-based (LoRA) [6] fine-tuning strategy to SAM. Chai *et al.* [1] proposed an auxiliary CNN encoder to achieve cost-efficient training of SAM. Motivated by these successes, in this paper, we propose Mammo-SAM and the contribution of which is listed as follows:

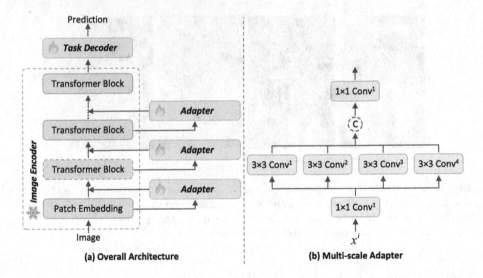

Fig. 2. (a) The overall architecture of our proposed Mammo-SAM, where tuneable parameters are shown in orange and frozen parameters are shown in blue. (b) The architecture of our proposed multi-scale adapter, where k × k $Conv^d$ denotes a Convolution-LayerNorm-GELU combination with kernel size k and dilation rate d. © denotes channel-wise concatenation. (Color figure online)

- To our best knowledge, this is the pioneering work focusing on how to tune vision foundation models for breast cancer diagnosis, which opens up opportunities for future exploration.
- We show that constructing better adapters and decoders are two promising directions that allow SAM to be well generalized to downstream tasks.
- Extensive experiments demonstrate that our proposed Mammo-SAM surpasses existing mass segmentation methods and other tuning paradigms designed for SAM, achieving new state-of-the-art performance.

2   Methodology

The overall architecture of our framework is illustrated in Fig. 2(a), which comprises a frozen SAM image encoder for feature extraction, multi-scale adapters for parameter-efficient fine-tuning, and a task decoder for final prediction. This work focuses on automatic segmentation, which does not involve pre-defined manual prompts, thus the SAM prompt encoder is removed. Next, we will illustrate how to enable adapters to explore multi-scale features and tailor a CNN-style decoder for effective mammogram analysis, harvesting accurate mass segmentation.

2.1   Multi-scale Adapter

In simple terms, the vanilla adapter [5] for natural language processing consists of a fully connected layer for down-projecting, a ReLU activation function for

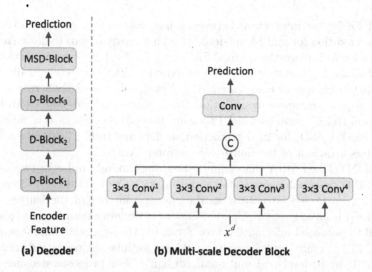

Fig. 3. (a) The architecture of multi-level decoder. (b) The architecture of multi-scale decoder block.

nonlinearity, and a fully connected layer for up-projecting. Jie *et al.* [7] have shown that convolution-based adapters, which benefit from the hard-coded inductive bias of convolution operations, are more suitable for vision adapting. However, it did not take advantage of utilizing multi-scale feature extraction, which proved to be very necessary for mass segmentation.

Motivated by the insights gained from the above-mentioned analysis, we introduce a convolutional multi-scale adapter whose architecture is illustrated in Fig. 2(b). Specifically, given an input feature x^i with c^i channels, our multi-scale adapter commences with an 1×1 Convolution-LayerNorm-GELU combination (abbreviated as 1×1 Conv) that down-projects the channels to $c^i/4$. Subsequently, the output is fed into four parallel 3 × 3 Conv branches, each with different dilation rates $d = 1, 2, 3, 4$, and reducing the channels to $c^i/16$ at the same time. These four features with diverse multi-scale perceptions are then aggregated through channel-wise concatenation. Finally, the obtained bottleneck feature is passed through an 1×1 Conv to up-project the channels back to c^i.

To encourage comprehensive adaptability, our adapters are inserted before each transformer block in the image encoder. The output of adapter is added to its input in a residual fashion [3], facilitating effective tuning. This design empowers our adapter to exploit appropriate feature representations for more effective fine-tuning, enhancing final mass segmentation performance.

2.2 Multi-level Decoder

As proven by U-Net [19], the cascaded pyramid decoder design is critical for accurate medical image segmentation, which is also widely applied in many mass segmentation methods [15,21]. However, the original SAM decoder is mainly designed for better interaction between image tokens and prompt tokens and lacks consideration for multi-level decoding. That motivated us to design a multi-level decoder to replace the original SAM decoder for better segmentation.

Specifically, the architecture of our proposed multi-level decoder is illustrated in Fig. 3(a). It comprises three cascaded U-Net decoder blocks, where each block consists of one transpose convolution for upsampling and two Convolution-BatchNorm-ReLU combinations. Following these three blocks is a multi-scale decoder block (MSD) for final prediction, as depicted in Fig. 3(b), which can be regarded as a variant of the multi-scale adapter (MSA).

While MSD and MSA share similarities, they mainly differ in two aspects. Firstly, MSD does not necessitate the initial 1×1 Conv for down-projecting to construct a bottleneck structure, as the channel number of the feature at the shallow level is already small. Conducting dimensionality reduction at this stage may lead to potential information loss. Secondly, the aggregated multi-scale features in MSD are sent into a convolutional layer for final segmentation prediction. Overall, the multi-level and multi-scale design of our proposed decoder allows for gradually recovering the detailed information of the mammogram and better segmenting breast masses of different sizes compared to the original SAM decoder.

3   Experiments

3.1   Datasets

Our experiments are conducted on the following two publicly available datasets:

CBIS-DDSM [9] dataset contains curated mammograms from the Digital Database for Screening Mammography (DDSM) [16]. The original DDSM dataset contains 2,620 cases in total. After data selection and curation by trained mammographers, the resulting CBIS-DDSM dataset contains 1,592 mass images in total. Different from previous work [10,28] that adapts a selected subset, we use the entire dataset for a more comprehensive evaluation, where the 75% images are divided into training set and the rest serve as testing set.

INbreast [13] dataset is built with full-field digital mammograms (FFDM). It has a total of 410 images, in which 107 images contains breast masses. Due to its relatively limited sample size and large modal differences with the CBIS dataset (digital mammograms versus digitized mammograms), we treat it as another test set to verify the generalization capability of different models.

3.2   Implementation Details

Our method is implemented by Pytorch framework and trained on quad NVIDIA Tesla A100 with 40GB memory. The SGD optimizer is employed, with an initial learning rate of 0.01, and the cosine decay is applied to stabilize training. The pixel-position aware loss [27] $\mathcal{L} = \mathcal{L}_{IoU}^w + \mathcal{L}_{BCE}^w$ is used for optimizing due to its satisfactory performance on segmenting medical objects. Two data augmentation strategies are employed, including random vertical and horizontal flips. We choose the ViT-L version of SAM to balance performance and training cost.

Table 1. Quantitative comparison on the CBIS-DDSM [9] and INbreast [13] datasets. All results are shown using percentages. Best results are in **bold**. All metrics are higher the better.

Methods	CBIS-DDSM				INbreast			
	Dice	*IoU*	*Prec*	*Rec*	*Dice*	*IoU*	*Prec*	*Rec*
U-Net [19]	44.32	35.73	48.69	50.61	65.79	55.11	73.93	69.38
UNet++ [31]	47.68	37.01	42.91	63.32	62.52	51.41	71.74	64.73
SCSGNet [10]	48.78	39.04	46.57	58.48	61.00	49.94	65.40	65.50
AUNet [21]	50.17	41.84	54.18	53.61	63.14	54.61	70.46	64.50
Zero-shot [30]	23.82	19.46	25.97	**86.84**	47.47	42.51	48.55	**93.65**
Freeze Encoder [11]	38.35	30.37	41.48	41.67	58.70	49.51	63.93	61.99
SAM Adapter [2]	56.12	45.35	52.22	68.20	72.67	62.62	71.86	81.69
SAMed [29]	56.79	47.29	55.16	64.37	72.59	62.45	74.18	78.18
Ours	**61.28**	**50.68**	**58.56**	71.56	**75.10**	**64.88**	**74.71**	83.35

3.3 Comparison with State-of-the-Art Methods

In order to assess the performance of different methods, we conduct a comparative analysis involving four traditional medical image segmentation methods, including U-Net [19], UNet++ [31], SCSGNet [10], and AUNet [21]. Further, to investigate how to adapt SAM for downstream tasks, we compared two baselines, namely zero-shot[1] and freeze encoder[2]; and two methods dedicated to SAM tuning, namely SAMed [29] and SAM Adapter [2]. For quantitative comparisons, four metrics are employed: dice similarity coefficient (*Dice*), intersection over Union (*IoU*), precision (*Prec*), and recall (*Rec*). Next, we will analyze the results of the quantitative comparison as well as the qualitative comparison.

Intra-Domain Comparison results are shown in the "DDSM" row of Table 1. Taking the Dice score as an example, among the different traditional methods, AUNet achieved the best result with a Dice value of 50.17%. Among the SAM-based methods, the performance of both zero-shot and freeze encoder is worse than that of the conventional method, which shows the necessity of designing parameter-efficient fine-tuning methods. It is worth noting that zero-shot achieves the highest recall because its predictions are manual selected. For the two competing tuning methods, SAM Adapter and SAMed, they both yielded better results than all conventional methods, demonstrating the great potential of the foundation model. In comparison, our method achieves the best results with a Dice value of 61.28%, an 11.11% and 4.49% improvement over the best traditional method (AUNet) and the second-best SAM-based method (SAMed), respectively.

---

[1] Since the original SAM produces multiple output masks to represent different segmented regions, following [24,30], we manually select the output mask that best matches ground truth to evaluate the upper bound performance of zero-shot.

[2] In this setting, only the SAM decoder is tunable, similar to [11].

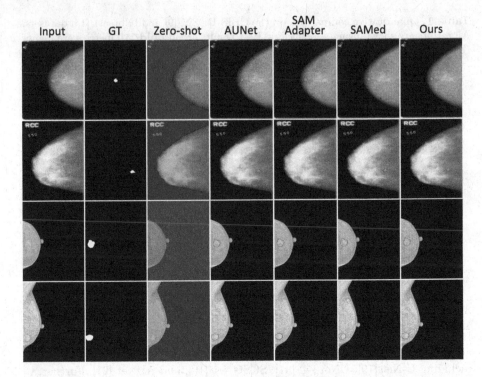

Fig. 4. Visual comparison on mammogram mass segmentation. Ground Truths are marked in blue and prediction results are marked in red. Zoom in for best view. (Color figure online)

Cross-Domain Validation results are shown in the "INbreast" row of Table 1. To analyze generalization ability, we leverage the above models trained on the CBIS-DDSM dataset to test on the INbreast dataset. This task is very challenging as there is a large domain gap between these two datasets due to differences in imaging styles. Our method still performs best with the Dice score of 75.10%, a 9.31% and 2.43% improvement over the best traditional method (U-Net) and the second-best SAM-based method (SAM Adapter), respectively.

Visual Comparison results are shown in Fig. 4. In various challenging scenarios, our Mammo-SAM can identify masses more accurately and obtain better prediction results on boundaries than competing methods.

3.4 Performance Under Limited Annotations

One major appealing aspect of the foundation model is its strong generalization capabilities. That is to say, it is also possible for SAM to achieve good results when the annotation budget is limited. This feature is important because expert annotations are often scarce in medical scenarios. To verify this, we compare the performance of MammoSAM with AUNet under different annotation budgets, and the results are shown in Fig. 5. It can be found that even though only 15% training samples are available, our Mammo-SAM beats AUNet under 100% budget, revealing that fine-tuning the foundation model is a promising choice under the annotation scarcity scenario such as medical imaging.

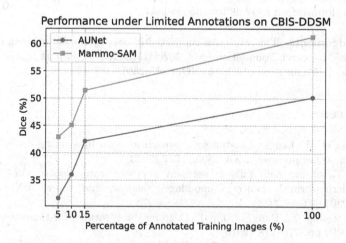

Fig. 5. Performance comparison under different annotation budgets.

3.5 Ablation Study

To verify the effectiveness of our key design choices, we conduct ablation experiments as presented as follows:

Effectiveness of Mulit-scale Adapter. We replace mulit-scale adapter with a simple convolutional adapter with an 1×1 Conv for down-projecting, a 3×3 Conv for feature learning and an 1×1 Conv for up-projecting, resulting in a 4.15% decrease in the Dice score. This suggests that a multi-scale design is necessary for adapters.

Effectiveness of Multi-level Decoder. We replace the multi-level decoder with the original SAM decoder, resulting in a 3.87% decrease in the Dice score. This demonstrates that designing multifarious decoders offers appealing possibilities for SAM to address a broader range of problems in medical imaging.

4   Conclusion

In this paper, we perform extensive experiments on automatic mammogram mass segmentation to assess the potential of the Segment Anything Model (SAM). Our findings reveal that directly applying SAM for zero-shot mass prediction yields unsatisfactory results. Fortunately, through efficient tuning, SAM demonstrates superior performance compared to existing solutions. Particularly, with the help of multi-scale adapters and multi-level decoder, our proposed Mammo-SAM shows impressive inter-domain learning ability and intra-domain generalization ability, achieving a new state-of-the-art level of performance. This result suggests that adapting foundation models is a promising choice for automatic mammogram analysis. We hope our work will inspire further research endeavours focusing on foundation models for medical AI.

Acknowledgements. This work was supported in part by the Guangdong Basic and Applied Basic Research Foundation (NO. 2020B1515020048), in part by the National Natural Science Foundation of China (NO. 61976250).

References

1. Chai, S., et al.: Ladder fine-tuning approach for sam integrating complementary network. arXiv preprint arXiv:2306.12737 (2023)
2. Chen, T., et al.: Sam fails to segment anything?-sam-adapter: Adapting sam in underperformed scenes: Camouflage, shadow, and more. arXiv preprint arXiv:2304.09148 (2023)
3. He, K., Zhang, X., Ren, S., Sun, J.: Deep residual learning for image recognition. In: CVPR, pp. 770–778 (2016)
4. He, S., Bao, R., Li, J., Grant, P.E., Ou, Y.: Accuracy of segment-anything model (sam) in medical image segmentation tasks. arXiv preprint arXiv:2304.09324 (2023)
5. Houlsby, N., et al.: Parameter-efficient transfer learning for NLP. In: ICML, pp. 2790–2799. PMLR (2019)
6. Hu, E.J., et al.: LoRA: Low-rank adaptation of large language models. In: ICLR (2022)
7. Jie, S., Deng, Z.H.: Convolutional bypasses are better vision transformer adapters. arXiv preprint arXiv:2207.07039 (2022)
8. Kirillov, A., et al.: Segment anything. arXiv preprint arXiv:2304.02643 (2023)
9. Lee, R.S., Gimenez, F., Hoogi, A., Miyake, K.K., Gorovoy, M., Rubin, D.L.: A curated mammography data set for use in computer-aided detection and diagnosis research. Sci. Data **4**(1), 1–9 (2017)
10. Li, Q., Xu, J., Yuan, R., Zhang, Y., Feng, R.: SCSGNet: Spatial-correlated and shape-guided network for breast mass segmentation. In: ICASSP, pp. 1–5. IEEE (2023)
11. Ma, J., Wang, B.: Segment anything in medical images. arXiv preprint arXiv:2304.12306 (2023)
12. Marmot, M.G., Altman, D., Cameron, D., Dewar, J., Thompson, S., Wilcox, M.: The benefits and harms of breast cancer screening: an independent review. Br. J. Cancer **108**(11), 2205–2240 (2013)
13. Moreira, I.C., Amaral, I., Domingues, I., Cardoso, A., Cardoso, M.J., Cardoso, J.S.: Inbreast: toward a full-field digital mammographic database. Acad. Radiol. **19**(2), 236–248 (2012)
14. OpenAI: Gpt-4 technical report (2023)
15. Pi, J., et al.: FS-UNet: Mass segmentation in mammograms using an encoder-decoder architecture with feature strengthening. Comput. Biol. Med. **137**, 104800 (2021)

16. PUB, M.H., Bowyer, K., Kopans, D., Moore, R., Kegelmeyer, P.: The digital database for screening mammography. In: Proceedings of the Third International Workshop on Digital Mammography, Chicago, IL, USA, pp. 9–12 (1996)

17. Radford, A., et al.: Learning transferable visual models from natural language supervision. In: ICML, pp. 8748–8763. PMLR (2021)

18. Ramesh, A., et al.: Zero-shot text-to-image generation. In: ICML, pp. 8821–8831. PMLR (2021)

19. Ronneberger, O., Fischer, P., Brox, T.: U-Net: convolutional networks for biomedical image segmentation. In: Navab, N., Hornegger, J., Wells, W.M., Frangi, A.F. (eds.) Medical Image Computing and Computer-Assisted Intervention – MICCAI 2015: 18th International Conference, Munich, Germany, October 5-9, 2015, Proceedings, Part III, pp. 234–241. Springer, Cham (2015). https://doi.org/10.1007/978-3-319-24574-4_28

20. Shi, P., Qiu, J., Abaxi, S.M.D., Wei, H., Lo, F.P.W., Yuan, W.: Generalist vision foundation models for medical imaging: a case study of segment anything model on zero-shot medical segmentation. Diagnostics **13**(11), 1947 (2023)

21. Sun, H., et al.: AUNet: attention-guided dense-upsampling networks for breast mass segmentation in whole mammograms. Phys. Med. Biol. **65**(5), 055005 (2020)

22. Sung, H., et al.: Global cancer statistics 2020: Globocan estimates of incidence and mortality worldwide for 36 cancers in 185 countries. CA: a cancer journal for clinicians **71**(3), 209–249 (2021)

23. Tabár, L., et al.: Swedish two-county trial: impact of mammographic screening on breast cancer mortality during 3 decades. Radiology **260**(3), 658–663 (2011)

24. Tang, L., Xiao, H., Li, B.: Can sam segment anything? when SAM meets camouflaged object detection. arXiv preprint arXiv:2304.04709 (2023)

25. Wang, C.R., Gao, F., Zhang, F., Zhong, F., Yu, Y., Wang, Y.: Disentangling disease-related representation from obscure for disease prediction. In: ICML, pp. 22652–22664. PMLR (2022)

26. Wang, C.R., et al.: Bilateral asymmetry guided counterfactual generating network for mammogram classification. IEEE Trans. Image Process. **30**, 7980–7994 (2021)

27. Wei, J., Wang, S., Huang, Q.: F^3net: fusion, feedback and focus for salient object detection. In: AAAI, pp. 12321–12328 (2020)

28. Xu, C., Qi, Y., Wang, Y., Lou, M., Pi, J., Ma, Y.: Arf-net: an adaptive receptive field network for breast mass segmentation in whole mammograms and ultrasound images. Biomed. Signal Process. Control **71**, 103178 (2022)

29. Zhang, K., Liu, D.: Customized segment anything model for medical image segmentation. arXiv preprint arXiv:2304.13785 (2023)

30. Zhou, T., Zhang, Y., Zhou, Y., Wu, Y., Gong, C.: Can sam segment polyps? arXiv preprint arXiv:2304.07583 (2023)

31. Zhou, Z., Rahman Siddiquee, M.M., Tajbakhsh, N., Liang, J.: UNet++: a nested u-net architecture for medical image segmentation. In: Stoyanov, D., et al. (eds.) DLMIA/ML-CDS -2018. LNCS, vol. 11045, pp. 3–11. Springer, Cham (2018). https://doi.org/10.1007/978-3-030-00889-5_1

Consistent and Accurate Segmentation for Serial Infant Brain MR Images with Registration Assistance

Yuhang Sun[1,2], Jiameng Liu[2], Feihong Liu[2,3], Kaicong Sun[2], Han Zhang[2], Feng Shi[2,4], Qianjin Feng[1], and Dinggang Shen[2,4,5(✉)]

[1] School of Biomedical Engineering, Southern Medical University, Guangzhou, China
fengqj99@smu.edu.cn
[2] School of Biomedical Engineering, ShanghaiTech University, Shanghai, China
dgshen@shanghaitech.edu.cn
[3] School of Information Science and Technology, Northwest University, Xi'an, China
[4] Shanghai United Imaging Intelligence Co., Ltd., Shanghai, China
[5] Shanghai Clinical Research and Trial Center, Shanghai, China

Abstract. The infant brain develops dramatically during the first two years of life. Accurate segmentation of brain tissues is essential to understand the early development of both normal and disease changes. However, the segmentation results of the same subject could demonstrate unexpectedly large variations across different time points, which may even lead to inaccurate and inconsistent results in charting infant brain development. In this paper, we propose a deep learning framework, which simultaneously exploits registration and segmentation for guaranteeing the longitudinal consistency among the segmentation results. Firstly, a manual label-guided registration model is designed to fast and accurately obtain the warped images from other time points. Secondly, a segmentation network with a longitudinal consistency constraint is developed to effectively obtain the temporal segmentation results. Thus, our proposed segmentation network could exploit the tissue information of warped intensity images from other time points to aid in segmenting the isointense phase (approximately 6–8 months) data, which is the most difficult case due to the low intensity contrast of tissues. Extensive experiments on infant brain images have shown improved performance achieved by our proposed method, compared with the existing state-of-the-art methods.

Keywords: Longitudinal Segmentation · Registration Assisted Segmentation

---

Y. Sun and J. Liu—These authors contributed equally to this work.

© The Author(s), under exclusive license to Springer Nature Switzerland AG 2024
X. Cao et al. (Eds.): MLMI 2023, LNCS 14348, pp. 186–195, 2024.
https://doi.org/10.1007/978-3-031-45673-2_19

1   Introduction

Longitudinal MRI studies of infants have enormous potential to reveal the pattern of early brain development, which could in turn help us early diagnose abnormal brain development and build longitudinal infant atlas [1–4]. To quantitatively measure the dynamic longitudinal brain development during the early postnatal stage, an accurate segmentation method is crucial to obtain the brain tissues, such as gray matter (GM), white matter (WM), and cerebrospinal fluid (CSF). However, applying the segmentation process independently for each infant image of different time points generally generate inconsistent results (the same brain tissues get different predicted label at different time points). In this work, we consider that when we segment a specific infant brain image, we use tissue labels from other time points of the same structure to constrain the segmentation. The segmentation results obtained in this way could be more longitudinally consistent. To do that, we develop a proper registration method to obtain the corresponding tissue labels of the same structure from all time points, which in turn contribute to the longitudinally consistent segmentation.

However, there is one main challenge for utilizing registration to help obtain the longitudinally consistent segmentation. As shown in Fig. 1, the infant's brain grows and develops significantly rapidly with ongoing myelination and maturation [5,6] in the first two years of life, and the intensity distributions of different tissues during this period also change a lot (Fig. 1(d-f)), which verifies the obvious appearance changes of different time point images (Fig. 1(a-c)) and brings difficulty for traditional intensity-based registration methods [7,8]. In addition, there is a special period, approximately 6–8 months (isointense phase), when the intensity distributions of GM and WM almost overlap (Fig. 1(d)), which illustrates the low intensity contrast at this time point (Fig. 1(a)). The low intensity contrast in 6 months yields a great challenge for both registration and segmentation tasks.

To date, some methods have been proposed for longitudinally consistent segmentation or atlas building. Specifically, longitudinally consistent terms, which

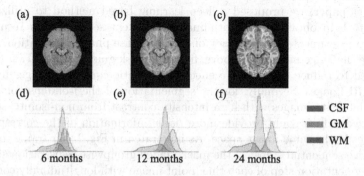

Fig. 1. Dynamic appearance and anatomical changes on a typical infant brain from 6 months to 2 years old.

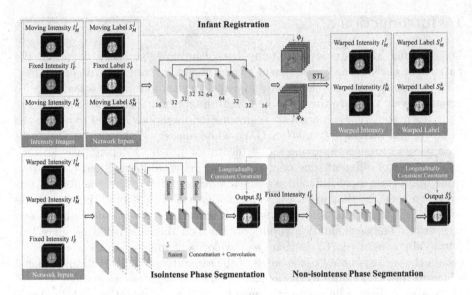

Fig. 2. Illustration of our proposed joint registration and segmentation framework, including a groupwise registration network (top panel), an isointense phase segmentation network (lower left panel), and a non-isointense phase segmentation (lower right panel).

define the distance between the different time point images, are added into the loss functions [9]. Segmentation methods combined with registration methods are also proposed [10–12]. However, these methods are almost traditional methods that are iterative and time-consuming. For the isointense phase infant brain segmentation problem, most methods fuse the information from multi-modalities [13,14], such as T1, T2, and fractional anisotropy images combined with designed fusion modules to boost the segmentation performance. However, most methods are designed for single month segmentation while ignoring the longitudinally consistent segmentation for longitudinal studies.

In this paper, we proposed a deep learning based method to utilize registration to help obtain the longitudinally consistent segmentation results and improve the segmentation accuracy of the isointense phase. In addition, to alleviate the challenge mentioned above, firstly, a mask guided registration network is utilized to reduce the negative effect of dramatic contrast changes in infant brain MRI images. Secondly, for the segmentation of the isointense phase, the warped intensity images with high intensity contrast from non-isointense phase are utilized, which could provide more tissue information of the corresponding structures in the same image space. As illustrated in Fig. 2, our framework consists of two sequential steps: 1) the mask guided groupwise registration step and 2) the segmentation step of each time point image with longitudinally consistent constraint. And the segmentation step can be further divided into isointense phase (lower left panel) and the non-isointense phase (lower right panel).

2   Method

Denote N as the number of time points in the longitudinal infant brain images. As shown in Fig. 2, the fixed tissue label S_F^i $(i \in \{1 \cdots N\})$ and the rest moving tissue labels $\{S_M^j | j = 1, \cdots, N\}$ of one subject are fed into the registration network to obtain the deformation fields $\{\phi_j | j = 1, \cdots, N\}$. In addition, we get the warped moving labels $\{S_M^j (\phi_j) | j = 1, \cdots, N\}$ and the corresponding warped intensity images $\{I_M^j (\phi_j) | j = 1, \cdots, N\}$. Then, the fixed intensity image I_F^i is passed to the segmentation network to obtain the corresponding segmentation result $\hat{S}_F^i$. Furthermore, to obtain longitudinally consistent segmentation results, the warped moving labels $\{S_M^j (\phi_j) | j = 1, \cdots, N\}$ are utilized to constrain the segmentation of the fixed intensity image I_F^i. In addition, if I_F^i is the isointense phase, the warped intensity images $\{I_M^j (\phi_j) | j = 1, \cdots, N\}$ are utilized by fusing the features extracted from the segmentation network to improve the performance.

2.1   Mask Guided Groupwise Registration

In the registration step, the infant brain manual labels are used for registration, instead of intensity images. The registration network is a U-Net based network combined with a spatial transformer layer (STL). The loss function of the pairwise (two images to be aligned) mask guided registration network is defined as follows:

$$\mathcal{L}_{pair} = \sum_{j=1(j \neq i)}^{N} -\mathcal{L}_{Dice}\left(S_M^j (\phi_j), S_F^i\right) + \lambda_1 \mathcal{L}_{smooth}(\phi_j), \ i \in \{1 \cdots N\}, \quad (1)$$

where $\mathcal{L}_{Dice}(\cdot, \cdot)$ measures the Dice similarity between $S_M^j (\phi_j)$ and S_F^i, $\mathcal{L}_{smooth}(\cdot)$ is a regularization term that constrains the deformation field to be smooth, and λ_1 is the regularization parameter.

As more than one images need to be registered to the fixed image in the longitudinal image series, we further consider the alignment of different time point images in a groupwise manner by calculating the similarity between each pair of images in the longitudinal infant brain images of one subject. The loss function of the mask guided groupwise registration network is defined as follows:

$$\mathcal{L}_{group} = \sum_{j=1(j \neq i)}^{N} \sum_{k=1(k \neq j)}^{N} -\mathcal{L}_{Dice}\left(S_M^j (\phi_j), S_M^k (\phi_k)\right), \quad (2)$$

And the overall objective function of the registration step is defined as follows:

$$\mathcal{L}_{reg} = \mathcal{L}_{pair} + \lambda_2 \mathcal{L}_{group}. \quad (3)$$

where λ_2 is a regularization parameter.

2.2   Longitudinally Consistent Segmentation

In the segmentation step, if the current time point to be segmented is not the isointense phase, then the fixed intensity image I_F^i (the current time point to be segmented) is inputted into the segmentation network and the network outputs the predicted segmentation results $\hat{S}_F^i$. The segmentation network for both the non-isointense phase and the isointense phase is U-Net based network.

Longitudinally Consistent Constraint (LCC). To obtain the longitudinally consistent segmentation results, the warped manual labels from other time points (obtained from the registration step) are utilized to constrain the segmentation step. Specifically, we calculate the similarity between the predicted segmentation mask and the warped masks of other time points, as the label information of the same structures from other time points could provide additional information to assist the segmentation. In this paper, the similarities of segmentation maps between the image to be segmented and the warped images of other time points are measured by the Dice score, and the loss function of the longitudinally consistent constraint is defined as follows:

$$\mathcal{L}_{LCC} = \sum_{j=1(j\neq i)}^{N} -\mathcal{L}_{Dice}\left(\hat{S}_F^i, S_M^j\left(\phi_j\right)\right), \tag{4}$$

The overall loss function for segmentation is defined as follows:

$$\mathcal{L}_{seg} = -\mathcal{L}_{Dice}\left(\hat{S}_F^i, S_F^i\right) + \lambda_3\mathcal{L}_{Focal}\left(\hat{S}_F^i, S_F^i\right) + \lambda_4\mathcal{L}_{LCC}, \ i \in \{1\cdots N\}. \tag{5}$$

where $\mathcal{L}_{Focal}(\cdot)$ is the Focal loss term. λ_3 and λ_4 are regularization parameters.

Segmentation of Isointense Phase. If the current time point to be segmented is the isointense phase, ($i = 1$ in our dataset), the low intensity contrast in this phase could seriously degrade the performance of segmentation. To mitigate the negative effect of low tissue contrast on the segmentation, we utilize a multi-level feature fusion strategy to improve the segmentation accuracy of the isointense phase. As the warped intensity images of other time points possess high tissue contrast compared with images of the isointense phase, the information of which could assist the segmentation of the isointense image. Specifically, as shown in Fig. 2 (lower left panel), the features of the intensity image of the isointense phase and the warped intensity images of other time points are extracted by three different branches, and the features of the warped image at different layers are concatenated to the corresponding layers of the branch of the fixed intensity image, then the information of the concatenated features are further extracted by the convolution layer, which fuses the information of the intensity images of all time points.

2.3   Detailed Implementation

Both networks for registration and segmentation are implemented based on Pytorch and trained on a workstation equipped with one NVIDIA A100 GPU. Data augmentation for the training set is conducted by randomly cropping. We first trained the registration network for 500 epochs with a batch size of 3. Then, we trained the segmentation network for 500 epochs with a batch size of 3. The patch size for both the registration and segmentation steps is set as $128 \times 128 \times 96$. Adam is adopted as the optimizer to train both models with a learning rate of 10^{-4} for the registration network and 10^{-3} for the segmentation network. And the hyperparameters λ_1, λ_2, λ_3, and λ_4 in the Eq. 1 and Eq. 3, and Eq. 5 are set as 1, 0.5, 10, and 0.5, respectively.

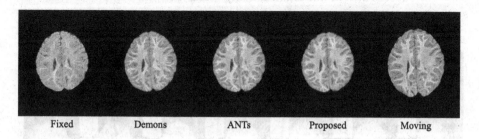

| Fixed | Demons | ANTs | Proposed | Moving |

Fig. 3. Visual performance of registration from 24 months image to 6 months image by Demons, SyN, and our proposed method.

Table 1. Quantitative comparison of three different registration methods for warping 12 months images to 6 months ones.

Method	Dice (%) ↑			HD (mm) ↓		
	GM	WM	CSF	GM	WM	CSF
Demons	75.78 ± 8.60	78.52 ± 6.71	78.91 ± 8.32	10.52 ± 8.98	5.08 ± 5.16	9.38 ± 3.65
ANTs	75.63 ± 6.09	77.47 ± 4.58	79.92 ± 6.88	10.10 ± 8.65	5.16 ± 5.16	9.69 ± 4.60
Proposed	$\mathbf{78.94 \pm 5.98}$	$\mathbf{81.82 \pm 3.61}$	$\mathbf{81.64 \pm 3.29}$	$\mathbf{9.62 \pm 8.49}$	$\mathbf{4.40 \pm 0.38}$	$\mathbf{9.02 \pm 1.59}$

3   Experiments

3.1   Dataset

We employ the publicly available dataset NDAR (National Database for Autism Research) [15] in our study. A total of 159 subjects are included with T1w MRI scans at three different time points (6 months, 12 months, and 24 months). For each image, three tissue labels of GM, WM, and CSF are obtained by clinical experts. All the images in the dataset are resampled to the size of $192 \times 224 \times 192$ with a resolution of $1 \times 1 \times 1\ mm^3$.

3.2   Experimental Setup

To validate the effectiveness of our method, we evaluate the registration performance and the segmentation performance, respectively. For registration performance, we compare our method with the traditional intensity based method Demons [8] and ANTs [7]. For segmentation performance, we first compare the segmentation accuracy of isointense phase images of different methods, i.e., HyperTransformer [16], DenseVoxelNet [17], 3D U-Net [18] (U-Net (single month)), 3D U-Net combined with Transformer (U-Net (Transformer)), and our proposed method. Similar to our multi-branch fusion segmentation network, HyperTransformer and U-Net (Transformer) also utilize warped intensity images to improve the segmentation accuracy of the isointense phase. Finally, we illustrate the effectiveness of our longitudinally consistent constraint with temporal segmentation results.

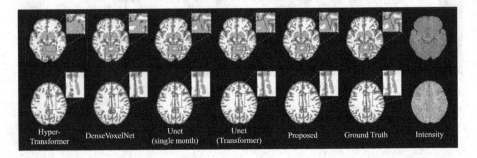

Fig. 4. Visual comparison of different infant segmentation methods for two representative examples.

Table 2. Quantitative comparison of five different segmentation methods for 6 months infant brain images. Our method achieves significant improvement ($p < 0.05$).

Method	Dice (%) ↑			HD (mm) ↓		
	GM	WM	CSF	GM	WM	CSF
HyperTransformer	88.54 ± 1.08	87.72 ± 1.15	89.85 ± 1.07	1.22 ± 0.29	1.06 ± 0.15	1.28 ± 0.26
DenseVoxelNet	91.11 ± 0.99	90.91 ± 1.16	91.99 ± 1.02	1.02 ± 0.07	1.08 ± 0.17	1.28 ± 0.28
U-Net (single month)	93.75 ± 1.27	92.82 ± 0.84	93.83 ± 0.72	1.65 ± 0.56	1.49 ± 0.40	1.41 ± 0.78
U-Net (Transformer)	94.99 ± 1.04	94.82 ± 1.08	94.73 ± 1.05	1.04 ± 0.18	1.04 ± 0.17	1.10 ± 0.20
Proposed	$\mathbf{95.89 \pm 0.92}$	$\mathbf{95.69 \pm 1.02}$	$\mathbf{95.83 \pm 1.01}$	$\mathbf{1.01 \pm 0.07}$	$\mathbf{1.04 \pm 0.13}$	$\mathbf{1.04 \pm 0.12}$

3.3   Experimental Results

Registration Performance. As shown in Fig. 3, our proposed method has the best visual performance. Note that our proposed method utilizes tissue manual

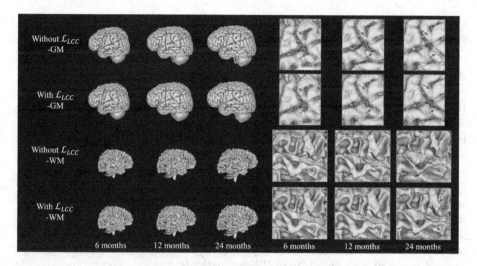

Fig. 5. Visual evidence of the proposed longitudinally consistent segmentation on the extracted surfaces of the GM and WM of infant brains.

labels to obtain the deformation fields while other methods do not, which implies the effectiveness of the utilization of manual labels. Quantitatively, we evaluate the registration performance through two metrics: (1) Dice coefficients of the GM, WM, and CSF; and (2) Hausdorff distance between the segmentations masks on the fixed image and the warped images. As shown in Table 1, we can see that our method is generally superior to the other methods, which is consistent with the visual comparison.

Segmentation Performance. The qualitative and quantitative segmentation results of isointense phase images are illustrated in Fig. 4 and Table 2, respectively. We can see that our proposed method generally achieves better results. Compared with segmentation methods without the assistance of warped intensity from other time points (DenseVoxelNet and U-Net (single month)), our method achieves better results, which illustrates the effectiveness of assistance of the high contrast warped intensity images. In addition, compared with the segmentation methods with different ways of fusion module (HyperTransformer and U-Net (Transformer)), the better results of our method illustrates the effectiveness of our multi-branch, multi-level fusion structure.

For an in-depth evaluation of the merit of our longitudinally consistent segmentation network, we illustrate the reconstructed surface of GM and WM based on the segmentation results at different time points in Fig. 5. The surfaces of different tissues are reconstructed through the marching cube algorithm. We compare the segmentation results with and without using the proposed LCC loss. We can easily see that the reconstructed surfaces based on the segmentation results with the longitudinally consistent constraint maintain more consistent anatomical structures than the ones without using LCC, which indicates the effectiveness of the proposed longitudinally consistent constraints.

4   Conclusion

In this paper, we propose a joint registration and segmentation method, which could help achieve longitudinally consistent segmentation for serial infant brain images and improve the segmentation accuracy of the isointense phase. Particularly, we first utilize a manual label guided registration network to obtain the warped masks and the corresponding warped intensity images from the other time points. Then the warped masks are utilized to constrain the longitudinally consistent segmentation. In addition, for the isointense phase segmentation, we utilize the warped intensity images with high intensity contrast to improve the segmentation accuracy of the isointense phase. Experimental results show that our proposed method achieves better performance for serial infant brain segmentation compared with the state-of-the-art methods.

Acknowledgment. This work was supported in part by National Natural Science Foundation of China (No. 62131015 and 62203355), and Science and Technology Commission of Shanghai Municipality (STCSM) (No. 21010502600), and The Key R&D Program of Guangdong Province, China (No. 2021B0101420006).

References

1. Sowell, E.R., Thompson, P.M., Leonard, C.M., Welcome, S.E., Kan, E., Toga, A.W.: Longitudinal mapping of cortical thickness and brain growth in normal children. J. Neurosci. **24**(38), 8223–8231 (2004)
2. Zhang, C., Adeli, E., Wu, Z., Li, G., Lin, W., Shen, D.: Infant brain development prediction with latent partial multi-view representation learning. IEEE Trans. Med. Imaging **38**(4), 909–918 (2018)
3. Chen, L., et al.: A 4D infant brain volumetric atlas based on the UNC/UMN baby connectome project (BCP) cohort. Neuroimage **253**, 119097 (2022)
4. Shi, F., et al.: Infant brain atlases from neonates to 1-and 2-year-olds. PLoS ONE **6**(4), e18746 (2011)
5. Hazlett, H.C., et al.: Brain volume findings in 6-month-old infants at high familial risk for autism. Am. J. Psychiatry **169**(6), 601–608 (2012)
6. Weisenfeld, N.I., Warfield, S.K.: Automatic segmentation of newborn brain MRI. Neuroimage **47**(2), 564–572 (2009)
7. Avants, B.B., Tustison, N., Song, G., et al.: Advanced normalization tools (ANTs). Insight j **2**(365), 1–35 (2009)
8. Thirion, J.P.: Image matching as a diffusion process: an analogy with maxwell's demons. Med. Image Anal. **2**(3), 243–260 (1998)
9. Wang, L., Shi, F., Yap, P.T., Gilmore, J.H., Lin, W., Shen, D.: 4D multi-modality tissue segmentation of serial infant images (2012)
10. Wu, G., Wang, L., Gilmore, J., Lin, W., Shen, D.: Joint segmentation and registration for infant brain images. In: Menze, B., et al. (eds.) MCV 2014. LNCS, vol. 8848, pp. 13–21. Springer, Cham (2014). https://doi.org/10.1007/978-3-319-13972-2_2

11. Shi, F., Yap, P.-T., Gilmore, J.H., Lin, W., Shen, D.: Spatial-temporal constraint for segmentation of serial infant brain MR images. In: Liao, H., Edwards, P.J.E., Pan, X., Fan, Y., Yang, G.-Z. (eds.) MIAR 2010. LNCS, vol. 6326, pp. 42–50. Springer, Heidelberg (2010). https://doi.org/10.1007/978-3-642-15699-1_5

12. Wang, L., Shi, F., Yap, P.-T., Gilmore, J.H., Lin, W., Shen, D.: Accurate and consistent 4D segmentation of serial infant brain MR images. In: Liu, T., Shen, D., Ibanez, L., Tao, X. (eds.) MBIA 2011. LNCS, vol. 7012, pp. 93–101. Springer, Heidelberg (2011). https://doi.org/10.1007/978-3-642-24446-9_12

13. Nie, D., Wang, L., Adeli, E., Lao, C., Lin, W., Shen, D.: 3-D fully convolutional networks for multimodal isointense infant brain image segmentation. IEEE Trans. Cybern. **49**(3), 1123–1136 (2018)

14. Qamar, S., Jin, H., Zheng, R., Ahmad, P., Usama, M.: A variant form of 3D-UNet for infant brain segmentation. Futur. Gener. Comput. Syst. **108**, 613–623 (2020)

15. Payakachat, N., Tilford, J.M., Ungar, W.J.: National database for autism research (NDAR): big data opportunities for health services research and health technology assessment. Pharmacoeconomics **34**(2), 127–138 (2016)

16. Bandara, W.G.C., Patel, V.M.: HyperTransformer: a textural and spectral feature fusion transformer for pansharpening. In: Proceedings of the IEEE/CVF Conference on Computer Vision and Pattern Recognition (CVPR), pp. 1767–1777 (2022)

17. Yu, L., et al.: Automatic 3D cardiovascular MR segmentation with densely-connected volumetric ConvNets. In: Descoteaux, M., Maier-Hein, L., Franz, A., Jannin, P., Collins, D.L., Duchesne, S. (eds.) MICCAI 2017. LNCS, vol. 10434, pp. 287–295. Springer, Cham (2017). https://doi.org/10.1007/978-3-319-66185-8_33

18. Çiçek, Ö., Abdulkadir, A., Lienkamp, S.S., Brox, T., Ronneberger, O.: 3D U-Net: learning dense volumetric segmentation from sparse annotation. In: Ourselin, S., Joskowicz, L., Sabuncu, M.R., Unal, G., Wells, W. (eds.) MICCAI 2016. LNCS, vol. 9901, pp. 424–432. Springer, Cham (2016). https://doi.org/10.1007/978-3-319-46723-8_49

Unifying and Personalizing Weakly-Supervised Federated Medical Image Segmentation via Adaptive Representation and Aggregation

Li Lin[1,2,3], Jiewei Wu[1], Yixiang Liu[1], Kenneth K. Y. Wong[2], and Xiaoying Tang[1,3](✉)

[1] Department of Electrical and Electronic Engineering, Southern University of Science and Technology, Shenzhen, China
tangxy@sustech.edu.cn
[2] Department of Electrical and Electronic Engineering, The University of Hong Kong, Hong Kong SAR, China
[3] Jiaxing Research Institute, Southern University of Science and Technology, Jiaxing, China

Abstract. Federated learning (FL) enables multiple sites to collaboratively train powerful deep models without compromising data privacy and security. The statistical heterogeneity (e.g., non-IID data and domain shifts) is a primary obstacle in FL, impairing the generalization performance of the global model. Weakly supervised segmentation, which uses sparsely-grained (i.e., point-, bounding box-, scribble-, block-wise) supervision, is increasingly being paid attention to due to its great potential of reducing annotation costs. However, there may exist label heterogeneity, i.e., different annotation forms across sites. In this paper, we propose a novel personalized FL framework for medical image segmentation, named FedICRA, which uniformly leverages heterogeneous weak supervision via adaptIve Contrastive Representation and Aggregation. Concretely, to facilitate personalized modeling and to avoid confusion, a channel selection based site contrastive representation module is employed to adaptively cluster intra-site embeddings and separate inter-site ones. To effectively integrate the common knowledge from the global model with the unique knowledge from each local model, an adaptive aggregation module is applied for updating and initializing local models at the element level. Additionally, a weakly supervised objective function that leverages a multiscale tree energy loss and a gated CRF loss is employed to generate more precise pseudo-labels and further boost the segmentation performance. Through extensive experiments on two distinct medical image segmentation tasks of different modalities, the proposed FedICRA demonstrates overwhelming performance over other state-of-the-art personalized FL methods. Its performance even approaches that of fully supervised training on centralized data. Our code and data are available at https://github.com/llmir/FedICRA.

L. Lin, J. Wu, Y. Liu—contributed equally to this work.

X. Cao et al. (Eds.): MLMI 2023, LNCS 14348, pp. 196–206, 2024.
https://doi.org/10.1007/978-3-031-45673-2_20

Keywords: Personalized federated learning · Heterogeneous weak supervision · Contrastive representation · Adaptive aggregation

1   Introduction

Deep learning methods have been widely adopted in many computer vision and medical image analysis tasks due to their impressive performance [3,15,16]. However, as a data-driven approach, its performance is highly reliant on the quantity of accessible data. Data sharing across institutions, particularly medical sites, is often infeasible due to regulatory constraints and privacy concerns regarding user/patient data [7]. In such cases, federated learning (FL) [8] has come to the fore and is receiving growing interest from the community as it allows different centers to collaboratively train a powerful global model without the need for sharing or centralizing data. As an illustration, FedAvg [22], a prevailing FL approach, averages models from all sites by their sample weights on the server side and then broadcasts the averaged result back to each site. Although FL has recently made promising progress in the medical image segmentation realm, most efforts still focus on improving the generalizability of a single global model via enhancing the aggregation strategy [10,11,17] or employing data augmentation [18,27]. Moreover, almost all existing FL studies are conducted under fully supervised paradigms.

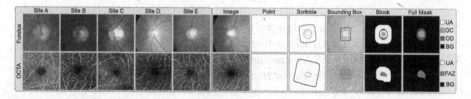

Fig. 1. Left: Data samples from different sites showcase the domain shifts in their distributions; Right: Examples of various types of sparse annotations and the corresponding full masks. UA, OC, OD, FAZ, and BG respectively represent unlabeled area, optic cup, optic disc, foveal avascular zone, and background.

A key obstacle in FL is statistical heterogeneity, mainly induced by non independent and identically distributed (non-IID) data, which makes it difficult for a single global model to perform well across all sites. FL's degradation could be even more severe in medical image analysis scenarios, as there may exist more diverse image domain shifts caused by differences in the imaging devices, protocols, patient populations, and physician expertise, as shown in Fig. 1 (Left). Personalized FL (pFL) has been proposed as a solution to these challenges, with the aim of training more customized local models for different sites. Existing pFL methods can be mainly classified into three categories: (1) fine-tuning the trained global model at each site, such as FedAvg with fine-tuning (FT) [32]; (2) partitioning the model into a global part and a personalized part, aggregating the global part on the server while retaining the locally trained parameters for the personalized part, such as FedBN and FedRep [4,12]; (3) combining information from other sites through local aggregation or knowledge distillation, such as FedAP and MetaFed [2,19,35]. Methods in category (1) risk forgetting common knowledge and overfitting local data, while those in (2) fail to leverage useful knowledge from other sites. Methods in category (3) relieve these issues, but local aggregation is typically performed at the model or layer level, which may result in excessive personalization and lead to performance degradation when the data heterogeneity is not that significant.

Recently, weakly supervised segmentation (WSS) has received significant research interest due to its ability to reduce annotation costs. WSS methods typically leverage sparsely-grained (e.g., point-, bounding box-, scribble-, block-wise) supervision through novel loss function designing, consistency learning, adversarial learning, or data synthesis [13,14,24]. Integrating WSS into FL can further reduce the annotation cost at each site while exploiting the benefits of FL. However, there is currently a lack of research on WSS in the context of pFL. A more challenging yet more practical setting is different sites have different forms of weak labels, as shown in Fig. 1 (Right), which further introduces label heterogeneity. In such context, designing a unified and easily deployable pFL framework for the WSS setting is highly desirable.

Here, we propose a novel pFL framework for medical image segmentation, named FedICRA, which leverages heterogeneous weak supervision via adapt**I**ve **C**ontrastive **R**epresentation and **A**ggregation. Specifically, to facilitate model personalization, a channel selection based **S**ite **C**ontrastive **R**epresentation (**SCR**) module is applied, adaptively characterizing the data distributions of different sites as being tightly self-related and different from each other. An **A**daptive **A**ggregation (**AA**) module is proposed to element-wisely aggregate each local model with the global one towards the local objective at each site. Moreover, a weakly supervised loss is designed to uniformly leverage heterogeneous labels.

Our main contributions are four-fold: (1) To our knowledge, we are the first to propose a pFL framework for heterogeneous WSS. (2) **SCR** enhances inter-site representation contrast via channel attention, while **AA** element-wisely aggregates the global model's generic knowledge and each local model's specific knowledge. Both modules effectively promote personalization but also alleviate over-personalization. (3) A WSS objective that leverages a multiscale tree energy loss and a gated CRF loss is proposed to uniformly leverage heterogeneous labels and generate high-quality pseudo-labels for better training. (4) We validate the effectiveness of our proposed method on two medical image segmentation tasks and our approach achieves state-of-the-art (SOTA) performance on both tasks.

2   Methodology

The overall pipeline of our FedICRA is provided in Fig. 2. We first overview our pFL paradigm, and then introduce details of the **SCR** and **AA** modules, as well as our weakly supervised objective in the following paragraphs.

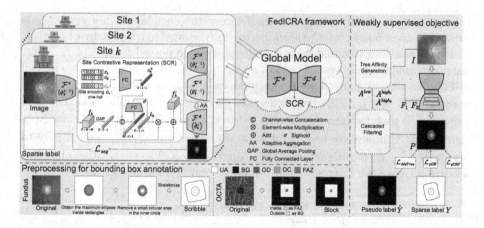

Fig. 2. Schematic representation of the proposed FedICRA framework.

Federation Paradigm Overview. Suppose there are K sites with private training data $D_1, ..., D_k$. These data are of distinct distributions and have different label types. With the help of a central server, FedICRA aims to collaboratively train a personalized model for each site without sharing data. The global and all local models share the same network architecture, namely UNet [26] with four encoding and decoding levels and an SCR module inserted in the middle. For more generalizable representations, we set the representation part of the network (encoder and SCR with parameter ϕ) as the global part and the task head (decoder with parameter θ) as the personalized part. FedICRA alternatively updates between local sites and the global server in each communication round. At round t, all sites receive the same parameters $(\phi_g^{t-1}, \theta_g^{t-1})$ from the server. The global part is initialized with ϕ_g^{t-1}, and the personalized part is initialized with $\hat{\theta}_k^t$ which is obtained from the AA module based on θ_g^{t-1} and the local parameters from the previous round (e.g., θ_k^{t-1} for site k). Each site updates its model by optimizing the local objective with its own data and its site encoding c_k utilized in SCR

$$\phi_k^t, \theta_k^t \leftarrow \text{GRD}\left(\phi_g^{t-1}, \hat{\theta}_k^t \mid D_k, c_k\right), \tag{1}$$

where $\text{GRD}(\cdot)$ denotes the local gradient-based update. As for the aggregation of the global model, we adopt the weighted averaging strategy in FedAvg [22].

Site Contrastive Based Channel Selection. pFL paradigms may suffer from confusion or over-personalization when data heterogeneity is low, performing even worse than traditional FL methods [35]. Inspired by the prompt-driven paradigms [31,36], the SCR module is designed to enhance the distance/contrast of inter-site data representations through site-contrastive learning based channel attention, which in turn facilitates personalization. Specifically, taking the k-th site as an example, a one-hot site encoding c_k (i.e., the k-th position is 1 and others are 0) and the output feature f_k from the encoder $\mathcal{F}^e$ are given. c_k is expanded to a length of C through two fully connected layers to obtain c_k^*, which is then concatenated with the global average pooled feature of f_k. After passing through a fully connected layer with Sigmoid activation, the site channel attention value $\hat{c}_k$ is obtained. The final feature f_k' is obtained in a residual manner: $f_k' = f_k + f_k \otimes \hat{c}_k$, where $\otimes$ denotes element-wise multiplication. The final feature f_k' is then fed into the decoder $\mathcal{F}^d$ to obtain the final segmentation. Note that during training, we assign different site encodings $\{c_i\}_{i=1}^K$ sequentially to obtain $\{\hat{c}_i\}_{i=1}^K$ with different site styles/distributions. The contrastive objective used to increase the inter-site embedding difference is formulated as

$$\mathcal{L}_{con} = -\frac{1}{K-1} \sum_{i=1}^K |\hat{c}_k - \text{StopGradient}(\hat{c}_i)|, \text{ s.t. } i \neq k. \tag{2}$$

Adaptive Head Aggregation. Following existing pFL paradigms, we globally share the representation part of the segmentation network (encoder $\mathcal{F}^e$ and the SCR module) and personalize the task head (decoder $\mathcal{F}^d$). To element-wisely aggregate two models without introducing multiple aggregation weight matrices, we adopt an adaptive learning based approach similar to residual learning for updating the weight matrices. The adaptive head aggregation is formulated as

$$\hat{\theta}_i^t := \theta_g^{t-1} + \left(\theta_i^{t-1} - \theta_g^{t-1}\right) \odot W_i, \tag{3}$$

where W_i is a learnable weight matrix. Inspired by [20], we clip and make $w \in [0,1]$, $\forall w \in W_i$ for regularization via $\sigma(w) = \max(0, \min(1, w))$. At the beginning, every element in W_i is initialized to be 1, and then iteratively gets updated via

$$W_i \leftarrow \text{GRD}\left(W_i^{t-1} \mid D_i, \hat{\theta}_i^t\right). \tag{4}$$

In the update process of W_i, Eqs. (3) and (4) are respectively utilized to alternatively update $\hat{\theta}_i^t$ and W_i. Upon convergence, the initialization parameter $\hat{\theta}_i^t$ for $\mathcal{F}^d$ in each federation round is obtained, after which local models are trained using Eq. (1). Note that when $t > 2$, only one epoch is trained to obtain W_i.

Weakly Supervised Objective. The WSS task aims to train a dense prediction model based on sparse annotations, wherein most pixels are unlabeled. To reduce the communication cost and enable all sites to cooperatively utilize different sparse labels in a uniform and easily deployable manner, we boost the WSS performance by optimizing the objective function rather than the network architecture nor the training strategy. For each image in the training set, it can be divided into a labeled set I_L and an unlabeled set I_U. The commonly used partial cross-entropy loss ($\mathcal{L}_{pCE}$) [30] is employed for I_L, which is formulated as

$$\mathcal{L}_{pCE} = -\frac{1}{|I_L|} \sum_{\forall i \in I_L} Y_i \log(P_i), \tag{5}$$

where Y_i and P_i respectively denote the sparse ground truth and the predicted probability of pixel i. The key to improving the WSS performance lies in how to effectively utilize I_U to generate accurate pseudo-labels. Since pixels belonging to the same object share similar patterns across different feature levels, we follow the design of tree filters [13,29]; we adapt the tree energy loss into a multi-scale recursive version that better accommodates medical image segmentation. Concretely, triple affinity matrices A^{low}, A^{high_1} and A^{high_2} are calculated via a *Borůvka* algorithm [6] based generation module respectively from the original image as well as feature maps from the second and third decoding stages. A^{low} contains object boundary information, while A^{high_1} and A^{high_2} maintain semantic consistency. By employing the same cascaded filtering operation $\mathcal{F}(\cdot)$ in [13], the pseudo-labels $\tilde{Y}$ can be generated through

$$\tilde{Y} = \mathcal{F}\left(\mathcal{F}\left(\mathcal{F}\left(P, A^{low}\right), A^{high_1}\right), A^{high_2}\right),\tag{6}$$

and the multi-scale recursive tree energy loss $(\mathcal{L}_{MsTree})$ goes as

$$\mathcal{L}_{\mathrm{MsTree}} = -\frac{1}{|I_U|}\sum_{\forall i \in I_U}\left|P_i - \tilde{Y}_i\right|.\tag{7}$$

Additionally, the gated CRF loss $(\mathcal{L}_{gCRF})$ [24] is employed as an extra regularization term to further increase the edge accuracy and reduce outliers in P and $\tilde{Y}$. Therefore, with trade-off parameters $\lambda_1, \lambda_2, \lambda_3$, the WSS objective and the total local objective can be respectively expressed as

$$\mathcal{L}_{seg} = \mathcal{L}_{pCE} + \lambda_1 \mathcal{L}_{\mathrm{MsTree}} + \lambda_2 \mathcal{L}_{gCRF},\tag{8}$$

$$\mathcal{L}_i = \mathcal{L}_{seg} + \lambda_3 \mathcal{L}_{con}.\tag{9}$$

3   Experiments and Results

Datasets and Preprocessing. Extensive experiments are conducted to verify the effectiveness of the proposed FedICRA on two medical image segmentation datasets, including optic disc/cup (ODOC) segmentation from fundus images and foveal avascular zone (FAZ) segmentation from optical coherence tomography angiography (OCTA) images. Both fundus images [5,25,28,34] and OCTA images [1,9,21,33] are obtained from publicly available datasets with heterogeneous distributions. For fundus images, we follow [18] to create the first four sites and add GAMMA [34] as the fifth site. The datasets for the five sites consist of {101, 159, 400, 400, 200} samples. For OCTA images, we partition the original dataset into five sites, respectively containing {304, 200, 300, 1012, 39} samples. Fundus images are center-cropped and then resized to 384 × 384 while OCTA images are directly resized to 256 × 256. We utilize automated algorithms to simulate point, scribble (two styles), bounding box, and block annotations for each site based on the original segmentation masks. The train-test splits of all datasets are in line with the original dataset partitioning; more information about the constructed datasets can be found in Table A1 in the appendix.

Data preprocessing includes normalizing all image intensities to between 0 and 1, while data augmentation includes randomly flipping images horizontally and vertically as well as rotation (spanning from -45° to 45°). Moreover, since bounding box annotations are not a direct sparse supervision signal in WSS, we perform certain preprocessing to covert them into scribble and block respectively for ODOC and FAZ segmentation, as shown in the lower left panel of Fig. 2.

Implementation Details. All compared FL methods and the proposed Fed-ICRA are implemented with PyTorch using NVIDIA A100 GPUs. We employ the vanilla UNet as the model architecture, with the number of channels progressively increasing from 16 to 256 from top to bottom. We use the AdamW optimizer with an initial learning rate of 1×10^{-2} to optimize the parameters and the polynomial policy with $power = 0.9$ to dynamically adjust the learning rate [23]. The trade-off coefficients $\lambda_1, \lambda_2, \lambda_3$ and batch size are respectively set to be 0.1, 0.1, 1 and 12. We train all FL methods for 500 federation rounds to ensure fair performance comparisons.

Evaluation Results. We compare FedICRA with several FL frameworks including traditional FL (FedAvg [22] and FedProx [10]) and SOTA pFL methods (i.e., FT [32], FedBN [12], FedAP [19], FedRep [4] and MetaFed [2]). Most of these methods are originally designed for classification, and we endeavor to maintain their design principles while adapting them to segmentation tasks. All the comparative methods are trained on the same datasets under the weakly supervised setting, consistent with FedICRA. We also compare with some baseline and ideal settings, including local training (LT) with weak labels and full labels, as well as centralized training (CT). All methods are evaluated using two metrics, i.e., the Dice similarity coefficient (DSC) and the 95% Hausdorff distance (HD95[px]).

Table 1 and Table A2 (in the appendix) display the quantitative results for the ODOC segmentation task. Compared to LT, all FL methods improve the overall segmentation performance of all sites. Sites C and D achieve significant performance improvement after participating in FL as their local data are distributed diversely, making it difficult for them to train sufficiently powerful models using local data alone. Due to the high data heterogeneity across sites, as shown in Fig. A2, pFL methods benefit from training personalized models for different sites and generally achieve higher performance than centralized FL methods. FedAP uses BN layers to measure the inter-site data distribution similarity and aggregate personalized models, thereby extracting useful information from other sites and achieving good performance. Notably, our FedICRA achieves further performance improvement with a DSC about 2% higher than FedAP, approaching centralized training with fully supervised labels.

Table 2 tabulates the quantitative results for the FAZ segmentation task. Different from the ODOC task, due to low data heterogeneity as shown in the left part of Fig. 4, most pFL methods (such as FedBN, FedAP and MetaFed) encounter confusion or over-personalization, performing even worse

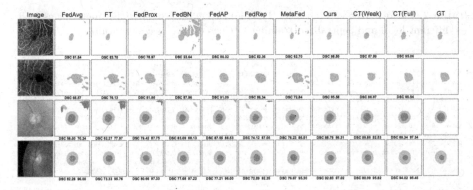

Fig. 3. Visualization results from FedICRA and other SOTA methods (**CT** indicates centralized training, with **Weak** and **Full** respectively denoting utilizing sparse annotations and full masks).

Table 1. Performance comparisons (DSC) of different FL methods as well as different local and centralized training settings on ODOC segmentation. The best results are highlighted in bold and the second-best ones are underlined.

Methods	DSC (OD) ↑						DSC (OC) ↑						Overall
	Site A	Site B	Site C	Site D	Site E	Avg.	Site A	Site B	Site C	Site D	Site E	Avg.	
FedAvg [22]	92.58±5.69	83.90±16.18	94.67±2.20	93.36±6.71	88.92±8.20	90.68	83.16±14.61	69.91±18.68	85.35±7.25	82.80±11.32	86.76±12.79	81.60	86.14
FedProx [10]	95.19±4.81	82.14±18.66	95.03±2.45	88.28±11.84	91.06±8.92	90.34	82.18±14.42	70.14±17.98	84.64±7.05	81.33±12.79	87.19±9.63	81.10	85.72
FT [32]	95.97±2.17	91.09±3.62	94.83±2.18	93.34±3.49	90.10±8.19	93.06	85.02±13.79	80.76±10.89	82.51±7.84	82.95±7.50	86.88±11.16	83.62	88.34
FedBN [12]	95.88±1.94	92.66±3.32	94.80±2.64	95.06±2.73	90.72±10.28	93.82	84.54±12.93	82.09±10.22	83.79±7.79	86.48±6.26	79.38±15.28	83.25	88.54
FedAP [19]	95.81±2.17	92.24±3.31	95.45±1.91	95.17±2.26	90.73±10.36	93.88	83.97±13.78	79.89±12.41	83.55±7.20	86.78±6.86	83.47±12.86	83.53	88.71
FedRep [4]	95.28±2.70	88.10±8.98	92.68±4.34	92.34±5.91	88.75±10.91	91.43	82.47±14.19	76.22±14.68	82.14±8.10	83.04±8.49	81.29±13.64	81.03	86.23
MetaFed [2]	91.71±8.77	91.48±7.55	88.70±9.20	91.76±6.96	82.29±11.94	89.19	80.10±18.55	79.43±11.13	83.14±8.02	85.95±6.87	81.38±15.27	82.01	85.59
FedICRA (Ours)	96.51±1.52	93.85±2.74	94.62±1.79	94.06±2.68	94.10±4.23	94.63	85.44±13.32	81.42±10.36	87.89±5.51	88.76±5.59	89.68±5.30	86.64	90.63
LT (Weak)	94.83±4.25	90.65±5.16	89.31±8.68	76.38±15.11	86.11±14.03	87.45	84.50±12.50	79.31±14.33	84.11±7.70	77.02±9.68	81.38±16.29	81.26	84.36
CT (Weak)	95.45±2.56	91.41±3.32	95.45±1.94	91.92±4.91	91.21±8.55	93.09	84.25±13.95	78.94±11.40	84.63±6.49	85.68±9.04	87.08±12.14	84.12	88.60
LT (Full)	96.28±2.69	96.21±2.50	95.21±3.95	95.56±1.80	94.33±10.64	95.52	87.94±8.08	81.72±12.28	87.22±6.41	88.60±7.37	86.53±12.24	86.40	90.96
CT (Full)	96.97±1.37	94.28±4.25	96.06±1.77	96.20±1.66	95.52±8.37	95.81	87.40±11.90	82.90±9.07	88.30±6.39	89.17±5.82	89.25±11.47	87.40	91.61

than FedAvg. And FT attains the second-best performance. Our FedICRA leverages the SCR module to enhance the inter-site data representation contrast (Fig. 4), thereby mitigating the aforementioned issue. As a result, it is still capable of training high-performance personalized model for each site and achieves the best performance across all sites. FedICRA outperforms FT by a large margin and achieves performance close to CT with full supervision. Visualization results for the two tasks are provided in Fig. 3 and Fig. A3.

We also calculate the p-values of the DSC values between FedICRA and the second-best method on OD, OC and FAZ tasks (respective 6.79×10^{-6}, 1.09×10^{-18} and 1.63×10^{-77}), identifying statistically significant performance improvements of FedICRA. Table 3 shows the ablation results on the four key components (i.e., **AA**, **SCR**, $\mathcal{L}_{MsTree}$, and $\mathcal{L}_{gCRF}$) of FedICRA. The first row represents the performance of FedAvg. Experimental results demonstrate the importance of the four key components. Even without the two WSS losses, incorporating the two FL modules still outperforms other FL methods on both segmentation tasks. It is worth noting that the last two rows of Table 1 and Table 2 show that centerlizing data or using FL may not provide significant

Table 2. Performance comparisons (DSC and HD95) of different FL methods as well as different local and centralized training settings on FAZ segmentation.

Methods	DSC↑						HD95[px]↓					
	Site A	Site B	Site C	Site D	Site E	Avg.	Site A	Site B	Site C	Site D	Site E	Avg.
FedAvg [22]	76.56±28.93	90.65±6.01	77.64±15.72	88.85±6.13	86.55±6.13	84.05	5.85±6.56	8.28±6.12	22.73±38.04	12.45±15.87	21.59±35.61	14.18
FedProx [10]	73.87±29.34	89.70±7.19	77.44±18.97	89.15±5.24	83.78±7.75	82.79	7.38±8.83	8.63±5.83	23.07±39.18	12.18±15.51	22.62±35.25	14.78
FT [32]	83.18±7.91	91.96±3.89	78.58±16.28	88.75±5.54	88.10±7.40	86.11	10.11±13.02	8.86±11.00	20.84±36.52	11.53±13.30	20.17±36.16	14.3
FedBN [12]	62.28±25.01	90.11±5.17	60.93±21.41	87.87±5.95	53.05±23.92	70.84	51.03±56.30	9.42±10.89	60.14±58.95	11.84±12.11	20.31±8.85	30.54
FedAP [19]	62.23±10.57	87.82±4.39	72.67±14.79	90.36±5.23	66.00±9.47	75.82	18.50±15.87	7.71±2.54	14.51±20.16	8.30±5.71	18.60±4.70	13.52
FedRep [4]	78.55±18.86	91.69±4.40	79.29±18.13	91.04±4.95	86.83±11.46	85.48	16.07±22.16	9.14±16.50	15.24±26.29	7.64±3.75	19.92±36.11	13.60
MetaFed [2]	70.99±29.03	86.22±9.19	74.15±21.40	88.77±6.24	74.62±21.15	78.95	9.18±12.04	10.37±6.43	24.61±38.99	13.91±16.95	39.61±47.34	19.54
FedICRA (Ours)	88.50±6.40	97.42±1.52	90.84±8.72	94.32±4.73	95.28±0.99	93.29	5.77±3.14	4.05±2.81	8.24±13.35	5.75±3.68	4.14±1.04	5.59
LT (Weak)	73.74±27.11	91.90±2.77	79.03±16.57	85.78±7.32	79.89±19.92	82.07	30.94±47.78	42.49±45.28	20.81±35.48	10.87±4.81	72.08±56.19	35.44
CT (Weak)	74.93±23.48	89.24±3.94	76.64±16.37	89.25±5.40	86.50±4.71	83.31	8.38±6.28	7.54±2.42	10.46±10.96	8.87±3.78	8.28±1.41	8.7
LT (Full)	90.88±5.06	97.75±2.13	89.22±17.63	95.14±5.46	95.11±3.00	93.62	5.31±6.00	3.36±3.49	7.84±13.20	4.80±3.08	4.54±2.61	5.17
CT (Full)	90.93±0.052	97.23±1.42	91.49±8.26	94.88±4.27	95.38±1.17	93.98	4.57±3.04	3.66±3.33	6.86±10.32	4.82±3.17	2.99±0.86	4.58

Table 3. Ablation study on the four key elements in FedICRA (DSC).

AA	SCR	$\mathcal{L}_{MsTree}$	$\mathcal{L}_{gCRF}$	OD	OC	FAZ
–	–	–	–	90.68	81.60	84.05
✓				92.98	84.67	88.96
	✓			93.19	84.07	88.27
✓	✓			93.65	85.10	89.20
✓	✓	✓		93.86	85.57	89.38
✓	✓		✓	94.63	85.82	92.47
✓	✓	✓	✓	94.63	86.64	93.29

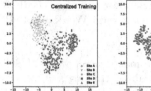

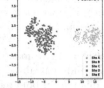

Fig. 4. Comparison of data embeddings on OCTA test sets via t-SNE visualization.

improvements over CT under the fully supervised setting. However, FL under weak supervision clearly highlights the importance of collaboration to reduce annotation costs and improve performance.

4 Conclusion

This paper presents a novel pFL framework, FedICRA, for personalized medical image segmentation. Through extensive experiments on two different modalities and two different segmentation tasks, the effectiveness of the proposed method has been demonstrated. Additionally, we have shown that under heterogeneous weakly-supervised conditions, FedICRA can achieve segmentation performance that is comparable to that of labor-intensive fully-supervised centralized training.

References

1. Agarwal, A., Raman, R., et al.: The foveal avascular zone image database (fazid). In: Applications of Digital Image Processing, pp. 507–512. SPIE (2020)
2. Chen, Y., Lu, W., et al.: MetaFed: Federated learning among federations with cyclic knowledge distillation for personalized healthcare. arXiv preprint arXiv:2206.08516 (2022)
3. Cheng, P., Lin, L., et al.: Prior guided fundus image quality enhancement via contrastive learning. In: 2021 IEEE 18th International Symposium on Biomedical Imaging (ISBI), pp. 521–525. IEEE (2021)

4. Collins, L., Hassani, H., Mokhtari, A., Shakkottai, S.: Exploiting shared representations for personalized federated learning. In: International Conference on Machine Learning, pp. 2089–2099. PMLR (2021)

5. Fumero, F., Alayón, S., Sanchez, J.L., Sigut, J., Gonzalez-Hernandez, M.: RIM-ONE: an open retinal image database for optic nerve evaluation. In: International Symposium on Computer-Based Medical Systems (CBMS), pp. 1–6. IEEE (2011)

6. Gallager, R.G., Humblet, P.A., Spira, P.M.: A distributed algorithm for minimum-weight spanning trees. ACM Trans. Programm. Lang. Syst. (TOPLAS) **5**(1), 66–77 (1983)

7. Goddard, M.: The EU general data protection regulation (GDPR): European regulation that has a global impact. Int. J. Mark. Res. **59**(6), 703–705 (2017)

8. Konečný, J., McMahan, et al.: Federated learning: Strategies for improving communication efficiency. arXiv preprint arXiv:1610.05492 (2016)

9. Li, M., Zhang, Y., et al.: IPN-V2 and OCTA-500: Methodology and dataset for retinal image segmentation. arXiv preprint arXiv:2012.07261 (2020)

10. Li, T., Sahu, A.K., et al.: Federated optimization in heterogeneous networks. In: Proceedings of Machine Learning and Systems, pp. 429–450. PMLR (2020)

11. Li, W., et al.: Privacy-preserving federated brain tumour segmentation. In: Suk, H.-I., Liu, M., Yan, P., Lian, C. (eds.) MLMI 2019. LNCS, vol. 11861, pp. 133–141. Springer, Cham (2019). https://doi.org/10.1007/978-3-030-32692-0_16

12. Li, X., Jiang, M., et al.: FedBN: Federated learning on non-IID features via local batch normalization. arXiv preprint arXiv:2102.07623 (2021)

13. Liang, Z., Wang, T., et al.: Tree energy loss: towards sparsely annotated semantic segmentation. In: Proceedings of the IEEE/CVF Conference on Computer Vision and Pattern Recognition (CVPR), pp. 16907–16916. IEEE (2022)

14. Lin, L., Peng, L., et al.: YoloCurvSeg: You only label one noisy skeleton for vessel-style curvilinear structure segmentation. arXiv preprint arXiv:2212.05566 (2022)

15. Lin, L., et al.: BSDA-Net: a boundary shape and distance aware joint learning framework for segmenting and classifying OCTA images. In: de Bruijne, M., et al. (eds.) MICCAI 2021. LNCS, vol. 12908, pp. 65–75. Springer, Cham (2021). https://doi.org/10.1007/978-3-030-87237-3_7

16. Lin, L., Wu, J., Cheng, P., Wang, K., Tang, X.: BLU-GAN: Bi-directional ConvL-STM U-Net with generative adversarial training for retinal vessel segmentation. In: Gao, W., et al. (eds.) FICC 2020. CCIS, vol. 1385, pp. 3–13. Springer, Singapore (2021). https://doi.org/10.1007/978-981-16-1160-5_1

17. Liu, D., Cabezas, M., et al.: MS lesion segmentation: Revisiting weighting mechanisms for federated learning. arXiv preprint arXiv:2205.01509 (2022)

18. Liu, Q., Chen, C., et al.: FedDG: federated domain generalization on medical image segmentation via episodic learning in continuous frequency space. In: Proceedings of the IEEE CVPR, pp. 1013–1023 (2021)

19. Lu, W., Wang, J., et al.: Personalized federated learning with adaptive batchnorm for healthcare. IEEE Transactions on Big Data, early access (2022)

20. Luo, L., Xiong, Y., et al.: Adaptive gradient methods with dynamic bound of learning rate. arXiv preprint arXiv:1902.09843 (2019)

21. Ma, Y., Hao, H., et al.: ROSE: a retinal OCT-angiography vessel segmentation dataset and new model. IEEE Trans. Med. Imaging **40**(3), 928–939 (2020)

22. McMahan, B., Moore, E., et al.: Communication-efficient learning of deep networks from decentralized data. In: Artificial Intelligence and Statistics, pp. 1273–1282. PMLR (2017)

23. Mishra, P., Sarawadekar, K.: Polynomial learning rate policy with warm restart for deep neural network. In: TENCON 2019–2019 IEEE Region 10 Conference (TENCON), pp. 2087–2092. IEEE (2019)

24. Obukhov, A., Georgoulis, S., Dai, D., Van Gool, L.: Gated CRF loss for weakly supervised semantic image segmentation. arXiv preprint arXiv:1906.04651 (2019)

25. Orlando, J.I., Fu, H., et al.: Refuge challenge: a unified framework for evaluating automated methods for glaucoma assessment from fundus photographs. Med. Image Anal. **59**, 101570 (2020)

26. Ronneberger, O., Fischer, P., Brox, T.: U-Net: convolutional networks for biomedical image segmentation. In: Navab, N., Hornegger, J., Wells, W.M., Frangi, A.F. (eds.) MICCAI 2015. LNCS, vol. 9351, pp. 234–241. Springer, Cham (2015). https://doi.org/10.1007/978-3-319-24574-4_28

27. Shen, C., et al.: Multi-task federated learning for heterogeneous pancreas segmentation. In: Oyarzun Laura, C., et al. (eds.) DCL/PPML/LL-COVID19/CLIP -2021. LNCS, vol. 12969, pp. 101–110. Springer, Cham (2021). https://doi.org/10.1007/978-3-030-90874-4_10

28. Sivaswamy, J., Krishnadas, S., et al.: A comprehensive retinal image dataset for the assessment of glaucoma from the optic nerve head analysis. JSM Biomed. Imaging Data Pap. **2**(1), 1004 (2015)

29. Song, L., Li, Y., et al.: Learnable tree filter for structure-preserving feature transform. Adv. Neural. Inf. Process. Syst. **32**, 12300–12311 (2019)

30. Tang, M., Djelouah, A., et al.: Normalized cut loss for weakly-supervised CNN segmentation. In: Proceedings of the IEEE Conference on Computer Vision and Pattern Recognition (CVPR), pp. 1818–1827 (2018)

31. Wang, J., Jin, Y., Wang, L.: Personalizing federated medical image segmentation via local calibration. In: Avidan, S., Brostow, G., Cissé, M., Farinella, G.M., Hassner, T. (eds.) Computer Vision – ECCV 2022. ECCV 2022. Lecture Notes in Computer Science, vol. 13681. Springer, Cham (2022). https://doi.org/10.1007/978-3-031-19803-8_27

32. Wang, K., Mathews, R., et al.: Federated evaluation of on-device personalization. arXiv preprint arXiv:1910.10252 (2019)

33. Wang, Y., Shen, Y., et al.: A deep learning-based quality assessment and segmentation system with a large-scale benchmark dataset for optical coherence tomographic angiography image. arXiv preprint arXiv:2107.10476 (2021)

34. Wu, J., Fang, H., et al.: GAMMA challenge: glaucoma grading from multi-modality images. arXiv preprint arXiv:2202.06511 (2022)

35. Zhang, J., Hua, Y., et al.: FedALA: Adaptive local aggregation for personalized federated learning. arXiv preprint arXiv:2212.01197 (2022)

36. Zhang, J., Xie, Y., et al.: DoDNet: learning to segment multi-organ and tumors from multiple partially labeled datasets. In: Proceedings of the IEEE Conference on Computer Vision and Pattern Recognition (CVPR), pp. 1195–1204 (2021)

Unlocking Fine-Grained Details with Wavelet-Based High-Frequency Enhancement in Transformers

Reza Azad[1]([envelope]), Amirhossein Kazerouni[2], Alaa Sulaiman[3], Afshin Bozorgpour[4], Ehsan Khodapanah Aghdam[5], Abin Jose[1], and Dorit Merhof[4]

[1] Faculty of Electrical Engineering and Information Technology, RWTH Aachen University, Aachen, Germany
rezazad68@gmail.com
[2] School of Electrical Engineering, Iran University of Science and Technology, Tehran, Iran
[3] Faculty of Information Science and Technology, Universiti Kebangsaan, Bangi, Malaysia
[4] Faculty of Informatics and Data Science, University of Regensburg, Regensburg, Germany
dorit.merhof@informatik.uni-regensburg.de
[5] Department of Electrical Engineering, Shahid Beheshti University, Tehran, Iran

Abstract. Medical image segmentation is a critical task that plays a vital role in diagnosis, treatment planning, and disease monitoring. Accurate segmentation of anatomical structures and abnormalities from medical images can aid in the early detection and treatment of various diseases. In this paper, we address the local feature deficiency of the Transformer model by carefully re-designing the self-attention map to produce accurate dense prediction in medical images. To this end, we first apply the wavelet transformation to decompose the input feature map into low-frequency (LF) and high-frequency (HF) subbands. The LF segment is associated with coarse-grained features, while the HF components preserve fine-grained features such as texture and edge information. Next, we reformulate the self-attention operation using the efficient Transformer to perform both spatial and context attention on top of the frequency representation. Furthermore, to intensify the importance of the boundary information, we impose an additional attention map by creating a Gaussian pyramid on top of the HF components. Moreover, we propose a multi-scale context enhancement block within skip connections to adaptively model inter-scale dependencies to overcome the semantic gap among stages of the encoder and decoder modules. Throughout comprehensive experiments, we demonstrate the effectiveness of our strategy on multi-organ and skin lesion segmentation benchmarks. The implementation code will be available upon acceptance. GitHub.

Keywords: Deep learning · High-frequency · Wavelet · Segmentation

Supplementary Information The online version contains supplementary material available at https://doi.org/10.1007/978-3-031-45673-2_21.

X. Cao et al. (Eds.): MLMI 2023, LNCS 14348, pp. 207–216, 2024.
https://doi.org/10.1007/978-3-031-45673-2_21

1   Introduction

In the field of computer vision, Convolutional Neural Networks (CNNs) have been the dominant architecture for various tasks for many years [12]. More recently, however, the Vision Transformer (ViT) [8] has been shown to achieve state-of-the-art (SOTA) results in diverse tasks with significantly fewer parameters than traditional CNN-based approaches. This has resulted in a shift in the field towards utilizing ViT, which is becoming increasingly popular for a wide range of computer vision tasks. The main success behind the ViTs is their ability to model long-range contextual dependencies by applying a grid-based self-affinities calculation on image patches (tokens). Unlike CNNs, which require stacked convolution blocks to increase the receptive field size, the ViT captures the global contextual representation within a single block. However, the ViT model usually suffers from a weak local description compared to the CNN models, which is crucial for semantic segmentation tasks in medical images.

To address the local feature deficiency of Transformer models, recent studies have explored the combination of CNN-Transformer models or pure Transformer-based designs with U-Net-like architectures [4,11]. The strength of the U-Net lies in its symmetrical hierarchical design with a large number of feature channels. However, a pure Transformer-based design involves quadratic computational complexity of the self-attention operation with respect to the number of patches, which makes a combination of U-Net and Transformer challenging. Furthermore, due to this fixed-size scale paradigm, ViT has no strong spatial inductive bias. Therefore, extensive research endeavors aim to overcome these issues by designing efficient and linear complexity self-attention mechanisms to make ViTs suitable for dense prediction tasks. Such designs either diminish the patch numbers (e.g., ATS [16] or A-ViT [28]), or apply downsampling or pooling operations, i.e., on images or key/value tensors (e.g., SegFormer [25], PVT [23], or MViT [9]). Furthermore, calculation on self-attentions is hindered by local windowing schemas as in studies such as Swin Transformer [14] or DW-ViT [15]. Swin-Unet [3] explored the linear Swin Transformer in a U-shaped structure as a Transformer-based backbone. MISSFormer [11] investigated the efficient self-attention from SegFormer as a main module for 2D medical image segmentation. In contrast, these methods endorse the ability of ViTs in segmentation tasks but still suffer from boundary mismatching and poor boundary localization due to the information dropping through their enhanced and efficient self-attention process. On the other hand, the Swin Transformer [14] utilizes non-overlapping windows to employ the self-attention mechanism which, however, may lead to the loss of detailed edges and other spatial information. Efficient self-attention [25] used in [11] decreases the dimensions of the input sequence in spatial dimensions that lose informative details and make the segmentation results error-prone. Moreover, recent studies [22] investigated how self-attention performs as a low-pass filter when Transformer blocks are stacked successively. Therefore, stacking Transformer blocks in a multi-scale paradigm (e.g., U-Net architecture) not only helps to model a multi-scale representation but also degrades the loss of local texture and localization features (high-frequency details) through the network.

High-frequency components are often critical in many real-world signals, such as speech and images, and they are usually associated with fine-grained details that can provide valuable information for many vision-based tasks. However, the Transformer model is known to consider low-frequency representations, making it challenging to capture these high-frequency components [22]. This limitation can result in vague and unsatisfactory feature extraction, leading to a suboptimal performance on the segmentation tasks, which requires a precise boundary extraction. Therefore, exerting wavelet analysis to enhance high-frequency representations in a Transformer can provide a multi-resolution decomposition of the input data, allowing us to identify and isolate high-frequency components that provide a more comprehensive representation.

In this paper, we propose a new Wavelet-based approach for medical image segmentation in a U-shaped structure with the help of efficient Transformers that modifies the quadratic self-attention map calculation by reformulating the self-attention map into a linear operation. We also propose incorporating a boundary attention map to highlight the importance of edge information further to distinguish overlapped objects, termed **Frequency Enhancement Transformer (FET)** block. Furthermore, we design an MSCE module within the skip connections to overcome the semantic gap among the encoder and decoder stages to build rich texture information transferring, which is otherwise limited by the multi-scale representation in a conventional encoder-to-decoder path. Our contributions are as ❶ We propose a novel FET block comprising a frequency-enhanced module and boundary-aware attention map to model both shape and texture representation in an adaptive way. ❷ Applying our proposed MSCE module to skip connections induces the informative texture information from the encoder to the decoder to enrich the missing localization information regarded as a low-frequency representation. ❸ In addition, our method leverages the high-frequency components after applying a Gaussian kernel to perform additional attention information that could effectively highlight the boundary and detailed information for dense prediction tasks, *e.g.* segmentation.

2   Proposed Method

As illustrated in Fig. 1, our proposed method trains in an end-to-end strategy that incorporates the frequency analysis in a multi-scale representation within the efficient Transformer paradigm. Therefore, this section first recapitulates the seminal vision Transformer's inner structure by investigating the multi-head self-attention (MHSA) general mathematical formulation. Assume $X \in \mathbb{R}^{H \times W \times D}$ to be the 2D input image (or feature map stream), then X can be reshaped as a sequence of patches consisting of $n = H \times W$ image patches, where D is the dimension of each patch. Afterward, three representations are learned from the X, namely $Q \in \mathbb{R}^{n \times D}$ Queries, $K \in \mathbb{R}^{n \times D}$ Keys, and $V \in \mathbb{R}^{n \times D}$ Values. The multi-head attention regime utilizes N_h diverse Queries, Keys, and Values, where $\{Q_j, V_j, K_j\} \in \mathbb{R}^{n \times D_h}$ depicts the j-th head information. Then, the MHSA follows and learns the final attention over calculated queries, keys, and values according to the following equations:

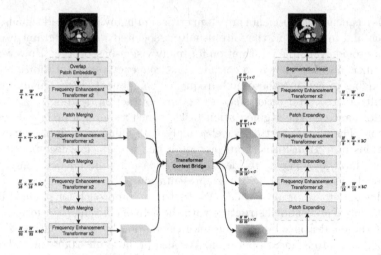

Fig. 1. The overview of the proposed **F**requency **E**nhanced **T**ransformer (**FET**) model. Each frequency-enhanced Transformer block comprises the sequential LayerNorm, FET block, LayerNorm, and Mix-FFN modules.

$$\mathbf{MHSA}(Q, K, V) = \mathbf{Concat}(head_0, head_1, ..., head_{N_h})W^O,$$
$$head_j = \mathbf{Attention}(Q_j, K_j, V_j),$$
$$\mathbf{Attention}(Q_j, K_j, V_j) = \mathbf{Softmax}(\frac{Q_j K_j^T}{\sqrt{D_h}})V_j, \tag{1}$$

where **Concat** and W^O denote the concatenation operation and the learnable transformation tensor, respectively. Thus, the conventional Transformer captures long-range dependencies but still suffers from several limitations that could affect the ViT's performance in dense segmentation tasks: first, the computational cost of multi-head self-attention is quadratic in patch numbers, $\mathcal{O}(n^2 D)$, making it unsuitable for high-resolution tasks. Second, the recent analytic work from Wang et al. [22] demonstrated the deficiency of a multi-head self-attention mechanism in capturing high-frequency details due to the included **Softmax** operation. Specifically, the lack of ability to capture high-frequency information degrades the segmentation performance with naive ViTs. Therefore, in the next section, we propose our FET module to address all aforementioned issues.

2.1   Efficient Transformer

Due to the quadratic computational complexity of seminal Transformers, a wide range of studies have been conducted to minimize this weakness. Shen et al. [20] revisited the dot production within the multi-head self-attention mechanism to circumvent redundant operations. From Eq. (1), it can be seen that the MHSA captures the similarity between each pair of patches, which is much more resource intensive. Efficient attention computes the self-attention as

$$\mathbf{Efficient\ Attention} = \rho_q(Q)(\rho_k(K)^T V), \tag{2}$$

where ρ_q and ρ_k denote the normalization functions for Q and K. In Eq. (2), instead of considering the keys as n feature vectors in $\mathbb{R}^D$, the module interprets them as d_k feature maps with only one channel. Efficient attention applies these feature maps as weights across all positions and combines the value features by weighted summation, resulting in a global context vector. This vector does not refer to any particular position, but rather represents a comprehensive overview of the input features, analogous to a global context vector.

2.2   Frequency Enhancement Transformer (FET)

As suggested by [27], we follow their intuition to preserve the high-frequency counterparts for medical image segmentation tasks. Discrete Wavelet Transform (DWT) is a mapping function from spatial resolution to spatial-frequency space. Wavelet decomposition is a powerful technique that decomposes images into high and low-frequency components, providing a multi-resolution analysis of the input signal. In medical image segmentation, high-frequency components of the image correspond to fine details such as edges and texture. In contrast, low-frequency components correspond to large-scale structures and background information. Thus, a wavelet decomposition which analyzes both high and low-frequency components of medical images may enhance the accuracy of segmentation models by capturing both local and global features of the image. While applying DWT on an image, there would be four distinct wavelet subbands, namely LL, LH, HL, and HH, demonstrating the texture, horizontal details, vertical details, and diagonal information, respectively.

The FET (visualized in Fig. 2a) is designed to address previous limitations by highlighting the boundary information (high-frequency details) for medical image segmentation. Motivated by [27], FET utilizes the DWT to account for the frequency analysis for focusing on high-frequency counterparts. First, the input 2D image (feature map) $X \in \mathbb{R}^{H \times W \times D}$ ($n = H \times W$) is linearly transformed into $\tilde{X} = \mathbb{R}^{n \times \frac{D}{4}}$ by reducing the channel dimension. Classical DWT applies pairs of low-pass and high-pass filters along rows and columns to extract frequency response subbands. Next, DWT is applied to $\tilde{X}$ to extract frequency responses and to downsample the input. As a result, the four subbands of input are $\tilde{X} = [\tilde{X}_{LL}, \tilde{X}_{LH}, \tilde{X}_{HL}, \tilde{X}_{HH}] \in \mathbb{R}^{n \times \frac{D}{4}}$. The high-frequency components ($\tilde{X}_{LH}$, $\tilde{X}_{HL}$, and $\tilde{X}_{HH}$) concatenate in a new dimension due to the underlying texture details at the fine-grained level. Then, a $3 \times 1 \times 1$ convolution is applied to the resulting feature map to recalibrate for a subsequent Gaussian hierarchical "Boundary Attention" mechanism. The process continues with another $3 \times 1 \times 1$ convolution, and then the encoded boundary features are concatenated in the channel dimension. Analogous to [27], another branch applies a 3×3 convolution for creating the keys and values. Furthermore, a global context results from incorporating keys and values. However, to compensate for the **Softmax** operation's destructive effect [22], we add the boundary attention to Value, to include the boundary preservation action when calculating attention. After boundary extraction, the FET block uses a query Q from the input X and key K and value V from the DWT to extract multi-disciplinary contextual correlations. While the firstmost left branch captures the spatial dependencies, the middle branch extracts the channel representation in an efficient concept. In addition, the most right branch highlights the boundary information within the value representation. Finally, the FET model in Fig. 1 is composed of a LayerNorm, FET block (see Fig. 2a), LayerNorm, and Mix-FFN [25] modules in sequence.

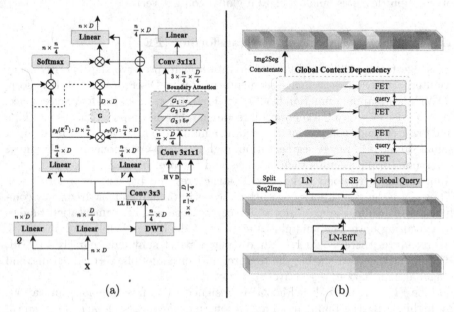

Fig. 2. (a) The **FET** Block. [LL, H, V, D] denotes the low-frequency, horizontal, vertical, and diagonal high-frequency counterparts. (b) The overview of **MSCE** skip connection enhancement module. LN, EffT, and SE are the LayerNorm, the efficient Transformer module, and the squeeze and excitation block, respectively.

2.3 Multi-Scale Context Enhancement (MSCE)

A multi-scale fusion paradigm is considered in our design for accurate semantic segmentation to alleviate the semantic gap between stages of U-shaped structures, as in Fig. 2b. Given the multi-level features that resulted from the hierarchical encoder, representations are flattened in spatial dimension and are reshaped to keep the same channel depth at each stage. Considering F_i as a hierarchical feature in each encoder stage $i \in \{1, \ldots, 4\}$, we flatten them spatially and reshape them to obtain the same channel depth for each stage before concatenating them in the spatial dimension. Following the LayerNorm and efficient Transformer, we create the hierarchical long-range contextual correlation. Afterward, the tokens are split and reshaped to their original shape of features in each stage and are fed to the FET block to capture the amalgamated hierarchical contextual representation. We capture the global information from the represented token space as a *Global Query* to the FET blocks.

3   Experiments

Our proposed method was implemented using the PyTorch library and executed on a single RTX 3090 GPU. A batch size of 24 and a SGD solver with a base learning rate of 0.05, a momentum of 0.9, and a weight decay of 0.0001 is used. The training was carried out for 400 epochs. For the segmentation task, both cross-entropy and Dice losses were utilized as the loss function. The segmentation task was performed using the combined loss ($Loss = 0.6 \cdot L_{dice} + 0.4 \cdot L_{ce}$ as used in [10]). **Datasets:** First, we evaluated our method on the *Synapse* dataset [13] that contains 30 cases of abdominal CT scans with 3,779 axial contrast-enhanced abdominal clinical CT images. Each CT data consists of 85 $\sim$ 198 slices of a consistent size 512×512 with the eight organ classes annotation. We followed the same preferences for data preparation as in [5]. Second, our study on skin lesion segmentation is based on the *ISIC 2018* [7] dataset, which was published by the International Skin Imaging Collaboration (ISIC) as a large-scale dataset of dermoscopy images. We follow the [2] for the evaluation setting.

Qualitative and Quantitative Results: In Table 1, we compare the performance of our proposed FET method with previous SOTA methods for segmenting abdominal organs using the DSC and the HD metrics. Our method surpasses existing CNN-based methods by a significant margin. FET exhibits superior learning ability on the DSC metric compared to other models, achieving an increase of 1.9% compared to HiFormer. The quantitative results highlight the FET superiority in segmenting kidney, pancreas, and spleen organs. The Table 2a also endorses the mentioned results qualitatively, and all other models suffer from organ deformations when segmenting the liver and suffer from under-segmentation while FET performs smoothly.

Table 1. Comparison results of the proposed method on the *Synapse* dataset. Blue indicates the best result, and red displays the second-best.

Methods	# Params (M)	DSC ↑	HD ↓	Aorta	Gallbladder	Kidney(L)	Kidney(R)	Liver	Pancreas	Spleen	Stomach
R50 U-Net [5]	30.42	74.68	36.87	87.74	63.66	80.60	78.19	93.74	56.90	85.87	74.16
U-Net [18]	14.8	76.85	39.70	89.07	69.72	77.77	68.60	93.43	53.98	86.67	75.58
Att-UNet [19]	34.88	77.77	36.02	89.55	68.88	77.98	71.11	93.57	58.04	87.30	75.75
TransUnet [5]	105.28	77.48	31.69	87.23	63.13	81.87	77.02	94.08	55.86	85.08	75.62
Swin-Unet [3]	27.17	79.13	21.55	85.47	66.53	83.28	79.61	94.29	56.58	90.66	76.60
LeVit-Unet [26]	52.17	78.53	16.84	78.53	62.23	84.61	80.25	93.11	59.07	88.86	72.76
DeepLabv3+ (CNN) [6]	59.50	77.63	39.95	88.04	66.51	82.76	74.21	91.23	58.32	87.43	73.53
HiFormer [10]	25.51	80.39	14.70	86.21	65.69	85.23	79.77	94.61	59.52	90.99	81.08
Baseline	27.36	80.39	20.56	85.69	69.68	83.83	80.07	94.20	60.72	90.92	77.98
FET (without MSCE bridge)	33.00	81.05	17.70	87.80	68.33	85.00	79.25	94.11	61.80	90.95	81.18
FET	47.01	81.92	18.41	85.31	69.67	86.66	80.06	94.43	67.08	91.85	80.34

Skin Lesion Segmentation: Table 2b also endorses the capability of FET compared to other well-known methods for skin lesion segmentation methods. Specifically, our method performs better than hybrid approaches such as TMU-Net [17]. Additionally, our method proves to be more resilient to noisy elements

Table 2. (a) Segmentation results of the proposed method versus SOTA methods on the *Synapse* dataset. (b) Quantitative results on ISIC2018 dataset.

(a) Segmentation visualization on Synapse dataset.

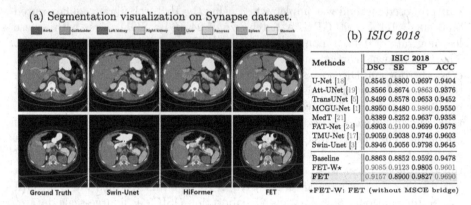

(b) *ISIC 2018*

Methods	ISIC 2018			
	DSC	SE	SP	ACC
U-Net [18]	0.8545	0.8800	0.9697	0.9404
Att-UNet [19]	0.8566	0.8674	0.9863	0.9376
TransUNet [5]	0.8499	0.8578	0.9653	0.9452
MCGU-Net [1]	0.8950	0.8480	0.9860	0.9550
MedT [21]	0.8389	0.8252	0.9637	0.9358
FAT-Net [24]	0.8903	0.9100	0.9699	0.9578
TMU-Net [17]	0.9059	0.9038	0.9746	0.9603
Swin-Unet [3]	0.8946	0.9056	0.9798	0.9645
Baseline	0.8863	0.8852	0.9592	0.9478
FET-W⋆	0.9085	0.9123	0.9805	0.9601
FET	0.9157	0.8900	0.9827	0.9690

⋆FET-W: FET (without MSCE bridge)

when compared to pure Transformer-based methods such as Swin-Unet [3], which suffer from reduced performance due to a lack of emphasis on local texture modeling. In addition, comparing qualitative results (presented as a supplementary file) on the ISIC 2018 dataset approves our method's capability to capture fine-grained boundary information.

To comprehensively evaluate the influence of our module on capturing high-frequency information in deeper layers, we conducted an extensive analysis of the spectrum response in Fig. 3. Our findings reveal that our method stands out from traditional self-attention modules by effectively preserving high-frequency information within the depths of the network.

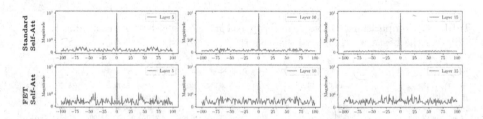

Fig. 3. Illustration of the spectral response of Standard Transformer (up) and FET (down) for capturing different frequency representation.

4   Conclusion

In this paper, we redesigned the Transformer block to recalibrate spatial and context representation adaptively. We further imposed a secondary attention map to highlight the importance of boundary information within the Transformer block. Moreover, we modeled the intra-scale dependency for further performance improvement by redesigning the skip connection path. The effectiveness of our module is illustrated through the experimental results.

Acknowledgments. This work was funded by the German Research Foundation (Deutsche Forschungsgemeinschaft, DFG) under project number 191948804.

References

1. Asadi-Aghbolaghi, M., Azad, R., Fathy, M., Escalera, S.: Multi-level context gating of embedded collective knowledge for medical image segmentation. arXiv preprint arXiv:2003.05056 (2020)
2. Azad, R., Asadi-Aghbolaghi, M., Fathy, M., Escalera, S.: Bi-directional convLSTM U-Net with densley connected convolutions. In: Proceedings of the IEEE/CVF International Conference on Computer Vision Workshops (2019)
3. Cao, H., et al.: Swin-UNet: UNet-like pure transformer for medical image segmentation. In: Proceedings of the European Conference on Computer Vision Workshops (ECCVW) (2022)
4. Chang, Y., Menghan, H., Guangtao, Z., Xiao-Ping, Z.: TransClaw U-Net: Claw U-Net with transformers for medical image segmentation. arXiv preprint arXiv:2107.05188 (2021)
5. Chen, J., et al.: TransUNet: Transformers make strong encoders for medical image segmentation. arXiv preprint arXiv:2102.04306 (2021)
6. Chen, L.C., Zhu, Y., Papandreou, G., Schroff, F., Adam, H.: Encoder-decoder with atrous separable convolution for semantic image segmentation. In: Proceedings of the European Conference on Computer Vision (ECCV), pp. 801–818 (2018)
7. Codella, N., et al.: Skin lesion analysis toward melanoma detection 2018: A challenge hosted by the international skin imaging collaboration (ISIC). arXiv preprint arXiv:1902.03368 (2019)
8. Dosovitskiy, A., et al.: An image is worth 16×16 words: Transformers for image recognition at scale. arXiv preprint arXiv:2010.11929 (2020)
9. Fan, H., et al.: Multiscale vision transformers. In: Proceedings of the IEEE/CVF International Conference on Computer Vision, pp. 6824–6835 (2021)
10. Heidari, M., et al.: HiFormer: hierarchical multi-scale representations using transformers for medical image segmentation. In: Proceedings of the IEEE/CVF Winter Conference on Applications of Computer Vision, pp. 6202–6212 (2023)
11. Huang, X., Deng, Z., Li, D., Yuan, X., Fu, Y.: MISSFormer: an effective transformer for 2D medical image segmentation. IEEE Trans. Med. Imaging **42**(5), 1484–1494 (2022)
12. Karimijafarbigloo, S., Azad, R., Merhof, D.: Self-supervised few-shot learning for semantic segmentation: An annotation-free approach. In: MICCAI 2023 workshop (2023)
13. Landman, B., Xu, Z., Igelsias, J., Styner, M., Langerak, T., Klein, A.: MICCAI multi-atlas labeling beyond the cranial vault-workshop and challenge. In: Proc. MICCAI Multi-Atlas Labeling Beyond Cranial Vault-Workshop Challenge. vol. 5, p. 12 (2015)
14. Liu, Z., et al.: Swin Transformer: hierarchical vision transformer using shifted windows. In: Proceedings of the IEEE/CVF International Conference on Computer Vision, pp. 10012–10022 (2021)

15. Ren, P., et al.: Beyond fixation: dynamic window visual transformer. In: Proceedings of the IEEE/CVF Conference on Computer Vision and Pattern Recognition, pp. 11987–11997 (2022)
16. Renggli, C., Pinto, A.S., Houlsby, N., Mustafa, B., Puigcerver, J., Riquelme, C.: Learning to merge tokens in vision transformers. arXiv preprint arXiv:2202.12015 (2022)
17. Reza, A., Moein, H., Yuli, W., Dorit, M.: Contextual attention network: Transformer meets U-Net. arXiv preprint arXiv:2203.01932 (2022)
18. Ronneberger, O., Fischer, P., Brox, T.: U-Net: convolutional networks for biomedical image segmentation. In: Navab, N., Hornegger, J., Wells, W.M., Frangi, A.F. (eds.) MICCAI 2015. LNCS, vol. 9351, pp. 234–241. Springer, Cham (2015). https://doi.org/10.1007/978-3-319-24574-4_28
19. Schlemper, J., et al.: Attention gated networks: learning to leverage salient regions in medical images. Med. Image Anal. **53**, 197–207 (2019)
20. Shen, Z., Zhang, M., Zhao, H., Yi, S., Li, H.: Efficient attention: attention with linear complexities. In: Proceedings of the IEEE/CVF Winter Conference on Applications of Computer Vision, pp. 3531–3539 (2021)
21. Valanarasu, J.M.J., Oza, P., Hacihaliloglu, I., Patel, V.M.: Medical transformer: gated axial-attention for medical image segmentation. In: de Bruijne, M., et al. (eds.) MICCAI 2021. LNCS, vol. 12901, pp. 36–46. Springer, Cham (2021). https://doi.org/10.1007/978-3-030-87193-2_4
22. Wang, P., Zheng, W., Chen, T., Wang, Z.: Anti-oversmoothing in deep vision transformers via the fourier domain analysis: From theory to practice. In: International Conference on Learning Representations (2022). https://openreview.net/forum?id=O476oWmiNNp
23. Wang, W., et al.: PVT V2: improved baselines with pyramid vision transformer. Comput. Vis. Media **8**(3), 415–424 (2022)
24. Wu, H., Chen, S., Chen, G., Wang, W., Lei, B., Wen, Z.: FAT-Net: feature adaptive transformers for automated skin lesion segmentation. Med. Image Anal. **76**, 102327 (2022)
25. Xie, E., Wang, W., Yu, Z., Anandkumar, A., Alvarez, J.M., Luo, P.: SegFormer: simple and efficient design for semantic segmentation with transformers. Adv. Neural. Inf. Process. Syst. **34**, 12077–12090 (2021)
26. Xu, G., Wu, X., Zhang, X., He, X.: LeViT-UNet: Make faster encoders with transformer for medical image segmentation. arXiv preprint arXiv:2107.08623 (2021)
27. Yao, T., Pan, Y., Li, Y., Ngo, C.W., Mei, T.: Wave-ViT: unifying wavelet and transformers for visual representation learning. In: Avidan, S., Brostow, G., Cissé, M., Farinella, G.M., Hassner, T. (eds.) Computer Vision – ECCV 2022. ECCV 2022. Lecture Notes in Computer Science, vol. 13685. Springer, Cham (2022). https://doi.org/10.1007/978-3-031-19806-9_19
28. Yin, H., Vahdat, A., Alvarez, J.M., Mallya, A., Kautz, J., Molchanov, P.: A-ViT: adaptive tokens for efficient vision transformer. In: Proceedings of the IEEE/CVF Conference on Computer Vision and Pattern Recognition, pp. 10809–10818 (2022)

Prostate Segmentation Using Multiparametric and Multiplanar Magnetic Resonance Images

Kuruparan Shanmugalingam[✉], Arcot Sowmya, Daniel Moses, and Erik Meijering

School of Computer Science and Engineering, University of New South Wales (UNSW), Sydney, Australia
kuruparan@unsw.edu.au

Abstract. Diseases related to the prostate and distal urethra, such as prostate cancer, benign prostatic hyperplasia and urinary incontinence, may be detected and diagnosed through noninvasive medical imaging. T2-weighted (T2W) magnetic resonance imaging (MRI) is the most commonly used modality for prostate and urethral segmentation due to its distinguishable features of anatomical texture. In addition to T2W multiplanar images, which capture information in the axial, sagittal and coronal planes, multiparametric MRI modalities such as dynamic contrast enhanced (DCE) and diffusion-weighted imaging (DWI) are usually also acquired in the scanning process to measure functional features. Feature fusion by combining multiparametric and multiplanar images is challenging due to the movement of the patient during image acquisition, the need for accurate image registration and the sheer volume of available scans. Here we propose a multi-encoder deep neural network named 3DDOSPyResidualUSENet to learn anatomical and functional features from multiparametric and multiplanar MRI images. Our extensive experiments on a public dataset show that combining T2W axial, sagittal and coronal images along with DCE information and apparent diffusion coefficient (ADC) maps computed from DWI images results in increased segmentation performance.

Keywords: Prostate segmentation · Magnetic resonance imaging · Deep learning · Multi-encoder-decoder network · Multiparametric imaging · Multiplanar imaging

1 Introduction

Benign prostatic hyperplasia (BPH), prostate cancer (PCa) and urinary incontinence (UI) are some of the known diseases related to the prostate and urethra. PCa itself resulted in 3.8% death in men in 2018 and is considered the second most common cancer to harm people [3,19]. Medical imaging helps to avoid tests such as prostate-specific antigen (PSA) and the invasive digital rectal examination (DRE). For surgery planning, it is important to segment the prostate and urethra accurately in the images to avoid post-operative complications [18,24].

X. Cao et al. (Eds.): MLMI 2023, LNCS 14348, pp. 217–226, 2024.
https://doi.org/10.1007/978-3-031-45673-2_22

Thanks to its good contrast of anatomical features, T2-weighted (T2W) magnetic resonance imaging (MRI) is most popular to perform segmentation and quantification of the prostate and urethra [5]. To further improve the accuracy, the use of multiple anatomical and functional images from different modalities and viewpoints may also be beneficial.

The prostate, which protects the sperm by producing prostatic fluid, is one of the organs in the male genital system. The urethra passes through the prostate and connects the bladder and the penis. The prostate can be divided into the peripheral zone (PZ), transitional zone (TZ) and central zone (CZ). The part of the urethra that passes through the prostate is known as the prostatic urethra and includes the proximal urethra (PU) and the distal prostatic urethra (DPU). The remaining part, from the prostate to the penis through smooth muscle, is known as the membranous urethra (MU) followed by the penile urethra.

MRI encompasses several modalities determined by the scanning protocols and acquisition parameters. T1-weighted and T2-weighted are the most common MRI modalities based on the relaxation times of the excited protons. In diffusion-weighted imaging (DWI), the image information is obtained by determining the sensitivity of the prostate tissue to the diffusion of water molecules. One of the images derived from DWI data is the apparent diffusion coefficient (ADC) map, which represents the magnitude of water diffusion within the tissues. Dynamic contrast-enhanced (DCE) images are acquired by time-lapse scanning, resulting in a 4D image. Here, sets of images are taken using gadolinium contrast, which perfuses through the blood vessels after intravenous injection. DCE is typically used to determine the vascularity of soft tissues.

In each MRI modality (Fig. 1), the prostate can be clearly identified by the naked eye, by focussing on local edges and texture features. This observation raises interest in combining several modalities to improve the accuracy of segmentation methods that use only T2W. However, despite several investigations using biparametric MRI (ADC+T2W) [20] and multiparametric MRI [7], this approach has been used mostly for cancer lesion segmentation or classification. To date, there have been few works related to multiparametric input feature fusion for accuracy improvement in prostate segmentation [12].

In this work, we propose a novel multi-encoder-decoder neural network architecture, named 3DDOSPyResidualUSENet, for feature learning from multiparametric and multiplanar MRI data for improved prostate segmentation. After a background literature review in Sect. 2, we present our network and experimental setup in Sect. 3, and discuss the quantitative and quality results in Sect. 4, supporting our conclusions in Sect. 5.

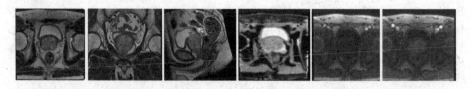

Fig. 1. Multiparametric MRI. Left to right: T2W axial, T2W coronal, T2W sagittal, ADC, DCE first frame (DCEF), DCE last frame (DCEL).

2   Literature Review

Multiparametric MRI (mpMRI) has been used in prostate cancer detection and classification tasks. Several works on mpMRI [2] focus on cancer lesion detection with combinations of multiparametric data such as T2W+ADC [6], T2W+ADC+DWI [12,25], T2W+ADC+DCE [1,22], and ADC+DWI+DCE [13]. By contrast, for the prostate segmentation task, mpMRI has not been thoroughly explored, possibly because of the challenges of multimodal image registration and feature fusion. Although the most commonly used MRI modality for prostate segmentation is T2W [15,16,23], some researchers have experimented with DWI [4], DCE [9], T2W+ADC [11], T2W+DWI [27], ADC+T2W+Ktrans [8], and T1W+T2W [21].

Clark et al. achieved a 0.89 Dice score by experimenting with 134 DWI images using U-Net [4]. Kang et al. worked with spatiotemporal DCE MRI volumes of 17 patients and achieved a Dice score of almost 0.87 using U-Net with ensemble LSTMs [9]. Litjens et al. used 48 mpMRI scans consisting of T2W and ADC and improved the central gland segmentation Dice score from 0.87 to 0.89 [11]. Zhu et al. reported a Dice score of nearly 0.93 using the classical U-Net on DWI and T2W images of 163 subjects [27]. Combining modalities such as T2W, ADC and DWI improved the prostate lesion segmentation accuracy in the I2CVB dataset [12]. Liu et al. concluded that adding DWI and ADC increased the Dice score of PCa lesion segmentation from about 0.78 to 0.82 [12].

Multiplanar prostate segmentation by combining axial, sagittal and coronal T2W images has resulted in improved accuracy compared to individually tested modalities [16]. The results were obtained on the public ProstateX dataset, an in-house private dataset and a merged dataset, resulting in a Dice score of 0.921 on 40 randomly selected ProstateX images and a score of 0.925 [15] on the combination of 66 randomly selected ProstateX images and 89 in-house images.

Although some studies have used multiparametric data for segmentation of prostate lesions or the prostate gland itself, and others have explored multiplanar prostate segmentation, a proper benchmark study has not been performed. To our knowledge, the combination of multiparametric and multiplanar MRIs for prostate segmentation has not been explored.

3   Materials and Methods

3.1   Dataset

The ProstateX challenge dataset [10] and corresponding image annotations were used in our study considering the availability of its multiparametric data. ProstateX has 349 studies of 346 participants consisting of T2W axial, T2W sagittal, T2W coronal, DWI, ADC maps, proton-density (PD) and several time-based DCE images. The dataset was originally acquired to study prostate cancer lesion grades. The ProstateX challenge was about classifying lesions as benign or malignant, and version 2 of the challenge focussed on estimating the grades and Gleason scores of the lesions. The annotation data was taken separately from elsewhere [14,17] and contains the PZ, TZ, anterior fibromuscular stroma (AFS)

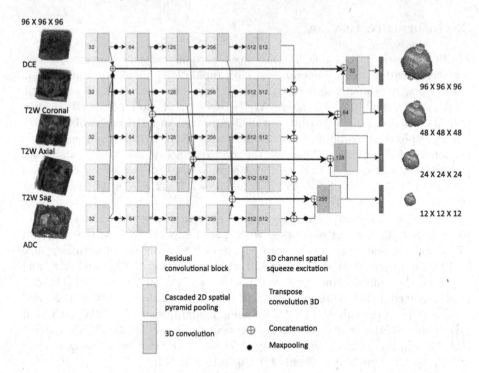

Fig. 2. Architecture of the proposed 3DDOSPyResidualUSENet. The network consists of five encoders and a single decoder with a four-stage design. DCEL, T2W axial, T2W coronal, T2W sagittal and ADC MRI images of size $96 \times 96 \times 96$ voxels are the input to one encoder each. The deep output supervision-based decoder is fed with four downsized masks of $96 \times 96 \times 96$, $48 \times 48 \times 48$, $24 \times 24 \times 24$, and $12 \times 12 \times 12$ voxels.

and DPU for 98 selected cases from the ProstateX dataset. For each case, all annotations were combined to obtain the whole prostate with DPU for that case. Since the annotated images have the same size, origin and spatial resolution as their corresponding T2W axial images, the images of all other modalities were registered to these axial images. All images were then resampled to $96 \times 96 \times 96$ voxels with unit spacing and normalised. The SimpleITK library [26] was used for registration and resampling with the nearest-neighbour algorithm.

3.2 Architecture

The proposed 3DDOSPyResidualUSENet (Fig. 2) has a multi-encoder single-decoder architecture. It is an extension of our recent 3DDOSPyUSENet architecture [23], named after its capability to efficiently process three-dimensional (3D) images using deep output supervision (DOS), pyramid pooling (Py) and a squeeze-and-excitation U-Net (USENet) design. In our new design, the dual sequential convolutional blocks were replaced by residual convolutional blocks to avoid losing memory from the previous blocks in the multi-encoder architecture. First, we trained and tested a single encoder-decoder 3DDOSPyResidualUSENet model for each modality separately (T2W axial, T2W coronal, T2W sagittal, ADC, DCEF, DCEL respectively). Next, a dual-encoder model was created, which takes images of two modalities of choice as inputs to produce a single segmentation output. This model was benchmarked to evaluate the potential added value of combining different modalities. Inspired by the triple encoder-decoder simple U-Net architecture for multiplanar segmentation [15,16], we have also trained and tested a triple-encoder 3DDOSPyResidualUSENet on multiplanar images as well as multiparametric images separately. Finally, we built a five-encoder single-decoder network (shown in Fig. 2), which can leverage the anatomical and functional information from all the available modalities to train the model. The proposed architecture was equipped with 3D channel and spatial squeeze-and-excitation blocks after each residual block and transpose convolutional block. Cascaded 2D pyramid pooling was used to handle all channels of the 3D blocks, and deep output supervision was employed to emphasise the label maps at each stage [23].

3.3 Experiments

The dataset was split into 80% for training and 20% for testing. Furthermore, the training data was split into 85% for actual training and 15% for validation. In each experiment, we performed five-fold cross-validation and calculated the mean and standard deviation across the five folds.

Single encoder-decoder based experiments were performed initially, using the six different modalities separately for training and testing in five folds, amounting to $6 \times 5 = 30$ experiments. Next, dual encoder-decoder based experiments were performed five-fold, adding up to another $5 \times 5 = 25$ experiments. Finally, two triple encoder-decoder based experiments and one experiment with our five encoder-decoder design were performed, taking another $3 \times 5 = 15$ experiments. Thus, in total, we ran 75 training-testing experiments to identify the configuration that would yield the highest segmentation accuracy.

Tensorflow 2.1 was used in the Anaconda environment along with all other supporting libraries to run the experiments. All models were trained with Nvidia V100 GPUs on the Gadi server of the National Computer Infrastructure (NCI) in Australia. Multistep learning was employed with a rate starting at 0.001 and using the stochastic gradient descent (SGD) optimiser. Each model was trained for 200 epochs.

3.4 Metrics

Segmentation performance was quantified in terms of both overlap-based and distance-based metrics. The overlap-based metric we used is the Dice similarity coefficient (DSC), which measures the overlap between the predicted (segmented) region and the reference (annotated) region. The distance-based metrics included the Hausdorff distance (HD), the average symmetric surface distance

Table 1. Mean and standard deviation of all considered performance metrics in the five-fold cross-validation experiments for various (combinations of) image types as input to 3DDOSPyResidualUSENet. To save space, we have abbreviated the naming of the axial (A), sagittal (S), and coronal (C) T2W planar images. Best results in bold.

Modality	DSC	HD	ASSD	RAVD
Axial T2W	0.8893 ± 0.0150*	8.5357 ± 1.7110	0.1372 ± 0.0278	0.0265 ± 0.0224
Sagittal T2W	0.8905 ± 0.0105*	10.0711 ± 3.1173	0.1329 ± 0.0193	0.0299 ± 0.0222
Coronal T2W	0.8835 ± 0.0150*	11.5274 ± 6.2334	0.1506 ± 0.0342	0.0354 ± 0.0349
ADC	0.8599 ± 0.0128*	10.8763 ± 2.8076	0.1844 ± 0.0162	0.0294 ± 0.1170
DCEF	0.8452 ± 0.0125*	11.2195 ± 4.0192	0.2264 ± 0.0408	0.0313 ± 0.0122
DCEL	0.8565 ± 0.0120*	10.1928 ± 2.6710	0.1936 ± 0.0436	0.0167 ± 0.0068
A + S T2W	0.8903 ± 0.0106*	9.7147 ± 3.8446	0.1332 ± 0.0241	0.0238 ± 0.0276
A + C T2W	0.8898 ± 0.0113*	8.4873 ± 1.8585	0.1385 ± 0.0266	0.0217 ± 0.0143
A T2W + ADC	0.8873 ± 0.0110*	8.4628 ± 1.9303	0.1383 ± 0.0217	0.0360 ± 0.0146
A T2W + DCEF	0.8884 ± 0.0106*	9.3180 ± 3.9647	0.1360 ± 0.0211	0.0366 ± 0.0201
A T2W + DCEL	0.8875 ± 0.0101*	11.7096 ± 5.4641	0.1401 ± 0.0238	0.0156 ± 0.0117
A + S + C T2W	0.8992 ± 0.0085	8.7781 ± 3.3059	0.1191 ± 0.0194	0.0323 ± 0.0135
A T2W + ADC+ DCEL	0.8922 ± 0.0101*	8.8972 ± 1.7456	0.1347 ± 0.0231	0.0241 ± 0.0198
Proposed	**0.9024 ± 0.0098**	**8.3430 ± 2.2318**	**0.1138 ± 0.0171**	**0.0112 ± 0.0106**

(ASSD) and the relative absolute volume difference (RAVD). HD measures the largest discrepancy between the two regions by calculating the maximum distance between the segmented region and the reference region. ASSD measures the average symmetrical distance between the surfaces of the segmented region and the reference region. RAVD assesses the volumetric agreement or discrepancy between a predicted segmented region and a reference volume.

4   Results and Discussion

From the DSC values of the single-encoder models (the first six entries of Table 1) column 1, we see that using the planar T2W images resulted in better segmentation performance than using the ADC maps or the DCE images (whether it be the first or the last frames of the DCE sequences). T2W MRI indeed captures the structural features of the prostate region with better visual clarity than all the other modalities. We also establish that the differences in DSC values of the three planar images are small and not statistically significant. Apparently, the relevant information is represented equally well along the three axes.

Using the DCEL images resulted in somewhat higher accuracy than the DCEF images. This is to be expected, as the prostate in the last frame of the DCE sequence is typically brighter than in the first frame. Over time, the gadolinium contrast perfuses through the blood vessels, and when the first image is taken the gadolinium usually has not yet reached the prostate, resulting in a darker prostate image than by the time the last image is taken (see Fig. 1 for a visual example). Using the ADC maps derived from DWI MRI gave a somewhat higher DSC score than either DCEF and DCEL. The prostate in ADC is indeed more clearly distinguishable due to the textures and gradients produced by the apparent diffusion of water molecules in the tissue.

To evaluate the performance of the dual-encoder architecture, we created various combinations of two out of the six types of images (planes and/or modalities), though we did not consider all possible combinations in order to keep the experiments feasible. Specifically, we used the axial T2W image as the first data type due to its widespread usage in clinical practice, and combined it with each of the other data types. The results obtained from these experiments are not statistically significantly better than those of the single-encoder architecture trained separately on the different T2W planar images.

Next, we made two combinations of three out of the six types of images, again using the axial T2W image as the starting point, to evaluate the performance of the tri-encoder architecture. Here, we did observe a notable increase in the segmentation performance according to the DSC score. In particular, the results in row 12 of Table 1 are in agreement with others [15,16] who obtained better segmentation performance by combining all three types of planar T2W images (multiplanar). In addition, our results add value to these findings by showing that better results are also obtained by combining images from different MRI modalities (multiparametric), corresponding to row 13.

Finally, we evaluated our proposed five-encoder single-decoder architecture 3DDOSPyResidualUSENet trained on T2W axial, T2W sagittal, T2W coronal, ADC and DCEL images, which achieved the highest DSC score (0.9024) compared to all other combinations described above. The differences between this model and all others were statistically significant according to a t-test with significance level $p < 0.05$, except for the model trained with all three T2W planar images. Apparently, the T2W images contribute most to the accuracy in both single-encoder and multi-encoder models.

While the comparison of segmentation performance so far was based on the DSC metric, other metrics such as HD, ASSD and RAVD provide further insight into the performance of our model not only in terms of overlap with the reference annotations but also in terms of distance. Based on the results (Table 1) we conclude that our proposed multi-encoder architecture yields superior segmentation results on all metrics.

This paper focuses on accurate segmentation of the prostate (PZ, TZ, AFS) and part of the urethra (DPU) that passes beyond PZ (annotation represented in the top-right corner of Fig. 2) with the aim to reduce UI, PCa, BPH related post-operative complications. Hence, our results are not directly comparable with prostate-only segmentation benchmarking results (such as the leaderboard of the Promise12 challenge), as the regions of interest are different.

Representative qualitative results (Fig. 3) illustrate the segmentation performance of our proposed architecture on a specific slice (depth level) of the 3D MRI scans. Since the images for each patient were registered and resampled to the

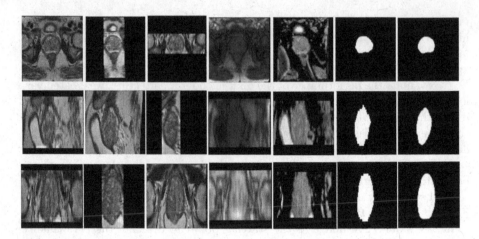

Fig. 3. Qualitative result of case #5, slice #35. Top to bottom row: axial, sagittal and coronal planes. Left to right in each row: T2W axial, T2W sagittal, T2W coronal, DCEL, ADC, reference annotation, prediction.

axial reference image, the reference annotation and prediction are representative of all three planes and all three modalities, facilitating visual interpretation.

5   Conclusion

Accurate segmentation of the prostate and urethra is vital to surgery planning and reducing post-operative complications. T2W MRI-based segmentation models are widely used in this domain due to the clear visibility of anatomical features in these images. However, multiparametric images including DCE and DWI may contain additional anatomical and functional information, and multiplanar views of the T2W images are typically available as well. We have proposed a novel multi-encoder single-decoder deep neural network architecture named 3DDOSPyResidualUSENet which can take advantage of these multiparametric and multiplanar images for prostate segmentation. We have presented several experiments demonstrating the performance of our model for different (combinations of) images. When using only a single modality, the best segmentation results are obtained with the T2W images, followed by ADC, the last frame of DCE, and the first frame of DCE, suggesting that there is no benefit in using modalities other than T2W alone. However, when combining multiple planes of T2W images, and combining multiple MRI modalities, as well as combining multiple planes with multiple modalities, the segmentation performance does improve, demonstrating their added value. Our proposed five-encoder model achieved a 0.9024 Dice score, which was significantly better than most simpler models. In future research, we plan to perform more experiments with additional (public and private) datasets, focussing on domain-generalised multiparametric learning and the use of surface-attention to further improve the segmentation performance closer the edges of the prostate.

References

1. Arif, M., et al.: Clinically significant prostate cancer detection and segmentation in low-risk patients using a convolutional neural network on multi-parametric MRI. Eur. Radiol. **30**(12), 6582–6592 (2020). https://doi.org/10.1007/s00330-020-07008-z
2. Bardis, M.D., et al.: Applications of artificial intelligence to prostate multiparametric MRI (mpMRI): current and emerging trends. Cancers **12**(5), 1204 (2020)
3. Bray, F., Ferlay, J., Soerjomataram, I., Siegel, R.L., Torre, L.A., Jemal, A.: Global cancer statistics 2018: GLOBOCAN estimates of incidence and mortality worldwide for 36 cancers in 185 countries. CA Can. J. Clin. **68**(6), 394–424 (2018)
4. Clark, T., Wong, A., Haider, M.A., Khalvati, F.: Fully deep convolutional neural networks for segmentation of the prostate gland in diffusion-weighted MR images. In: International Conference on Image Analysis and Recognition (ICIAR), pp. 97–104 (2017)
5. Cuocolo, R., et al.: Machine learning applications in prostate cancer magnetic resonance imaging. Eur. Radiol. Exp. **3**(1), 1–8 (2019)
6. Dai, Z., et al.: Segmentation of the prostatic gland and the intraprostatic lesions on multiparametic magnetic resonance imaging using mask region-based convolutional neural networks. Adv. Radiat. Oncol. **5**(3), 473–481 (2020)
7. Fedorov, A., Vangel, M.G., Tempany, C.M., Fennessy, F.M.: Multiparametric magnetic resonance imaging of the prostate: repeatability of volume and apparent diffusion coefficient quantification. Invest. Radiol. **52**(9), 538–546 (2017)
8. Hoar, D., et al.: Combined transfer learning and test-time augmentation improves convolutional neural network-based semantic segmentation of prostate cancer from multi-parametric MR images. Comput. Methods Programs Biomed. **210**, 106375 (2021)
9. Kang, J., Samarasinghe, G., Senanayake, U., Conjeti, S., Sowmya, A.: Deep learning for volumetric segmentation in spatio-temporal data: application to segmentation of prostate in DCE-MRI. In: IEEE International Symposium on Biomedical Imaging (ISBI), pp. 61–65 (2019)
10. Litjens, G., Debats, O., Barentsz, J., Karssemeijer, N., Huisman, H.: Computer-aided detection of prostate cancer in MRI. IEEE Trans. Med. Imaging **33**(5), 1083–1092 (2014)
11. Litjens, G., Debats, O., van de Ven, W., Karssemeijer, N., Huisman, H.: A pattern recognition approach to zonal segmentation of the prostate on MRI. In: International Conference on Medical Image Computing and Computer-Assisted Intervention (MICCAI), pp. 413–420 (2012)
12. Liu, Y., Zhu, Y., Wang, W., Zheng, B., Qin, X., Wang, P.: Multi-scale discriminative network for prostate cancer lesion segmentation in multiparametric MR images. Med. Phys. **49**(11), 7001–7015 (2022)
13. Mehrtash, A., et al.: Classification of clinical significance of MRI prostate findings using 3D convolutional neural networks. In: SPIE Medical Imaging: Computer-Aided Diagnosis, vol. 10134, pp. 589–592 (2017)
14. Meyer, A., Schindele, D., von Reibnitz, D., Rak, M., Schostak, M., Hansen, C.: PROSTATEx zone segmentations [dataset]. The Cancer Imaging Archive (2020)
15. Meyer, A., et al.: Anisotropic 3D multi-stream CNN for accurate prostate segmentation from multi-planar MRI. Comput. Methods Programs Biomed. **200**, 105821 (2021)

16. Meyer, A., et al.: Automatic high resolution segmentation of the prostate from multi-planar MRI. In: IEEE International Symposium on Biomedical Imaging (ISBI), pp. 177–181 (2018)

17. Meyer, A., et al.: Towards patient-individual PI-Rads v2 sector map: CNN for automatic segmentation of prostatic zones from T2-weighted MRI. In: IEEE International Symposium on Biomedical Imaging (ISBI), pp. 696–700 (2019)

18. Muñoz-Calahorro, C., García-Sánchez, C., Barrero-Candau, R., García-Ramos, J.B., Rodríguez-Pérez, A.J., Medina-López, R.A.: Anatomical predictors of long-term urinary incontinence after robot-assisted laparoscopic prostatectomy: a systematic review. Neurol. Urodyn. **40**(5), 1089–1097 (2021)

19. Rawla, P.: Epidemiology of prostate cancer. World J. Oncol. **10**(2), 63 (2019)

20. de Rooij, M., Israel, B., Bomers, J.G., Schoots, I.G., Barentsz, J.O.: Can biparametric prostate magnetic resonance imaging fulfill its PROMIS? Eur. Urol. **78**(4), 512–514 (2020)

21. Rundo, L., et al.: Automated prostate gland segmentation based on an unsupervised fuzzy c-means clustering technique using multispectral T1w and T2w MR imaging. Information **8**(2), 49 (2017)

22. Seah, J.C., Tang, J.S., Kitchen, A.: Detection of prostate cancer on multiparametric MRI. In: SPIE Medical Imaging: Computer-Aided Diagnosis, vol. 10134, pp. 585–588 (2017)

23. Shanmugalingam, K., Sowmya, A., Moses, D., Meijering, E.: Attention guided deep supervision model for prostate segmentation in multisite heterogeneous MRI data. In: International Conference on Medical Imaging with Deep Learning, pp. 1085–1095 (2022)

24. Tienza, A., Robles, J.E., Hevia, M., Algarra, R., Diez-Caballero, F., Pascual, J.I.: Prevalence analysis of urinary incontinence after radical prostatectomy and influential preoperative factors in a single institution. Aging Male **21**(1), 24–30 (2018)

25. Xu, H., Baxter, J.S.H., Akin, O., Cantor-Rivera, D.: Prostate cancer detection using residual networks. Int. J. Comput. Assist. Radiol. Surg. **14**(10), 1647–1650 (2019). https://doi.org/10.1007/s11548-019-01967-5

26. Yaniv, Z., Lowekamp, B.C., Johnson, H.J., Beare, R.: SimpleITK image-analysis notebooks: a collaborative environment for education and reproducible research. J. Digit. Imaging **31**(3), 290–303 (2018)

27. Zhu, Y., et al.: Fully automatic segmentation on prostate MR images based on cascaded fully convolution network. J. Magn. Reson. Imaging **49**(4), 1149–1156 (2019)

SPPNet: A Single-Point Prompt Network for Nuclei Image Segmentation

Qing Xu[1,2]([⊠]), Wenwei Kuang[2], Zeyu Zhang[2], Xueyao Bao[2], Haoran Chen[2], and Wenting Duan[1]

[1] University of Lincoln, Brayford Pool, UK
xq141839@connect.hku.hk, wduan@lincoln.ac.uk
[2] University of Hong Kong, Pok Fu Lam, Hong Kong
{kuangww,zeyu2022,baogela,rickchen}@connect.hku.hk

Abstract. Image segmentation plays an essential role in nuclei image analysis. Recently, the segment anything model has made a significant breakthrough in such tasks. However, the current model exists two major issues for cell segmentation: (1) the image encoder of the segment anything model involves a large number of parameters. Retraining or even fine-tuning the model still requires expensive computational resources. (2) in point prompt mode, points are sampled from the center of the ground truth and more than one set of points is expected to achieve reliable performance, which is not efficient for practical applications. In this paper, a single-point prompt network is proposed for nuclei image segmentation, called SPPNet. We replace the original image encoder with a lightweight vision transformer. Also, an effective convolutional block is added in parallel to extract the low-level semantic information from the image and compensate for the performance degradation due to the small image encoder. We propose a new point-sampling method based on the Gaussian kernel. The proposed model is evaluated on the MoNuSeg-2018 dataset. The result demonstrated that SPPNet outperforms existing U-shape architectures and shows faster convergence in training. Compared to the segment anything model, SPPNet shows roughly 20 times faster inference, with 1/70 parameters and computational cost. Particularly, only one set of points is required in both the training and inference phases, which is more reasonable for clinical applications. The code for our work and more technical details can be found at https://github.com/xq141839/SPPNet.

Keywords: Single-point prompt · Feature combination · Nuclei segmentation

1 Introduction

Automatic nuclei segmentation has received significant attention from pathologists and become an indispensable part of computer-aid diagnosis in the future

X. Cao et al. (Eds.): MLMI 2023, LNCS 14348, pp. 227–236, 2024.
https://doi.org/10.1007/978-3-031-45673-2_23

[15]. In the last few years, deep neural networks displayed outstanding performance in various computer vision tasks. Specifically. U-Net, proposed by Ronneberger et al. [12], has achieved great success in biomedical image segmentation. It adopts a convolutional encoder and decoder to extract image features and recover the original image shape respectively. A skip connection strategy is used to incorporate low-level semantic information with high-level semantic information in each layer. Zhou et al. [19] designed a nested U-Net (Unet++). It adds a series of nested and dense skip pathways between the encoder and decoder so that each layer can receive multi-scale semantic information. Schlemper et al. [14] developed an attention U-Net that uses a bottom-up attention gate to weight the concatenated feature map of the encoder and decoder. It highlights target regions and improves the sensitivity of the model. Instead of using a solely convolutional neural network (CNN) to encode image features, Chen et al. [2] proposed TransUNet that combines CNN with vision transformer (ViT) [4] for image patch extraction and global context collection. With the increasing influence of transformer, a growing number of modern medical segmentation models embed ViT or even adopt fully-transformer architecture [1,17]. Recently, segment anything model (SAM) [10] was proposed as a foundation model for natural image segmentation. It consists of three modules: a large masked autoencoders (MAE) [5] pre-trained ViT as image encoder. A prompt encoder is used to fetch tokens from box, point, or text inputs. A two-layer mask decoder leverages cross-attention for updating both the image embedding and prompt tokens. However, it can be difficult to retrain or fine-tune SAM when limited training sources are available. For nuclei image segmentation, although SAM can present remarkable performance when provided sufficient cell box or point prompts [6], it is not efficient for clinical applications.

In this paper, we propose a single-point prompt network (SPPNet) for nuclei segmentation in microscopic images. In order to improve the model robustness in practical scenarios, a center neighbour selection algorithm is established using the combination of distance transform and Gaussian kernel. To reduce the parameters of image encoder, the MAE ViT of SAM is replaced with a Tiny-ViT [16]. This small network may be not able to provide feature representations as complex as the large one. Therefore, we introduce a helpful shadow network, named low-level semantic information extractor (LLSIE). We evaluate the proposed on the small dataset: MoNuSeg-2018 [11], and train the model with only one RTX2080Ti graphic card. The experimental result shows that SPPNet performs better than other state-of-the-art (SOTA) models and takes more than 20% less training time. Compared to SAM, our proposed model requires much lower parameters, computational cost, and faster inference speed. Overall, SPPNet shows promising potential as a new cell segmentation method for pathological images in clinical settings.

2   Method

2.1   Center Neighborhood Point Sampling

The SAM model supports box and point prompts as inputs to direct the network focus on target regions and provide precise segmentation masks. The former requires gathering all potential cell bounding boxes in images. The latter expects to acquire the center point of cells. Both methods are difficult to be implemented in practice as one whole-slide microscopic image may include thousands of nuclei and medical experts cannot guarantee to always label the center of cells. To address this issue, we design a center neighborhood point sampling algorithm, which is presented in Fig. 1.

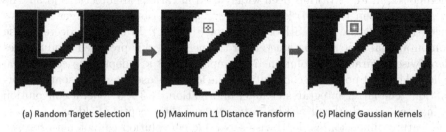

(a) Random Target Selection   (b) Maximum L1 Distance Transform   (c) Placing Gaussian Kernels

Fig. 1. Algorithm of center neighborhood point selection.

Firstly, we randomly select one cell from the ground truth as the target region x. Secondly, to find the center point M of the nuclei, the L1 distance transform method is applied to this area, which can be represented as:

$$M(x) = \arg\max \left[\min_{y \in \Omega} |x - y| \right] \tag{1}$$

The transform calculates the minimum L1 distance between each point in x and the cell border Ω. Then, we select the point with the maximum value in this set as the center of the cell. If more than one point satisfies the condition, we choose the first point in order. Thridly, a Gaussian filter is operated on this center to capture its neighborhood points. They can be defined as:

$$G_\sigma(m_x, m_y) = C \cdot e^{-\frac{m_x^2 + m_y^2}{2\sigma^2}}, m_x, m_y \in \{-K, \ldots, 0, \ldots, K\} \tag{2}$$

where $G_\sigma(m_x, m_y) \in \mathbb{R}^{(2K+1) \times (2K+1)}$ stands for a 2D Gaussian kernel and $\sum_{m_x=-K}^{K} \sum_{m_y=-K}^{K} G_\sigma(m_x, m_y) = 1$. σ^2 is the isotropic covariance, the kernel size can be defined as $(2K + 1) \times (2K + 1)$ and C is a constant. Finally, we randomly pick one from these neighbors as the positive input, which mimics the human behaviour of point selection. The parameter K is set to 2. For the negative point, we randomly choose one from the background.

2.2 Patch Up Low-Level Semantic Information

For the image encoder, SAM adopts a large-sized MAE [5] pre-trained ViT to extract features from input images. Although point prompt-based SAM has performed amazing zero-shot ability in natural image segmentation, medical image segmentation still can be a challenging task due to its complexity of interpretation. To retrain or fine-tune SAM on Nuclei datasets, 256 A100 GPUS are required in the training phase [10]. It can be a non-trivial burden for many researchers. Therefore, we replace SAM image encoder with a Tiny-ViT [16], which drops the number of parameters by about 99.1%. However, such an operation somehow diminishes the feature extraction ability of the image encoder. Inspired by previous U-shape segmentation networks, we construct a low-level semantic information extractor (LLSIE) to patch up shallow features. The U-Net module [12] in Fig. 2 (a) has been widely used in various architectures [8,9]. The stem block [3] in Fig. 2 (b) is usually developed to obtain the larger receptive field. The structure of LLSIE is shown in Fig. 2 (c). In order to decrease the computational cost and parameters of the network, our proposed module uses depthwise separable convolution [7], composed of 3×3 depthwise convolution followed by 1×1 pointwise convolution. The former is used to merge spatial information. The latter integrates channel information. A 3×3 general convolution is inserted into the head of LLSIE to preserve the receptive field and increase the feature dimension as depthwise separable convolution cannot perform well on low-channel feature maps [13]. In the end, we involve a residual connection to mitigate the potential effect of the gradient vanish.

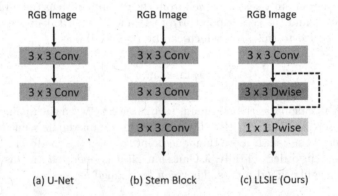

Fig. 2. Comparison of our design and other common structures for low-level feature extraction from input images.

2.3 SPPNet Architecture

For nuclei image segmentation, we establish the SPPNet based on the proposed point sampling strategy and LLSIE block. Figure 3 displays an overview of our

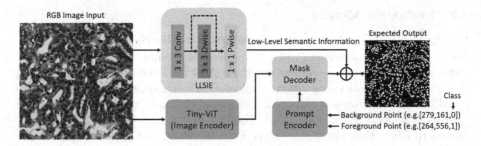

Fig. 3. Overview of our single-point prompt network with efficient vision transformer.

SSPNet. The input RGB microscopic images are first fed to the Tiny-ViT for feature extraction. A single set of point prompts, including one cell point and one background point, will be generated using center neighborhood point sampling, and then used to collect tokens by prompt encoder. The mask decoder leverages multilayer perceptrons and a cross-attention mechanism to upgrade the image embedding as well as prompt, and raise the width and height of feature maps to 256×256. In addition, the LLSIE block is in parallel with the image encoder for capturing low-level semantic information from inputs. This block is followed by a 2×2 max pooling with stride 2 for aligning with the shape of the feature map produced by the mask decoder. A connection operation fuses both two features and one 1×1 convolution declines the channel to the same number of classes as prediction. Finally, we reuse the post-processing method of SAM to recover the image size and generate masks.

3   Experiments and Results

3.1   Implementation Details

All experiments are conducted with PyTorch 1.10.0 framework on a single NVIDI-A RTX2080Ti Tensor Core GPU, 4-core CPU, and 28 GB RAM. A standard dice loss function and an Adam optimizer with a learning rate of $5e-4$ are used to train all models. We set batch sizes and epochs to 4 and 200 respectively. The early stopping strategy is implemented to avoid overfitting. During training our SSPNet, we input original microscopic images to the image encoder and resized images (256×256) to the LLSIE block. These resized images are also used to train other SOTA methods for comparison. Meanwhile, Tiny-ViT loads a pre-trained checkpoint based on knowledge distillation [18]. The prompt encoder and mask decoder of SPPNet reuse the pre-trained weight of SAM [10]. We randomly select 80% samples as a training set, 10% as a validation set, and 10% as a test set. Moreover, data augmentation methods have been widely used to extend image diversity and improve model robustness. In the experiment, horizontal flip, rotation, and cutout with a probability of 0.25 are randomly applied to the training set.

3.2   Evaluation Metrics

In order to evaluate and quantify mode performance. Mean intersection over union (mIoU) and dice coefficient (DSC), the standard medical image segmentation metrics, are computed on the test set. Both show the similarity between prediction and ground truth using different calculation methods, where mIoU tends to penalise single instances of bad classification. Additionally, we report parameters (Params), floating point operations (FLOPs), and frames per second (FPS) of models to reveal their sizes, computational complexity, and inference speed.

3.3   Comparison on MoNuSeg-2018 Dataset

Microscopic images usually include a large number of cells. Manual labeling is expensive and time-consuming. Therefore, we select a small dataset to evaluate all models, which is more feasible in clinical applications. MoNuSeg-2018 is an open-access dataset for MICCAI 2018 multi-organ nuclei segmentation challenge. It only contains 51 fully-labeled pathological images. We trained SPPNet and four SOTA models. The quantitative result of the test set is provided in Table 1. We can observe that SPPNet achieves a DSC of 79.77% and a mIoU of 66.43%, which outperforms Attention U-Net by 2.09% in terms of DSC and 2.58% in mIoU. Particularly, our proposed model demonstrates a significant enhancement over the two previous transformer-based architectures, where the mIoU of SPPNet is 8.56% and 8.22% higher than TransUNet and Swin-UNet, and the DSC of SPPNet is 6.68% and 6.41% higher than these two models respectively. Also, SPPNet costs considerably fewer parameters. Consequently, our proposed model reveals the highest score in standard evaluation metrics of medical image segmentation.

Table 1. Performance comparison between our method (SSPNet) and other SOTA models on MoNuSeg-2018 dataset.

Method	mIoU(%)	DSC(%)	Params(M)	FLOPs	FPS
U-Net [12]	61.64 ± 8.28	75.92 ± 6.79	13.40	23.83	93.22
Unet++ [19]	62.28 ± 6.54	76.05 ± 5.06	9.16	26.73	69.88
Attention U-Net [14]	63.85 ± 7.23	77.68 ± 5.82	34.88	51.02	55.19
TransUNet [2]	57.87 ± 6.41	73.09 ± 5.51	61.82	32.63	66.67
Swin-UNet [1]	58.21 ± 6.50	73.36 ± 5.50	27.15	5.91	43.12
SPPNet	**66.43±4.32**	**79.77±3.11**	9.79	39.90	22.61

3.4   Ablation Study

In this section, a detailed ablation study is conducted on SPPNet. LLSIE block is an essential module to capture low-level semantic information from input images. We first compare LLSIE with other prevalent blocks, which is shown in Table 2.

Table 2. Performance comparison between our LLSIE block and other general methods for low-level semantic information extraction on MoNuSeg-2018 dataset.

Block	mIoU(%)	DSC(%)	Params(M)	FLOPs	FPS
U-Net [12]	64.95 ± 7.94	77.70 ± 6.37	9.80	40.41	22.12
Stem	65.13 ± 6.69	78.93 ± 5.19	9.81	41.02	21.55
LLSIE	$\mathbf{66.43 \pm 4.32}$	$\mathbf{79.77 \pm 3.11}$	9.79	39.90	22.61

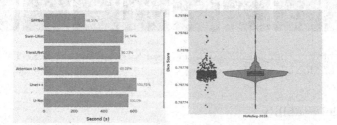

Fig. 4. Training time cost to achieve SOTA performance with the U-Net baseline (left side) and presentation of model stability with violin plot (right side).

Table 3. Detailed ablation study of our single-point prompt network architecture on dataset. IE: image encoder, PE: prompt encoder, MD: mask decoder, CNPS: center neighborhood point sampling

SAM			SPPNet							
IE	PE	MD	IE	CNPS	LLSIE	mIoU(%)	DSC(%)	Params(M)	FLOPs	FPS
✓	✓	✓				60.18 ± 8.15	74.76 ± 7.00	635.93	2736.63	1.39
	✓	✓	✓			62.33 ± 4.05	76.22 ± 2.98	9.78	39.73	23.78
	✓	✓	✓	✓		63.77 ± 5.77	77.73 ± 4.26	9.78	39.73	23.64
	✓	✓	✓	✓	✓	$\mathbf{66.43 \pm 4.32}$	$\mathbf{79.77 \pm 3.11}$	9.79	39.90	22.61

It can be demonstrated that the LLSIE block not only performs better than U-Net and Stem blocks in metrics of mIoU and DSC, but also has faster inference speed and costs a lower number of parameters as well as computational resources. Furthermore, we evaluate the efficiency of Tiny-ViT, center neighborhood point sampling strategy, and LLSIE block, which is presented in Table 3. We make a comparison to SAM with fine-tuned prompt encoder and mask decoder. From Table 3, we can first conclude that it is necessary to retrain the image encoder when applied to the nuclei segmentation task as retrained Tiny-ViT outperforms pre-trained MAE ViT. Secondly, our center neighborhood point sampling strategy can improve model performance without any extra parameters and computational costs. Thirdly, incorporating low-level semantic information can help the model predict a more precise segmentation mask. Overall, compared to SAM, SPPNet shows better performance, considerably fewer parameters, lower computational costs, and faster inference.

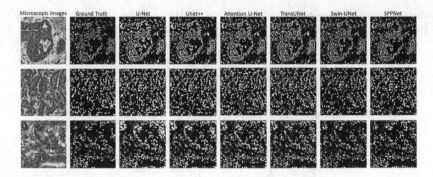

Fig. 5. Qualitative comparison results between SPPNet and other SOTA models on the test set of MoNuSeg-2018.

3.5 Model Stability

When applying the center neighborhood point sampling strategy to determine a positive point, the target cell and neighbors are randomly selected by our program. To evaluate our model stability, we use the same trained model to implement the test set with 500 iterations. The result is provided in Fig. 4 (right side) with a violin plot. It can be demonstrated that the dice score of the proposed SPPNet is clustered at 79.776% to 79.778%. The maximum performance gap is only about 0.01%. As a result, we can argue that the random point selection based on our sampling strategy does not have a significant impact on the model performance.

3.6 Dicussion

Nuclei image segmentation is an important step for pathological analysis for patients. From the above experimental results, our approach shows an outstanding improvement compared to SAM in training resource costs and better performance than existing SOTA models of medical image segmentation. Also, SPPNet can take about 50% less time than others in the training phase, which is shown in Fig. 4 (left side). To further demonstrate the efficiency of SPPNet in clinical applications, some segmentation masks using all models on the test set are visualised in Fig. 5. Compared to other architectures, SPPNet is able to provide more accurate masks and fewer false positive cases. In addition, SAM displays a great zero-shot ability in natural image segmentation as it is trained on an extremely sufficient dataset. Therefore, we will also explore the zero-shot performance of SPPNet for nuclei image segmentation in the future. Achieving a such foundation model can considerably alleviate the paucity of pathologists.

4   Conclusion

In this paper, we propose a single-point prompt network for nuclei image segmentation. The LLSIE block is used to patch up lower-level semantic information in the mask decoder. A center neighborhood point sampling strategy makes inference more feasible in clinical applications. The experimental results demonstrate that our SPPNet can be considered a new benchmark for cell segmentation.

References

1. Cao, H., et al.: Swin-unet: unet-like pure transformer for medical image segmentation. In: Karlinsky, L., Michaeli, T., Nishino, K. (eds.) ECCV 2022. LNCS, vol. 13803, pp. 205–218. Springer, Cham (2022). https://doi.org/10.1007/978-3-031-25066-8_9

2. Chen, J., et al.: Transunet: transformers make strong encoders for medical image segmentation. arXiv preprint arXiv:2102.04306 (2021)

3. Chen, K., et al.: MMDetection: Open MMLAB detection toolbox and benchmark. arXiv preprint arXiv:1906.07155 (2019)

4. Dosovitskiy, A., et al.: An image is worth 16x16 words: transformers for image recognition at scale. arXiv preprint arXiv:2010.11929 (2020)

5. He, K., Chen, X., Xie, S., Li, Y., Dollár, P., Girshick, R.: Masked autoencoders are scalable vision learners. In: Proceedings of the IEEE/CVF Conference on Computer Vision and Pattern Recognition, pp. 16000–16009 (2022)

6. He, S., Bao, R., Li, J., Grant, P.E., Ou, Y.: Accuracy of segment-anything model (SAM) in medical image segmentation tasks. arXiv preprint arXiv:2304.09324 (2023)

7. Howard, A.G., et al.: Mobilenets: efficient convolutional neural networks for mobile vision applications. arXiv preprint arXiv:1704.04861 (2017)

8. Huang, H., et al.: Unet 3+: a full-scale connected Unet for medical image segmentation. In: ICASSP 2020–2020 IEEE International Conference on Acoustics, Speech and Signal Processing (ICASSP), pp. 1055–1059. IEEE (2020)

9. Jha, D., Riegler, M.A., Johansen, D., Halvorsen, P., Johansen, H.D.: Doubleu-net: a deep convolutional neural network for medical image segmentation. In: 2020 IEEE 33rd International Symposium on Computer-Based Medical Systems (CBMS), pp. 558–564. IEEE (2020)

10. Kirillov, A., et al.: Segment anything. arXiv preprint arXiv:2304.02643 (2023)

11. Kumar, N., Verma, R., Sharma, S., Bhargava, S., Vahadane, A., Sethi, A.: A dataset and a technique for generalized nuclear segmentation for computational pathology. IEEE Trans. Med. Imaging 36(7), 1550–1560 (2017)

12. Ronneberger, O., Fischer, P., Brox, T.: U-Net: convolutional networks for biomedical image segmentation. In: Navab, N., Hornegger, J., Wells, W.M., Frangi, A.F. (eds.) MICCAI 2015. LNCS, vol. 9351, pp. 234–241. Springer, Cham (2015). https://doi.org/10.1007/978-3-319-24574-4_28

13. Sandler, M., Howard, A., Zhu, M., Zhmoginov, A., Chen, L.C.: Mobilenetv 2: Inverted residuals and linear bottlenecks. In: Proceedings of the IEEE Conference on Computer Vision and Pattern Recognition, pp. 4510–4520 (2018)

14. Schlemper, J., et al.: Attention gated networks: learning to leverage salient regions in medical images. Med. Image Anal. 53, 197–207 (2019)

15. Wang, R., Lei, T., Cui, R., Zhang, B., Meng, H., Nandi, A.K.: Medical image segmentation using deep learning: a survey. IET Image Proc. 16(5), 1243–1267 (2022)

16. Wu, K., et al.: TinyViT: fast pretraining distillation for small vision transformers. In: Avidan, S., Brostow, G., Cissé, M., Farinella, G.M., Hassner, T. (eds.) ECCV 2022. LNCS, vol. 13681, pp. 68–85. Springer, Cham (2022)

17. Yan, X., Tang, H., Sun, S., Ma, H., Kong, D., Xie, X.: After-Unet: axial fusion transformer Unet for medical image segmentation. In: Proceedings of the IEEE/CVF Winter Conference on Applications of Computer Vision, pp. 3971–3981 (2022)
18. Zhang, C., et al.: Faster segment anything: towards lightweight SAM for mobile applications. arXiv preprint arXiv:2306.14289 (2023)
19. Zhou, Z., Siddiquee, M.M.R., Tajbakhsh, N., Liang, J.: Unet++: redesigning skip connections to exploit multiscale features in image segmentation. IEEE Trans. Med. Imaging **39**(6), 1856–1867 (2019)

Automated Coarse-to-Fine Segmentation of Thoracic Duct Using Anatomy Priors and Topology-Guided Curved Planar Reformation

Puyang Wang[1,2], Panwen Hu[3], Jiali Liu[4(✉)], Hang Yu[3], Xianghua Ye[5],
Jinliang Zhang[6], Hui Li[6], Li Yang[6], Le Lu[1], Dakai Jin[1],
and Feng-Ming (Spring) Kong[4]

[1] DAMO Academy, Alibaba Group, Hangzhou, China
[2] Hupan Lab, Hangzhou 310023, China
[3] The Chinese University of Hong Kong, Shenzhen, China
[4] The University of Hong Kong, Pok Fu Lam, Hong Kong
liujiali1995@hotmail.com
[5] Zhejiang University, Hangzhou, China
[6] The University of Hong Kong, Shenzhen Hospital, Ap Lei Chau, Hong Kong

Abstract. Recent studies have emphasized the importance of protecting thoracic duct during radiation therapy (RT), as dose distributions in thoracic duct may be associated with the development radiation-induced lymphopenia. Because of its thin/slim size, curved geometry and extremely poor (intensity) contrast of thoracic duct, manual delineation of thoracic duct in RT planning CT is time-consuming and with large inter-observer variations. In this work, we aim to automatically and accurately segment thoracic duct in RT planning CT, as the first attempt to tackle this clinically critical yet under-studied task. A two-stage coarse-to-fine segmentation approach is proposed. At the first stage, we automatically segment six chest organs and combine these organ predictions with the input planning CT to better infer and localize the thoracic duct. Given the coarse initial segmentation from first stage, we subsequently extract the topology-corrected centerline of initial thoracic duct segmentation at stage two where curved planar reformation (CPR) is applied to transform the planning CT into a new 3D volume representation that provides a spatially smoother reformation of thoracic duct in its elongated medial axis direction. Thus the CPR-transformed CT is employed as input to the second stage deep segmentation network, and the output segmentation mask is transformed back to the original image space, as the final segmentation. We evaluate our approach on 117 lung cancer patients with RT planning CT scans. Our approach significantly outperforms a strong baseline model based on nnUNet, by reducing 57% relative Hausdorff distance error (from 49.9 mm to 21.2 mm) and improving 1.8% absolute Jaccard Index.

P. Wang, P. Hu, J. Liu—Equal contribution.
D. Jin, F.-M. S. Kong—Co-senior author.

Keywords: Thoracic Duct Segmentation · Radiation Therapy · Centerline Extraction · Curved Planar Reformation

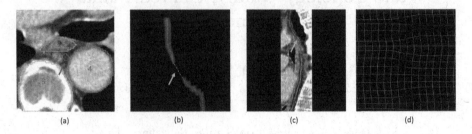

Fig. 1. (a) Thoracic duct indicated by red arrow (1: esophagus, 2: azygos vein, 3: spine, 4: aorta). (b) Segmentation breakage produced (indicated by yellow arrow) using the state-of-the-art segmentation method nnUNet [5] and the result of our topology-corrected centerline extraction. (c) CPR of CT scan based on extracted centerline. (d) CPR displayed in grid transform. (Color figure online)

1   Introduction

Thoracic cancer is a significant public health challenge in the United States, with breast, lung, and esophageal cancers accounting for over 540,000 confirmed patient cases in 2022 alone [18]. Out of these patient populations, approximately 50–60% will receive radiation therapy (RT) as part of their treatment procedure [22]. While radiation is an effective treatment option for thoracic cancer, it also can cause severe toxic effects, e.g., radiation-induced lymphopenia (RIL), which is a condition that radiation dosage damages circulating immune cells and significantly impairs tumor control and patient survival [2,20]. Several factors, such as mean lung dose, mean heart dose, and the effective dose to circulating immune cells (EDIC), are associated with the development of RIL [10,19,24]. Recent studies have showed the importance of protecting the thoracic duct during RT [15]. Thus it becomes essential for the accurate delineation of thoracic duct to minimize the risk of RIL in RT planning.

Thoracic duct, as the main collecting vessel of the lymphatic system, is a continuous tubular structure with a mean diameter of 2–3 mm in the axial slices, and a mean intensity attenuation value of 15.3 HU (ranging between 4.5 to 38 HU) in CT scans that is slightly lower than that of arteries and veins [13,16]. It is very challenging for accurate thoracic duct delineation due to its slim and curved 3D structure and the extremely poor contrast with surrounding adipose tissue in CT scans (see Fig. 1 (a, b) for an illustration). Moreover, the low spatial resolution of RT planning CT may not adequately capture intricate details of the thoracic duct because the planning CT scans typically have the slice thickness between 3 and 5 mm with thé pixel spacing of $1 \sim 1.2$ mm in axial plane. The imaging quality can become even worse for some medical institutions when the planning CT is an average intensity projection CT generated from respiratory four-dimensional (4D-CT) scanning. All these factors compound the difficulty of manual delineation of thoracic duct, making it time-consuming with large inter-observer variability. This could ultimately lead to sub-optimal RT planning and produce potential radiation toxicity.

Recent advancements in deep learning have shown great promise in auto-mated segmentation of organs at risk (OARs) and tumors in various body parts [1,3,4,6–8,23,26]. UaNet adopts a segmentation-by-detection strategy to achieve 28 head & neck OAR segmentation [21], while SOARS [25] achieves a comprehensive of 42 head & neck OAR segmentation using stratified learning and neural architecture search [3,25]. RTP-Net [17] develops a cascade coarse-to-fine segmentation scheme with organ size adaptive module and attention mechanisms for organ boundaries to segment 67 whole-body OARs. For thoracic OARs, a DeepStationing model has segmented 22 chest anatomical structures to support the mediastinal lymph node station segmentation [4], where high accuracy is achieved for OARs such as lungs, heart, esophagus and spinal cord. These stud-ies demonstrate the capability of deep learning models to improve the OAR seg-mentation accuracy, consistency and reproducibility to benefit the RT planning in clinical practice. Nevertheless, none of the previous work have tackled tho-racic duct segmentation. To address this challenging task, important anatomic knowledge and clinical insights can be leveraged. 1) Anatomy of thoracic duct is closely related to several key organs, e.g., near the level of the fifth thoracic vertebra, thoracic duct passes through the space between esophagus and spine. Physicians often utilize these reference organs to locate the spatial regions that thoracic duct may appear. 2) Considering that the relative low spatial resolution of planning CT, physicians often zoom in the potential region of interest (ROI) to better visualize the 3D extension and boundary of thoracic duct.

In this work, we propose a two-stage coarse-to-fine thoracic duct segmenta-tion framework based on the anatomy prior and topology guidance. At stage one, using a recent multi-organ deep segmentation model [4], we first segment six key chest organs that are spatially related to the thoracic duct. These organ predictions are then used as anatomy guidance to better localize and segment the thoracic duct. Specifically, 3D mask image consisting six key organs is concate-nated with the planning CT to serve as input to the stage one deep segmentation network. We then extract the centerline from the initial thoracic duct segmen-tation and use a minimum-cost path approach [9] to connect the discontinuous ones if there exists. Based on the topology-corrected centerline, CT scan and key organ mask can be resampled around the initial thoracic duct segmentation using the curved planar reformation (CPR) technique [11,12] to generate newly reformatted 3D volumes (refer Fig. 1(c,d) as an example). The resampled CT volume provides a spatially smoother representation of thoracic duct along its medial axial direction, and used as input to the stage two segmentation network. The segmentation result of stage two is later transformed back into the origi-nal image space using inverse CPR transform. We evaluate our approach on an

in-house dataset, including 117 lung cancer planning CT scans with manual thoracic duct annotations. Our approach significantly outperforms a strong baseline
of nnUNet [5] by reducing 57% relative Hausdorff distance error (from 49.9 mm
to 21.2 mm) and improving 1.8% absolute Jaccard Index.

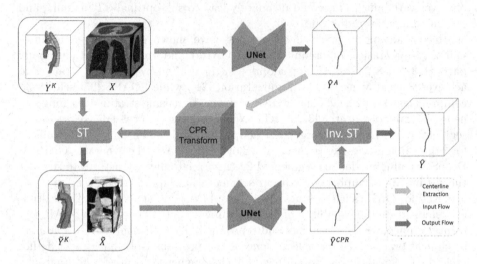

Fig. 2. Overall workflow of our proposed 2-stage Anatomy and Topology Guided
Coarse-to-fine Segmentation of Thoracic Duct. ST and Inv. ST denotes Spatial Transform and Inverse Spatial Transform.

2   Methods

The proposed coarse-to-fine thoracic duct segmentation framework consists of
two main stages. At stage-1, anatomy-guided coarse segmentation is conducted.
At stage-2, the centerline of coarse segmentation is extracted first, which is utilized to perform the CPR transformation. After that, the fine-scale segmentation
is executed in the CPR space and the segmentation output is transformed back
in the original image space to get the final thoracic duct segmentation. Figure 2
depicts an overview of our proposed method.

2.1   Anatomy-Guided Coarse Segmentation

To better localize and segment thoracic duct at stage one, we first segment a set
of six key organs using a recent multi-organ deep segmentation mode [4]. Six key
organs include: *esophagus, aorta, spine, azygos vein, subclavian vein* and *internal
jugular vein*. Their predictions are used as anatomy prior to guide the thoracic
duct segmentation. Let a dataset of N instances denoted as $\mathbf{D} = \{X_n, Y_n{}^S, Y_n\}^N$,
where X_n, Y_n^S, Y_n denote the input CT image, prediction mask of six supporting
organs and ground truth mask of thoracic duct. Dropping n for clarity, the

anatomy-guided segmentation model at stage one predicts a coarse thoracic duct Y^C given X and Y^S:

$$\hat{Y}^C = f^C(X, Y^S | \mathbf{W}^C) \tag{1}$$

where $f^{(*)}(.)$ and $\mathbf{W}^{(*)}$ denote the network function and the corresponding network parameters, respectively, and $\hat{Y}^C$ represents the predicted coarse thoracic duct output. As demonstrated in the experiment, using six supporting organs leads to more accurate segmentation results with less false positives.

2.2   Segmentation Refinement Through Topology-Guided CPR

Although the prediction $\hat{Y}^C$ achieves overall reasonable segmentation results, however, discontinuous segmentation along vertical axis and inaccurate boundary in xy-planes may still exist. We aim to solve these issues at stage two by refining the segmentation.

It is noticed that physicians often use the curved planar reformation (CPR) to visualize vascular abnormalities for small vessels [11,12], where CPR can generate longitudinal cross-sections of a tubular structure in a curved plane. Inspired by that, we apply the CPR to transform the planning CT into a new 3D volume representation that provides a spatially smoother reformation of thoracic duct in its elongated medial axis direction. Then, the CPR-transformed CT along with CPR-transformed organ mask are employed as input to the second stage deep segmentation network for fine-scale thoracic duct segmentation. This refinement stage two consists of the following steps: 1) extract a single and continuous centerline $\hat{C}$ from predicted thoracic duct $\hat{Y}^C$ mask even if $\hat{Y}^C$ has segmentation breakage; 2) compute CPR transformation map based on $\hat{C}$; 3) apply CPR transformation to CT and support organ mask and train a deep segmentation network using transformed CT and organ mask in CPR space, 4) apply inverse CPR transformation to get the final thoracic duct segmentation in original CT space. The centerline extraction and CPR transformation steps are described as follows.

Topology-Corrected Centerline Extraction. To compute the CPR transformation, it first requires to extract a complete single centerline regardless of the number of components that $\hat{Y}^C$ has. To achieve that, consider $\hat{Y}^C$ has M connected components $\hat{Y}_m^C$ because of the breakage in the coarse segmentation. Centerline of each component $\hat{Y}_m^C$ can be extracted by a thinning algorithm [14], i.e., $\hat{C}_m = \text{Thinning}(\hat{Y}_m^C)$. Then, the gap between components are connected by iteratively computing a minimum-cost path [9] between each two adjacent components. This leads to a complete connected centerline of the same topology as original thoracic duct. The process is illustrated in Fig. 1 (b).

CPR Transformation. The goal of CPR transformation is to make a tubular structure visible in it's entire length within one single image. In particular, we use the stretched CPR. To do this, thoracic duct centerline is required. Assume that the extracted centerline $\hat{C}$ of $\hat{Y}^C$ is a sequence of points at sub-voxel resolution. By processing all points successively, the corresponding lines-of-interest are mapped to the image. This is done by rotating the consecutive point around the current line-of-interest. The point is rotated in a way that the resulting plane is coplanar to the viewing plane. Let P_i to be the last processed point and point P_{i+1} the currently processed point of the centerline. The vector $d_i = P_i P_{i+1}$ and l represent the path direction at position i and the normalized direction of the line-of-interest respectively. The offset Δ_i in image space is derived as:

$$\Delta_i = \sqrt{|d_i|^2 - l \cdot d_i}. \tag{2}$$

The image position (y-coordinates) y_{i+1} of the line-of-interest related to point P_{i+1} is given by $y_{i+1} = y_i + \Delta_i$ where $y_0 = 0$. The resampling map is computed by consecutive viewing planes perpendicular to derived line-of-interest. An example of CPR transformed CT using centerline of thoracic duct is shown in Fig. 1 (c) with its grid transform map in (d).

3   Experiments and Results

Dataset. After obtaining approval by the appropriate institutional review board, we retrospectively collected patients with primary lung cancer treated by radiotherapy from May 2005 to February 2020 at The University of Hong Kong, Shenzhen Hospital. A total of 117 patients with RT planning CT were included, with an average CT volume size of $514 \times 514 \times 139$ voxels and an average voxel resolution of $1.2 \times 1.2 \times 3.0 \, \text{mm}^3$. The thoracic duct is manually delineated by an experienced radiation oncologist (10 yr) with the guidance of a second senior radiation oncologist (25 yr), while the segmentation of six supporting organs, including azygos vein, aorta, esophagus, spine, left internal jugular vein and left subclavian vein, are provided using a recent multi-organ segmentation model [4].

Implementation Details. We adopt '3d-fullres' version of nnUNet [5] with Dice+CE losses as our backbone modules. Each encoder is the same as the default nnUNet encoder. We use the default nnU-Net data augmentation settings for our model training, and set the patch size to $192 \times 192 \times 48$ and $96 \times 96 \times 208$ for 1st and 2nd stages. We implemented our framework using PyTorch and trained on an NVIDIA Tesla V100. The total training epochs is 500. The average training time is 0.5 GPU days. For CPR transform, we set the field of view to 6.4×6.4 cm and adopt a resampling spacing of $0.5 \times 0.5 \times 1.5 \, \text{mm}$.

Comparing Method and Evaluation Metrics. We employ five-fold cross-validation protocol split at the patient level. As there is no previous works solving the thoracic duct segmentation, we compare our method with nnUNet [5], which represents the current leading organ segmentation approach and use it as our segmentation backbone. Three quantitative metrics are reported to evaluate the thoracic duct segmentation performance: Jaccard Index (Jac.), Dice score (DSC), and Hausdorff distance (HD) in "mm". We further divide the whole thoraic duct (TD) into two anatomical segments, upper TD and lower TD based on the top of aortic arch, and report their corresponding quantitative metrics, respectively.

Table 1. Quantitative ablation results for proposed 2-stage Thoracic Duct (TD) segmentation framework using nnUNet [5] as backbone. AG represents the anatomy-guided coarse segmentation (with six key supporting organs). CPR refers to the topology-guided curved planar reformation based segmentation. DSC and Jacarrd (Jac.) Index are shown in "%" and Hausdorff distance (HD) in "mm".

AG	CPR	Upper TD			Lower TD			Whole TD		
		Jac.	DSC	HD	Jac.	DSC	HD	Jac.	DSC	HD
−	−	21.31	32.85	34.89	47.45	63.96	29.71	42.48	59.18	49.79
✓	−	27.88	41.90	28.32	47.67	64.17	20.58	43.24	60.04	38.19
−	✓	26.78	41.70	28.11	47.38	63.68	27.56	43.27	58.06	40.44
✓	✓	**30.21**	**43.99**	**12.22**	**48.87**	**65.34**	**14.46**	**44.15**	**60.09**	**21.18**

Quantitative Results

Our quantitative ablation results which demonstrate the effectiveness of each component in the proposed framework and comparison to leading general organ segmentation approach nnUNet [5] are tabulated in Table 1. The volumetric index of Jac. or DSC scores are low in general. This indicates the difficulty of this task. However, we can observed that by applying the proposed anatomy guidance using six key support organs to nnUNet newtork (row 2 vs. row 1), HD drops from 49.79 mm to 38.19 mm, meanwhile improves the absolute Jac. score by 0.8%. Especially, for upper TD region, the Dice score increase around 30%, from 32.85% to 41.90% which clearly demonstrated the importance of incorporating anatomy priors for segmenting small and hard organs in CT scans. The effectiveness of proposed topology-guided CPR based segmentation refinement is first validated by comparing row 3 vs. row 1 where one can observe a similar improvement as using anatomy guidance. Note that, since our proposed framework is a two-stage pipeline, the performance of stage 2 relies on the output of stage 1. Thus, although the AG or CPR alone can help better segment the thoracic duct, their combination further significantly improves the results (row 4 vs. row 2/3). For instance, even if AG is already utilized in the nnUNet model to guide the segmentation, combining it with the second stage CPR segmentation refinement still leads to an additional 16 mm, 6 mm and 17 mm Hausdorff distance error reduction in upper, lower and whole TD respectively. When compared with original nnUNet, our complete 2-stage framework (AG + CPR) showed a significant improvement in terms of all metrics (row 4 vs. row 1). Furthermore, we examined the inter-observer variation in 117 patients. The consistency between two physicians following our internal delineation guideline is 64.64% in Dice score. As comparison, our method achieves 60.09% Dice score.

Qualitative Results. Apart from the quantitative comparison, we also compare our method with nnUNet qualitatively by the visual inspection. Three qualitative results are shown in Fig. 3. As can be observed, the proposed 2-stage method can generate topology desired and more complete thoracic duct segmentation. In contrast, nnUNet yields several clear segmentation breakages. While our stage

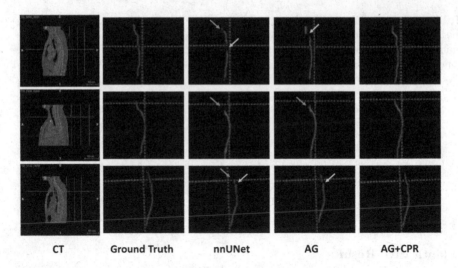

| CT | Ground Truth | nnUNet | AG | AG+CPR |

Fig. 3. Qualitative results of thoracic duct segmentation in three patients using different methods. nnUNet is the leading comparison method. AG is our stage one segmentation method, while AG+CPR represents the proposed two-stage method. Segmentation breakages and missing thoracic duct are indicted by yellow and blue arrows, respectively. (Color figure online)

one segmentation model (nnUNet with AG) can reasonably improve the performance, breakages or under-segmentations still exist as shown in the figure. With AG+CPR as our proposed 2 stage framework, all previous defects are absent in all three example cases.

4 Conclusion

In this work, we present a two stage coarse-to-fine thoracic duct segmentation approach. Recent studies have emphasized the importance of protecting thoracic duct during radiation therapy, and we are the first to tackle this clinically critical yet under-studied task using an automated method. Because of its thin/slim size, curved geometry and extremely poor (intensity) contrast of thoracic duct, we propose to use anatomy priors and topology guidance in a two stage framework to address these challenges. At stage one, six key chest organs are automatically segmented and combined with the planning CT to better infer and localize the thoracic duct. At stage two, we subsequently extract the centerline of initial thoracic duct segmentation, where curved planar reformation is applied to transform the planning CT into a new 3D volume representation that provides a spatially smoother reformation of thoracic duct in its elongated medial axis direction. Then, the CPR-transformed CT is employed as input to the stage two deep segmentation network, and the output prediction is transformed back to the original image space, as the final thoracic duct segmentation. Experimental results demonstrate the effectiveness of our approach, as it significantly outperforms a strong baseline nnUNet by reducing 57% relative HD error (from 49.9 mm to 21.2 mm) and improving 1.8% absolute Jaccard Index.

References

1. Chen, X., et al.: A deep learning-based auto-segmentation system for organs-at-risk on whole-body computed tomography images for radiation therapy. Radiother. Oncol. **160**, 175–184 (2021)
2. Davuluri, R., et al.: Absolute lymphocyte count nadir during chemoradiation as a prognostic indicator of esophageal cancer survival outcomes. Int. J. Radiat. Oncol. Biol. Phys. **96**(2), E177 (2016)
3. Guo, D., et al.: Organ at risk segmentation for head and neck cancer using stratified learning and neural architecture search. In: IEEE/CVF Conference on Computer Vision and Pattern Recognition, pp. 4223–4232 (2020)
4. Guo, D., et al.: DeepStationing: thoracic lymph node station parsing in CT scans using anatomical context encoding and key organ auto-search. In: de Bruijne, M., et al. (eds.) MICCAI 2021. LNCS, vol. 12905, pp. 3–12. Springer, Cham (2021). https://doi.org/10.1007/978-3-030-87240-3_1
5. Isensee, F., Jaeger, P.F., Kohl, S.A., Petersen, J., Maier-Hein, K.H.: NNU-net: a self-configuring method for deep learning-based biomedical image segmentation. Nat. Methods **18**(2), 203–211 (2021)
6. Ji, Z., et al.: Continual segment: towards a single, unified and accessible continual segmentation model of 143 whole-body organs in CT scans. In: IEEE International Conference on Computer Vision (2023)
7. Jin, D., Guo, D., Ge, J., Ye, X., Lu, L.: Towards automated organs at risk and target volumes contouring: defining precision radiation therapy in the modern era. J. Natl. Can. Center **2**, 306–313 (2022)
8. Jin, D., et al.: Deeptarget: gross tumor and clinical target volume segmentation in esophageal cancer radiotherapy. Med. Image Anal. **68**, 101909 (2021)
9. Jin, D., Iyer, K.S., Chen, C., Hoffman, E.A., Saha, P.K.: A robust and efficient curve skeletonization algorithm for tree-like objects using minimum cost paths. Pattern Recogn. Lett. **76**, 32–40 (2016)
10. Jin, J.Y., et al.: A framework for modeling radiation induced lymphopenia in radiotherapy. Radiother. Oncol. **144**, 105–113 (2020)
11. Kanitsar, A., Fleischmann, D., Wegenkittl, R., Felkel, P., Groller, E.: CPR-curved planar reformation. IEEE (2002)
12. Kanitsar, A., Wegenkittl, R., Fleischmann, D., Groller, M.E.: Advanced curved planar reformation: flattening of vascular structures. IEEE (2003)
13. Kiyonaga, M., Mori, H., Matsumoto, S., Yamada, Y., Sai, M., Okada, F.: Thoracic duct and cisterna chyli: evaluation with multidetector row CT. Br. J. Radiol. **85**(1016), 1052–1058 (2012)
14. Lee, T.C., Kashyap, R.L., Chu, C.N.: Building skeleton models via 3-d medial surface axis thinning algorithms. CVGIP Graphical Models Image Process. **56**(6), 462–478 (1994)
15. Liu, J., et al.: Integrate sequence information of dose volume histogram in training LSTM-based deep learning model for lymphopenia diagnosis. Int. J. Radiat. Oncol. Biol. Phys. **111**(3), e112–e113 (2021)
16. Schnyder, P., et al.: CT of the thoracic duct. Eur. J. Radiol. **3**(1), 18–23 (1983)

17. Shi, F.: Deep learning empowered volume delineation of whole-body organs-at-risk for accelerated radiotherapy. Nat. Commun. **13**(1), 6566 (2022)

18. Siegel, R.L., Miller, K.D., Fuchs, H.E., Jemal, A.: Cancer statistics, 2022. CA Can. J. Clin. **72**(1), 7–33 (2022)

19. So, T.H., et al.: Lymphopenia and radiation dose to circulating lymphocytes with neoadjuvant chemoradiation in esophageal squamous cell carcinoma. Adv. Radiat. Oncol. **5**(5), 880–888 (2020)

20. Tang, C., et al.: Lymphopenia association with gross tumor volume and lung v5 and its effects on non-small cell lung cancer patient outcomes. Int. J. Radiat. Oncol. Biol. Phys. **89**(5), 1084–1091 (2014)

21. Tang, H., et al.: Clinically applicable deep learning framework for organs at risk delineation in CT images. Nat. Mach. Intell. **1**(10), 480–491 (2019)

22. Tyldesley, S., Boyd, C., Schulze, K., Walker, H., Mackillop, W.J.: Estimating the need for radiotherapy for lung cancer: an evidence-based, epidemiologic approach. Int. J. Radiat. Oncol. Biol. Phys. **49**(4), 973–985 (2001)

23. Wang, P., et al.: Accurate airway tree segmentation in ct scans via anatomy-aware multi-class segmentation and topology-guided iterative learning. arXiv preprint arXiv:2306.09116 (2023)

24. Xu, C., et al.: The impact of the effective dose to immune cells on lymphopenia and survival of esophageal cancer after chemoradiotherapy. Radiother. Oncol. **146**, 180–186 (2020)

25. Ye, X., et al.: Comprehensive and clinically accurate head and neck cancer organs-at-risk delineation on a multi-institutional study. Nat. Commun. **13**(1), 6137 (2022)

26. Zhu, Z., et al.: Lymph node gross tumor volume detection and segmentation via distance-based gating using 3D CT/PET imaging in radiotherapy. In: Martel, A.L., et al. (eds.) MICCAI 2020. LNCS, vol. 12267, pp. 753–762. Springer, Cham (2020). https://doi.org/10.1007/978-3-030-59728-3_73

Leveraging Self-attention Mechanism in Vision Transformers for Unsupervised Segmentation of Optical Coherence Microscopy White Matter Images

Mohamad Hawchar(✉) [iD] and Joël Lefebvre(✉) [iD]

Laboratoire d'Imagerie Numérique, Neurophotonique et Microscopie (LINUM),
Université du Québec à Montréal (UQAM), Montréal, QC H2L 2C4, Canada
`hawchar.mohamad@courrier.uqam.ca`, `lefebvre.joel@uqam.ca`

Abstract. A new microscope has been created to capture detailed images of the brain using a technology called optical coherence microscopy (OCM). However, there is still much to discover and understand about this valuable data. In this paper, we focus on the important task of segmenting the white matter in these high-resolution OCM images. A closed-up accurate segmentation of white matter tracts has the potential to enhance our knowledge of brain connections. In this paper, we propose an unsupervised segmentation approach that leverages the self-attention mechanism of Vision Transformers (ViT). Our approach uses the output attention weights from a ViT pre-trained with Masked Image Modeling (MIM) to generate binary segmentations that we use as Pseudo-Ground-Truth (PGT) to train an additional segmentation model. Our method achieved superior performance when compared with classical unsupervised computer vision methods and common unsupervised deep learning architectures designed for natural images. Additionally, we compared our results with those of a supervised U-Net model trained on different numbers of labels and a semisupervised approach where we selected the best-performing model based on labeled data. Our model achieved comparable results to the U-Net model trained on 30% of the labeled data. Furthermore, through finetuning, our model demonstrated an improvement of 3% over the supervised U-Nets. The code and data are available on GitHub repository https://github.com/linum-uqam/ViT-OCM-WMSegmentation.

Keywords: Semantic Segmentation · Self-supervised Learning · Deep Learning · Unsupervised Learning · Vision Transformers · White Matter · Optical Coherence Microscopy

1 Introduction

In recent years, we have developed a high-resolution optical microscope capable of capturing high-resolution images of ex vivo mouse brain samples [14]. This

---

Supported by an FRQNT and an NSERC grant.

new microscope is able to acquire both a low-resolution overview of the brain and a high-resolution view of a selected region of interest (ROI) at the same time through Optical Coherence Microscopy (OCM), displaying a close-up view of white fiber tracts, cell nuclei, and other components of the brain, enabling us to gain valuable insights into the brain microstructures [14]. However, to fully leverage the potential of this imaging technology, it is essential to develop robust and efficient tools that facilitate a comprehensive analysis of the acquired data. These tools will play a crucial role in extracting meaningful information from the high-resolution optical data and advancing our knowledge in the field of brain analysis. Segmenting the white matter connections in these high-resolution OCM images holds immense potential and presents a crucial task with wide-ranging implications. It can serve as a fundamental step for developing various brain analysis algorithms, including optical tractography, 3D reconstruction of white matter connections, analysis of white matter density in specific brain regions, evaluation of white matter orientation, analysis of fiber crossing regions, and more. However, the complexity and heterogeneity of the brain make it challenging to acquire labeled data for supervised segmentation or to develop hand-crafted unsupervised methods. Therefore, developing a self-supervised method for white matter segmentation that uses the data itself as guidance is a crucial step. This work aims to perform segmentation on 40× OCM white matter acquisitions obtained through serial blockface histology [14].

Vision Transformer (ViT) [6] has gained significant attention for its success in natural image analysis. ViT leverages the multi-head self-attention mechanism to learn long-range contextual information [6,21]. Moreover, some recent works have shown that self-supervised pretraining of ViT leads to improved performance compared to Convolutional Neural Networks (CNNs) in medical image analysis [19]. Self-supervised pretraining enables ViT to learn contextual image representation features through pretext tasks such as self-distillation [4], colorization [24], denoising [24], Masked Image Modeling (MIM), etc. Various approaches for applying MIM for ViT pretraining have been proposed, including discrete token prediction [2], auto-encoder [10], or appending a simple raw pixel prediction layer [27]. The ViT attention mechanism is a powerful tool that can be used for unsupervised segmentation. Previous works have attempted to leverage the attention for supervised downstream medical tasks [8,11,15,25], as well as unsupervised downstream natural-image tasks such as object detection [23], object segmentation [26], and semantic segmentation [9]. In this work, we tackle the task of white matter segmentation in high-resolution OCM data for the first time. We introduce a technique that leverages the self-attention mechanism feature of ViTs by MIM pretraining [27] and use it to produce binary segmentation masks that serve as a Pseudo-Ground-Truth (PGT) to train a final U-Net [22] model.

2   Methods

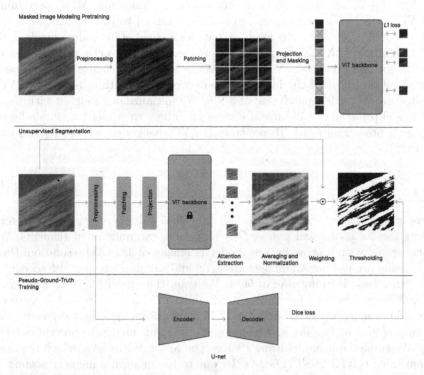

Fig. 1. Overview of our model. The proposed architecture consists of a two-step training process. In the first step, a Vision Transformer (ViT) backbone is pre-trained using Masked Image Modeling (MIM). In the second step, unsupervised labels are generated using the pre-trained model, and a U-Net model is trained to predict these Pseudo-Ground-Truth (PGT) labels.

2.1   Dataset Description and Preprocessing

The OCM white matter dataset consists of 457 high resolution (40×) OCM images of size 334 × 334, along with 254 high resolution (40×) OCM mosaics with a resolution of 1.5 micron/pixel and width and height values ranging between 1500 and 6000 pixels, which we will resize and use solely for pre-training. All the OCM data are from ex vivo mouse brain samples, and the specific tissue preparation and image acquisition procedures are described in detail in [14]. In total, We have 711 images for pre-training and 44 annotated 40× images for evaluation (25 training, 5 validation, and 14 testing). To preprocess the OCM images, we start by applying a Gaussian blur filter of $\sigma = 1$, followed by a top hat filter with a 20-pixel radius to remove the gray background. The annotation process began with an initial segmentation using Ilastik [3], followed by manual curation to include only regions of white matter.

2.2 Masked Image Modeling

The first step of our approach is the Masked Image Modeling (MIM) pertaining of a ViT model as shown in Fig. 1. ViTs are designed to process images in the form of patches, which is why we mask patches instead of individual pixels. This approach aligns with the natural processing mechanism of ViTs, making it more effective for image analysis. We use by default a mask patch size of 16×16 and a mask ratio of 50% of the image. For our encoder, we utilize a standard ViT architecture [6] with a patch size of 8×8. We initialize our weights with those of a model pre-trained on natural images [4]. Then, we append a simple linear layer as a prediction head. To pre-train the ViT backbone, we use L_1 loss just like in [27].

$$L = \frac{1}{\Omega(x_M)}||y_M - x_M||_1 \tag{1}$$

where x and y represent the true and predicted values, respectively; M represents the set of masked pixels; $\Omega(.)$ represents the number of elements. We experimented with pretraining our model on images of 384×384 resolution. During training, we use the Adam optimizer [12] and a cosine learning rate scheduler [17] with a base learning rate of 5e–4. We train the model for 30 epochs, with the first 20 epochs used for warm-up, where we gradually increase the learning rate. Additionally, we use a batch size of 16 and perform light data augmentation techniques such as flipping and cropping on the input images to prevent overfitting. We trained our model using PyTorch on 384×384 images, which required 20 min using 6 RTX 2080 11 GB GPUs due to the quadratic memory scaling of ViTs with data.

2.3 Model Selection

Our model selection method for the MIM pretraining step shown in Fig. 1 is based on monitoring the behavior of the loss function during training as in [13]. When the change in the loss function falls below a predefined threshold (in our case, 1e–3) for a consecutive number of epochs (in our case, 3 epochs), we stop the training process and use the current model to generate segmentation results. This strategy allows us to select the most suitable model based on the stability and convergence of the loss function. This is the approach we adopt by default.

For experimental purposes, we also try a semi-supervised model selection approach where we aim to determine the optimal hyperparameters for pre-training by trying different Mask ratio ranges (10–90%) and Mask sizes (8, 16, 32) on 10 labeled images, selecting the model with the highest DC score.

2.4 Attention Map Weighting and Thresholding

We extract the output attention from the attention heads of the last layer of the pre-trained ViT, and we average them. The resulting attention map will have a lower resolution than the input image due to the ViT computing the attention between input patches [6]. To address this issue, we interpolate the average attention map to the original image size using bilinear interpolation. We then utilize a min-max normalization technique to scale the attention map within the range of 0 to 1. This normalization process can be expressed mathematically as:

$$A_{norm} = \frac{A - \min(A)}{\max(A) - \min(A)} \tag{2}$$

where A is the attention map and A_{norm} is the normalized attention map.

Lastly, we multiply the resulting attention map with the input image to get the weighted image to which the Otsu [20] thresholding method is applied to obtain the segmentation. As shown in Fig. 2.

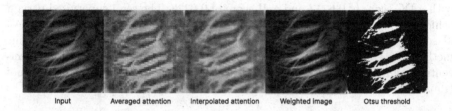

Input   Averaged attention   Interpolated attention   Weighted image   Otsu threshold

Fig. 2. Label generation steps of the output attention.

2.5 Pseudo-Ground-Truth Training, Fine-Tuning, and Linear Prob

Following the weighted image thresholding, as explained in the previous section, we obtain binary labeled images, which act as Pseudo-Ground-Truth (PGT), to train the additional U-Net model [22] using the default architecture. The U-Net model is trained for 150 epochs, with a learning rate of 1e–4 and a batch size of 8. The input images fed into the model have a resolution of 512 × 512 pixels.

We conducted two additional experiments to explore the capabilities of the MIM-trained ViT model in supervised applications. In the Fine-Tuning experiment, we added linear layers and allowed weight modifications to the backbone. In the Linear prob experiment, we used the same process without modifying the backbone weights. Both experiments used the same ViT backbone as in our unsupervised experiment and were trained with the same configuration as our PGT U-Net.

3 Experimental Results

We evaluated our approach by measuring the Dice coefficient (DC) and the Jaccard index (JI) using the following formulas:

$$\text{DC} = \frac{2|A \cap B|}{|A| + |B|}, \ \text{JI} = \frac{|A \cap B|}{|A \cup B|} \tag{3}$$

A and B represent the ground truth segmentation and our method's segmentation, respectively. $\cap$ and $\cup$ denote the intersection and union operations. Refer to Fig. 3 for an example of our method's results.

Table 1 provides a comprehensive performance comparison of various classical unsupervised methods (Otsu [20], K-means [1], Chan-Vese [7][1]) and unsupervised deep learning methods (DINO [4], STEGO [9], PiCIE [5]). Additionally, it showcases the results of supervised U-Nets trained on a subset of the OCM dataset using the same settings as our PGT U-Net model. The table also includes our unsupervised (Unsup), semi-supervised (Semi-sup), PGT, Linear prob, and Fine-Tuning experiments.

Among the classical unsupervised methods, Otsu achieves the highest Jaccard Index (JI) score of 0.6120, while K-means obtains a similar Dice Coefficient (DC) score of 0.5235. Our proposed unsupervised approach outperforms these methods with a DC score of 0.6213 and a JI score of 0.6329. The semi-supervised variant achieves even better results, with the highest DC score of 0.6863 and the highest JI score of 0.7445. Lastly, the performance of the unsupervised deep learning methods lags behind, mainly since these techniques are developed for object-centric segmentation tasks and require huge amounts of data, which do not translate well to our dataset.

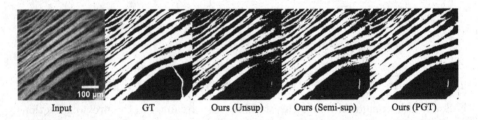

| Input | GT | Ours (Unsup) | Ours (Semi-sup) | Ours (PGT) |

Fig. 3. Results of our method for an example from the OCM datasets. (Contrast adjusted for representation purposes)

The performance of U-Net models is influenced by the amount of supervised data used. As supervised data decreases, performance generally declines, with a few exceptions. For instance, the U-Net trained with 70% supervised data outperforms the 80% supervised data model, indicating potential challenges in generalization due to limited training data.

Our semi-supervised and unsupervised experiments demonstrate promising results. The semi-supervised variant surpasses the fully unsupervised version, emphasizing the benefits of utilizing some labeled data (10 images in our case) to select the best-performing model. Its performance lands between the U-Nets trained on 10% and 20% of the data, if we compare both the DC and JI scores.

Our PGT (Unsup) model demonstrates exceptional performance, surpassing the Semi-sup approach and eliminating the need for label guidance. Additionally,

---

[1] K-means and Chan-Vese are iterated for 200 iterations, while Chan-Vese uses parameters $\mu = 0.25$, $lambda1 = 1$, $lambda2 = 1$, and $tol = 1e - 3$.

Table 1. Comparison of all conducted experiments on the same test dataset. Best-performing experiments and our PGT method are in bold.

Method	Dice Coefficient (DC)	Jaccard Index (JI)
U-Nets		
U-Net (Sup 100%)	0.8373	0.8013
U-Net (Sup 90%)	0.8183	0.7749
U-Net (Sup 80%)	0.8133	0.7728
U-Net (Sup 70%)	0.8248	0.7916
U-Net (Sup 40%)	0.7848	0.7635
U-Net (Sup 30%)	0.7661	0.7537
U-Net (Sup 20%)	0.7085	0.7402
U-Net (Sup 10%)	0.6835	0.7378
Unsupervised Deep Learning Techniques		
STEGO [9]	0.4792	0.3054
PiCIE [5]	0.4842	0.3134
DINO [4]	0.4649	0.2170
Our Experiments		
Ours ViT + Fine-tuning (Sup)	**0.8673**	**0.8220**
Ours ViT + Linear prob (Sup)	0.8577	0.7641
Ours PGT U-Net (Unsup)	**0.7489**	**0.7487**
Ours (Semi-sup)	0.6863	0.7445
Ours (Unsup)	0.6213	0.6329
Classical Image Processing		
Otsu [20]	0.5247	0.6120
K-means [1,18]	0.5235	0.6050
Chan-Vese [7]	0.4582	0.3120

the PGT performance falls just 1% short of a U-Net model trained with 30% of labeled data. This demonstrates the effectiveness of using PGT labels for training when labeled data is scarce.

The fine-tuned model achieves the highest DC and JI scores, followed closely by the linear probing model, demonstrating the effectiveness of self-supervision and the learned features from MIM training. This highlights the superiority of the MIM-trained features, as they outperform the U-Net architecture in generating segmentations. Especially in the linear probing case, where the model is prohibited from adjusting the ViT backbone weights.

To summarize, while supervised U-Net models and fine-tuning methods still exhibit superior performance, our method, especially when trained on PGT data, offers a robust alternative in scenarios where acquiring labeled data is difficult and has the potential to compete with models trained on small datasets.

4   Discussion

The combination of ViTs and MIM in our study successfully generated attention maps that captured important information in our medical image datasets. MIM was found to be a suitable general pre-training task for medical images, because it does not rely on color or object shape, and instead encourages the model to capture higher-level semantic information by predicting missing parts of an image using the available information.

The limitations of our approach come mainly from the ViT's computational nature, which exhibits quadratic scaling and requires large amounts of data to train [21]. To address these limitations, weights learned on natural images were used to initialize our model.

Our model has the potential for future enhancements by addressing the output of the different heads of the ViT. By selecting only the relevant attention heads and disregarding those that focus solely on the background, we can potentially improve the overall performance of the model. Additionally, exploring alternative ViT backbones such as SWIN [16], or designing customized backbone architectures with specific layers and heads may yield higher-resolution attention maps without significantly increasing computational demands. For example, a custom model with a 4×4 patch size, two layers, and four heads could offer more detailed attention maps while keeping computational requirements manageable. However, it should be noted that the absence of pre-trained weights for these custom architectures could pose a challenge.

One of our future research directions includes investigating the analysis of white matter density using the segmentation of white matter in high-resolution images. This work aims to contribute to the understanding of brain structure, function, and neurological disorders, particularly in the context of neurodegenerative diseases such as Alzheimer's, Parkinson's, and multiple sclerosis. By accurately quantifying white matter concentration and distribution, we aim to provide valuable insights into brain health and enhance the interpretation of neuroimaging data for clinical and research purposes. Another future application is the detection of white fiber crossing areas within the brain. Accurate segmentation of white matter in OCM volumes enables us to identify these crossing regions and study the intricate network of neural connections. By quantifying the crossings and constructing a dataset of volumes and images, we aim to train a model that can assist the microscope in detecting fiber-crossing regions. This will improve the ROI selection method used by the microscope [14] and make it accurate and automatic instead of the inaccurate statistical and manual methods that are being used. This will also help optimize memory usage during the acquisition process, addressing the challenge of storing large amounts of data.

5   Conclusion

This paper introduces a straightforward unsupervised segmentation method based on self-supervised learning. By utilizing the learned attention from ViT backbones trained with masked image modeling, we generate binary segmentations using ViT's attention maps and employ them as Pseudo-Ground-Truth data to train an additional U-Net model. Our proposed approach outperformed classical methods and achieved comparable results to U-Nets trained on 30% of the labeled data, highlighting its effectiveness in the task of segmentation.

Acknowledgments. This work and M.H. scholarship was supported by an FRQNT team research project grant (2021-PR-282231) and by an NSERC Discovery grant (RGPIN-2020-06109).

References

1. Achanta, R., Shaji, A., Smith, K., Lucchi, A., Fua, P., Süsstrunk, S.: Slic superpixels compared to state-of-the-art superpixel methods. IEEE Trans. Pattern Anal. Mach. Intell. **34**(11), 2274–2282 (2012)
2. Bao, H., Dong, L., Piao, S., Wei, F.: Beit: bert pre-training of image transformers. arXiv preprint arXiv:2106.08254 (2021)
3. Berg, S.: Ilastik: interactive machine learning for (bio) image analysis. Nat. Methods **16**(12), 1226–1232 (2019)
4. Caron, M., et al.: Emerging properties in self-supervised vision transformers. In: Proceedings of the IEEE/CVF International Conference on Computer Vision, pp. 9650–9660 (2021)
5. Cho, J.H., Mall, U., Bala, K., Hariharan, B.: Picie: unsupervised semantic segmentation using invariance and equivariance in clustering. In: Proceedings of the IEEE/CVF Conference on Computer Vision and Pattern Recognition, pp. 16794–16804 (2021)
6. Dosovitskiy, A., et al.: An image is worth 16×16 words: transformers for image recognition at scale. arXiv preprint arXiv:2010.11929 (2020)
7. Getreuer, P.: Chan-vese segmentation. Image Process. Line **2**, 214–224 (2012)
8. Gulzar, Y., Khan, S.A.: Skin lesion segmentation based on vision transformers and convolutional neural networks-a comparative study. Appl. Sci. **12**(12), 5990 (2022)
9. Hamilton, M., Zhang, Z., Hariharan, B., Snavely, N., Freeman, W.T.: Unsupervised semantic segmentation by distilling feature correspondences. arXiv preprint arXiv:2203.08414 (2022)
10. He, K., Chen, X., Xie, S., Li, Y., Dollár, P., Girshick, R.: Masked autoencoders are scalable vision learners. In: Proceedings of the IEEE/CVF Conference on Computer Vision and Pattern Recognition, pp. 16000–16009 (2022)
11. Jiang, J., Tyagi, N., Tringale, K., Crane, C., Veeraraghavan, H.: Self-supervised 3d anatomy segmentation using self-distilled masked image transformer (smit). In: Medical Image Computing and Computer Assisted Intervention-MICCAI 2022: 25th International Conference, Singapore, 18–22 September 2022, Proceedings, Part IV, pp. 556–566. Springer, Heidelberg (2022). https://doi.org/10.1007/978-3-031-16440-8_53
12. Kingma, D.P., Ba, J.: Adam: a method for stochastic optimization. arXiv preprint arXiv:1412.6980 (2014)
13. Krug, C., Rohr, K.: Unsupervised cell segmentation in fluorescence microscopy images via self-supervised learning. In: Pattern Recognition and Artificial Intelligence: Third International Conference, ICPRAI 2022, Paris, France, 1–3 June 2022, Proceedings, Part I, pp. 236–247. Springer, Heidelberg (2022). https://doi.org/10.1007/978-3-031-09037-0_20

14. Lefebvre, J., Delafontaine-Martel, P., Pouliot, P., Girouard, H., Descoteaux, M., Lesage, F.: Fully automated dual-resolution serial optical coherence tomography aimed at diffusion mri validation in whole mouse brains. Neurophotonics 5(4), 045004–045004 (2018)

15. Liu, Q., Kaul, C., Anagnostopoulos, C., Murray-Smith, R., Deligianni, F.: Optimizing vision transformers for medical image segmentation. ArXiv arXiv:2210.08066 (2022)

16. Liu, Z., et al.: Swin transformer: hierarchical vision transformer using shifted windows. In: Proceedings of the IEEE/CVF International Conference on Computer Vision, pp. 10012–10022 (2021)

17. Loshchilov, I., Hutter, F.: Sgdr: stochastic gradient descent with warm restarts. arXiv preprint arXiv:1608.03983 (2016)

18. MacQueen, J.: Classification and analysis of multivariate observations. In: 5th Berkeley Symposium on Mathematical Statistics Probability, pp. 281–297. University of California, Los Angeles (1967)

19. Matsoukas, C., Haslum, J.F., Söderberg, M., Smith, K.: Is it time to replace CNNs with transformers for medical images? arXiv preprint arXiv:2108.09038 (2021)

20. Otsu, N.: A threshold selection method from gray-level histograms. IEEE Trans. Syst. Man Cybern. 9(1), 62–66 (1979)

21. Raghu, M., Unterthiner, T., Kornblith, S., Zhang, C., Dosovitskiy, A.: Do vision transformers see like convolutional neural networks? Adv. Neural. Inf. Process. Syst. 34, 12116–12128 (2021)

22. Ronneberger, O., Fischer, P., Brox, T.: U-net: convolutional networks for biomedical image segmentation. In: Navab, N., Hornegger, J., Wells, W.M., Frangi, A.F. (eds.) MICCAI 2015. LNCS, vol. 9351, pp. 234–241. Springer, Cham (2015). https://doi.org/10.1007/978-3-319-24574-4_28

23. Siméoni, O., et al.: Localizing objects with self-supervised transformers and no labels. arXiv preprint arXiv:2109.14279 (2021)

24. Taleb, A., et al.: 3d self-supervised methods for medical imaging. Adv. Neural. Inf. Process. Syst. 33, 18158–18172 (2020)

25. Valanarasu, J.M.J., Oza, P., Hacihaliloglu, I., Patel, V.M.: Medical transformer: gated axial-attention for medical image segmentation. ArXiv arxiv:2102.10662 (2021)

26. Van Gansbeke, W., Vandenhende, S., Van Gool, L.: Discovering object masks with transformers for unsupervised semantic segmentation. arXiv preprint arXiv:2206.06363 (2022)

27. Xie, Z., et al.: Simmim: a simple framework for masked image modeling. In: Proceedings of the IEEE/CVF Conference on Computer Vision and Pattern Recognition, pp. 9653–9663 (2022)

PE-MED: Prompt Enhancement for Interactive Medical Image Segmentation

Ao Chang[1,2,3,4], Xing Tao[1,2,3], Xin Yang[1,2,3], Yuhao Huang[1,2,3], Xinrui Zhou[1,2,3], Jiajun Zeng[1,2,3], Ruobing Huang[1,2,3], and Dong Ni[1,2,3(✉)]

[1] National-Regional Key Technology Engineering Laboratory for Medical Ultrasound, School of Biomedical Engineering, Health Science Center, Shenzhen University, Shenzhen, China
nidong@szu.edu.cn
[2] Medical Ultrasound Image Computing (MUSIC) Lab, Shenzhen University, Shenzhen, China
[3] Marshall Laboratory of Biomedical Engineering, Shenzhen University, Shenzhen, China
[4] Shenzhen RayShape Medical Technology Co., Ltd., Shenzhen, China

Abstract. Interactive medical image segmentation refers to the accurate segmentation of the target of interest through interaction (e.g., click) between the user and the image. It has been widely studied in recent years as it is less dependent on abundant annotated data and more flexible than fully automated segmentation. However, current studies have not fully explored user-provided prompt information (e.g., points), including the knowledge mined in one interaction, and the relationship between multiple interactions. Thus, in this paper, we introduce a novel framework equipped with prompt enhancement, called PE-MED, for interactive medical image segmentation. First, we introduce a Self-Loop strategy to generate warm initial segmentation results based on the first prompt. It can prevent the highly unfavorable scenarios, such as encountering a blank mask as the initial input after the first interaction. Second, we propose a novel Prompt Attention Learning Module (PALM) to mine useful prompt information in one interaction, enhancing the responsiveness of the network to user clicks. Last, we build a Time Series Information Propagation (TSIP) mechanism to extract the temporal relationships between multiple interactions and increase the model stability. Comparative experiments with other state-of-the-art (SOTA) medical image segmentation algorithms show that our method exhibits better segmentation accuracy and stability.

Keywords: Interactive Segmentation · Prompt Learning

1 Introduction

Medical image segmentation is a pivotal aspect of research in medical image analysis, intended for the extraction of specific targets or regions in medical

---

A. Chang and X. Tao—Contribute equally to this work.

© The Author(s), under exclusive license to Springer Nature Switzerland AG 2024
X. Cao et al. (Eds.): MLMI 2023, LNCS 14348, pp. 257–266, 2024.
https://doi.org/10.1007/978-3-031-45673-2_26

images for further analysis and diagnosis [22]. Traditional segmentation methods heavily rely on image processing techniques and machine learning algorithms, which can be computationally intensive, time-consuming, and require a high level of expertise. Deep learning-based methods have achieved state-of-the-art (SOTA) performance in automatic segmentation of medical images [17]. However, most current automatic methods lack the learning of informative prompts, resulting in inaccurate and inflexible segmentation.

Interactive segmentation methods offer a promising solution to these challenges, utilizing limited user guidance to extract the target object and combining this with image features to yield finely segmented results [21]. While user interaction leads to more precise segmentation results, the interaction process should be efficient and time-saving to reduce the burden on the user.

Traditional interactive methods for image segmentation use low-level features like edge information or color distribution, such as GraphCuts and Random Walk [2,6,7,9]. These methods often require multiple user interactions and are time-consuming to produce satisfactory results as low-level features may not always distinguish the desired object from the background. To reduce user interactions and improve segmentation accuracy, machine learning techniques are leveraged. For instance, GrabCut [16] uses a Gaussian mixture model to estimate foreground and background distributions. The initial results can be obtained via a user-provided bounding box, and refined through additional interactions.

Recently, **deep learning-based methods** have achieved SOTA performance in medical image segmentation [4,5,8,11], thanks to the neural networks for automatically capturing high-level semantic features [17]. Thus, deep models have been proposed to integrate with interactive methods for medical image segmentation. There are two main streams of current approaches, as introduced below:

-The first type of method ignores prompt learning during training [3,14,19,20]. These methods require pre-training a semantic segmentation network, followed by fine-tuning the predicted mask through user interactions. However, they may not be suitable for multi-class segmentation tasks that require accurate delineation of different targets. Besides, they cannot ensure the quality of feature extraction when dealing with unfamiliar data patterns or categories, causing a performance drop.

-The second type of method involves prompt learning during training. Prompts are leveraged to guide the learning of deep models in these methods, including iSegFormer [13], Segment Anything Model (SAM) [10,12], etc. During testing, users need to interactively click on the foreground or background to achieve an accurate target segmentation. Such prompt learning techniques have the potential to make interactive segmentation more flexible, accurate, and general to complex scenarios. However, it is still challenging to deeply mine sparse prompts information and improve network response to user-provided prompts.

In this study, we propose a novel interactive approach with prompt enhancement to improve medical image segmentation performance, named PE-MED. Our contribution is three-fold. First, we employ a simple yet effective self-loop method to address the issue of insufficient information during the first interac-

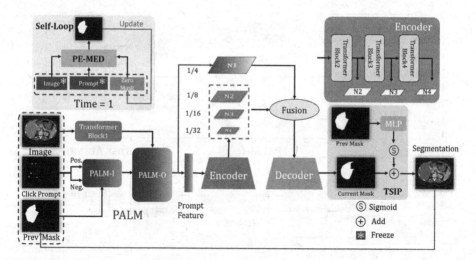

Fig. 1. Overview of our proposed interactive segmentation framework. The green and red dots in the Click Prompt represent positive and negative prompt points, respectively. Sequential processing through four Transformer blocks (block1–4) enables the generation of multi-level features at $\{1/2^n, n = 2, 3, 4, 5\}$ of the original image resolution. The fusion module is composed of a convolutional layer and a normalization layer. The decoder module is a simple multilayer perceptron. (Color figure online)

tion. Second, we propose a novel Prompt Attention Learning Module (PALM) that explores the relationship between user interactions and image features to extract essential interaction details, enhancing the network's response to user input. Third, we introduce a Time Series Information Propagation (TSIP) mechanism to model the continuity between multiple interactions for improving stability. Extensive experiments validated that, compared with the SOTA methods, our PE-MED can achieve accurate results with less user interaction.

2   Methodology

Figure 1 shows the schematic view of our proposed method. We propose the PE-MED, a novel iterative refinement framework equipped with prompt-enhanced modules, for medical image segmentation. PE-MED consists of two stages: 1) The initial stage (see *Time = 1*) serves as the foundation for the subsequent clicks, 2) The main stage represents the subsequent clicks after the initial one. In the first stage, we first introduce a Self-Loop strategy to obtain a good initialized mask. Then, the original image, previous mask and prompts (positive and negative points) are taken as inputs, and transmitted to the PALM and TSIP for prompt enhancement. Last, the network will output the refined segmentation iteratively.

2.1   Self-Loop Strategy for Warm Start

For interactive segmentation methods, the initial segmentation generated by the user's first click plays a fundamental role in the subsequent interactions. Most

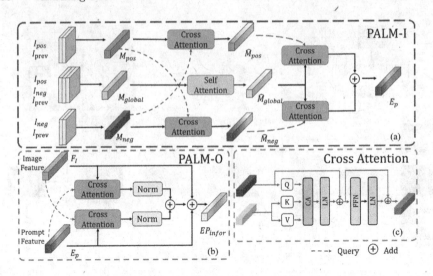

Fig. 2. Overview of our proposed PALM.

of the current method lacks the optimization of the first interaction, resulting in poor initialization with inadequate prompt information. This will have a negative impact on the subsequent network module optimization and even segmentation failure. To address this issue, inspired by [23,24], we introduce a Self-Loop strategy to obtain a warm start for iterative segmentation (left-upper yellow block in Fig. 1). Specifically, after the user's first click, fully empty masks are sent to the network, obtaining the rough segmentation ($M0$). Then, the coarse prediction will enter the loop, and output the information-enhanced mask ($M1$) without any interaction. Compared to $M0$, $M1$ contains richer prompt information, and makes the learning of subsequent modules easier.

2.2 PALM for Prompt Feature Enhancement

Extracting rich information from sparse user hints is a challenging task. In our study, we developed PALM to effectively leverage and enhance the prompt information. As shown in Fig. 2, PALM consists of two modules: PALM-I (see Fig. 2(a)) and PALM-O (see Fig. 2(b)). Specifically, PALM-I primarily focuses on augmenting the intrinsic features of prompts, whereas PALM-O enhances the interplay between prompts and images.

PALM takes four inputs, including the image I_{image}, positive and negative prompts (I_{pos} and I_{neg}), and previous mask (I_{prev}). After the users input is converted to a disk map following a click, I_{pos} and I_{neg} are encoded and utilized in the PALM module. As shown in Fig. 2(a), in the PALM-I part, we combine I_{pos}, I_{neg}, and I_{prev} to form the concatenated feature map M_{pos}, M_{global}, and M_{neg} using patch embeddings. The Cross Attention Module (Fig. 2(c)) is then applied to enhance these feature maps, resulting in augmented features $\hat{M}_{pos} = \theta_c(M_{neg}, M_{pos}, M_{pos})$ and $\hat{M}_{neg} = \theta_c(M_{pos}, M_{neg}, M_{neg})$, with

$\theta_c(Q, K, V) = Softmax(QK^T/d_k)V$. Simultaneously, the Self Attention Module improves M_{global}, yielding the global feature map $\hat{M}_{global}$:

$$\hat{M}_{global} = \theta_s(M_{global}) = \theta_c(M_{global}, M_{global}, M_{global}), \tag{1}$$

After obtaining enriched interaction and global information, we compute the prompt feature E_p by summing the features resulting from cross-attentive mechanisms using $\hat{M}_{pos}$ and $\hat{M}_{neg}$ as query vectors, E_p can be defined as:

$$E_p = \theta_c(\hat{M}_{pos}, \hat{M}_{global}, \hat{M}_{global}) + \theta_c(\hat{M}_{neg}, \hat{M}_{global}, \hat{M}_{global}), \tag{2}$$

Besides, we propose the PALM-O to bridge the information gap between prompt and image features. It takes E_p and the image features F_I (after Transformer Block 1) as input. Then, with the normalized layer ($norm$), the enhanced mixture feature EP_{infor} can be illustrated as:

$$EP_{infor} = F_I + norm(\theta_c(F_I, E_p, E_p))$$
$$+E_p + norm(\theta_c(E_p, F_I, F_I)). \tag{3}$$

2.3 TSIP Mechanism for Stability Enhancement Among Prompts

Most of the existing interactive segmentation algorithms often ignore the relationship between consecutive interactions. It may lead to poor segmentation stability, especially when dealing with multiple interactions. To overcome this limitation, we propose to integrate TSIP mechanism into the framework to enhance the stability among multiple prompts.

The TSIP mechanism enables the extraction of dynamic and continuous interaction information, with the entire network functioning as a cohesive unit to convey temporal information. Specifically, the previous network output serves as a candidate memory for the current moment using a simple multilayer perceptron (MLP), while the current output serves as a candidate memory for the next moment. This mechanism is mathematically represented by:

$$O_t = F(I_{input}) + Sigmoid(\theta(O_{t-1})), \tag{4}$$

where F refers to PE-MED without TSIP, θ refers a MLP. I_{input} denotes the input image, click prompts and previous mask, O_{t-1} and O_t represent the output of the previous and current network, respectively.

3 Experimental Results

Materials and Implementation Details. We validate our proposed framework on the public multi-organ dataset (named **Synapse** [1]) proposed in the 2015 MICCAI Multi-Atlas Abdomen Labeling Challenge. **Synapse** includes 30 cases with a total of 3779 axial abdomen 2D images where each CT volume

Table 1. Method comparison on **Synapse** dataset. $-p^*$ denotes the number of point prompts.

Methods	DSC(%) ↑								
	AVG	Aorta	Gallbladder	Kidney(L)	Kidney(R)	Liver	Pancreas	Spleen	Stomach
U-Net [15]	76.85	89.07	69.72	77.77	68.60	93.43	53.98	86.67	75.58
TransUnet [5]	77.48	87.23	63.13	81.87	77.02	94.08	55.86	85.08	75.62
Swin-Unet [4]	79.13	85.47	66.53	83.28	79.61	94.29	56.58	90.66	76.60
HiFormer [8]	80.69	87.03	68.61	84.23	78.37	94.07	60.77	90.44	82.03
GrabCut [16]	42.46	27.87	0.00	67.46	59.87	72.39	3.42	41.08	33.85
iSegFormer(2D)-p1 [13]	74.62	82.88	58.08	75.05	72.23	78.46	62.89	79.13	64.67
iSegFormer(2D)-p3 [13]	86.43	89.07	70.53	89.60	86.92	89.73	77.25	88.71	83.22
iSegFormer(2D)-p5 [13]	89.74	91.43	79.14	91.56	90.61	92.20	82.42	90.57	88.61
SAM-p1 [12]	75.33	88.98	52.74	87.05	85.98	72.41	41.19	78.06	64.44
SAM-p3 [12]	78.61	87.93	52.22	86.76	86.04	81.59	48.27	79.29	71.21
SAM-p5 [12]	79.64	87.37	53.26	86.71	85.96	84.26	51.18	81.23	74.53
Ours-p1	80.76	82.67	57.05	86.46	85.85	84.31	67.97	87.96	73.86
Ours-p3	90.51	89.43	77.99	91.57	91.75	94.53	83.45	94.54	90.66
Ours-p5	92.76	91.68	84.81	92.88	92.88	96.00	87.52	95.43	94.15

Table 2. Method comparison on **OL12** dataset, evaluated by DSC.

Method	Point:1	Point:2	Point:3	Point:5	Point:10	NoC@85	NoC@90
iSegFormer(2D) [13]	47.03(20.95)	58.27(17.66)	65.87(14.37)	76.17(9.59)	85.83(4.89)	7.51(2.19)	9.56(1.13)
SAM [12]	51.07(34.85)	56.90(32.96)	58.49(32.79)	61.31(32.23)	57.28(33.31)	6.83(4.12)	7.67(3.80)
Ours	86.49(12.25)	91.35(7.41)	92.91(5.65)	94.24(3.97)	95.18(2.97)	1.65(1.45)	2.39(2.37)

involves 85–198 slices. The dataset is divided randomly into 24 cases for training, and 6 cases for testing. Eight organs are annotated by experts, including Aorta, Gallbladder, Left Kidney, Right Kidney, Liver, Pancreas, Spleen and Stomach. To further test the performance of our proposed PE-MED, we built another dataset including different modalities (CT&MRI) and 12 common Organs/Lesions, named **OL12**. We randomly split **OL12** into 5050, 2041 and 4912 images for training, validation and testing at the case level.

We implemented our framework using PyTorch using one NVIDIA 3090 GPU with 24 GB of memory. The input image sizes are 224×224 and 256×256 in **Synapse** and **OL12** datasets, respectively. The models are trained for 100 epochs using a batch size of 128 and 64 for each dataset, using the normalized focal loss function [18]. We optimized our model using the Adam optimizer, starting with a learning rate of 5×10^{-3}, and reducing the learning rate by a rate factor of 0.6 every 20 epochs. The dice score coefficient (DSC) was adopted to quantitatively evaluate segmentation performance. We also evaluated the methods using *number of clicks* (NoC@†) metric, to measure the number of interactions required to achieve a predefined DSC (†).

Quantitative and Qualitative Analysis. For **Synapse** dataset, we compared the proposed PE-MED with the fully supervised image segmentation methods,

Table 3. Ablation study under the setting of five point prompts.

Methods	DSC(%) ↑								
	AVG	Aorta	Gallbladder	Kidney(L)	Kidney(R)	Liver	Pancreas	Spleen	Stomach
Baseline	89.11	90.25	74.17	89.14	88.40	93.71	78.03	92.00	91.25
Baseline-SL	91.70	93.12	73.02	91.77	91.92	95.13	83.79	94.45	93.11
Baseline-I	91.84	93.79	73.00	93.07	92.10	95.40	82.69	94.93	91.70
Baseline-O	91.22	92.29	71.63	91.84	91.91	95.06	83.12	93.62	92.56
Baseline-IO	92.24	91.15	83.00	92.15	92.54	95.46	87.35	94.85	93.84
Baseline-T	91.01	91.90	72.62	91.17	91.15	94.79	82.65	93.73	93.18
Ours	92.76	91.68	84.81	92.88	92.88	96.00	87.52	95.43	94.15

including TransUnet [5], Swin-Unet [4], and HiFormer [8]. Moreover, we also evaluated the performance of PE-MED with interactive segmentation methods such as GrabCut [16], iSegFormer (2D version) [13], and SAM [12]. For **OL12** dataset, we only compared PE-MED with two SOTA interactive segmentation methods, i.e., iSegFormer [13] and SAM [12]. The ablation study was conducted on **Synapse** dataset, by comparing the PE-MED with different components, including Self-Loop (Baseline-SL), PALM-I (BaseLine-I), PALM-O (BaseLine-O), PALM (BaseLine-IO), and TSIP (BaseLine-T). Quantitative results are presented in Table 1-3, with best results shown in Blue.

For Table 1, it can be observed that our method achieves an average DSC of 80.76% with a single point prompt, slightly higher than the current SOTA fully automated segmentation algorithm, i.e., HiFormer (80.69%). With more prompts (-$p5$), our method achieve a DSC of 92.76%, outperforming all the reported fully automatic and interactive segmentation methods. It is also notable that our proposed method shows the best performance compared to interactive methods under the same number of prompts. Specifically, the DSC are 6.14%/5.43% ($p1$), 4.08%/11.90% ($p3$), 3.02%/13.15% ($p5$) higher than iSegFormer and SAM, respectively. Results on **OL12** dataset are reported in Table 2. It can be seen that iSegFormer and SAM struggle to obtain satisfactory DSC even with 5 or 10 point prompts (<86%). However, our method achieved the DSC of 86.49% even with one user interaction, and can outperform iSegFormer with 5&10 prompts by about 20%&10% DSC, respectively. The last two columns (NoC@†) reveals that our method only require average 1.65 and 2.39 to reach a DSC of 85% and 90%, respectively. While the other two methods require more user interaction to satisfy the corresponding requirements.

Results of ablation study can be found in Table 3. Experiments validated that by adding our proposed modules separately on the basis of baseline, the DSC performance can be improved. Simultaneously, integrating all the modules (**Ours**) can further enhance the performance of the network. It can also be observed that PE-MED achieves the highest DSC for six out of the eight organs, with only a slight deviation from the best results for the remaining two organs. Specifically, for the most challenging organ, i.e., *Gallbladder* with DSC of 74.17% in Baseline, our proposed PE-MED can increase it by 10.64%.

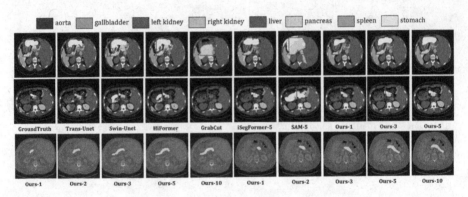

Fig. 3. Segmentation performance on the **Synapse** and **OL12** dataset. The red rectangles highlight organ regions where the superiority of *Ours* is evident.

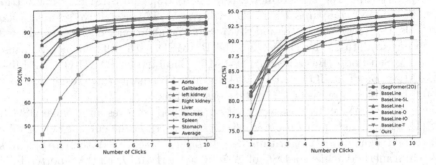

Fig. 4. DSC variation tendency curve on **Synapse** dataset.

Figure 3 depicts the visualization results of different methods (rows 1–2), demonstrating our method as the closest to the Ground truth. Row 3 of Fig. 3 presents the visualization results for various numbers of interactions. Furthermore, increasing the number of interactions in our methodology leads to a progressive improvement in performance. We also test the number of clicks and the results are shown in Fig. 4. In the left figure, we observe the rising trend of DSC for different organs and the average value, from clicks 1 to 10. Specifically, see the *Average* curve, clicks 1–5 gain a higher improvement than clicks 6–10. Thus, we consider 5 clicks a suitable choice, since it is a trade-off between the NoC and DSC performance. In the right figure, compared with *Ours* and the ablation results, the contribution of our proposed prompt enhancement techniques can be validated. Besides, it can be found that *Ours* outperforms *iSegFormer* at every click (red and blue curves), which further illustrates the power of our method.

4   Conclusion

In this work, we introduce an interactive framework for medical image segmentation, named PE-MED. Via the click user prompts, PE-MED can progressive optimize the segmentation results. We proposed three techniques for enhancing the prompt information, including 1) Self-Loop strategy for providing warm initialization at the first interaction, 2) PALM for feature aggregation at one click, and 3) TSIP for temporal modeling among multiple interactions. Extensive experiments on two large datasets validate that PE-MED is general and efficient, achieving the best DSC results among all the strong competitors. In the future, we will extend a 3D version PE-MED to directly handle the volumetric data.

Acknowledgement. This work was supported by the grant from National Natural Science Foundation of China (Nos.62171290, 62101343, 62101342), Shenzhen-Hong Kong Joint Research Program (No.SGDX20201103095613036), Shenzhen Science and Technology Innovations Committee (No.20200812143441001), and Guangdong Basic and Applied Basic Research Foundation (No.2023A1515012960).

References

1. MICCAI 2015 Multi-Atlas Abdomen Labeling Challenge. Synapse multi-organ segmentation dataset (2015). https://www.synapse.org/#!Synapse:syn3193805/wiki/217789. Accessed 10 July 2023
2. Boykov, Y.Y., Jolly, M.P.: Interactive graph cuts for optimal boundary & region segmentation of objects in nd images. In: Proceedings Eighth IEEE International Conference on Computer Vision, ICCV 2001, vol. 1, pp. 105–112. IEEE (2001)
3. Bredell, G., Tanner, C., Konukoglu, E.: Iterative interaction training for segmentation editing networks. In: Shi, Y., Suk, H.-I., Liu, M. (eds.) MLMI 2018. LNCS, vol. 11046, pp. 363–370. Springer, Cham (2018). https://doi.org/10.1007/978-3-030-00919-9_42
4. Cao, H., et al.: Swin-unet: unet-like pure transformer for medical image segmentation. In: Proceedings of the European Conference on Computer Vision Workshops (ECCVW) (2022)
5. Chen, J., et al.: Transunet: transformers make strong encoders for medical image segmentation. arXiv preprint arXiv:2102.04306 (2021)
6. Criminisi, A., Sharp, T., Blake, A.: GeoS: geodesic image segmentation. In: Forsyth, D., Torr, P., Zisserman, A. (eds.) ECCV 2008. LNCS, vol. 5302, pp. 99–112. Springer, Heidelberg (2008). https://doi.org/10.1007/978-3-540-88682-2_9
7. Grady, L.: Random walks for image segmentation. IEEE Trans. Pattern Anal. Mach. Intell. **28**(11), 1768–1783 (2006)
8. Heidari, M., et al.: Hiformer: hierarchical multi-scale representations using transformers for medical image segmentation. In: Proceedings of the IEEE/CVF Winter Conference on Applications of Computer Vision, pp. 6202–6212 (2023)
9. Hu, Y., Soltoggio, A., Lock, R., Carter, S.: A fully convolutional two-stream fusion network for interactive image segmentation. Neural Netw. **109**, 31–42 (2019)
10. Huang, Y., et al.: Segment anything model for medical images? arXiv preprint arXiv:2304.14660 (2023)
11. Huang, Y., et al.: Flip learning: erase to segment. In: de Bruijne, M., et al. (eds.) MICCAI 2021. LNCS, vol. 12901, pp. 493–502. Springer, Cham (2021). https://doi.org/10.1007/978-3-030-87193-2_47
12. Kirillov, A., et al.: Segment anything (2023)

13. Liu, Q., Xu, Z., Jiao, Y., Niethammer, M.: iSegFormer: interactive segmentation via transformers with application to 3D knee MR images. In: Wang, L., Dou, Q., Fletcher, P.T., Speidel, S., Li, S. (eds.) MICCAI 2022. LNCS, vol. 13435, pp. 464–474. Springer, Heidelberg (2022). https://doi.org/10.1007/978-3-031-16443-9_45

14. Liu, W., Ma, C., Yang, Y., Xie, W., Zhang, Y.: Transforming the interactive segmentation for medical imaging. In: Wang, L., Dou, Q., Fletcher, P.T., Speidel, S., Li, S. (eds.) MICCAI 2. LNCS, vol. 13434, pp. 704–713. Springer, Heidelberg (2022). https://doi.org/10.1007/978-3-031-16440-8_67

15. Ronneberger, O., Fischer, P., Brox, T.: U-net: convolutional networks for biomedical image segmentation. In: Navab, N., Hornegger, J., Wells, W.M., Frangi, A.F. (eds.) MICCAI 2015. LNCS, vol. 9351, pp. 234–241. Springer, Cham (2015). https://doi.org/10.1007/978-3-319-24574-4_28

16. Rother, C., Kolmogorov, V., Blake, A.: "grabcut" interactive foreground extraction using iterated graph cuts. ACM Trans. Graph. (TOG) **23**(3), 309–314 (2004)

17. Shen, D., Wu, G., Suk, H.I.: Deep learning in medical image analysis. Ann. Rev. Biomed. Eng. **19**, 221–248 (2017)

18. Sofiiuk, K., Petrov, I.A., Konushin, A.: Reviving iterative training with mask guidance for interactive segmentation. In: 2022 IEEE International Conference on Image Processing (ICIP), pp. 3141–3145. IEEE (2022)

19. Wang, G., et al.: Interactive medical image segmentation using deep learning with image-specific fine tuning. IEEE Trans. Med. Imaging **37**(7), 1562–1573 (2018)

20. Wang, G., et al.: Deepigeos: a deep interactive geodesic framework for medical image segmentation. IEEE Trans. Pattern Anal. Mach. Intell. **41**(7), 1559–1572 (2018)

21. Zhou, L., Wang, Y., Chen, D., Zeng, W., Zhang, Q., Yang, J.: Embracing imperfect datasets: a review of deep learning solutions for medical image segmentation. Med. Image Anal. **63**, 101693 (2020)

22. Zhou, L., Wang, S., Zhang, Q., Shen, D.: A review of deep learning in medical imaging: imaging traits, technology trends, case studies with progress highlights, and future promises. Proc. IEEE **109**(5), 820–838 (2021)

23. Zhou, W., Tao, X., Wei, Z., Lin, L.: Automatic segmentation of 3D prostate MR images with iterative localization refinement. Dig. Signal Process. **98**, 102649 (2020)

24. Zhou, Y., et al.: Multi-task learning for segmentation and classification of tumors in 3D automated breast ultrasound images. Med. Image Anal. **70**, 101918 (2021)

A Super Token Vision Transformer and CNN Parallel Branch Network for mCNV Lesion Segmentation in OCT Images

Xiang Dong[1], Hai Xie[1], Yunlong Sun[1], Zhenquan Wu[2], Bao Yang[1], Junlong Qu[1], Guoming Zhang[2(✉)], and Baiying Lei[1(✉)]

[1] School of Biomedical Engineering, Health Science Center, National-Regional Key Technology Engineering Laboratory for Medical Ultrasound, Guangdong Key Laboratory for Biomedical Measurements and Ultrasound Imaging, Shenzhen University, Shenzhen, China
leiby@szu.edu.cn

[2] Shenzhen Eye Hospital, Jinan University, Shenzhen Eye Institute, Shenzhen, Guangdong, China
zhangguoming@sz-eyes.com

Abstract. Myopic choroidal neovascularization (mCNV) is a vision-threatening complication of high myopia characterized by the growth of abnormal blood vessels in the choroid layer of the eye. In OCT images, mCNV typically presents as a highly reflective area within the subretinal layer. Therefore, accurate segmentation of mCNV in OCT images can better assist clinicians in assessing the disease status and guiding treatment decisions. However, accurate segmentation in OCT images is highly challenging due to the presence of noise interference, complex lesion areas, and low contrast. Consequently, we propose a parallel-branch network architecture that combines super token vision transformer (STViT) and CNN to more efficiently capture global dependency and low-level feature details. The super token attention mechanism (STA) in STViT reduces the number of tokens in self-attention and preserves global modeling. Additionally, we create a novel feature fusion module that utilizes depth-wise separable convolutions to efficiently fuse multi-level features from two pathways. We conduct extensive experiments on an in-house OCT dataset and a public OCT dataset, and the results demonstrate that our proposed method achieves state-of-the-art segmentation performance.

Keywords: mCNV lesion segmentation · Super token vision transformer · Feature fusion module

1 Introduction

Myopic choroidal neovascularization (mCNV) [1] is a common complication of high myopia. Its occurrence is related to choroidal atrophy and thinning caused by high myopia. The severity of mCNV depends on the number and location of blood vessels, and some mCNV lesions can regress spontaneously, while others can lead to lesion expansion and vision loss. The lesion involves the macular region of the retina and has

X. Cao et al. (Eds.): MLMI 2023, LNCS 14348, pp. 267–276, 2024.
https://doi.org/10.1007/978-3-031-45673-2_27

a serious impact on vision, making it one of the leading causes of irreversible blindness. Currently, drug therapy is one of the main treatments for mCNV, which involves intraocular injection of anti-vascular endothelial growth factor (VEGF) [2] drugs to inhibit the growth and leakage of neovascularization. Multiple injections are usually required to convert the lesion from active to inactive. Optical coherence tomography (OCT) is a non-invasive imaging technique that provides high-resolution images of the fine structures of the retina and choroid. OCT images can be used to observe the size of mCNV lesions, to judge the effectiveness of drug injection treatment and the possibility of disease recurrence. Accurate segmentation of mCNV lesions in OCT images can effectively evaluate their development status, assist doctors in assessing the condition, and guide the formulation of appropriate follow-up treatment plans. However, accurate segmentation in OCT images is highly challenging due to the presence of noise interference, complex lesion areas, and low contrast.

Currently, several researchers have conducted research on the aforementioned challenges and proposed some effective OCT image segmentation algorithms. For instance, Wilkins et al. proposed a fully automated retinal cyst segmentation technique using OCT images, which employed a computationally fast bilateral filter to remove speckle noise while maintaining CME boundaries [3]. Xiang et al. proposed a supervised method for automatic retinal layer segmentation on OCT scans of eyes with neovascularization, which not only introduced multi-scale bright and dark layer detection filters to describe and enhance retinal layers but also proposed a constrained graph search algorithm to detect the retinal surface accurately [4]. However, the drawback of these traditional algorithms is their inability to extract deep features from images efficiently.

Convolutional neural networks (CNNs) have shown remarkable performance in many medical image segmentation tasks due to their powerful feature extraction capability. In 2015, Ronneberger et al. designed the U-Net [5], which consists of an encoder-decoder structure with operations such as up-sampling, down-sampling, and skip connections, to perform pixel-level predictions on images. U-Net has been widely used in various medical image segmentation tasks. In addition, some variants of the U-Net structure, such as U-Net++ [6], Attention U-Net [7], and ResUNet++ [8], have achieved superior performance in medical image segmentation through various improvements. However, due to the unique convolutional operation of CNNs, it is difficult to capture global feature information. Dosovitskiy et al. made a breakthrough by proposing the Vision Transformer (ViT) [9], which applies the Transformer to computer vision tasks and can better capture global feature information. The ViT model has also been applied in the field of medical image segmentation. For example, Gao et al. proposed a hybrid transformer architecture called UTNet in 2021 [10], which uses an efficient self-attention mechanism and relative position encoding to capture long-range dependencies of features at different scales. Some studies have also attempted to combine CNNs and Transformers to leverage their respective strengths and maximize the combination of detailed local information with global context, such as TransFuse [11] and TMUNet [12], both of which have shown promising performance in medical image segmentation tasks. Although the above-mentioned Transformer networks can extract contextual information and capture global features, the computational complexity and sacrifice of the ability to capture global dependencies are still significant problems.

In order to address the aforementioned issues, we propose a parallel branch network based on STViT [13] and CNN for mCNV lesion segmentation in OCT images. The transformer branch based on the super token attention mechanism can significantly reduce computational complexity and learn global contextual information. The CNN branch can acquire rich low-level spatial features. We also design a novel feature fusion module that integrates the features of the two paths and introduces depth-wise separable convolutions [14], which can adapt to features of different scales and improve the model's expression ability. We conducted extensive comparative experiments on two different OCT datasets, demonstrating the effectiveness of our proposed method.

2 Method

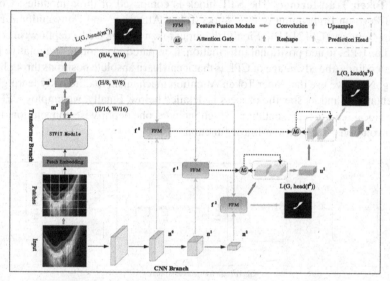

Fig. 1. Overall architecture of the proposed network. STViT Module: Super Token Vision Transformer.

As shown in Fig. 1, the overall architecture of the proposed network is composed of five parts: the transformer branch, CNN branch, feature fusion module, attention gate, and up-sampling. The transformer branch corresponds to the purple branch and follows a typical encoder-decoder structure. The input image is partitioned into patches of the same size using a stem consisting of four 3×3 convolutions, which are then fed into the patch embedding layer and STViT module. We adopt a progressive up-sampling method in the decoder, and reshape the output feature map before using two standard up-sampling convolutions to restore the spatial resolution. Additionally, we save three different-scale feature maps m^0, m^1, and m^2, for fusion with the corresponding output feature maps of the CNN branch. For the CNN branch, we directly use ResNet-based models and only retain the first four blocks, each of which down-samples the feature map by a factor of two. The output feature maps of the second, third, and fourth blocks are saved as n^0, n^1, and n^2, respectively. The feature maps m^i and n^i with the same size are then input into our FFM module for feature fusion. Finally, the attention gate [15] and up-sampling are used to combine the multi-scale fusion features f^i and generate the segmentation.

2.1 STViT Module

Figure 2 provides an overview of the STViT module. Tokens that have been subjected to patch embedding undergo hierarchical representation extraction through four stacked Super Token Transformer (STT) blocks. To reduce the number of tokens, a 3×3 convolution with a stride of 2 is used between every two blocks. Afterwards, the feature map is outputted by passing through a 1×1 convolution layer.

Super Token Transformer. The STT block is composed of three modules: Convolutional Position Embedding (CPE), Super Token Attention, and Convolutional Feed-Forward-Network (CFFN)[16]. Firstly, the input token is subjected a depth-wise convolution, i.e., CPE, to add positional information. Compared to absolute and relative positional encodings, the advantage of CPE is that it can learn absolute positions through zero padding. Next, we use the Super Token Attention mechanism to efficiently learn global representations, and its specific process is detailed below. Finally, we employ CFFN to enhance local feature representation, which includes one depth-wise convolution, two 1×1 convolutions, and one non-linear function.

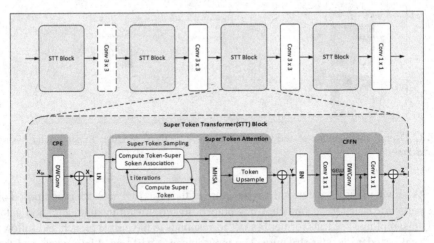

Fig. 2. The architecture of STViT Module.

2.2 Super Token Attention

The super token attention module, as shown in Fig. 2, consists of Super Token Sampling (STS), Multi-Head Self-Attention (MHSA), and Token Upsample (TU). First, we use

STS to aggregate vision tokens into a super token, then model the global dependency between super tokens using MHSA, and finally convert the super tokens back into visual tokens using TU.

Super Token Sampling. In the process of STS, we adapt the super-pixel algorithm based on k-means clustering from SSN [17]. Given the visual tokens $G \in \mathbb{R}^{N \times C}$ (where $N = H \times W$ is the number of tokens), assuming that each token $G_i = \mathbb{R}^{1 \times C}$ belongs to one of the e super tokens $S \in \mathbb{R}^{e \times C}$, it is necessary to compute the association map $Q \in \mathbb{R}^{N \times e}$ between G and S to implement the mapping. First, we sample the initial super tokens S^0 by averaging the tokens in the regular grid regions. Assuming the grid size is $h \times w$, the number of super tokens is $e = H/h \times W/w$. Then, we iteratively run the sampling algorithm for t times according to the following two steps:

1) **Token & Super Token Association.** In contrast to SSN, we use a method that is more similar to attention manner to compute the token & super token association map at iteration t, defined as

$$Q^t = Softmax\left(\frac{GS^{t-1^T}}{\sqrt{d}}\right), \tag{1}$$

   where d represents the channel number C and T represents transposition.
2) **Super Token Update.** The super tokens are updated as a weighted sum of the tokens, defined as

$$S = \left(\widehat{Q}^t\right)^T X, \tag{2}$$

where $\widehat{Q}^t$ is column-normalized Q^t.

Multi-head Self-Attention. As the super tokens are continuous and compact representations of visual content, we apply MHSA to learn global context dependencies more efficiently and further enhance the model's representational capacity. The MHSA applied to the sampled super tokens can be defined as follows:

$$Attn(S) = Softmax\left(\frac{q(S)k^T(S)}{\sqrt{d}}\right)v(S) = A(S)v(S), \tag{3}$$

where $A(S) = Softmax\left(\frac{q(S)k^T(S)}{\sqrt{d}}\right) \in \mathbb{R}^{m \times m}$ is the attention map, $q(S) = SW_q, k(S) = SW_k$ and $v(S) = SW_v$ are linear functions with parameters W_q, W_k and W_v. For clarity, we omit the multi-head setting here.

Token Upsample. Although the super tokens capture better global context information with MHSA, it also loses many local details during the sampling process. Therefore, we use the association map Q from the previous step to map the super tokens back into visual tokens and add them to the original tokens G, which are then used as inputs for subsequent layers. The process is defined as follows:

$$TU(Attn(S)) = QAttn(S). \tag{4}$$

2.3 Feature Fusion Module

To better fuse the multi-scale features extracted by the Transformer branch and the CNN branch, which contain rich semantic texture and global features, we propose a novel FFM that adapts spatial attention, channel attention, and depth-wise separable convolution. The overall framework is shown in Fig. 3.

We employ the channel attention from the SE block [18] to extract global features from the Transformer branch. The spatial attention is adopted from the CBAM block [19] as a spatial filter to enhance local details and suppress irrelevant regions in the features extracted from the CNN branch. In addition, we add the features from the two branches and use depth-wise separable convolution to extract their interaction features D^i. By concatenating the attention features M^i, N^i and the interaction features D^i, and passing them through a residual block, we obtain the final fusion feature f^i, which contains rich local and global information at the current spatial resolution. Furthermore, as shown in Fig. 1, we use attention gates and upsampling to generate the final mCNV lesion segmentation mask.

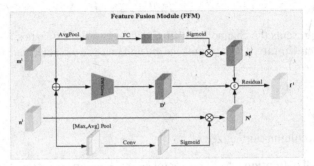

Fig. 3. Overall architecture of the proposed FFM.

3 Experiments and Results

3.1 Experimental Setup

To assess the effectiveness of our model, we performed experiments using both an in-house OCT dataset of mCNV from Zhongshan Ophthalmic Center of Sun Yat-sen University and a public OCT dataset available from Duke University [20]. The in-house OCT dataset consists of 2402 OCT images with corresponding ground truth annotations completed by experienced ophthalmologists. The public OCT dataset is commonly used for irregular area segmentation of diabetic macular edema (DME), which includes 110 OCT images from 10 patients. To increase the number of images in this dataset, we augmented it to 880 images using mirroring, rotation, horizontal and vertical flipping. For the experiments, we split both datasets into training, validation, and testing sets in an 8:1:1 ratio, and resized all images to 256×256.

All experiments were conducted using the PyTorch framework, and the model was trained on an NVIDIA TITAN XP GPU. We adopt Adam as training optimizer with the initial learning rate of le-4 and all models were trained for 200 epochs as well as batch size of 16.

3.2 Results

In order to evaluate the segmentation performance of the proposed method, we choose five popular models in the field of medical image segmentation in recent years for comparison, including U-Net, Atten U-Net, UTNet, Res_UTNet, TMUNet. The results of the comparative experiments are illustrated in Table 1.

From Table 1, we can see that our proposed method achieved the best performance in all evaluation metrics except for SP. In the task of in-house dataset, the Dice and IoU were **1.70%** and **1.01%** higher than the second best-performing method, and **2.87%** and **2.86%** higher than U-Net, respectively. By comparing U-Net and UTNet, we observed a significant improvement in segmentation performance using the hybrid network. All results indicate that our proposed model shows superior performance and demonstrate that the Super Token Vision Transformer combined with CNN can more fully extract local semantic features and global features with long-range dependencies and generate more accurate lesion segmentation maps.

Table 1. Segmentation performance camparison of popular methods on the two test sets (%).

Model	Task									
	in-house dataset					public dataset				
	Dice	IoU	Acc	SP	SE	Dice	IoU	Acc	SP	SE
U-Net	88.65	79.66	98.99	99.36	90.92	77.82	63.69	99.76	99.97	66.60
Atten U-Net	89.20	80.50	99.06	99.45	90.28	78.32	64.36	99.77	**99.98**	66.89
UTNet	89.70	81.33	99.10	99.43	91.67	77.95	63.87	99.76	99.95	68.73
Res_UTNet	89.80	81.48	99.11	99.48	90.90	78.26	63.93	99.74	99.95	67.42
TMUNet	89.82	81.51	99.12	**99.54**	89.82	78.45	64.54	99.77	99.97	67.28
Proposed	**91.52**	**82.52**	**99.30**	99.52	**91.68**	**80.25**	**66.34**	**99.78**	99.96	**69.54**

As shown in Fig. 4, we also demonstrate the results of mCNV lesion segmentation prediction for OCT images using the above methods. We use red boxes to highlight the segmented details for better comparison. By observing the first three rows, we can see that these comparative methods basically overlook some specific lesion details. In the third row, Atten U-Net and TMUNet even suffer from extensive missing lesions in the segmentation results. In contrast, our proposed method exhibits excellent performance and preserves lesion details well.

In order to further demonstrate the effectiveness of our method, we used the OCT dataset of DME as a comparative validation dataset and the experimental results are

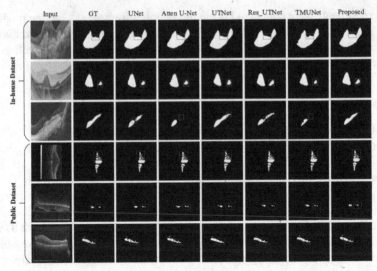

Fig. 4. Qualitative results of the comparison methods in two datasets. Each column denotes the segmentation mask of a method. The red boxes are to reflect the differences in segmentation details between different methods.

shown in Table 1. It can be seen that our method outperforms the second-best method by **1.8%** in Dice and IoU metrics, achieving **80.25%** and **66.34%**, respectively. In addition, our method also achieves the best performance in terms of Acc and SE metrics, but other methods also have high values in these two metrics, mainly because most of the lesion areas are small. From the fifth row in Fig. 4, it can be seen that only our method does not segment the small lesion on the right side of the image into two parts. In addition, thanks to the parallel branch feature extraction, our method is very effective in capturing the details of lesion areas. This further demonstrates the superior performance of our proposed method in OCT image segmentation tasks.

3.3 Ablation Studies

Table 2. Ablation studies of different modules on the in-house test set (%).

Setting	STViT	FFM	Dice	IoU	Acc	SP	SE
E.1			89.90	81.14	99.11	99.41	90.95
E.2		√	90.54	81.65	99.27	99.46	90.98
E.3	√		90.87	81.96	99.26	99.45	91.43
E.4	√	√	**91.52**	**82.52**	**99.30**	**99.52**	**91.68**

To validate the effectiveness of our modules, we conducted some ablation experiments, and the results are shown in Table 2. Four sets of comparative experiments (E1, E2,

E3, and E4) were designed, in which the STViT module was not selected, representing the use of a regular ViT instead, and the FFM module was not selected, representing the use of a feature fusion module without depth-wise separable convolution. By comparing E1 and E2, as well as E3 and E4 in Table 2, we observed a significant improvement in segmentation performance with our proposed FFM module. Furthermore, comparing E1 and E3, we observed that the use of our STViT module resulted in better performance, with a 0.97% increase in dice and a 0.82% increase in IoU. Finally, our proposed network achieved the best performance in all evaluation metrics.

4 Conclusion

In this paper, we propose a parallel branch network structure for mCNV lesion segmentation in OCT images. The transformer branch adopts super tokens vision transformer, which exhibits powerful capability in modeling global dependencies. The CNN branch effectively preserves local details in the image. A novel FFM module is used to fuse different features from both branches, enhancing the overall representation ability of the model. The experimental results on our private dataset and the comparative dataset confirm the effectiveness of our method, which can assist clinicians in evaluating the disease condition more efficiently.

Acknowledgements. This work is supported in part by the National Natural Science Foundation of China (No. U22A2024, 62106153, 82271103), Guangdong Basic and Applied Basic Research Foundation (No. 2020A1515110605, 2022A1515012326) and Natural Science Foundation of Shenzhen (No. JCYJ20220818095809021).

References

1. Cheung, C.M.G., et al.: Myopic choroidal neovascularization: review, guidance, and consensus statement on management. Ophthalmology **124**, 1690–1711 (2017)
2. Ohno-Matsui, K., Ikuno, Y., Lai, T.Y., Cheung, C.M.G.: Diagnosis and treatment guideline for myopic choroidal neovascularization due to pathologic myopia. Progr. Retinal Eye Res. **63**, 92–106 (2018)
3. Wilkins, G.R., Houghton, O.M., Oldenburg, A.L.: Automated segmentation of intraretinal cystoid fluid in optical coherence tomography. IEEE Trans. Biomed. Eng. **59**(4), 1109–1114 (2012)
4. Xiang, D., et al.: Automatic segmentation of retinal layer in OCT images with choroidal neovascularization. IEEE Trans. Image Process. **27**(12), 5880–5891 (2018)
5. Ronneberger, O., Fischer, P., Brox, T.: U-net: convolutional networks for biomedical image segmentation. In: Navab, N., Hornegger, J., Wells, W.M., Frangi, A.F. (eds.) Medical Image Computing and Computer-Assisted Intervention – MICCAI 2015: 18th International Conference, Munich, Germany, October 5-9, 2015, Proceedings, Part III, pp. 234–241. Springer International Publishing, Cham (2015). https://doi.org/10.1007/978-3-319-24574-4_28
6. Zhou, Z., Rahman Siddiquee, M.M., Tajbakhsh, N., Liang, J.: Unet++: A nested u-net architecture for medical image segmentation. In: Stoyanov, D., et al. (ed.) DLMIA/ML-CDS -2018. LNCS, vol. 11045, pp. 3–11. Springer, Cham (2018). https://doi.org/10.1007/978-3-030-008 89-5_1

7. Oktay, O., et al.: Attention u-net: learning where to look for the pancreas (2018)
8. Jha, D., et al.: Resunet++: an advanced architecture for medical image segmentation. In: 2019 IEEE International Symposium on Multimedia (ISM), pp. 225–2255. IEEE (2019)
9. Dosovitskiy, A., et al.: An image is worth 16×16 words: transformers for image recognition at scale (2020)
10. Gao, Y., Zhou, M., Metaxas, D.N.: UTNet: a hybrid transformer architecture for medical image segmentation. In: de Bruijne, M., et al. (ed.) MICCAI 2021. LNCS, vol. 12903, pp. 61–71. Springer, Cham (2021). https://doi.org/10.1007/978-3-030-87199-4_6
11. Zhang, Y., Liu, H., Qiang, H.: Transfuse: fusing transformers and cnns for medical image segmentation. In: de Bruijne, M., et al. (ed.) Medical Image Computing and Computer Assisted Intervention – MICCAI 2021: 24th International Conference, Strasbourg, France, September 27–October 1, 2021, Proceedings, Part I, pp. 14–24. Springer International Publishing, Cham (2021). https://doi.org/10.1007/978-3-030-87193-2_2
12. Azad, R., Heidari, M., Yuli, W., Merhof, D.: Contextual attention network: transformer meets u-net. In: Lian, C., Cao, X., Rekik, I., Xuanang, X., Cui, Z. (eds.) Machine Learning in Medical Imaging: 13th International Workshop, MLMI 2022, Held in Conjunction with MICCAI 2022, Singapore, September 18, 2022, Proceedings, pp. 377–386. Springer Nature Switzerland, Cham (2022). https://doi.org/10.1007/978-3-031-21014-3_39
13. Huang, H., Zhou, X., Cao, J., He, R., Tan, T.: Vision transformer with super token sampling. In: Proceedings of the IEEE/CVF Conference on Computer Vision and Pattern Recognition, pp. 22690–22699 (2023)
14. Chollet, F.: Xception: deep learning with depthwise separable convolutions. In: Proceedings of the IEEE Conference on Computer Vision and Pattern Recognition, pp. 1251–1258 (2019)
15. Schlemper, J., et al.: Attention gated networks: learning to leverage salient regions in medical images. Med. Image Anal. **53**, 197–207 (2019)
16. Huang, H., Zhou, X., He, R.: Orthogonal transformer: an efficient vision transformer backbone with token orthogonalization. Adv. Neural Inf. Process. Syst. **35**, 14596–14607 (2022)
17. Jampani, V., Sun, D., Liu, M.-Y., Yang, M.-H., Kautz, J.: Superpixel sampling networks. In: Proceedings of the European Conference on Computer Vision (ECCV), pp. 352–368 (2018)
18. Hu, J., Shen, L., Sun, G.: Squeeze-and-excitation networks. In: Proceedings of the IEEE Conference on Computer Vision and Pattern Recognition, pp. 7132–7141 (2018)
19. Woo, S., Park, J., Lee, J.-Y., Kweon, I.S.: CBAM: convolutional block attention module. In: Proceedings of the European Conference on Computer Vision (ECCV), pp. 3–19 (2018)
20. Rashno, A., et al.: Fully automated segmentation of fluid/cyst regions in optical coherence tomography images with diabetic macular edema using neutrosophic sets and graph algorithms. IEEE Trans. Biomed. Eng. **65**(5), 989–1001 (2017)

Boundary-RL: Reinforcement Learning for Weakly-Supervised Prostate Segmentation in TRUS Images

Weixi Yi[1,2(✉)], Vasilis Stavrinides[3,4], Zachary M. C. Baum[1,2], Qianye Yang[1,2],
Dean C. Barratt[1,2], Matthew J. Clarkson[1,2], Yipeng Hu[1,2],
and Shaheer U. Saeed[1,2]

[1] Wellcome/EPSRC Centre for Interventional and Surgical Sciences; Centre for
Medical Image Computing, University College London, London, UK
`weixi.yi.22@ucl.ac.uk`
[2] Department of Medical Physics and Biomedical Engineering, University College
London, London, UK
[3] Division of Surgery and Interventional Science, University College London, London,
UK
[4] Department of Urology, UCL Hospital NHS Foundation Trust, London, UK

Abstract. We propose Boundary-RL, a novel weakly supervised segmentation method that utilises only patch-level labels for training. We envision segmentation as a boundary detection problem, rather than a pixel-level classification as in previous works. This outlook on segmentation may allow for boundary delineation under challenging scenarios such as where noise artefacts may be present within the region-of-interest (ROI) boundaries, where traditional pixel-level classification-based weakly supervised methods may not be able to effectively segment the ROI. Particularly of interest, ultrasound images, where intensity values represent acoustic impedance differences between boundaries, may also benefit from the boundary delineation approach. Our method uses reinforcement learning to train a controller function to localise boundaries of ROIs using a reward derived from a pre-trained boundary-presence classifier. The classifier indicates when an object boundary is encountered within a patch, serving as weak supervision, as the controller modifies the patch location in a sequential Markov decision process. The classifier itself is trained using only binary patch-level labels of object presence, the only labels used during training of the entire boundary delineation framework. The use of a controller ensures that sliding window over the entire image is not necessary and reduces possible false-positives or -negatives by minimising number of patches passed to the boundary-presence classifier. We evaluate our approach for a clinically relevant task of prostate gland segmentation on trans-rectal ultrasound images. We show improved performance compared to other tested weakly supervised methods, using the same labels e.g., multiple instance learning.

Keywords: TRUS · Weak Supervision · Reinforcement Learning

X. Cao et al. (Eds.): MLMI 2023, LNCS 14348, pp. 277–288, 2024.
https://doi.org/10.1007/978-3-031-45673-2_28

1   Introduction

Automated segmentation of anatomical structures and other regions-of-interest (ROI) plays a vital role in the field of medical imaging [18,26]. Accurate delineation and contouring of ROI boundaries is an essential step in procedures such as treatment planning e.g., for radiotherapy [24], longitudinal diagnostic or prognostic planning e.g., by tracking fetal dimensions [35] or organ sizes such as prostate volume [32], for locating abnormalities within imaged structures e.g., for lung abnormality detection [5], and for trans-rectal ultrasound (TRUS) guided prostate biopsy or brachytherapy [12]. In particular, of interest in this work, TRUS-guided biopsy has recently been aided by automated segmentation of the gland boundaries for precise sampling along the gland for accurate diagnoses [12,14,29]. Recent advancements in the field have been driven by the introduction of fully supervised learning-based segmentation methods [9,30] that require expert annotated data for learning. However, due to the time-consuming nature of pixel-level annotation of medical images, the requirement for specialized knowledge and significant inter-observer variability, the acquisition of ample annotated data for model training may be challenging [1,2,4,21,23].

Weakly-supervised semantic segmentation (WSSS) may address some of the above-mentioned issues with fully-supervised learning by employing weak labels, e.g., image-level classification labels, for learning segmentation [10,15,36]. However, segmenting the boundaries of the prostate from TRUS images with weak labels remains significantly challenging [7] due to: (1) often indistinct boundaries between prostate and surrounding tissues caused by the low contrast of soft tissues, (2) frequently missing segments of prostate boundaries due to the presence of shadowing artefacts, (3) considerable patient-dependent variability in prostate shape and size, (4) uneven intensity distribution of the prostate, and (5) lack of informative signals, e.g., in full supervision, where in commonly used approaches for weak supervision, presence of object to learn a valid prior such as global correlation between missing and distinct boundaries becomes challenging.

Within the literature on WSSS, methods based on multiple instance learning (MIL) have been proposed, with varying definitions of bags and patches used for training. For instance, MILinear [19], with its bag-splitting-based mechanism, is capable of iteratively generating new negative bags from positive ones. Xu et al. [33] integrated the concept of clustering and introduced a multi-cluster instance learning technique for segmenting cancerous and non-cancerous tissues. CAMEL [31] treated lattice patches as instances and histopathological images as bags. Increasing research efforts are oriented towards considering the image as a bag and treating each pixel within the image as an instance, for improved efficiency at inference, thereby transforming the weakly supervised segmentation task with only image-level annotations into an instance prediction task based on bag-level annotations. Jia et al. [8] proposed DWS-MIL, based on a fully convolutional network, for segmenting carcinogenic areas within histopathological images. Li et al. [13] introduced a self-attention mechanism to capture the correlation between instances within MIL. Additionally, some methods employ patch-level labels (i.e., binary labels of ROI presence within a patch or crop of a larger image) to achieve pixel-level segmentation with a comparatively lower annotation cost [6]. Most previous methods are application-specific with no general solutions.

We present a novel WSSS approach that only utilises patch-level binary classification labels of object-presence for training. These are more-time efficient to obtain than full segmentation maps traditionally used in fully-supervised learning. The WSSS in this work envisions the segmentation task as a boundary detection problem rather than a pixel-level classification, as in most previous works. The boundary detection follows a Markov Decision Process (MDP) whereby a controller function, trained using reinforcement learning (RL) based on reward function derived from a boundary presence classifier (similar to previously proposed task-based rewards [3,22,34,37]), aims to localise the ROI boundary.

The use of RL for boundary delineation allows us to pose this as a sequential MDP whereby the patch-location refinement is done until the boundary of the object is found, as determined by the boundary presence classifier. As opposed to previously proposed WSSS methods that pose the segmentation task as a pixel-level classification, our method does not require a sliding window to be passed over the entire image to generate pixel-level labels for the segmentation map, which may be more efficient at inference, reducing the number of forward passes through the classifier. Rather, the boundary delineation may simply be generated using a trajectory that the controller follows within an image. Moreover, this approach of boundary delineation has the potential to perform well even when artefacts such as shadowing are present within the ROI but not occluding the boundary, where traditional weakly supervised pixel-wise classification approaches to segmentation may produce inaccurate results [11,16,17,19,27].

Contributions of this work are summarised: First, we propose a RL-based framework, Boundary-RL, trained only using patch-level labels of object presence, capable of effectively identifying object boundaries; Second, we evaluate Boundary-RL for a clinically relevant task of prostate gland boundary delineation on TRUS images using data from real prostate cancer patients; And lastly, we compare Boundary-RL to commonly used algorithms such as fully-supervised learning and MIL-based WSSS, which we adapt for our application, from previous work.

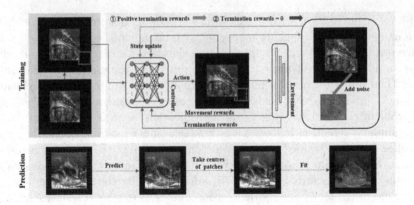

Fig. 1. The workflow of our proposed framework.

2   Methods

In our work (Fig. 1), the boundary presence classifier, trained using only binary patch-level labels of ROI presence within the patch, serves as a weak signal that may be used to train the boundary detecting controller. The patch-localising controller ensures that the location of the patch is refined, starting from the edge of the image, until part of the ROI exists within the patch, as determined by the boundary presence classifier. This may be repeated with randomised starting locations in order to get an effective boundary delineation for the ROI.

2.1   Boundary Presence Classifier

The boundary presence classifier $f(\cdot; w) : \mathcal{X} \to \{0, 1\}$, with weights w, predicts object presence, given an image or image patch $x \in \mathcal{X}$, where $\mathcal{X}$ is the domain of images or image patches. Here, 0, corresponds to object absence, and 1, corresponds to object presence. Image-label pairs $\{x_k, y_k\}_{k=1}^N$, where $y \in \{0, 1\}$, may be used to train the classifier f using a binary cross-entropy loss function by means of gradient descent following $w^* = arg\min_w \mathbb{E}[L(y, f(x; w))]$ where $L(y, f(x; w))$ is the binary cross-entropy loss function and w^* are the optimal weights of the classifier which may be fixed after the initial pre-training stage. In practice the threshold for prediction probability is set as 0.9, so any classifier-predicted probability greater than 0.9 is rounded to 1, and 0 otherwise. The prediction probability of the classifier is treated as a boundary-presence probability prediction, which is used to inform the controller function, as outlined below.

2.2   Boundary-RL - Delineating Boundaries of Objects Using RL

We define an image patch x_c, of fixed dimensions, from image x as having coordinates $c = (i, j)$. The aim is for the controller function is to refine c until the ROI or its boundary are within the bounds of the patch.

The controller function $g(\cdot; \theta) : \mathcal{X} \times \mathcal{X} \rightarrow \mathcal{A}$, with weights θ, takes in the image $x \in \mathcal{X}$, the image patch $x_c \in \mathcal{X}$ and outputs an action $a \in \mathcal{A}$ where $\mathcal{A}$ is the discrete action space defined as $a \in \mathcal{A} \in \{-1, 0, +1\}^2$. Here the two values for a correspond to the change in coordinates for the patch, in each axis i.e., $a = (\delta i, \delta j)$ which may be applied to the coordinates $c_t = (i, j)$ at time-step t, to translate the patch to the new coordinates $c_{t+1} = (i + \delta i, j + \delta j)$. In practice these actions are modelled as four discrete actions of up, down, left or right, that is to say that if $\delta i \neq 0$ then $\delta j = 0$ and if $\delta j \neq 0$ then $\delta i = 0$.

Interacting with an Environment to Learn Boundary Delineation: The image x_t and pre-trained presence classifier $f(\cdot; w^*)$ may be considered part of a MDP environment with which the controller interacts, i.e., as a MDP. The environment may be defined as $(\mathcal{S}, \mathcal{A}, p, r, \pi, \gamma)$.

States: The observed state $s_t \in \mathcal{S}$ for this environment consists of the image x_t and the image patch localised by the controller $x_{c,t}$, i.e., $s_t = \{x_t, x_{c,t}\}$, where $\mathcal{S}$ is the state space.

Actions: The action $a_t \in \mathcal{A}$, where $\mathcal{A}$ is the action space, impacts the observed state at the subsequent time-step $t + 1$ as $a_t = (\delta i, \delta j)$, which leads to $c_{t+1} = (i + \delta i, j + \delta j)$, which in-turn results in an updated patch, $x_{c,t+1}$, which forms part of the updated state.

State Transition Distribution: The state-action conditioned state transition distribution is given by $p : \mathcal{S} \times \mathcal{S} \times \mathcal{A} \rightarrow [0, 1]$. Assuming next state s_{t+1} as input, given the current state s_t and action a_t, the probability of the next state is denoted as $p(s_{t+1}|s_t, a_t)$.

Rewards: The reward function $r : \mathcal{S} \times \mathcal{A} \rightarrow \mathbb{R}$ takes the state-action pair as input and produces a reward $R_t = r(s_t, a_t)$, at time-step t. The reward consists of two components. 1) The first component is the movement reward r_{mov}, which is computed with the help of the estimated centroid of the prostate gland, given by coordinates $c_p = (i_p, j_p)$. The centroid of the prostate may be practically easy to obtain e.g., by means of pixel-intensity values or by manually labelling a subset of samples in roughly aligned images potentially utilising intensity-based registration. Then for time-step t, the L^2-norm between the location of the patch and the centroid of the prostate gland is given by $||c_t - c_p||$. We then define $r_{mov} = +1$ if $||c_t - c_p|| - ||c_{t-1} - c_p|| \leq 0$ and $r_{mov} = -1$ otherwise. 2) The second component is a termination reward r_{term}, given when the patch contains the

ROI or its boundary, as determined by the boundary presence classifier function $f(\cdot; w^*)$, with pre-trained fixed weights w^*. We compute $r_{term} = f(x_{c,t}; w^*)$ for the termination reward, which is $f(x_{c,t}; w^*) = 1$ when the ROI is in the patch and $f(x_{c,t}; w^*) = 0$ otherwise. The two rewards may then be combined into our reward function $r(s_t, a_t) = r_{mov} + 100 r_{term}$, where 100 is a scalar that allows weighting the two components. In this formulation, r_{mov} may be considered as reward shaping and r_{term} as the final sparse reward.

Policy: The probability of performing an action a_t is given by the policy $\pi : \mathcal{S} \times \mathcal{A} \rightarrow [0, 1]$, e.g., for time-step t the probability of performing action a_t is $\pi(a_t | s_t)$. Sampling an action according to a policy is denoted as $a_t \sim \pi(\cdot)$. Following the state transition distribution for sampling next states, the policy for sampling actions and the reward function for generating corresponding rewards, we can collect states, actions and rewards over multiple time-steps $(s_1, a_1, R_1, \ldots, s_T, a_T, R_T)$, also called trajectories. These are used to update our controller.

Policy Optimisation: The policy is modelled as a parametric neural network $\pi(\cdot; \theta)$ with weights θ. This neural network predicts parameters of a distribution from which to sample actions, which are the mean and standard deviation of a Gaussian distribution for continuous actions in this work [25]. The action is sampled from this predicted distribution $a \sim \pi(\cdot)$. We compute a cumulative reward $Q^{\pi(\cdot;\theta)}(s_t, a_t) = \sum_{k=0}^{T} \gamma^k R_{t+k}$ using a discount factor γ for future rewards. We can then use gradient ascent to obtain optimal policy parameters: $\theta^* = arg\max_\theta \mathbb{E}[Q^{\pi(\cdot;\theta)}(s_t, a_t)]$.

Episodic Training and Terminal Signals: In practice, we collect trajectories by defining episode termination as finding a patch containing the organ boundary. More concretely, we terminate an episode when $f(x_{c,t}; w^*) = 1$, we may denote this time-step at termination as t_{term}. At t_{term}, we add random Gaussian noise over our patch within the image such that finding the same patch-location on the next episode within the same image does not lead to $f(x_{c,t}; w^*) = 1$. This encourages finding unique patches as boundary points for the ROI. We conduct M episodes for each image until M patches have been found as the object boundary. At each episode the patch location is randomly selected as an edge-pixel on one of the four edges, so the controller learns to move from the edge of the image to the boundary of the object within the image. In practice, we also terminate the episode after T time-steps in case the boundary cannot be found after T steps, where in our formulation, $T = 1000$.

Using Controller-Predicted Patches for Object Boundary Delineation:
The centre-points of all M patches are taken to be the estimated boundary points
for the object, outliers are removed via a simple distance-based outlier removal
i.e., by rejecting points beyond a distance of 10 pixels from the mean of the
boundary point locations. A polygon is constructed using the remainder of the
point points which serves as the delineated boundary. With sufficiently densely
sampled points, the nearest points are connected without and topological issues
found in practice. The entire training algorithm is summarised in Algorithm 1,
where at inference, the only difference is that rewards are not computed and the
controller is not updated.

3   Experiments

3.1   Dataset

During the (NCT02290561, NCT02341677) clinical trials, TRUS images from
249 patients were collected using a bi-plane transperineal ultrasound probe during manual/rotational positioning of a digital transperineal stepper, resulting in
a total of 5185 prostate-containing images with the size of 403×361, which were
center-cropped to 360×360. Ground truth labels were acquired through pixel-
wise consensus among three researchers and verified by an expert radiologist.
From 203 patients, we randomly selected 4000 images for controller training,
and further cropped 80000 patches sized 90×90 for MIL and Boundary Presence Classifier training. From the remaining 46 patients, we randomly chose 200
images each for validation and testing.

3.2   Architecture and Implementation Details

Boundary Presence Classifier: EfficientNet [28] architecture was used, with
the Adam optimiser with a learning rate of 1e–4 and batch size 16.

Boundary RL Controller: We trained the controller, which has 3 convolu-
tional layers followed by 2 dense layers, for Boundary-RL using Proximal Policy
Optimization (PPO) [25], which took about 27 h on a single Nvidia Tesla V100
GPU, with a learning rate set at 3e–4 and a batch size of 4096.

Data: Images to construct $s_t \in \mathcal{S}$
Result: Trained RL policy π_{θ^*}.

while *not converged* **do**

    Randomly sample an image $x_t \in \mathcal{X}$;

    for $m \leftarrow 1$ **to** M **do**

        Start at $t = 0$;

        Randomly sample patch coordinates $c_0 = (i, j)$;

        Construct the state $s_t = \{x_t, x_{c,0}\}$;

        Sample the action a_0 according to the policy $a_0 \sim \pi_\theta(a_0|s_0)$;

        Compute object presence-based reward $R_0 = r(s_0, a_0)$;

        for $t \leftarrow 1$ **to** T **do**

            Note: t is now iterating starting at $t = 1$;

            Given $a_{t-1} = (\delta i, \delta j)$, update patch coordinates $c_t = (i + \delta i, j + \delta j)$;

            Construct the state $s_t = \{x_t, x_{c,t}\}$;

            Sample the action a_t according to the policy $a_t \sim \pi_\theta(a_t|s_t)$;

            Compute object presence-based reward $R_t = r(s_0, a_0)$;

            End if object presence detected i.e., $t = t_{term}$ if $f(x_{c,t}; w^*) = 1$; add Gaussian noise to patch to update x_t;

        end

    end

    Once M trajectories of $R_{t=1:T}$ or $R_{t=1:t_{term}}$ collected, update RL function using gradient ascent

end

Algorithm 1: Training procedure for Boundary RL.

3.3 Comparisons and Baselines

MIL: MIL is adapted from previous works [11,16,17,19,27], for our application of prostate gland segmentation. We treat each patch as a bag, and each pixel within a patch is considered an instance. We use a modified U-Net [20], with global average pooling added before the full connection layer to aggregate instance probability distribution vectors into bag features. We adopt both partition (MIL-P) and sliding window (MIL-S) approaches for prediction. For overlapping predictions within the sliding window, we take their average.

Fully Supervised Learning (FSL): We chose the commonly used U-Net, trained using segmentation labels, to compare with our proposed WSSS method.

4 Results

Results presented in Table 1, show that our proposed Boundary-RL attained highest performance amongst the tested WSSS methods. While performance was statistically significantly better than the commonly used MIL with image partitioning (MIL-P) for inference, significance was not observed for the comparison with MIL using sliding windows (MIL-S) at inference. The fully supervised (FSL) method performed better than all tested WSSS methods, with statistical significance. Samples from each of the methods are presented in Fig. 2.

Table 1. Table of results and corresponding statistical tests.

Algorithm	Dice	Statistical tests (t-value/ p-value)			
		Proposed	MIL - P	MIL - S	FSL
Proposed	0.751 ± 0.161	N/A	–	–	–
MIL - P	0.663 ± 0.156	5.53/<0.01	N/A	–	–
MIL - S	0.737 ± 0.152	0.88/0.38	4.80/<0.01	N/A	–
FSL	0.846 ± 0.167	5.76/<0.01	11.26/<0.01	6.79/<0.01	N/A

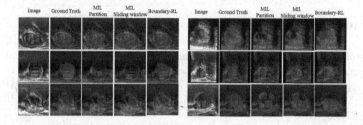

Fig. 2. Samples with segmentation from the tested methods.

5   Discussion and Conclusion

Results show that an effective boundary delineation policy may be learnt using our proposed Boundary-RL training scheme. Proposed adaptations to the commonly used MIL approach, also yielded comparable performance, with Boundary-RL being potentially more time-efficient at inference due to a fewer passes through the classifier due to the controller-learnt localisation policy, compared to sliding windows in MIL which may be akin to brute-search. All WSSS methods under-performed fully-supervised learning as expected due to the use of pixel-level segmentation maps for training, compared to only binary classification labels used in WSSS. Boundary-RL outperformed all tested WSSS methods but statistical significance was not observed compared to MIL-S at inference. We maintain that our method produces an effective boundary delineation, evidenced from qualitative results and high observed Dice. Qualitative samples show that Boundary-RL can handle cases where artefacts exist within ROI boundaries, where MIL was unable to effectively segment since it relies on pixel-level classification which may be impacted by presence of artefacts within ROI boundaries. Future studies could explore augmenting rewards with other weak supervision signals such as self re-construction instead of just the patch-level classification.

We proposed a WSSS method based on RL that envisions segmentation as a boundary detection problem. A MDP is formulated to effectively learn to localise ROI boundaries using a reward based on boundary-presence within the localised region. The reward, generated using a classifier trained using binary patch-level labels of object presence, serves as a weak signal to inform boundary detection. We evaluate our approach for a clinically relevant task of boundary delineation of the prostate gland on TRUS, used during navigation in TRUS guided biopsy.

Acknowledgements. This work was supported by the EPSRC grant [EP/T02940 4/1], Wellcome/EPSRC Centre for Interventional and Surgical Sciences [203145Z/ 16/Z], and the International Alliance for Cancer Early Detection, an alliance between Cancer Research UK [C28070/A30912; 73666/A31378], Canary Center at Stanford University, the University of Cambridge, OHSU Knight Cancer Institute, University College London and the University of Manchester.

References

1. Chalcroft, L.F., et al.: Development and evaluation of intraoperative ultrasound segmentation with negative image frames and multiple observer labels. In: Noble, J.A., Aylward, S., Grimwood, A., Min, Z., Lee, S.-L., Hu, Y. (eds.) ASMUS 2021. LNCS, vol. 12967, pp. 25–34. Springer, Cham (2021). https://doi.org/10.1007/978-3-030-87583-1_3

2. Chen, M.Y., Woodruff, M.A., Dasgupta, P., Rukin, N.J.: Variability in accuracy of prostate cancer segmentation among radiologists, urologists, and scientists. Cancer Med. **9**(19), 7172–7182 (2020)

3. Cubuk, E.D., Zoph, B., Mane, D., Vasudevan, V., Le, Q.V.: Autoaugment: learning augmentation policies from data. arXiv preprint arXiv:1805.09501 (2018)

4. Czolbe, S., Arnavaz, K., Krause, O., Feragen, A.: Is segmentation uncertainty useful? In: Feragen, A., Sommer, S., Schnabel, J., Nielsen, M. (eds.) IPMI 2021. LNCS, vol. 12729, pp. 715–726. Springer, Cham (2021). https://doi.org/10.1007/978-3-030-78191-0_55

5. Dertkigil, S., Appenzeller, S., Lotufo, R., Rittner, L.: A systematic review of automated segmentation methods and public datasets for the lung and its lobes and findings on computed tomography images. Yearbook Med. Inf. **31**(01), 277–295 (2022)

6. Han, C., Lin, J., Mai, J., Wang, Y., Zhang, Q., et al.: Multi-layer pseudo-supervision for histopathology tissue semantic segmentation using patch-level classification labels. MedIA **80**, 102487 (2022)

7. Hulsmans, F.J.J., Castelijns, J.A., Reeders, J.W., Tytgat, G.N.: Review of artifacts associated with transrectal ultrasound: understanding, recognition, and prevention of misinterpretation. J. Clin. Ultrasound **23**(8), 483–494 (1995)

8. Jia, Z., Huang, X., Eric, I., Chang, C., Xu, Y.: Constrained deep weak supervision for histopathology image segmentation. IEEE TMI **36**(11), 2376–2388 (2017)

9. Karimi, D., Zeng, Q., Mathur, P., Avinash, A., et al.: Accurate and robust deep learning-based segmentation of the prostate clinical target volume in ultrasound images. MedIA **57**, 186–196 (2019)
10. Kervadec, H., Dolz, J., Tang, M., Granger, E., Boykov, Y., et al.: Constrained-cnn losses for weakly supervised segmentation. MedIA **54**, 88–89 (2019)
11. Kots, M., Chukanov, V.: U-net adaptation for multiple instance learning. In: Journal of Physics: Conference Series, vol. 1236. IOP Publishing (2019)
12. Lei, Y., Tian, S., He, X., Wang, T., et al.: Ultrasound prostate segmentation based on multidirectional deeply supervised v-net. Med. Phys. **46**(7), 3194–3206 (2019)
13. Li, K., Qian, Z., Han, Y., Eric, I., Chang, C., et al.: Weakly supervised histopathology image segmentation with self-attention. MedIA **86**, 102791 (2023)
14. Liu, D., Wang, L., Du, Y., Cong, M., Li, Y.: 3-d prostate MR and TRUS images detection and segmentation for puncture biopsy. IEEE Trans. Instrument. Meas. **71**, 1–13 (2022)
15. Pathak, D., Krahenbuhl, P., Darrell, T.: Constrained convolutional neural networks for weakly supervised segmentation. In: ICCV (2015)
16. Pathak, D., Shelhamer, E., Long, J., Darrell, T.: Fully convolutional multi-class multiple instance learning. arXiv preprint arXiv:1412.7144 (2014)
17. Pinheiro, P.O., Collobert, R.: From image-level to pixel-level labeling with convolutional networks. In: CVPR (2015)
18. Ramesh, K., Kumar, G.K., Swapna, K., Datta, D., Rajest, S.S.: A review of medical image segmentation algorithms. EAI PHAT **7**(27), e6 (2021)
19. Ren, W., Huang, K., Tao, D., Tan, T.: Weakly supervised large scale object localization with multiple instance learning and bag splitting. IEEE Trans. Pattern Anal. Mach. Intell. **38**(2), 405–416 (2015)
20. Ronneberger, O., Fischer, P., Brox, T.: U-net: convolutional networks for biomedical image segmentation. In: Navab, N., Hornegger, J., Wells, W.M., Frangi, A.F. (eds.) MICCAI 2015. LNCS, vol. 9351, pp. 234–241. Springer, Cham (2015). https://doi.org/10.1007/978-3-319-24574-4_28
21. Saeed, S.U., et al.: Adaptable image quality assessment using meta-reinforcement learning of task amenability. In: Noble, J.A., Aylward, S., Grimwood, A., Min, Z., Lee, S.-L., Hu, Y. (eds.) ASMUS 2021. LNCS, vol. 12967, pp. 191–201. Springer, Cham (2021). https://doi.org/10.1007/978-3-030-87583-1_19
22. Saeed, S.U., Fu, Y., Stavrinides, V., Baum, Z.M., Yang, Q., et al.: Image quality assessment for machine learning tasks using meta-reinforcement learning. MedIA **78**, 102427 (2022)
23. Saeed, S.U., Yan, W., Fu, Y., Giganti, F., et al.: Image quality assessment by overlapping task-specific and task-agnostic measures: application to prostate multiparametric MR images for cancer segmentation. In: Machine Learning for Biomedical Imaging (IPMI 2021), vol. 1 (2022)
24. Savjani, R.R., Lauria, M., Bose, S., Deng, J., et al.: Automated tumor segmentation in radiotherapy. In: Seminars in Radiation Oncology, vol. 32. Elsevier (2022)
25. Schulman, J., Wolski, F., Dhariwal, P., Radford, A., Klimov, O.: Proximal policy optimization algorithms. arXiv preprint arXiv:1707.06347 (2017)
26. Sharma, N., Aggarwal, L.M.: Automated medical image segmentation techniques. J. Med. Phys. **35**(1), 3 (2010)
27. Shi, X., Xing, F., Xie, Y., Zhang, Z., Cui, L., et al.: Loss-based attention for deep multiple instance learning. In: AAAI 2020, vol. 34 (2020)
28. Tan, M., Le, Q.: Efficientnet: rethinking model scaling for convolutional neural networks. In: International Conference on Machine Learning. PMLR (2019)

29. Wang, X., Chang, Z., Zhang, Q., Li, C., Miao, F., et al.: Prostate ultrasound image segmentation based on dsu-net. Biomedicines **11**(3), 646 (2023)

30. Wang, Y., Dou, H., Hu, X., Zhu, L., Yang, X., et al.: Deep attentive features for prostate segmentation in 3D transrectal ultrasound. IEEE TMI **38**(12), 2768–2778 (2019)

31. Xu, G., Song, Z., Sun, Z., Ku, C., et al.: Camel: a weakly supervised learning framework for histopathology image segmentation. In: IEEE/CVF ICCV (2019)

32. Xu, R.S., Michailovich, O., Salama, M.: Information tracking approach to segmentation of ultrasound imagery of the prostate. IEEE Trans. Ultrasonics Ferroelectr. Freq. Control **57**(8), 1748–1761 (2010)

33. Xu, Y., Zhu, J.Y., Eric, I., Chang, C., Lai, M., et al.: Weakly supervised histopathology cancer image segmentation and classification. MedIA **18**(3), 591–604 (2014)

34. Yoon, J., Arik, S., Pfister, T.: Data valuation using reinforcement learning. In: International Conference on Machine Learning. PMLR (2020)

35. Zeng, Y., Tsui, P.H., Wu, W., Zhou, Z., Wu, S.: Fetal ultrasound image segmentation for automatic head circumference biometry using deeply supervised attention-gated v-net. J. Dig. Imaging **34**, 134–148 (2021)

36. Zhang, M., Zhou, Y., Zhao, J., Man, Y., Liu, B., et al.: A survey of semi-and weakly supervised semantic segmentation of images. AIRE **53**, 4259–4288 (2020)

37. Zhang, X., Wang, Q., Zhang, J., Zhong, Z.: Adversarial autoaugment. arXiv preprint arXiv:1912.11188 (2019)

A Domain-Free Semi-supervised Method for Myocardium Segmentation in 2D Echocardiography Sequences

Wenming Song, Xing An, Ting Liu, Yanbo Liu, Lei Yu, Jian Wang, Yuxiao Zhang, Lei Li, Longfei Cong, and Lei Zhu[✉]

Shenzhen Mindray BioMedical Electronics, Co., Ltd., Shenzhen, China
zhulei@mindray.com

Abstract. Many deep learning methods have been applied in myocardium segmentation, however, the robustness of these algorithms is relatively low, especially when dealing with datasets from different domains, such as machines. In this paper, we propose a domain-free semi-supervised deep learning algorithm to improve the model robustness between different machines. Two domain-free factors (the shape of the myocardium and the motion tendency between adjacent frames) are adopted. Specifically, an optical flow field-based segmentation network is proposed for enhancing the performance by combining the motion tendency of myocardium between adjacent frames. Moreover, a shape-based semi-supervised adversarial network is presented to utilize the shape of the myocardium for the purposes of improving the segmentation robustness. Experiments on our private and public datasets show that the proposed method not only improves the segmentation performance, but also decreases the performance gap when applied to different machines, thus demonstrating the effectiveness of the proposed method.

Keywords: Domain-free · Robustness · Semi-supervising · Segmentation · Myocardium · Optical flow

1 Introduction

Cardiovascular diseases (CVDs) remain the leading cause of death worldwide with more than 17 million fatalities each year [1]. Echocardiography is one of the chief imaging modalities to assist cardiologists in assessing the functional status of the heart, due to the method's availability, portability, and real-time imaging ability [2]. Accurate estimation of ejection fraction (EF) from echocardiography images is of great importance for the evaluation of cardiac function, which depends on the delineations of the left ventricular (LV) myocardium boundaries at both end-diastole (ED) and end-systole (ES). Manual annotation of the delineations is time-consuming and has inter- and intra-variations. Therefore, exploring effective, accurate, and automatic segmentation algorithms has great clinical significance in improving the work efficiency of cardiologists.

Over the last decade, various computer-aided diagnosis algorithms have been developed to conduct segmentation of LV myocardium in echocardiography and effectively

© The Author(s), under exclusive license to Springer Nature Switzerland AG 2024
X. Cao et al. (Eds.): MLMI 2023, LNCS 14348, pp. 289–298, 2024.
https://doi.org/10.1007/978-3-031-45673-2_29

relieves the workload of physicians [3–5]. [3] proposed a machine learning method based on the Structured Random Forest algorithm to fully automate the segmentation of the myocardium and LV on heterogeneous clinical data. [4] designed a novel framework to obtain the cardiac structure segmentation from spatial semantic information and optical flow features. [5] proposed a multi-task semi-supervised framework for precise EF estimation of echocardiographic sequences from two cardiac views.

Many methods have achieved optimal performances, thus demonstrating the great potential of deep learning in LV myocardium segmentation. However, the robustness of a deep learning segmentation network on datasets acquired from different machines is particularly poor due to the various imaging parameters and image qualities of the latter. For example, the Dice of myocardium decreased by approximately 10% when a model was applied to an external dataset from a different machine [4]. Additionally, as shown in Table 2 (Row 3 and 7), a network trained on a dataset acquired by GE achieved a Dice of 0.875 on GE's dataset during testing, yet achieved only 0.721 on a dataset from Mindray's machine (Fig. 1 shows the domain difference between the two datasets). In this paper, we aim to optimize this situation.

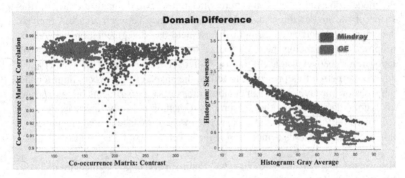

Fig. 1. Domain difference between datasets from GE and Mindray.

To improve the robustness and machine adaptability, a domain-free semi-supervised method is proposed. First, an optical flow field-based segmentation network is proposed for enhancing the segmentation performance by combining the motion tendency of myocardium between adjacent frames. Second, a shape-based semi-supervised adversarial network is presented to utilize the outline of the myocardium to improve semantic segmentation.

2 Methodology

As illustrated in Fig. 2, we proposed a domain-free semi-supervised architecture to effectively segment LV myocardium in 2D echocardiography sequences. These innovations consist of two parts: 1) Optical flow-based segmentation network, which fuses the semantic feature and the optical flow feature to improve the segmentation performance. 2) Shape-based semi-supervised adversarial module, which utilizes the shape of the myocardium to improve semantic segmentation. Details are described as follows.

2.1 Optical Flow-Based Segmentation Network (OFSNet)

Although echocardiography images acquired by different machines are various, the motion tendency between the adjacent frames is consistent in the imaging of a heart. Based on this fact, we propose OFSNet which incorporates optical flow field into the U2net [6] to fuse the machine-free feature to the semantic feature, thus improving the general segmentation performance and robustness of machines.

In recent years, many methods have demonstrated that fusing optical flow field can improve the segmentation performance [7–9]. The normal networks which combine optical flow and semantic segmentation are divided into three categories: 1) Networks with two branches, which are redundant and come at a cost to memory [7]. 2) Images which have been stacked as input, thus preventing the encoder stage to build the sequence relevance [8]. 3) Combination of optical flow and semantic features solely at the final output layer, disregarding the multiple scales [4]. Considering the above weaknesses, we propose OFSNet to combine the optical flow and semantic segmentation more effectively.

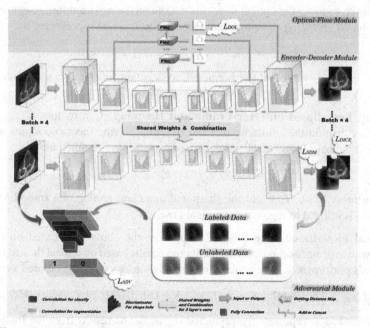

Fig. 2. Illustration of our method. L_{ADV}: Loss of adversarial network. L_{SDM}: Loss of signed distance map. L_{OF}: Loss of optical flow field. L_{DICE}: Loss of the segmentation results.

The overall architecture is shown in Fig. 2. It takes N $(N > 2)$ consecutive images, which contain labeled and unlabeled images, as input. The input images go through the shared encoder and decoder to extract the semantic features. Between the encoder and decoder, the PWC block [9] is adopted to estimate the optical field in multiple scales, which are concatenated with features in the encoder and then are fed into decoder stages, thus fusing the motion tendency and semantic features. Considering the consistency of the global photometric and local structure, the smoothness of optical flow, and the segmentation consistency, the optical flow estimation is trained in an unsupervised manner based on the following loss:

A. Photometric Consistency Loss Based on Cross-Frame and Heatmap. Photometric consistency loss [10] takes a current frame (I_c), a next frame (I_{c+1}), a forward optical flow ($F_{c \to c+1}$) and a reverse optical flow ($F_{c+1 \to c}$) as input, and constrains the photometric consistency between the two frames, where:

$$L_{pho} = F_{pho}(I_c, I_{c+1}, F_{c \to c+1}, F_{c+1 \to c}) \tag{1}$$

We argue that the frame rate in echocardiography imaging is various due to the differences of heartbeat frequency. Therefore, we propose a cross frame-based photometric consistency loss (L'_{pho}), namely:

$$L'_{pho} = F_{pho}(I_c, I_{c+t}, F_{c \to c+t}, F_{c+t \to c}) \tag{2}$$

where t is a single value and refers to the frame intervals.

Moreover, to make the optical flow concentrate on the myocardium, a segmentation heatmap-based loss (L''_{pho}) is proposed according to L'_{pho}, as represented by:

$$L''_{pho} = F_{pho}(I'_c, I'_{c+t}, F_{c \to c+t}, F_{c+t \to c}) \tag{3}$$

where I'_c and I'_{c+t} represent the predicted myocardium in frame c and frame $c + t$, respectively.

B. Census Loss Based on Over-Frame and Heatmap. Census loss (L_{cen}) [11] is utilized to constrain the similarity of local structure, having the same inputs as L_{pho}. Like L''_{pho}, the over-frame and heatmap-based census loss is defined as:

$$L''_{cen} = F_{cen}(I'_c, I'_{c+t}, F_{c \to c+t}, F_{c+t \to c}) \tag{4}$$

C. Smoothness Loss. To avoid the sharp and unnatural motion, the smoothness loss (L_{smo}) [11] is utilized to regulate the optical flow.

D. Optical Flow-Based Dice Loss. Except from the above loss based on original images, we constrain the predicted segmentation results based on optical flow to improve the model's performance. The optical flow-based dice loss (L_{ofdc}) is defined as:

$$L_{ofdc} = Dice(w(M_c, F_{c \to c+t}), M_{c+t}) + Dice(w(M_{c+t}, F_{c+t \to c}), M_c) \tag{5}$$

where M_c and M_{c+t} represent the predicted segmentation results from the frame c and the frame $c + t$, respectively. $F_{c \to c+t}$ and $F_{c+t \to c}$ refer to the forward and reverse optical flows between M_c and M_{c+t}, separately. $w(A, B)$ [12] signifies warping A based on B.

The total loss of this part is summarized as:

$$L_{OF} = \gamma_{pho} \bullet L''_{pho} + \gamma_{cen} \bullet L''_{cen} + \gamma_{smo} \bullet L_{smo} + \gamma_{ofdc} \bullet L_{ofdc} \tag{6}$$

where $\gamma_{pho}, \gamma_{cen}$, and γ_{ofdc} are balance parameters and are set to 1.0, respectively. γ_{smo} is a function of epoch, defined as:

$$\gamma_{smo} = 1.0 + 0.01 * epoch \tag{7}$$

2.2 Shape-Based Semi-supervised Adversarial Module (SSAM)

Regarding the echocardiography images, the parameters in various machines result in significant image differences [13]. However, for people, the shape of the heart ventricle and shape of the myocardium are similar, the shape being a domain free factor. As such, a shape-based semi-supervised adversarial module is proposed.

In this module, the OFSNet is regarded as the generator (G), and the network in [14] which consists of 5 convolution layers followed by an MLP is adopted as the discriminator (D). The signed distance map [15] (SDM) is utilized to represent the shape of the myocardium. As shown in Fig. 2, the output of OFSNet consists of two parts. One is the predicted segmentation result of the myocardium, the other is the predicted SDM of it.

For the generative model, the loss is defined as:

$$L_{gen} = L_{DICE} + L_{SDM} + \lambda_d \bullet L_d \tag{8}$$

where L_{DICE} is the Dice loss [16] of the predicted segmentation results. L_{SDM} is the mean square error loss of the predicted SDM. L_d is defined as:

$$L_d = -log(D(x_u, G(x_u))) \tag{9}$$

where x_u represents unlabeled images. $G(x_u)$ is the predicted SDM by OFSNet. The discriminator (D) takes the original unlabeled image and predicted SDM as input. The purpose of optimizing L_d is to make $G(x_u)$ closer to SDM generated according to manual annotation. λ_d is a dynamic function to adjust the weight of L_d. λ_d is defined in [14].

For the adversarial part, the main purpose of the loss is to train the adversarial network in recognizing which SDMs are generated according to manual annotations and which are generated by OFSNet. The adversarial loss is defined as:

$$L_{ADV} = -\left[\log(D(x_l, G(x_l))) + \log(1 - D(x_u, G(x_u)))\right] \tag{10}$$

where x_l and x_u represent labeled and unlabeled images, separately.

3 Experiments

Dataset. The proposed method was assessed on both a public dataset (CAMUS [17]) and on our private dataset. The CAMUS dataset was provided by the University Hospital of St Etienne (France), containing 450 patient cases acquired by GE Vivid E95. Each case contained the apical two-chamber (A2C) and the four-chamber (A4C) sequences from ED to ES. Three highly experienced cardiologists annotated the myocardium, LV endocardium and epicardium at ED and ES for the dataset following the same annotation scheme. The private dataset was used solely for testing the robustness in a different machine. It contained 50 cases acquired by Probe SP5-1U, Mindray Resona 7. Each case also contained A2C and A4C sequences from ED to ES. Two highly experienced cardiologists annotated the ED and ES for the private dataset.

Implementation Details. All images were converted to square by cropping equal pixels of the longer edges, and then resized to 208 × 208, with pixel values being normalized to between 0 and 1. For CAMUS dataset, 360 cases (720 sequences of 1440 labeled and 12524 unlabeled images), 45 cases (90 sequences of 180 labeled and 1474 unlabeled images), and 45 cases (90 sequences of 180 labeled and 1466 unlabeled images) were randomly divided into training, validation, and test datasets. All 50 cases (100 sequences of 200 labeled images) from Mindray were adopted as the external test set to assess the robustness of the model. During the training process, the U2net was adopted as the backbone network. Random translating, scaling, and brightness and contract transformation were utilized as data augmentation.

In further detail, the optimizer was set as Adam with a learning rate of 0.0001 for the unsupervised optical flow training, while the optimizer was set as SGD with a learning rate of 0.001 for the adversarial module. The entire network was trained for 100 epochs with a batch size of 4. The model that performed with the highest Dice on validation set was chosen as the final model for testing.

Result Analysis. We measured Dice [19], HD [20], and JAC [21] of the predicted myocardium segmentation results to evaluate the proposed method. As shown in Table 1, our method outperforms others in most metrics on CAMUS dataset.

Table 1. Result comparison on CAMUS dataset. (LV_{Endo}: Endocardium; LV_{Epi}: Epicardium)

Method	ED				ES			
	LV_{Endo}		LV_{Epi}		LV_{Endo}		LV_{Epi}	
	Dice	HD	Dice	HD	Dice	HD	Dice	HD
Unet [22]	0.934 ± 0.042	5.5 ± 2.9	0.951 ± 0.024	5.9 ± 3.4	0.905 ± 0.063	5.7 ± 3.7	0.943 ± 0.035	6.1 ± 4.1
Unet + + [23]	0.927 ± 0.046	6.5 ± 3.9	0.945 ± 0.026	7.2 ± 4.5	0.904 ± 0.060	6.3 ± 4.2	0.939 ± 0.034	7.1 ± 5.1
BASnet [24]	0.939 ± 0.021	3.6 ± 2.4	0.949 ± 0.019	3.2 ± 2.1	0.919 ± 0.036	3.8 ± 2.7	0.944 ± 0.032	3.2 ± 2.2
U2net [6]	0.943 ± 0.018	3.3 ± 1.7	0.951 ± 0.016	3.3 ± 2.1	0.919 ± 0.033	3.3 ± 1.9	0.946 ± 0.027	3.2 ± 1.8
SOCOF [4]	0.946 ± 0.027	3.3 ± 0.8	**0.958** ± 0.020	4.1 ± 1.2	0.924 ± 0.051	**2.9** ± 0.7	0.951 ± 0.029	3.9 ± 1.0
Ours	**0.947** ± 0.016	**2.6** ± 0.7	0.957 ± 0.012	**2.7** ± 0.9	**0.932** ± 0.026	3.0 ± 1.1	**0.955** ± 0.018	**2.5** ± 0.8

The ablation experiment results on both the CAMUS dataset and the inhouse dataset are shown in Table 2. The U2net was utilized as the baseline network. By either fusing the optical flow (OF) or by introducing the shape-based semi-supervised adversarial module (SSAM) built on U2net, the predictions improved approximately 1% over all metrics in the CAMUS dataset. The improvement was notably higher in the private dataset, demonstrating that the proposed method can effectively improve the segmentation performance and robustness.

Table 2. Ablation results on public and private datasets. (LV_{myo}: Myocardium)

Dataset	LV_{myo}			
	Method	Dice	HD	JAC
CAMUS (GE)	U2net	0.875	4.53	0.736
	U2net + OF	0.882	3.82	0.761
	U2net + SSAM	0.887	3.39	0.751
	U2net + OF + SSAM	**0.893**	**3.11**	**0.786**
Private (Mindray)	U2net	0.721	8.36	0.568
	U2net + OF	0.767	6.23	0.624
	U2net + SSAM	0.771	6.68	0.631
	U2net + OF + SSAM	**0.786**	**5.88**	**0.638**

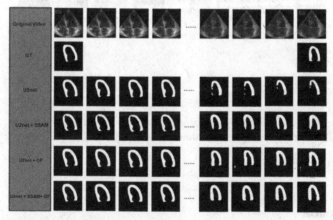

Fig. 3. Result comparison of different methods on the CAMUS dataset. Row 1 is a sequence from ED to ES. Only the ED and ES frames are annotated, as shown in Row 2.

Figure 3 compares the segmentation results of different methods on the CAMUS dataset. As shown in Row 3, the original U2net did not handle the high noise areas sufficiently, such as the intervals of the heart. The adversarial module successfully improved the situation (Row 4), by utilizing the unlabeled datasets and the shapes of myocardium to constrain the model. The optical flow-based U2net not only improved the predictions, but also enhanced the segmentation relevancies between adjacent frames (Row 5). However, the optical flow field was easily affected by noise, resulting in unneeded tissues in the segmentation results. Finally, by combining the shape-based adversarial network

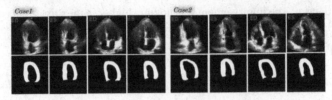

Fig. 4. Results on our private dataset. Row 1 is the original images. Row 2 is the predicted results by the proposed method.

and the optical flow-based U2net, the impact of noise was eliminated and the overall segmentation performance was significantly improved.

Figure 4 shows the predicted segmentation results on our private dataset. Although the private dataset (Mindray) is significantly different from the CAMUS dataset (GE) in grey distribution, the proposed method can obtain an ideal result on it.

Figure 5 illustrates the color map and vector field of the predicted optical flow. The direction of the vector field is consistent with the movement trend of the myocardium, indicating the potential of the optical flow in improving the segmentation performance.

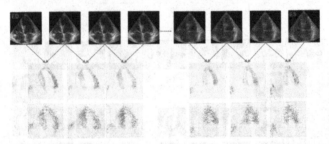

Fig. 5. Row 1 is an echocardiography sequence from ED to ES. Row 2 and Row 3 represent the color maps and vector fields of the optical flows between adjacent frames, respectively.

4   Conclusion

In conclusion, a domain-free semi-supervised deep learning algorithm was proposed to improve the segmentation performance and robustness for the myocardium segmentation task in 2D echocardiography sequences. Both the shape of the myocardium and the motion tendency between adjacent frames were utilized as the domain-free factors. Firstly, we proposed an optical flow field-based segmentation network to enhance performance by combining the motion tendency of myocardium between adjacent frames. Secondly, a shape-based semi-supervised adversarial network, which utilized the shapes of the left ventricle and myocardium for improving the segmentation robustness, was presented. The experiments on two different datasets demonstrated the effectiveness of the proposed method and its potential for clinical application.

Acknowledgements. This work was supported in part by Major Technical Research Projects in Shenzhen (Project No.: JSGGZD20220822095200002, Project Name: Development and Application of Key Technologies for 2022D009 Area Array Ultrasonic Transducers).

References

1. Virani, S.S., et al.: American Heart Association Council on Epidemiology and Prevention Statistics Committee and Stroke Statistics Subcommittee. Heart Disease and Stroke Statistics-2020 Update: A Report from the American Heart Association. Circulation. (2020)
2. Marwick, T.H.: The future of echocardiography. Eur. J. Echocardiogr. **10**(5), (2009)
3. Leclerc, S., Grenier, T., Espinosa, F., Bernard, O.: A fully automatic and multi-structural segmentation of the left ventricle and the myocardium on highly heterogeneous 2D echocardiographic data. In: 2017 IEEE International Ultrasonics Symposium (IUS), Washington (2017)
4. Xue, W., Cao, H., Ma, J., Bai, T., Wang, T., Ni, D.: Improved segmentation of echocardiography with orientation-congruency of optical flow and motion-enhanced segmentation. IEEE J. Biomed. Health Inf. **26**(12), 6105–6115 (2022)
5. Wei, H., Ma, J., Zhou, Y., Xue, W., Ni, D.: Co-learning of appearance and shape for precise ejection fraction estimation from echocardiographic sequences. Med. Image Anal. (2023)
6. Qin, X., Zhang, Z., Huang, C., Dehghan, M., Zaiane, O.R., Jagersand, M.: U2-net: going deeper with nested U-structure for salient object detection. Pattern Recognit. **106**, 107404 (2022)
7. Ta, K., Ahn, S.S., Stendahl, J.C., Sinusas, A.J., Duncan, J.S.: A semi-supervised joint network for simultaneous left ventricular motion tracking and segmentation in 4D echocardiography. In: Martel, A.L., et al. (eds.) MICCAI 2020. LNCS, vol. 12266, pp. 468–477. Springer, Cham. (2020)
8. Ding, M., Wang, Z., Zhou, B., Shi, J., Lu, Z., Luo, P.: Every frame counts: joint learning of video segmentation and optical flow. In: The Thirty-Fourth AAAI Conference on Artificial Intelligence, AAAI 2020, The Thirty-Second Innovative Applications of Artificial Intelligence Conference, IAAI 2020, The Tenth AAAI Symposium on Educational Advances in Artificial Intelligence, EAAI 2020, New York, 7–12 February 2020, pp. 10713–10720 (2020)
9. Sun, D., Yang, X., Liu, M. -Y., Kautz, J.: PWC-net: CNNs for optical flow using pyramid, warping, and cost volume. In: 2018 IEEE/CVF Conference on Computer Vision and Pattern Recognition, Salt Lake City (2018)
10. Yu, J.J., Harley, A.W., Derpanis, K.G.: Back to basics: Unsupervised learning of optical flow via brightness constancy and motion smoothness. In: Hua, G., Jégou, H. (eds.) ECCV 2016. LNCS, vol. 9915, pp. 3–10. Springer, Cham (2016). https://doi.org/10.1007/978-3-319-494 09-8_1
11. Hafner, D., Demetz, O., Weickert, J.: Why is the census transform good for robust optic flow computation? In: International Conference on Scale Space and Variational Methods in Computer Vision, pp. 210–221 (2013)
12. Jaderberg, M., et al.: Spatial transformer networks. In: Advances in Neural Information Processing Systems, pp. 2017–2025 (2015)
13. Fabiszewska, E., Pasicz, K., Grabska, I., Skrzyński, W., Ślusarczyk-Kacprzyk, W., Bulski, W.: Evaluation of imaging parameters of ultrasound scanners: baseline for future testing. Pol. J. Radiol. **82**, 773–782 (2017)
14. Li, S., Zhang, C., He, X.: Shape-aware semi-supervised 3D semantic segmentation for medical images. In: Martel, A.L., et al. (eds.) MICCAI 2020. LNCS, vol. 12261, pp. 552–561. Springer, Cham (2020). https://doi.org/10.1007/978-3-030-59710-8_54

15. Park, J.J., Florence, P., Straub, J., Newcombe, R., Lovegrove, S.: Deepsdf: learning continuous signed distance functions for shape representation. In: CVPR, pp. 165–174 (2019)
16. Milletari, F., Navab, N., Ahmadi, S.-A.: V-net: fully convolutional neural networks for volumetric medical image segmentation. In: 2016 Fourth International Conference on 3D Vision (3DV), Stanford (2016)
17. Leclerc, S., et al.: Deep learning for segmentation using an open large-scale dataset in 2d echocardiography. IEEE Trans. Med. Imaging **38**(9), 2198–2210 (2019)
18. Ronneberger, O., Fischer, P., Brox, T.: U-net: convolutional networks for biomedical image segmentation. In: Navab, N., Hornegger, J., Wells, W.M., Frangi, A.F. (eds.) MICCAI 2015. LNCS, vol. 9351, pp. 234–241. Springer, Cham (2015). https://doi.org/10.1007/978-3-319-24574-4_28
19. Dice, L.R.: Measures of the amount of ecologic association between species. Ecology **26**(3), 297–302 (1945)
20. Aspert, N., Santa-Cruz, D., Ebrahimi, T.: Mesh: measuring errors between surfaces using the hausdorff distance. In: Proceedings of the IEEE International Conference on Multimedia and Expo, vol. 1, pp. 705–708, IEEE (2002)
21. Dharavath, R., Singh, A.K.: Entity resolution-based jaccard similarity coefficient for heterogeneous distributed databases. In: Satapathy, S.C., Raju, K.S., Mandal, J.K., Bhateja, V. (eds.) Proceedings of the Second International Conference on Computer and Communication Technologies. AISC, vol. 379, pp. 497–507. Springer, New Delhi (2016). https://doi.org/10.1007/978-81-322-2517-1_48
22. Ronneberger, O., Fischer, P., Brox, T.: U-net: convolutional networks for biomedical image segmentation. In: International Conference on Medical Image Computing and Computer-Assisted Intervention, pp. 234–241. Springer, Cham (2015)
23. Zhou, Z., Rahman Siddiquee, M.M., Tajbakhsh, N., Liang, J.: Unet++: a nested u-net architecture for medical image segmentation. In: Stoyanov, D., et al. (eds.) DLMIA/ML-CDS -2018. LNCS, vol. 11045, pp. 3–11. Springer, Cham (2018). https://doi.org/10.1007/978-3-030-00889-5_1
24. Qin, X., Zhang, Z., Huang, C., Gao, C., Dehghan, M., Jagersand, M.: BASNet: boundary-aware salient object detection. In: 2019 IEEE/CVF Conference on Computer Vision and Pattern Recognition (CVPR), Long Beach (2019)

Self-training with Domain-Mixed Data for Few-Shot Domain Adaptation in Medical Image Segmentation Tasks

Yongze Wang, Maurice Pagnucco, and Yang Song[✉]

School of Computer Science and Engineering, University of New South Wales, Sydney, Australia
yang.song1@unsw.edu.au

Abstract. Deep learning has shown significant progress in medical image analysis tasks such as semantic segmentation. However, deep learning models typically require large amounts of annotated data to achieve high accuracy; often a limiting factor in medical applications where labeled data is scarce. Few-shot domain adaptation (FSDA) is one approach to address this problem. It adapts a model trained on a source domain to a target domain which includes a few labeled data. In this paper, we present an FSDA method adapting pre-trained models to the target domain via domain-mixed data in the self-training framework. Our network follows the traditional encoder-decoder structure, which consists of a Transformer encoder, a DeeplabV3+ decoder for segmentation tasks and an auxiliary decoder for boundary-supervised learning. Our approach fine-tunes the source-domain pre-trained model with a few labeled examples from the target domain and by including unlabeled target domain data as well. We evaluate our method on two commonly used publicly available datasets for optic disc/cup and polyp segmentation, and show that it outperforms other state-of-the-art FSDA methods with only 5 labeled examples in the target domain. Overall, our FSDA method shows promising results and has potential to be applied to other medical imaging tasks with limited labeled data in the target domain.

Keywords: Few-shot domain adaptation · Self-training · Pseudo-labeling · Optic disc/cup segmentation · Polyp segmentation

1 Introduction

Deep learning models have achieved remarkable success in medical image segmentation tasks, such as identifying regions of interest in fundus and polyp images. However, one of the main challenges in applying these models to real-world scenarios is the lack of labeled data for new domains. Traditional methods for addressing this issue involve collecting more data, which can be time-consuming and costly. To overcome this restriction, most domain adaptation

Supplementary Information The online version contains supplementary material available at https://doi.org/10.1007/978-3-031-45673-2_30.

(DA) methods are in source-free [2,16,19,34], test-time [15,28] or unsupervised domain adaptation (UDA) settings, which do not rely on any labeled target data. Common adaptive modules are self-training and adversarial training. Self-training-related methods focus on optimizing the quality of pseudo-label [2,4,34], while adversarial training-related methods are developed based on the assumption that the cross-domain features can be extracted when the discriminators are confused [9,18,29,30]. For instance, BEAL [30] tries to confuse the discriminator by using entropy maps and boundary prob, while ADVENT [29] uses self-information maps.

While UDA has been most commonly studied due to its advantage of not requiring any labeled target domain data, we suggest that having a few annotated target samples could be widely affordable. Few-shot domain adaptation (FSDA) can thus be a suitable approach that aims to adapt a model to new domains with only a few labeled samples in the target domain. Moreover, in recent years, U-net [24], a convolutional neural network, has dominated DA tasks of medical segmentation [4,15,28,32,36] due to its performance and simplicity. Meanwhile, transformer-based models have shown remarkable performance in natural language processing and computer vision tasks [7,20], raising the question of whether they can improve FSDA in medical image segmentation.

Our motivation is to leverage prior knowledge from related source domains and maximize the benefits derived from a limited number of target examples with segmentation labels and the remaining unlabeled target samples to improve the model's generalization ability for medical image segmentation. In this work, we first pre-train a transformer-based model on source domain images to learn the underlying distribution of the data, then fine-tune it for the target domain with *domain mixed* images in a self-training framework. The domain-mixed images are generated based on ClassMix [21], which produces augmented images by mixing target and source domain images. The overall framework contains a lightweight network customized from DAFomer [14], where the segmentation component consists of a SegFormer [33] MIT-B1 backbone and a modified DeeplabV3+ decoder [3]. It also contains the Mean-Teacher architecture to realize the domain adaptation. By fine-tuning the pre-trained model with the domain mixed data, the model will gradually adapt to a new target domain in the self-training process, consequently narrowing the domain gap between the pre-trained model and the target domain. In addition, during our experiments, we noticed that the predicted boundaries were rough in certain cases, and the missing areas corresponding to the input images were concealed by blood vessels (in fundus images) or reflection (in polyp images). We thus incorporated an auxiliary branch for boundary-supervised learning, allowing our model to concentrate on the object's shape during training.

Our method has been evaluated on two segmentation tasks: 1) REFUGE challenge [22] as the source domain dataset and RIM-ONE-r3 [10] as the target domain dataset for disc/cup segmentation in retinal fundus images; and, 2) CVC-612 [1] and Kvasir [23] as the source domain dataset and CVC-300 [8] as the target domain dataset for polyp segmentation in colonoscopy images. The results

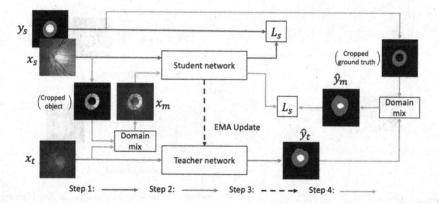

Fig. 1. The pipeline of our framework. Step 1: train the student network with L_s on source domain dataset $\mathcal{D}_s = \{\mathcal{X}_s, \mathcal{Y}_s\}$, where $\{x_s, y_s\} \in \{\mathcal{X}_s, \mathcal{Y}_s\}$. Step 2: input target domain sample $x_t \in \mathcal{D}_t$ to the teacher network and get pseudo-label $\hat{y}_t$. Step 3: the teacher network weights are updated as an exponential moving average (EMA) of the weights in the student network. Step 4: randomly select and crop one object (disc/cup in fundus) from x_s, and paste the object on top of x_t to form x_m. Similarly, crop the corresponding ground truth of the selected object from y_s, and paste it on top of pseudo-label $\hat{y}_t$ to form $\hat{y}_m$. The synthesized data $\{x_m, \hat{y}_m\}$ will be used to train the student network as augmented data.

demonstrate that our approach achieves state-of-the-art performance on both datasets, outperforming existing FSDA methods. Our experiments also show that our approach is robust to variations in the number of labeled examples, indicating its potential for practical applications.

2   Methodology

Figure 1 illustrates the overall Mean-Teacher based framework, in which the student network has a Transformer-based segmentation model. Given the target domain including mostly unlabeled images and a few images with segmentation labels, we aim to fine-tune the student network with the help of the teacher network and domain mix module. The teacher network generates pseudo-labels, whereas the domain mix module generates domain mixed images and their corresponding semi-pseudo-labels, as described in Sect. 2.3. The shape-guided module is incorporated into the student network, while the teacher network is a lightweight student network without the shape-guided module.

2.1   Mean-Teacher Architecture

In our framework, we apply the Mean-Teacher architecture [25] in the self-training model. The student network is pre-trained on the source domain dataset. The teacher network generates pseudo-labels for unlabeled target domain samples. Compared to the traditional self-training framework, we demonstrate that

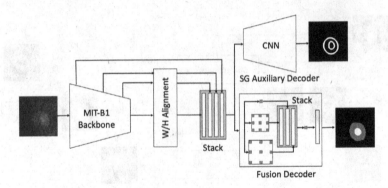

Fig. 2. The student network overview, consisting of three main parts: a SegFormer MIT-B1 backbone, fusion decoder modified from DeeplabV3+ decoder and an SG auxiliary decoder. This figure shows the feed-forward pass.

the teacher network can produce more reliable predictions mitigating the effect of negative transfer caused by the low-confident pseudo-labels during the adaptation (in Table 2). The few-shot annotated target domain samples will be augmented by the domain mix module in the last step to fine-tune the student network and boost its performance.

In the teacher network, the network generates pseudo-labels for unlabeled target domain images. The pseudo-labels will be denoised via a pixel-level threshold that linearly increases along with the training iterations. The prediction probability of each pixel will be set to 0 if it is less than the threshold, otherwise will be set to 1. The network structure contains two parts: MIT-B1 encoder and DeeplabV3+ decoder, where the parameters are all updated from the student network via EMA after each training iteration.

2.2 The Student Network Design

As shown in Fig. 2, the overall structure of the student network is the encoder-decoder model, which consists of three components: a Transformer-based backbone (encoder), a fusion decoder and a shape-guided (SG) auxiliary decoder. The encoder structure is a lightweight version of [14], which utilizes a SegFormer MIT-B1 backbone [33]. The fusion (main) decoder is a modified DeeplabV3++ decoder containing multi-size dilated convolutions.

Moreover, we added an SG auxiliary decoder for boundary-supervised learning as we observed the boundaries of the masks were rugged in some cases. The problematic regions corresponded to vessels or reflections in the fundus or polyp images, which caused difficulties in segmentation. We suggest this could be due to texture bias, where the model prefers texture over shape during the segmentation process [12]. This discovery motivated us to guide the model to rely on both shape and texture concurrently. The SG auxiliary decoder only contains one layer of the convolutional network with kernel size 1. We keep the convolutional layer as simple as possible to avoid the overfitting problem in this branch, as the auxiliary decoder has a simpler task compared to the fusion decoder.

To train the student network, we combine the dice loss (L_{dice}) and the cross entropy (L_{ce}) as the total loss function for both the fusion decoder and SG auxiliary decoder.

2.3   Domain Mix Module

We design a domain-mix module based on Domain Adaptation via Cross-domain mixed Sampling (DACS) [26] as the core method to deal with the domain shift problem. DACS applies ClassMix [21] to cut out objects from source domain images and paste them onto target domain images along with the corresponding ground truth and pseudo-labels. In our work, we customize DACS for the FSDA setting and extend the rule of the ClassMix [21] algorithm. Compared to the standard DACS, we apply ClassMix on three groups of data: the few-shot labeled target domain images, the source domain dataset and the unlabeled target domain dataset. Any two images from these three groups can be applied to build the augmented image via ClassMix. Their corresponding ground truth will be augmented similarly when the ground truths are both available. We use pseudo-labels as the replacement only when the image is unlabeled.

On the one hand, the augmented data contains features from both domains, adapting the network to a new domain between the source and target domain, narrowing the domain gap between the model and the target domain. On the other hand, it can also increase the diversity of the training data, resulting in a more robust model with higher generalization.

2.4   Training Strategies for FSDA

Our overall training strategy is to prepare the pre-trained model first, then fine-tune it with the few-shot labeled target domain images and their domain-mixed augmented data.

Prepare Pre-trained Models: Our pre-trained model is trained on the source domain dataset without adaptation (*Step 1* for both Setting 1 and 2 in Table 3). The checkpoint with the highest Dice score is selected as the pre-trained model. We use the student network in Fig. 2 for our pre-trained model.

Settings for Domain Adaptation: Our approach follows the baseline [18], which consists of two settings: (1) during the whole training process, the model can access the source domain dataset and a few-shot annotated target domain dataset; (2) the model can access unlabeled target domain dataset as the additional dataset apart from (1).

Fine-Tune Pre-trained Model: To meet the above two settings, we have two types of domain adaptation. For Setting 1, we fine-tune the pre-trained model with few-shot annotated target domain images and the corresponding augmented data generated by the domain mix module (*Step 2* for Settings 1 in Table 3). Compared to Setting 1, we have an extra step in Setting 2. We first fine-tune the pre-trained model with domain-mix data built from the source and unlabeled target domain data (*Step 2* for Setting 2 in Table 3). Then, similar to Setting 1, we fine-tune the model with the few-shot annotated target domain images and the domain-mix data (*Step 3* for Setting 2 in Table 3).

3   Experiments

3.1   Dataset and Evaluation Metrics

Following [18], We conduct experiments on two medical segmentation tasks: optic disc/cup and polyp segmentation tasks on publicly available fundus and colonoscopy image datasets respectively. For optic disc/cup segmentation, we use the REFUGE challenge dataset (1200 images) [22] as the source domain dataset and RIM-ONE-r3 (159 images) [10] as the target domain dataset. For polyp segmentation, we use CVC-612 (612 images) [1] and Kvasir (1000 images) [23] as the source domain dataset, and CVC-300 (60 images) [8] as the target domain dataset. We crop a 576×576 pixels optic disc region as input and resize each image in the polyp dataset to 320×320 pixels as input.

In the FSDA settings, we usually have 5 to 20 samples [18] in the few-shot annotated target dataset. In this work, we mainly focus on the 5-shot scenario. Other scenarios will be discussed in the supplementary file. We follow [18] and randomly select 5 samples from the target domain datasets. Testing results are evaluated on the target domain dataset, excluding these 5 samples. We repeat the experiments 3 times and each time, we have different random samples from the target domain datasets. We use the commonly used Dice coefficient to measure the performance and report the mean and standard deviation results.

3.2   Experimental Results

Strategies of Domain Mix: To explore the best way of exploiting the information in the few-shot annotated target dataset, we tested some potential combinations of datasets for the data augmentation via domain mix, and the Dice scores are listed in Table 1. The augmented images can be built by the images from three datasets: 1) few-shot annotated target domain dataset (FSTDD); 2) source domain dataset (SDD); and 3) unlabeled target domain dataset (UTDD). For example, in the first case with 1⇒2, we crop an object from an image in FSTDD and paste it onto every image in SDD. We use these domain-mix images to fine-tune both Setting 1 (SDD → FSTDD) and Setting 2 (SDD → FSTDD+UTDD), which are the last step for both settings (the *Step 2* and *Step 3* respectively in Table 3). Also, for Setting 2, there is an additional fine-tuning step using domain mixed data between SDD and UTDD, which is consistently applied to all combinations under Setting 2 in Table 1. The results show that with different combinations for domain mix, the results can be quite different and the results are generally better when the domain mix is only applied on datasets 1 and 2 but not 3. This is because when we build domain mix images based on dataset 3, we use the pseudo-labels of objects to mix with the ground truth of images in datasets 1 and 2. The pseudo-labels normally contain noise which may negatively affect the network's performance.

Table 1. Ablation study for the strategy of domain mix.

From ⇒ To		Setting 1 (SDD → FSTDD)				Setting 2 (SDD → FSTDD+UTDD)			
		RIM-ONE			CVC-300	RIM-ONE			CVC-300
		Disc	Cup	Avg		Disc	Cup	Avg	
1	2	93.43	83.53	$88.48_{\pm0.39}$	$87.93_{\pm0.92}$	95.33	84.85	$90.09_{\pm0.24}$	$88.45_{\pm0.98}$
2	1	89.63	83.76	$86.7_{\pm0.23}$	$86.58_{\pm0.56}$	94.91	85.71	$\mathbf{90.31}_{\pm0.19}$	$87.95_{\pm0.69}$
1	(1,2)	93.57	83.3	$88.43_{\pm0.36}$	$\mathbf{88.91}_{\pm1.0}$	95.28	84.72	$90.00_{\pm0.24}$	$88.61_{\pm0.78}$
1	1	94.44	83.3	$\mathbf{88.87}_{\pm0.46}$	$88.36_{\pm0.22}$	95.78	84.72	$90.25_{\pm0.17}$	$\mathbf{88.98}_{\pm0.33}$
(1,2)	1	89.86	82.84	$86.35_{\pm0.25}$	$85.87_{\pm0.68}$	95.02	85.22	$90.12_{\pm0.17}$	$87.47_{\pm0.67}$
1	3					95.41	85.13	$90.27_{\pm0.19}$	$87.44_{\pm0.77}$
1	(1,3)					95.59	84.93	$90.26_{\pm0.15}$	$87.74_{\pm0.89}$
(1,2)	3					94.35	84.12	$89.24_{\pm0.07}$	$86.03_{\pm0.31}$
(1,2)	(1,3)					94.38	84.75	$89.56_{\pm0.10}$	$86.74_{\pm0.32}$
Fine-tune on 1		93.39	83.71	$88.55_{\pm0.59}$	$86.26_{\pm0.58}$	95.84	84.56	$90.20_{\pm0.10}$	$87.99_{\pm0.57}$
w/o domain mix						81.97	70.84	$76.41_{\pm0.91}$	$84.49_{\pm0.01}$

Network Architecture: Compared to the standard self-training framework applied in [2,16,19,34], which uses the training network to produce pseudo-labels, our model uses the teacher network instead. To verify the assumption that the teacher network can produce high-quality pseudo-labels, we conducted ablation experiments, fine-tuning our pre-trained model by the pseudo-labels produced by different networks, and then comparing the performances of fine-tuned models. Figure 1 shows the pseudo-labels generated by the teacher network within the Mean-Teacher model. For the self-training framework without the Mean-Teacher model, we can use the training network, the student network, to generate pseudo-labels. Furthermore, we also investigated the potential contribution of the shape-guided module to the overall performance. We fine-tune the pre-trained model with domain mix images built from FSTDD and UTDD with different modules in the framework (*Step 2* for Setting 2 in Table 3) and evaluate the models on UTDD. The Dice scores in Table 2 show that pseudo-labels generated within the Mean-Teacher model can endow the model with better performance. Meanwhile, the shape-guided (SG) module can cooperate with the Mean-Teacher module and further boost performance.

Table 2. Ablation study for the network architecture.

	RIM-ONE			CVC-300		RIM-ONE			CVC-300
	Disc	Cup	Avg			Disc	Cup	Avg	
Without Mean-teacher Model					**With** Mean-teacher Model				
w/o SG	92.82	80.12	86.47	81.4	w/o SG	94.06	80.21	87.14	85.99
w/ SG	93.23	82.35	87.79	83.38	w/ SG	94.29	83.5	88.90	85.46

Domain Adaptation Techniques: Table 3 shows the performance of our models in different steps. Notably, the performance of Setting 1 (SDD → FSTDD) in Step 2 is comparable to that of Setting 2 (SDD → FSTDD+UTDD) in Step 2. This may reflect that without collecting many unlabeled target domain samples, networks may still achieve similar performance within only 5 labeled target domain samples.

Table 3. Dice scores evaluated on UTDD for different settings in different steps.

	RIM-ONE			CVC-300
	Disc	Cup	Avg	
For both **Setting 1** and **Setting 2**				
Step 1: Pre-trained model	86.02	79.49	82.76	82.46
Setting 1 (SDD → FSTDD)				
Step 2: Setting 1 model	94.44	83.30	$88.87_{\pm0.46}$	$88.91_{\pm1.0}$
Setting 2 (SDD → FSTDD+UTDD)				
Step 2: Setting 2 model	94.29	83.5	88.90	85.46
Step 3: Setting 2 model	**94.91**	**85.71**	$\mathbf{90.31_{\pm0.19}}$	$\mathbf{88.98_{\pm0.33}}$
Supervised learning on target domain (50%)				
Oracle	96.76	86.87	91.82	83.93

Comparison with State-of-the-Art Methods: The results of the compared approaches in Table 4 are directly copied from [18]. Compared with their performances, our approach achieves the best scores in both settings. All other methods use U-net in the adversarial training framework. The reasons could be manifold. First, self-training has been proven to be more stable and outperformed adversarial training [26,35]. Second, compared to CNN, the Transformer has a larger receptive field [33] and less inductive bias [5]. Furthermore, the self-training framework is enhanced by incorporating the Mean-Teacher model.

Table 4. Comparison to the Dice scores of the state-of-the-art methods from [18].

		Setting 1		Setting 2						
		PF [18]	Ours	CG [17,37]	RG [11]	AD [27]	DA [6,31]	CSDA [13]	PF [18]	Ours
RIM-ONE	Disc	90.0	94.44	74.7	86.0	87.4	88.5	86.9	91.3	94.91
	Cup	75.3	83.30	69.0	73.2	72.6	72.5	75.6	75.8	85.71
	Avg	82.65	$88.87_{\pm 0.46}$	71.85	79.6	80	80.5	81.25	83.55	$90.31_{\pm 0.19}$
CVC-300		83.0	$88.91_{\pm 1.0}$	70.9	81.3	83.6	83.0	80.5	83.4	$88.98_{\pm 0.33}$

4   Conclusion

In this work, we propose a lightweight self-training segmentation model for few-shot domain adaptation in medical images. Our method demonstrates improved performance over the state-of-the-art methods by more than 6% on both target domain datasets. We have demonstrated that (1) even with just 5 annotated target domain samples, the network can still adapt to the new domain and achieve performance comparable to that under UDA setting; (2) the combination of Transformer backbone and self-training framework can be more powerful than traditional methods like U-net in adversarial training frameworks.

References

1. Bernal, J., et al.: Comparative validation of polyp detection methods in video colonoscopy: results from the MICCAI 2015 endoscopic vision challenge. IEEE Trans. Med. Imaging **36**(6), 1231–1249 (2017)
2. Chen, C., Liu, Q., Jin, Y., Dou, Q., Heng, P.-A.: Source-free domain adaptive fundus image segmentation with denoised pseudo-labeling. In: de Bruijne, M., et al. (eds.) MICCAI 2021. LNCS, vol. 12905, pp. 225–235. Springer, Cham (2021). https://doi.org/10.1007/978-3-030-87240-3_22
3. Chen, L.C., et al.: Encoder-decoder with atrous separable convolution for semantic image segmentation. In: ECCV, pp. 801–818 (2018)
4. Cho, H., Nishimura, K., Watanabe, K., Bise, R.: Cell detection in domain shift problem using pseudo-cell-position heatmap. In: de Bruijne, M., et al. (eds.) MICCAI 2021. LNCS, vol. 12908, pp. 384–394. Springer, Cham (2021). https://doi.org/10.1007/978-3-030-87237-3_37
5. Cordonnier, J.B., Loukas, A., Jaggi, M.: On the relationship between self-attention and convolutional layers. arXiv preprint arXiv:1911.03584 (2019)
6. Dong, N., Kampffmeyer, M., Liang, X., Wang, Z., Dai, W., Xing, E.: Unsupervised domain adaptation for automatic estimation of cardiothoracic ratio. In: Frangi, A.F., Schnabel, J.A., Davatzikos, C., Alberola-López, C., Fichtinger, G. (eds.) MICCAI 2018. LNCS, vol. 11071, pp. 544–552. Springer, Cham (2018). https://doi.org/10.1007/978-3-030-00934-2_61
7. Dosovitskiy, A., et al.: An image is worth 16x16 words: Transformers for image recognition at scale. arXiv preprint arXiv:2010.11929 (2020)
8. Fan, D.-P., et al.: PraNet: parallel reverse attention network for polyp segmentation. In: Martel, A.L., et al. (eds.) MICCAI 2020. LNCS, vol. 12266, pp. 263–273. Springer, Cham (2020). https://doi.org/10.1007/978-3-030-59725-2_26

9. Feng, W., et al.: Unsupervised domain adaptive fundus image segmentation with category-level regularization. In: Wang, L., Dou, Q., Fletcher, P.T., Speidel, S., Li, S. (eds.) Medical Image Computing and Computer Assisted Intervention – MICCAI 2022: 25th International Conference, Singapore, September 18–22, 2022, Proceedings, Part II, pp. 497–506. Springer, Cham (2022). https://doi.org/10.1007/978-3-031-16434-7_48

10. Fumero, F., Alayón, S., Sanchez, J.L.: Rim-one: an open retinal image database for optic nerve evaluation. In: CBMS, pp. 1–6. IEEE (2011)

11. Ganin, Y., Lempitsky, V.: Unsupervised domain adaptation by backpropagation. In: International Conference on Machine Learning, pp. 1180–1189. PMLR (2015)

12. Geirhos, R., et al.: Imagenet-trained CNNs are biased towards texture; increasing shape bias improves accuracy and robustness. arXiv preprint arXiv:1811.12231 (2018)

13. Haq, M.M., Huang, J.: Adversarial domain adaptation for cell segmentation. In: Medical Imaging with Deep Learning, pp. 277–287. PMLR (2020)

14. Hoyer, L., Dai, D., Van Gool, L.: Daformer: improving network architectures and training strategies for domain-adaptive semantic segmentation. In: CVPR, pp. 9924–9935 (2022)

15. Hu, M., et al.: Fully test-time adaptation for image segmentation. In: de Bruijne, M., et al. (eds.) MICCAI 2021. LNCS, vol. 12903, pp. 251–260. Springer, Cham (2021). https://doi.org/10.1007/978-3-030-87199-4_24

16. Lee, J., et al.: Confidence score for source-free unsupervised domain adaptation. In: ICML, pp. 12365–12377. PMLR (2022)

17. Li, K., Wang, S., Yu, L., Heng, P.-A.: Dual-teacher: integrating intra-domain and inter-domain teachers for annotation-efficient cardiac segmentation. In: Martel, A.L., et al. (eds.) MICCAI 2020. LNCS, vol. 12261, pp. 418–427. Springer, Cham (2020). https://doi.org/10.1007/978-3-030-59710-8_41

18. Li, S., et al.: Few-shot domain adaptation with polymorphic transformers. In: de Bruijne, M., et al. (eds.) MICCAI 2021. LNCS, vol. 12902, pp. 330–340. Springer, Cham (2021). https://doi.org/10.1007/978-3-030-87196-3_31

19. Liang, J., Hu, D., Feng, J.: Do we really need to access the source data? source hypothesis transfer for unsupervised domain adaptation. In: ICML, pp. 6028–6039. PMLR (2020)

20. Liu, Z., et al.: Swin transformer: hierarchical vision transformer using shifted windows. In: ICCV, pp. 10012–10022 (2021)

21. Olsson, V., et al.: Classmix: segmentation-based data augmentation for semi-supervised learning. In: WACV, pp. 1369–1378 (2021)

22. Orlando, J.I., et al.: Refuge challenge: a unified framework for evaluating automated methods for glaucoma assessment from fundus photographs. Med. Image Anal. **59**, 101570 (2020)

23. Pogorelov, K., et al.: Kvasir: a multi-class image dataset for computer aided gastrointestinal disease detection. In: Proceedings of the 8th ACM on Multimedia Systems Conference, pp. 164–169 (2017)

24. Ronneberger, O., Fischer, P., Brox, T.: U-net: convolutional networks for biomedical image segmentation. In: Navab, N., Hornegger, J., Wells, W.M., Frangi, A.F. (eds.) MICCAI 2015. LNCS, vol. 9351, pp. 234–241. Springer, Cham (2015). https://doi.org/10.1007/978-3-319-24574-4_28

25. Tarvainen, A., Valpola, H.: Mean teachers are better role models: weight-averaged consistency targets improve semi-supervised deep learning results. In: NeurIPS **30** (2017)

26. Tranheden, W., et al.: Dacs: domain adaptation via cross-domain mixed sampling. In: WACV, pp. 1379–1389 (2021)

27. Tzeng, E., Hoffman, J., Saenko, K., Darrell, T.: Adversarial discriminative domain adaptation. In: Proceedings of the IEEE Conference on Computer Vision and Pattern Recognition, pp. 7167–7176 (2017)

28. Varsavsky, T., et al.: Test-time unsupervised domain adaptation. In: Martel, A.L., et al. (eds.) MICCAI 2020. LNCS, vol. 12261, pp. 428–436. Springer, Cham (2020). https://doi.org/10.1007/978-3-030-59710-8_42

29. Vu, T.H., Jain, H., Bucher, M., Cord, M., Pérez, P.: Advent: adversarial entropy minimization for domain adaptation in semantic segmentation. In: CVPR, pp. 2517–2526 (2019)

30. Wang, S., Yu, L., Li, K., Yang, X., Fu, C.-W., Heng, P.-A.: Boundary and entropy-driven adversarial learning for fundus image segmentation. In: Shen, D., et al. (eds.) MICCAI 2019. LNCS, vol. 11764, pp. 102–110. Springer, Cham (2019). https://doi.org/10.1007/978-3-030-32239-7_12

31. Wang, S., Yu, L., Yang, X., Fu, C.W., Heng, P.A.: Patch-based output space adversarial learning for joint optic disc and cup segmentation. IEEE Trans. Med. Imaging **38**(11), 2485–2495 (2019)

32. Wu, S., Chen, C., Xiong, Z., Chen, X., Sun, X.: Uncertainty-aware label rectification for domain adaptive mitochondria segmentation. In: de Bruijne, M., et al. (eds.) MICCAI 2021. LNCS, vol. 12903, pp. 191–200. Springer, Cham (2021). https://doi.org/10.1007/978-3-030-87199-4_18

33. Xie, E., et al.: Segformer: simple and efficient design for semantic segmentation with transformers. NeurIPS **34**, 12077–12090 (2021)

34. Xu, Z., et al.: Denoising for relaxing: unsupervised domain adaptive fundus image segmentation without source data. In: Wang, L., Dou, Q., Fletcher, P.T., Speidel, S., Li, S. (eds.) Medical Image Computing and Computer Assisted Intervention – MICCAI 2022: 25th International Conference, Singapore, September 18–22, 2022, Proceedings, Part V, pp. 214–224. Springer, Cham (2022). https://doi.org/10.1007/978-3-031-16443-9_21

35. Zhang, P., Zhang, B., Zhang, T., Chen, D., Wang, Y., Wen, F.: Prototypical pseudo label denoising and target structure learning for domain adaptive semantic segmentation. In: CVPR, pp. 12414–12424 (2021)

36. Zhao, Z., Xu, K., Li, S., Zeng, Z., Guan, C.: MT-UDA: towards unsupervised cross-modality medical image segmentation with limited source labels. In: de Bruijne, M., et al. (eds.) MICCAI 2021. LNCS, vol. 12901, pp. 293–303. Springer, Cham (2021). https://doi.org/10.1007/978-3-030-87193-2_28

37. Zhu, J.Y., Park, T., Isola, P., Efros, A.A.: Unpaired image-to-image translation using cycle-consistent adversarial networks. In: ICCV, pp. 2223–2232 (2017)

Bridging the Task Barriers: Online Knowledge Distillation Across Tasks for Semi-supervised Mediastinal Segmentation in CT

Muhammad F. A. Chaudhary[1(✉)], Seyed Soheil Hosseini[1], R. Graham Barr[2], Joseph M. Reinhardt[1,3], Eric A. Hoffman[1,3], and Sarah E. Gerard[1(✉)]

[1] The Roy J. Carver Department of Biomedical Engineering, The University of Iowa, Iowa, IA 52242, USA
{muchaudhary,segerard}@uiowa.edu
[2] Department of Medicine, Columbia University Irving Medical Center, Columbia University, New York, NY 10027, USA
[3] Department of Radiology at The Roy J. and Lucille A. Carver College of Medicine, The University of Iowa, Iowa, IA 52242, USA

Abstract. Segmentation of the mediastinal vasculature in computed tomography (CT) enables automated extraction of important biomarkers for cardiopulmonary disease characterization and outcome prediction. However, the limited contrast between blood and surrounding soft tissue makes manual segmentation of mediastinal structures challenging in non-contrast CT (NCCT) images, resulting in limited annotations for training deep learning models. To overcome this challenge, we propose a semi-supervised mediastinal vasculature segmentation method that utilizes knowledge distillation from unlabeled training data of contrast-enhanced dual-energy CT to achieve segmentation of the main pulmonary artery, main pulmonary veins, and aorta in NCCT. Our framework incorporates multitask learning with attention feature fusion bridges for online knowledge transfer from a related image-to-image translation task to the target segmentation task. Experimental evaluations demonstrate superior segmentation accuracy of our approach compared to fully supervised methods as well as two sequential approaches that do not leverage distillation between tasks. The proposed approach achieves a Dice similarity coefficient of 0.871 for the main pulmonary artery, 0.920 for the aorta, and 0.824 for the main pulmonary veins. By leveraging a large dataset without annotations through multitask learning and knowledge distillation, our approach improves performance in the target task of mediastinal segmentation with limited annotated training data.

Keywords: Knowledge Distillation · Image Segmentation · Multitask Learning · Vision Transformers · Semi-Supervised Learning

X. Cao et al. (Eds.): MLMI 2023, LNCS 14348, pp. 310–319, 2024.
https://doi.org/10.1007/978-3-031-45673-2_31

1 Introduction

The vascular structures within the mediastinum play an important role in sustaining the homeostatic function of the cardiovascular system. Malformations of the mediastinal vasculature have been demonstrated to serve as indicators of various cardiovascular conditions including, pulmonary hypertension, arterial stenosis, and pulmonary embolism [12,22]. To better characterize these abnormalities, contrast-enhanced computed tomography (CECT) is employed to enhance the vasculature, enabling improved delineation of key structures such as the main pulmonary artery (MPA), aorta (AA), and the main pulmonary veins (MPV). This has led to the development of important biomarkers for pulmonary hypertension [12], pulmonary artery aneurysm, and exacerbation susceptibility in chronic obstructive pulmonary disease [22]. Acquiring a CECT, however, is expensive, takes more time, and may not be included in the routine CT acquisition protocol. Additionally, certain patients may experience side effects from contrast agents [6]. Often, when quantitative measures are used as an indicator of pathology, the measurements are provided by a radiologist estimating diameters by placing calipers on a two-dimensional cross section of the structure of interest. Furthermore, quantitative measures are often desired from non-contrast CT (NCCT) studies.

To accurately assess the quantitative relationships between the aortic and pulmonary vascular structures of the mediastinum, it is necessary to fully segment these structures from a volumetric CT, and this is particularly challenging if the structures are not contrast enhanced since the radiodensities of tissue and blood are similar. Therefore, it has been challenging to acquire reliable voxelwise annotations of the MPA, AA, and the MPV on NCCT scans. Recognizing this limitation, several studies have proposed deep generative models to translate NCCT scans into CECT scans, facilitating the identification and visualization of vessel structures. A cascaded framework of two generative models, called Dye-FreeNet, was proposed for NCCT to CECT translation [14]. Hu *et al.* proposed an aorta-aware GAN for synthesizing CECT scans from NCCT scans. They demonstrated that synthetic CECT scans could be used for detecting aortic aneurysms [10]. Recently, a cycle-consistent vision transformer architecture for translation of NCCT to CECT scans, called CyTran, was developed [19].

A few studies have recently used the synthetic image priors for improved segmentation of vascular structures in NCCT scans [18,21]. Wang *et al.* used a GAN to first synthesize CECT scans from NCCT scans, that were subsequently used for aorta and main pulmonary artery segmentation [21]. Pang *et al.* used a similar approach for lung vessel tree segmentation [18], where a synthesis model was trained to convert NCCT to sythetic CECT and independently a segmentation model was trained to segment the vessel tree from CECT. At test time, a synthetic CECT image was generated and used as input to the segmentation model. On one dataset, this approach showed superior performance compared to training a segmentation model on NCCT. However, evaluation on a second dataset contradicted these results and revealed worse performance on the CECT

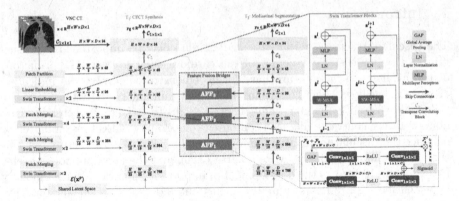

Fig. 1. The proposed multitask knowledge distillation framework. The encoder maps the input **x**, a 3D VNC CT image, to a latent space $\mathcal{E}(\mathbf{x})$ using Swin Transformer blocks. The latent space branches into two decoders for multitask learning: T_1 performs CECT synthesis and T_2 performs mediastinal vasculature segmentations. The two decoder branches interact through attentional feature fusion (AFF) modules for knowledge transfer. The right upper panel show details of the two consecutive Swin Transformer blocks with alternating windowed (W-MSA) and shifted-windowed (SW-MSA) multi-head self-attention. The lower right panel shows the multiscale channel attention module for feature sharing across decoder backbones.

segmentation model, which was attributed to increased segmentation vulnerability to underlying synthesis errors when generating the CECT input from NCCT.

There is a significant overlap in the higher-level knowledge required to enhance the vasculature structures in a NCCT image and the knowledge required to segment these structures; therefore, learning both tasks can be mutually beneficial. However, previous strategies have failed to harness the potential of incorporating synthesis and segmentation into a unified multitask framework with online knowledge sharing. We hypothesize that mediastinal segmentation can be improved by leveraging knowledge from a similar task of translating NCCT to CECT. We propose a semi-supervised framework with knowledge sharing to mitigate small labeled training set by harnessing knowledge from the image-to-image translation of NCCT to CECT. We utilize dual-energy CT (DECT) scans and perform material decomposition to generate a unique and abundant source of virtual non-contrast (VNC) and CECT image pairs that are intrinsically spatially aligned.

2   Methods

2.1   Overview

We propose an online knowledge distillation framework for semi-supervised segmentation of the MPA, AA, and the MPV from the VNC CT scans. Our goal is to be able to distill useful representations from a generative task T_1, with large paired data to the segmentation task T_2, where voxelwise annotations are hard to acquire. We propose a dual-decoder architecture fed jointly by a shifted window hierarchical vision transformer, Swin Transformer [16], as shown in Fig. 1. Each decoder branch caters to a different task and the overall model is trained in two stages. We first train a single encoder-decoder network on image-to-image translation (T_1) task, of VNC CT scans to CECT scans, where paired data is abundant. The next stage involves re-training of the pre-trained, generative model with an additional decoder branch for low-resource semantic segmentation task (T_2). The second stage involves multitask learning where the decoder branches interact through multiscale channel attention bridges for sharing useful representations [3]. Our model allows efficient online knowledge transfer from one task to the other at both the encoder and decoder levels (see Fig. 1).

2.2   Dataset

We used $N = 600$ iodine contrast-enhanced DECT scans acquired at functional residual capacity (FRC) from phase 6 of the Multi-Ethnic Study of Atherosclerosis (MESA) Lung Study, which is a prospective cohort study of the prevalence of cardiovascular disease [9]. Material decomposition was performed on the DECT scans to subtract out the iodine contrast and generate VNC CT scans [4]. The VNC and low-energy DECT scans provided VNC-CECT pairs for training a generative model for the task of translating non-contrast images to contrast-enhanced images (T_1). Being derived from DECT, the VNC and CECT were intrinsically spatially aligned and provided abundant paired training data without image registration or manual annotation. A subset of $N_s = 49$ DECT scans were manually annotated to label the MPA, AA, and MPV by a consensus of two trained experts. We define the training set of VNC-CECT scan pairs as $\mathcal{Q} = \{(\mathbf{x}^i, \mathbf{y}_g^i)\}_{i=1}^{N}$ and the subset with annotations as $\mathcal{Q}_s = \{(\mathbf{x}^i, \mathbf{y}_g^i, \mathbf{y}_s^i)\}_{i=1}^{N_s}$ where $\mathbf{x}^i$, $\mathbf{y}_g^i$, and $\mathbf{y}_s^i$ correspond to 3D images: VNC, CECT, and ground truth segmentation, respectively.

2.3   Swin Transformer Encoder

For the encoder of the proposed architecture we used a pyramidal Swin Transformer backbone with four stages $\{s_1, s_2, s_3, s_4\}$, each operating at a different resolution with different number of transformer blocks set as $\{2, 4, 2, 2\}$. Unlike other transformer architectures, the Swin Transformer block reduces spatial resolution at each stage through patch merging and performs multi-head self-attention (MSA) within local windows for computational efficiency. To model dependencies across windows, a shifted window partitioning alternates between two consecutive blocks of a Swin Transformer [16], as illustrated in upper right panel in Fig. 1. With shifted window partitioning in every other block, we can express two consecutive Swin Transformer blocks as:

$$\hat{\mathbf{z}}^l = W - MSA(LN(\mathbf{z}^{l-1})) + \mathbf{z}^{l-1},$$
$$\mathbf{z}^l = MLP(LN(\hat{\mathbf{z}}^l)) + \hat{\mathbf{z}}^l,$$
$$\hat{\mathbf{z}}^{l+1} = SW - MSA(LN(\mathbf{z}^l)) + \mathbf{z}^l, \tag{1}$$
$$\mathbf{z}^{l+1} = MLP(LN(\hat{\mathbf{z}}^{l+1})) + \hat{\mathbf{z}}^{l+1},$$

where $\mathbf{z}^l$ and $\mathbf{z}^{l+1}$ denotes outputs from the regular (W-MSA) and shifted window (SW-MSA) multi-head self-attention, respectively. The input to the encoder was a VNC CT image $\mathbf{x}$ mapped to the shared latent space $\mathcal{E}(\mathbf{x})$ that is fed to the two decoder branches for learning T_1 and T_2 (see Fig. 1).

2.4   SwinUNETR GAN for CECT Synthesis

We first pretrained a GAN for synthesizing CECT images from VNC CT images using the larger dataset $\mathcal{Q}$. We used a Swin Transformer-based, UNet-like conditional GAN called SwinUNETR GAN. The generator, $\mathcal{G} : \mathbf{x} \rightarrow \mathbf{y}_g$, consisted of the Swin Transformer encoder described in Sect. 2.3 followed by a decoder branch for T_1 illustrated in Fig. 1. We used the least squares GAN framework, with a patch GAN discriminator $\mathcal{D}$, for the adversarial training of our generative model [11]. The min-max least squares objective was expressed as:

$$\mathcal{L}_{LSGAN} = -\mathbb{E}_{\mathbf{x}, \mathbf{y}_g}[(\mathcal{D}(\mathbf{x}, \mathbf{y}_g) - 1)^2] - \mathbb{E}_{\mathbf{x}}[\mathcal{D}(\mathbf{x}, \mathcal{G}(\mathbf{x}))^2]. \tag{2}$$

We used a combination of multiview perceptual similarity and ℓ_1-distance to stabilize training and improve image synthesis. The details of the volumetric image synthesis method can be found in [2].

2.5   Multitask Learning with Attention Feature Fusion

The pretrained generator from the SwinUNETR GAN was used for distilling useful representations from the CECT synthesis task to the segmentation task. A second decoder branch for T_2 was added stemming from the latent space $\mathcal{E}(\mathbf{x})$, with the same structure as the decoder for T_1. In this phase, multitask learning was employed to simultaneously train T_1 and T_2.

To efficiently transfer useful representations from the synthesis branch to the segmentation branch, we used multiscale attentional feature fusion (AFF) blocks [3] that acted as information bridges between two tasks. The multiscale AFF block computes attention on a local and a global scale, and its architecture is shown in lower right panel of Fig. 1. For global channel attention, the input tensor $\mathcal{F}$ is spatially collapsed using global average pooling. Meanwhile, the local attention path only reduces the channel dimensions by a factor of r before computing local channel attention. The two paths are subsequently combined and followed up by a sigmoid function to compute overall attention. For knowledge distillation, feature maps from generation $\mathcal{F}_g$ and segmentation $\mathcal{F}_s$ branch are fused through the following expression:

$$\mathcal{F}' = \mathrm{AFF}(\mathcal{F}_g, \mathcal{F}_s) = \Psi(\mathcal{F}_g + \mathcal{F}_s) \odot \mathcal{F}_g + (1 - \Psi(\mathcal{F}_g + \mathcal{F}_s)) \odot \mathcal{F}_s, \qquad (3)$$

where $\mathcal{F}'$ is the fused feature map and Ψ is the multiscale channel attention module shown in Fig. 1, and $\odot$ denotes the Hadamard product. The feature fusion module learns to selectively emphasize useful representations that could improve segmentation task. The hyperparameter r was set to 4.

We used a combination of focal [13], Dice [17], and Tversky loss [20] for segmentation. The Tversky loss was added to penalize false negatives in the predicted segmentation masks. At this stage, the output from the generative branch was optimized using just the ℓ_1-distance. All terms within the segmentation cost function were weighted equally and the weight for ℓ_1-distance was set to 0.1.

3 Experiments

3.1 Preprocessing

Lung segmentations were generated using a multiresolution convolutional network [5] and all images were cropped to the bounding box of the lungs. Images were spatially resampled to $1\mathrm{mm}^3$ isotropic voxels and intensities were clipped between -1024 Hounsfield units (HU) and 1024 HU and then linearly mapped to -1 and 1. Random crops were sampled during training with dimensions $128 \times 128 \times 128$ voxels.

3.2 Evaluation

For the generative model, we split the data into disjoint training and validation sets of size 450 and 150, respectively. We used five different metrics for evaluating synthesized images: peak-signal-to-noise-ratio (PSNR), structural similarity metric (SSIM), learned perceptual image similarity (LPIPS) [23], normalized mean squared error (NMSE), and mean absolute error (MAE) in HU. For the segmentation model, we used 5-fold cross-validation and quantitatively evaluated performance using the Dice similarity coefficient (DSC) and average symmetric surface distance (ASSD) [8]. Although the model was trained on 3D patches, we computed evaluation metrics on complete 3D volumes obtained by combining the patches with an overlap of 32 voxels. We compared our method to different state-of-the-art supervised segmentation methods, including the UNet and UNETR [7]. We also compared our method to two sequential models that first learn a model for CECT synthesis, similar to [18,21], without sharing knowledge between tasks. The first model trained a UNet using synthetic CECT (sCECT) image priors obtained from SwinUNETR GAN used in this study. The second model by Pang et al. trained separate networks for synthesizing CECT images and performing segmentation from CECT volumes. During inference, their method used sCECT images instead of the original CECT images [18]. To assess model generalizability, we evaluated the qualitative performance of our model on true NCCT scans at different lung volumes, acquired using a totally different CT acquisition protocol. These multiple volume CT scans were acquired as part of the smoking cessation study [15].

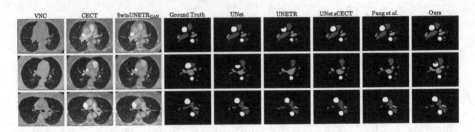

Fig. 2. Qualitative evaluation of the proposed model on axial slices. We also show synthetic CT slice generated by SwinUNETR GAN.

Table 1. Quantitative model evaluation and comparison with other state-of-the-art methods. MPA = main pulmonary artery, MPV = main pulmonary vein, AA = aorta. We report means of the metric values (standard deviations) across five folds. Difference between the metric means of proposed and other methods were assessed using Mann-Whitney's U test where †: not significant; *: $p < 0.05$; **: $p < 0.01$; ***: $p < 0.001$.

	DSC_{MPA}	DSC_{MPV}	DSC_{AA}	$ASSD_{MPA}$	$ASSD_{MPV}$	$ASSD_{AA}$
UNet – VNC	$0.847^{***}_{0.043}$	$0.807^{*}_{0.038}$	$0.912^{†}_{0.029}$	$2.14^{***}_{0.98}$	$2.59^{**}_{1.88}$	$1.41^{*}_{0.96}$
UNETR – VNC	$0.826^{***}_{0.058}$	$0.782^{***}_{0.055}$	$0.865^{***}_{0.059}$	$2.49^{***}_{1.91}$	$2.33^{**}_{0.71}$	$2.15^{***}_{1.35}$
UNet – sCECT	$0.852^{***}_{0.034}$	$0.803^{**}_{0.041}$	$0.914^{†}_{0.025}$	$2.35^{***}_{1.62}$	$3.24^{***}_{2.79}$	$1.50^{†}_{1.36}$
Pang *et al.* [18]	$0.825^{***}_{0.053}$	$0.798^{**}_{0.043}$	$0.896^{**}_{0.047}$	$2.17^{***}_{0.85}$	$2.47^{***}_{0.72}$	$1.57^{***}_{0.81}$
Ours – VNC	$0.871_{0.034}$	$0.824_{0.039}$	$0.920_{0.028}$	$1.62_{0.65}$	$1.94_{0.51}$	$1.07_{0.50}$

3.3   Implementation Details

We implemented our networks using the open source frameworks PyTorch and MONAI [1]. All models were trained with a learning rate of 0.0002, with the exception of the discriminator in the GAN framework which was trained with a learning rate of 0.00005. All models were training using a single NVIDIA A100 GPU with 80 GB memory.

4   Results

The SwinUNETR GAN model for synthesis of CECT from VNC achieved an overall PSNR of 27.36 dB, SSIM of 0.83, LPIPS of 0.19, and NMSE of 4.18 on a testing cohort of 150 subjects. In Table 1, we present quantitative evaluation of the proposed mediastinal segmentation method and the comparison methods. Our model showed superior performance with an overall DSC_{MPA} of 0.871 and $ASSD_{MPA}$ of 1.62 for MPA segmentation. The performance was better than supervised learning UNet and UNETR models that were trained with VNC CTs as inputs. Our model also performed better than the approach by Pang *et al.* [18], which showed a DSC_{MPA} of 0.825 and $ASSD_{MPA}$ of 2.17 for MPA segmentation (see Table 1). A similar trend was observed for MPV and AA segmentation. Qualitative model evaluation is shown in Fig. 2. We show 3D renderings of different methods in Fig. 3. Qualitative evaluation on multiple true NCCT volumes from a different cohort is shown in Fig. 4.

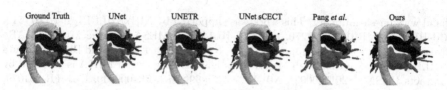

Fig. 3. Three-dimensional surface renderings of the MPA, MPV, and AA segmentations from different methods.

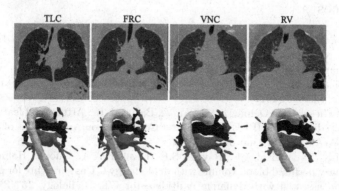

Fig. 4. Qualitative evaluation of the proposed method on an external dataset of true NCCTs acquired at total lung capacity (TLC), functional residual capacity (FRC), and residual volume (RV).

5   Discussion and Conclusion

We proposed a multitask framework for online knowledge distillation to improve mediastinal vasculature segmentation in non-contrast images when voxelwise annotations are not abundantly available. We demonstrated that channel attention could be used to select and distill useful representations from a similar task of synthesizing contrast-enhanced images from non-contrast images. Our model performed better than two sequential models that used generative priors, which highlighted the benefit of representation transfer between related tasks given limited annotations. A limitation of this study was that training and evaluation were performed with VNC CT scans instead of true NCCTs, which could potentially lead to degradation of performance on true NCCT scans. However, qualitative evaluation on an external cohort of true NCCTs demonstrated the proposed method could successfully segment the mediastinal vasculature in true NCCT images at different lung volumes that were not included in training. These results suggest high potential for generalizability and broader applicability of our model to large datasets of NCCTs acquired in clinical settings and large multicenter imaging studies.

Acknowledgements. The authors thank the other investigators, the staff, and the participants of the Multi-Ethnic Study of Atherosclerosis (MESA) for their valuable contributions. A full list of participating MESA investigators and institutions can be found www.mesa-nhlbi.org. This work was supported by NIH/NHLBI R01-HL077612, R01-HL093081, R01-HL121270, and R01-HL142028. MESA and the MESA SHARe project are conducted and supported by the National Heart, Lung, and Blood Institute (NHLBI) in collaboration with MESA investigators. Support for MESA is provided by contracts HHSN268201500003I, N01-HC-95159-69, UL1-TR-000040, UL1-TR-001079, UL1-TR-001420, UL1-TR-001881, and DK063491.

References

1. Cardoso, M.J., et al.: MONAI: an open-source framework for deep learning in healthcare. arXiv preprint arXiv:2211.02701 (2022)

2. Chaudhary, M.F., et al.: Lung2Lung: volumetric style transfer with self-ensembling for high-resolution cross-volume computed tomography. arXiv preprint arXiv:2210.02625 (2022)

3. Dai, Y., Gieseke, F., Oehmcke, S., Wu, Y., Barnard, K.: Attentional feature fusion. In: Proceedings of the IEEE/CVF Winter Conference on Applications of Computer Vision (WACV), pp. 3560–3569 (2021)

4. Fuld, M.K., Halaweish, A.F., Haynes, S.E., Divekar, A.A., Guo, J., Hoffman, E.A.: Pulmonary perfused blood volume with dual-energy CT as surrogate for pulmonary perfusion assessed with dynamic multidetector CT. Radiology **267**(3), 747–756 (2013)

5. Gerard, S.E., Herrmann, J., Kaczka, D.W., Musch, G., Fernandez-Bustamante, A., Reinhardt, J.M.: Multi-resolution convolutional neural networks for fully automated segmentation of acutely injured lungs in multiple species. Med. Image Anal. **60**, 101592 (2020)

6. Hagan, J.B.: Anaphylactoid and adverse reactions to radiocontrast agents. Immunol. Allergy Clin. **24**(3), 507–519 (2004)

7. Hatamizadeh, A., et al.: UNETR: transformers for 3D medical image segmentation. In: Proceedings of the IEEE/CVF Winter Conference on Applications of Computer Vision (WACV), pp. 574–584 (2022)

8. Heimann, T., et al.: Comparison and evaluation of methods for liver segmentation from CT datasets. IEEE Trans. Med. Imaging **28**(8), 1251–1265 (2009)

9. Hermann, E.A., et al.: Pulmonary blood volume among older adults in the community: the MESA lung study. Circul. Cardiovas. Imaging **15**(8), e014380 (2022)

10. Hu, T., et al.: Aorta-aware GAN for non-contrast to artery contrasted CT translation and its application to abdominal aortic aneurysm detection. Int. J. Comput. Assist. Radiol. Surg. 1–9 (2022)

11. Isola, P., Zhu, J.Y., Zhou, T., Efros, A.A.: Image-to-image translation with conditional adversarial networks. In: Proceedings of the IEEE Conference on Computer Vision and Pattern Recognition, pp. 1125–1134 (2017)

12. Iyer, A.S., Wells, J.M., Vishin, S., Bhatt, S.P., Wille, K.M., Dransfield, M.T.: CT scan-measured pulmonary artery to aorta ratio and echocardiography for detecting pulmonary hypertension in severe COPD. Chest **145**(4), 824–832 (2014)

13. Lin, T.Y., Goyal, P., Girshick, R., He, K., Dollár, P.: Focal loss for dense object detection. In: Proceedings of the IEEE International Conference on Computer Vision (ICCV), pp. 2980–2988 (2017)

14. Liu, J., et al.: DyeFreeNet: deep virtual contrast CT synthesis. In: Burgos, N., Svoboda, D., Wolterink, J.M., Zhao, C. (eds.) SASHIMI 2020. LNCS, vol. 12417, pp. 80–89. Springer, Cham (2020). https://doi.org/10.1007/978-3-030-59520-3_9

15. Liu, Y., et al.: An incentive-based program coupled with sildenafil provides enhanced success of smoking cessation associated with an accelerated loss of CT assessed smoking-associated lung density (inflammation) and improved DLCO. In: D76. COPD: Clinical Studies, pp. A7556–A7556. American Thoracic Society (2020)
16. Liu, Z., et al.: Swin transformer: hierarchical vision transformer using shifted windows. In: Proceedings of the IEEE/CVF International Conference on Computer Vision (ICCV), pp. 10012–10022 (2021)
17. Milletari, F., Navab, N., Ahmadi, S.A.: V-Net: fully convolutional neural networks for volumetric medical image segmentation. In: 2016 Fourth International Conference on 3D Vision (3DV), pp. 565–571. IEEE (2016)
18. Pang, H., et al.: NCCT-CECT image synthesizers and their application to pulmonary vessel segmentation. Comput. Methods Prog. Biomed. **231**, 107389 (2023)
19. Ristea, N.C., et al.: CyTran: a cycle-consistent transformer with multi-level consistency for non-contrast to contrast CT translation. Neurocomputing **538**, 126211 (2023)
20. Salehi, S.S.M., Erdogmus, D., Gholipour, A.: Tversky loss function for image segmentation using 3D fully convolutional deep networks. In: Wang, Q., Shi, Y., Suk, H.-I., Suzuki, K. (eds.) MLMI 2017. LNCS, vol. 10541, pp. 379–387. Springer, Cham (2017). https://doi.org/10.1007/978-3-319-67389-9_44
21. Wang, H.J., et al.: Automated 3D segmentation of the aorta and pulmonary artery on non-contrast-enhanced chest computed tomography images in lung cancer patients. Diagnostics **12**(4), 967 (2022)
22. Wells, J.M., et al.: Pulmonary arterial enlargement and acute exacerbations of COPD. N. Engl. J. Med. **367**(10), 913–921 (2012)
23. Zhang, R., Isola, P., Efros, A.A., Shechtman, E., Wang, O.: The unreasonable effectiveness of deep features as a perceptual metric. In: Proceedings of the IEEE Conference on Computer Vision and Pattern Recognition, pp. 586–595 (2018)

RelationalUNet for Image Segmentation

Ivaxi Sheth[1,2(✉)], Pedro H. M. Braga[1,2,4], Shivakanth Sujit[1,2],
Sahar Dastani[1,2], and Samira Ebrahimi Kahou[1,2,3]

[1] ÉTS Montréal, Montreal, Canada
ivaxi.sheth16@imperial.ac.uk
[2] Mila, Quebec AI Institute, Montreal, Canada
{shivakanth.sujit.1,sahar.dastani-oghani}@ens.etsmtl.ca
[3] CIFAR AI Chair, Montreal, Canada
samira.ebrahimi-kahou@etsmtl.ca
[4] Universidade Federal de Pernambuco, Recife, Brazil
pedromagalhaes.hb@gmail.com
https://cifar.ca/ai/canada-cifar-ai-chairs/

Abstract. Medical image segmentation is one of the most classic applications of machine learning in healthcare. A variety of Deep Learning approaches, mostly based on Convolutional Neural Networks (CNNs), have been proposed to this end. In particular, U-Shaped Network (UNet) have emerged to exhibit superior performance for medical image segmentation. However, some properties of CNNs, such as the stationary kernels, may limit them from capturing more in-depth visual and spatial relations. The recent success of transformers in both language and vision has motivated dynamic feature transforms. We propose RelationalUNet (RelationalUNet) which introduces relational feature transformation to the UNet architecture. RelationalUNet models the dynamics between visual and depth dimensions of a 3D medical image by introducing Relational Self-Attention blocks in skip connections. As the architecture is mainly intended for the semantic segmentation of 3D medical images, we aim to learn their long-range depth relations. Our method was validated on the Multi-Atlas Labeling Beyond The Cranial Vault (BTCV) dataset for multi-organ segmentation. Robustness to distribution shifts is a particular challenge in safety-critical applications such as medical imaging. We further test our model performance on realistic distributional shifts on the Shifts 2.0 White Matter Multiple Sclerosis Lesion Segmentation. Experiments show that our architecture leads to competitive performance. The code is available at https://github.com/ivaxi0s/runet.

Keywords: Medical Image Segmentation · Attention

I. Sheth, P.H.M. Braga, and S. Sujit—Equal Contribution.

Supplementary Information The online version contains supplementary material available at https://doi.org/10.1007/978-3-031-45673-2_32.

1 Introduction

The field of medical imaging has made tremendous advancements in recent times with the aid of computer vision algorithms, assisting radiologists and pathologists in accurately diagnosing diseases. Medical image segmentation allows the extraction of detailed information from medical images by partitioning an image into sub-groups and classifying each pixel of the image. This information can aid in identifying illnesses, supporting treatment plans, and tracking the progression of an organ failure. However, the various artifacts present in medical images and their complexity make segmentation a challenging task. Various CNN models [3,14,23] have been adapted for segmentation. But the inability of CNNs to manage different input sizes have limited their performance [18] for medical segmentation. Furthermore, they are prone to overfitting when trained on small datasets, resulting in poor generalization performance [11]. Current state-of-the-art models follow an encoder-decoder, UNet-like structure as proposed by Ronneberger et al. [21] to automatically segment medical image lesions. The UNet is a U-shaped network consisting of an encoder for learning contextual representations connected to a decoder for up-sampling the extracted representations.

Current UNet models do not explicitly incorporate dynamic feature transforms as part of their learning process [6]. Therefore, the relational composition of each dimension within a 3D volumetric medical image has not been thoroughly explored. Alom et al. [1] utilized Recurrent Neural Networks (RNNs) to model depth-dependent characteristics in images for segmentation purposes. However, they encountered limitations in capturing deeper relationships present in medical images, such as Computed Tomography (CT) scans and Magnetic Resonance Imagings (MRIs) [25]. Motivated by limited research in dynamic learning of volumetric images, our work proposes a RelationalUNet that aims to learn relational attention between spatial and depth dimensions of medical images.

2 Related Work

Since its introduction, UNet [21] has served as a backbone for numerous state-of-the-art models in biomedical image segmentation. Its encoder-decoder framework provides the ability to project discriminative features at different stages. Additionally, the skip connections concatenate encoder and decoder features facilitating gradient flow during backpropagation. A variety of UNet-based models have been proposed to achieve state-of-the-art performance for both 2D images and 3D volumes [5,7,17]. Popular UNet-based models either extend the basic architecture [20,24], or modify its components, such as the encoder [9]. Roth et al. [22] introduced a hierarchical approach as a multi-scale framework for multi-organ segmentation.

Transformers have started to gain prevalence in computer vision tasks after the seminal success in natural language processing. The self-attention mechanism in transformers has been the key to its success, as it allows the model to identify the important features of the input even with long dependencies. Attention UNet [20] was one of the first methods to use attention to segment medical images. After the successful introduction of Vision Transformerss (ViTs) [4], the UNEt TRansformers (UNETR) [9] was presented as a transformer-based model for 3D image segmentation. It employs a ViT encoder linked to the decoder through skip connections, drawing inspiration from the UNet architecture. In a related approach, Hatamizadeh et al. [8] proposed the utilization of a Swin encoder (hierarchical vision transformers) paired with a CNN decoder, creating a UNet-like structure tailored for multi-modal segmentation.

3   Methodology

3.1   Relational Self Attention

Relational Self-Attention (RSA). [13] is a relational dynamic feature transform initially proposed to address the limitation of existing methods for video understanding. It leverages structures of space and time to dynamically generate aggregating relational kernels and contexts. In essence, RSA combines these two main components with regular dynamic kernels and context features to capture and learn the desired dynamics.

The first component, called the relational kernel, is responsible for predicting the weights of kernels based on the relevance of context information. This information is determined by the structure of content-to-content interactions and is computed as the projection of the Hadamard product vector between query and key using a learnable matrix $H \in \mathbb{R}^{M \times M}$. The Hadamard product is employed to leverage channel-wise correlations and to avoid the loss of semantic information that can occur when contracting all channels in regular dot-product computation.

$$\kappa_n^R = \text{vec}(\mathbf{1}(x_n^Q)^T \odot X_n^K)H. \tag{1}$$

H is crucial for transforming the target into a dynamic kernel. It achieves this by making predictions based on the Hadamard product vector. This enables it to adapt to various visual structures located at different spatial positions. Instead of having a regular convolutional operation, involution is employed. Note that the main point of involution [16] is to leverage from such potential interactions.

The second component, the relational context, uses self-correlation to describe volumetric information and thus provides relational patterns of content-to-content interactions, as defined by Eq. 2.

$$X_n^R = X_n^V (X_n^V)^T G, \tag{2}$$

where G is a learnable matrix that tries to map any content-to-content interaction patterns revealed by the self-correlation of X_n^V to the relational context.

Both relational components, κ_n^R and X_n^R, are then combined with their basic counterparts, κ_n^V and X_n^V, into the RSA transform as per Eq. 3.

$$y_n = (\kappa_n^V + \kappa_n^R)(X_n^V + X_n^R). \tag{3}$$

Figure 1 provides an illustration of the overall architecture. RSA is a strong model for video understanding and has been shown to outperform self-attention mechanisms. For this reason, we investigate the power of RSA layers in medical image segmentation to learn relations across depth in 3D medical volumes.

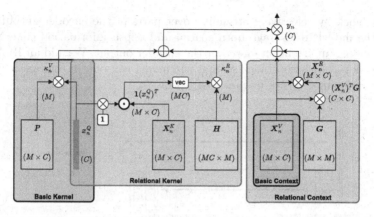

Fig. 1. Relational Self Attention Mechanism used by our model. For further details about the working, please see Sect. 3.1

3.2   RelationalUNet

RSA [13] has proven to outperform other attention mechanisms, such as self-attention in the domain of video understanding, where regular convolutional layers are replaced by RSA layers. Continuing along a similar path, but now within the realm of medical image segmentation, we begin by utilizing the widely recognized and firmly established 3D U-Shaped Networks [21] as our backbone. We then extend this framework to leverage relational information in the combination of encoder and decoder features. While the original application of RSA centered around temporal relationships, our approach attempts to harness depth-based relationships within 3D medical imaging volumes.

Our proposed model's architecture is shown in Fig. 2. We adapt the UNet structure from Ronneberger et al. [21], consisting of a contracting encoder and an expanding decoder path. Each encoder layer transforms the features produced from the previous layer through two convolutional layers and then downsamples the output by a factor of 2 before passing it to the next encoder layer. Each decoder layer processes two inputs, one from the preceding decoder layer and one from the encoder at the current layer through a skip connection. These inputs are concatenated before being transformed through two consecutive convolutional layers. The output is up-sampled by a factor of 2 before passing it to the subsequent decoder layer. The convolutional layers use Rectified Linear Unit (ReLU) for the activation function and down-sampling is done using max pooling.

In traditional UNet architectures, each decoder (except the first) block receives as input the concatenation of the corresponding encoder block's output and the output of the previous decoder block. However, this does now allow the decoder block to decide which parts of the encoder output are relevant to the task. That is, the expressivity of the network for combining these two inputs is low. Instead, we propose that RSA is used to provide additional context for the decoder block by selectively attending over parts of the encoder output. RSA allows the model to leverage both spatial and depth information concurrently when building an attention map over the encoder output. We add an RSA layer to every skip-connection block so that the model can obtain relevant attention maps at different levels of feature abstraction.

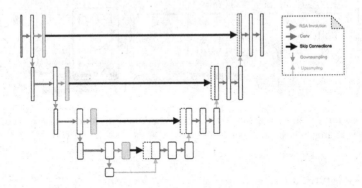

Fig. 2. Overview of the proposed Relational-UNet architecture.

4   Experiments

To evaluate the effectiveness of our model, RelationalUNet, we perform segmentation across 2 benchmark datasets. BTCV is an abdominal organ segmentation task while Shifts 2.0 involves segmentation of brain MRIs. The experiments were conducted on a single NVIDIA V100 GPU. Please find further implementation details in the appendix.

4.1   Datasets

BTCV: The BTCV dataset [15] is a medical imaging dataset that consists of abdominal CT scans of 30 patients. It contains annotations for 13 different body organs that were segmented by expert radiologists. Each CT scans were taken using a special dye that makes the images clearer. These scans are made up of 80 to 225 thin slices, where the thickness of each slice can vary between 1 and 6 mm. We follow the train, validation and test splits as introduced in the codebase of [9]. We train each model on 34 images and test the model on the remaining 6 images, a similar training regime to [2].

Shifts 2.0: The Shifts dataset [19] is a publicly available dataset including Fluid Attenuated Inversion Recovery (FLAIR) scans and manual white matter lesions (WML) annotations. The FLAIR scan is utilized to resample the images to a standard resolution, remove noise, and correct for biases. The training (33 scans) and validation datasets (7 scans) contain data from 4 different centers, and the testing dataset (58 patients) contains both in-domain and out-of-domain data from different centers and scanners. The Dev_{in} and Dev_{out} are development validation sets and $Eval_{in}$ is in-domain test set.

4.2 Metrics

We use the Dice Similarity Coefficient (DSC) to evaluate performance on BTCV in order to be aligned with current benchmarks. It is calculated as a geometric ratio between precision and recall:

$$DSC = 2TP/(FP + 2TP + FN), \tag{4}$$

where TP, FP, and FN correspond to true positive, false positive, and false negative, respectively.

The Shifts 2.0 dataset uses a new metric, Normalized Dice Similarity Coefficient (nDSC). Therefore in our experiments, we report the nDSC score. NDSC is robust across patients with different sizes of lesions. It keeps the recall fixed but adjusts the precision by scaling FP according to a factor, k_p, for each patient p. k_p is defined as the FP rate at 100% scaled by a reference value, r selected as 0.1% as it is the estimated average of the fraction of lesion voxels. Precision is then adapted at the given threshold, r^*, as per

$$\overline{Pr}_{r^*} = \text{TP}_{r^*}/(\text{TP}_{r^*} + k_p\text{FP}_{r^*}). \tag{5}$$

Finally, nDSC is computed as the geometric mean (as DSC in Eq. 4) of $\overline{Pr}_{r^*}$ and the recall at r^* (which is not affected by the change). A higher nDSC indicates better performance.

4.3 Results

We compare the performance of RelationalUNet, with current popular and state-of-the-art medical segmentation models. The baselines for comparison include UNet, a modified version of UNet [21], Attention UNet [20], nn-UNet [10], Vision Transformer based UNETR [9], and Residual UNet [12]. We used Dice score for BTCV and normalized Dice score for Shifts 2.0 as metrics for model evaluation.

From Table 1, we observe that our proposed model RelationalUNet achieves the overall highest Dice score of 81.94% for the BTCV dataset compared to the baselines. We also report the dice score for each organ. While the dice score is comparable for almost every organ, it must be noted that using dynamic self-attention over the 3D images, leads to a massive improvement in the Dice score for segmenting gall bladders. Transformers are known to be strong models, from our experiments we observe that ViT based encoder model does perform on par with our model. From the qualitative example in Fig. 3, we observe that only RelationalUNet is able to segment the finer details of an organ as compared to other baselines.

To evaluate the robustness of our proposed RelationalUNet model, we explore the more challenging out-of-domain experimental setup of Shifts 2.0, where data

Table 1. Results on the BTCV dataset. The DSC score is multiplied by 100 %. Results for some of the organs are also detailed.

Model	Spl	LKid	RKid	Gall	Eso	Liv	Sto	Aor	Avg (%)
UNet [21]	93.26	92.94	93.11	50.20	74.21	95.89	78.58	89.10	77.96
UNet* [21]	91.57	92.15	91.04	62.82	74.15	94.96	77.83	88.94	78.23
Att-UNet [20]	**95.62**	93.58	94.04	63.58	74.45	96.65	81.97	89.24	80.67
nn-UNet [10]	95.57	94.07	94.03	62.14	75.15	96.40	80.12	88.70	80.36
UNETR [8]	93.67	90.93	91.29	58.39	70.40	95.92	76.44	87.39	77.39
ResUNet [12]	93.45	94.04	93.91	59.96	75.59	96.28	**82.61**	89.93	80.15
RelationalUNet (ours)	94.52	**94.53**	**94.52**	**63.71**	**76.32**	**96.82**	82.07	**89.97**	**81.94**

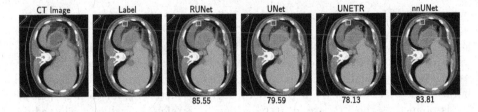

Fig. 3. Qualitative comparison of segmentation of a slice of a BTCV image.

is acquired from different scanners and hospitals. Table 2 illustrates the results in terms of nDSC. Overall, our RelationalUNet model achieved competitive accuracy in out-of-domain settings, Dev_{out} among baselines. On the in-domain test set, our model portrays competitive nDSC performance with nn-UNet for the in-domain $Eval_{in}$ setting. Evaluation in a realistic medical setting is extremely important for the deployment of a segmentation model.

Table 2. Results on Shifts 2.0 dataset.

Model	Dev_{in}	Dev_{out}	$Eval_{in}$
UNet [21]	68.97	50.61	70.40
UNet* [21]	71.41	54.19	73.66
nn-UNet [10]	73.22	53.73	**74.87**
ResUNet [12]	72.07	52.06	72.55
UNETR [9]	72.25	53.41	72.45
RelationalUNet (ours)	**73.36**	**57.61**	74.24

Model Parameters vs Accuracy. From Fig. 5, we observe that while the addition of a relational self-attention module does increase the number of parameters, it provides a significant boost in performance. We observe that RelationalUNet achieves dominant performance with a relatively small number of parameters.

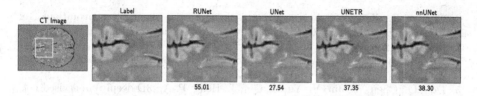

Fig. 4. Qualitative comparison of segmentation of a slice of a Shifts 2.0 image. Our model RelationalUNet has the best performance, as evident from the qualitative comparison.

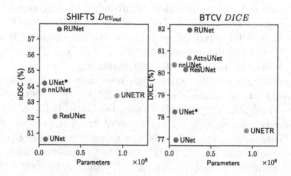

Fig. 5. Comparison of model performance versus model size (number of parameters).

5   Conclusion

This paper introduces a new architecture, a UNet based model for semantic segmentation of medical images. We used RSA layers in a UNet based architecture to capture relations between spatial and depth dimensions. We also illustrated that using RelationalUNet could lead to more generalization performance. To evaluate our contribution, we performed segmentation on CT and MRI images - BTCV and Shifts 2.0, which allows us to have both in-domain and out-of-domain settings. We hope that our work leads to further research in exploiting relational features in volumetric images for segmentation.

Acknowledgements. The authors would like to thank the Digital Research Alliance of Canada for compute resources and CIFAR, Google, NSERC and Imagia Canexia Health for funding.

References

1. Alom, M.Z., Hasan, M., Yakopcic, C., Taha, T.M., Asari, V.K.: Recurrent residual convolutional neural network based on u-net (r2u-net) for medical image segmentation. arXiv preprint arXiv:1802.06955 (2018)
2. Chen, J., et al.: Transunet: transformers make strong encoders for medical image segmentation. arXiv preprint arXiv:2102.04306 (2021)

3. Chollet, F.: Xception: deep learning with depthwise separable convolutions. In: Proceedings of the IEEE Conference on Computer Vision and Pattern Recognition, pp. 1251–1258 (2017)

4. Dosovitskiy, A., et al.: An image is worth 16x16 words: transformers for image recognition at scale. arXiv preprint arXiv:2010.11929 (2020)

5. Dou, Q., Chen, H., Jin, Y., Yu, L., Qin, J., Heng, P.-A.: 3D deeply supervised network for automatic liver segmentation from CT volumes. In: Ourselin, S., Joskowicz, L., Sabuncu, M.R., Unal, G., Wells, W. (eds.) MICCAI 2016. LNCS, vol. 9901, pp. 149–157. Springer, Cham (2016). https://doi.org/10.1007/978-3-319-46723-8_18

6. Futrega, M., Milesi, A., Marcinkiewicz, M., Ribalta, P.: Optimized U-net for brain tumor segmentation. In: Crimi, A., Bakas, S. (eds.) Brainlesion: Glioma, Multiple Sclerosis, Stroke and Traumatic Brain Injuries: 7th International Workshop, BrainLes 2021, Held in Conjunction with MICCAI 2021, pp. 15–29. Springer, Cham (2022). https://doi.org/10.1007/978-3-031-09002-8_2

7. Gibson, E., et al.: Automatic multi-organ segmentation on abdominal CT with dense v-networks. IEEE Trans. Med. Imaging 37(8), 1822–1834 (2018)

8. Hatamizadeh, A., Nath, V., Tang, Y., Yang, D., Roth, H.R., Xu, D.: Swin unetr: swin transformers for semantic segmentation of brain tumors in MRI images. In: Brainlesion: Glioma, Multiple Sclerosis, Stroke and Traumatic Brain Injuries: 7th International Workshop, BrainLes 2021, Held in Conjunction with MICCAI 2021, pp. 272–284. Springer, Cham (2022). https://doi.org/10.1007/978-3-031-08999-2_22

9. Hatamizadeh, A., et al.: Unetr: transformers for 3d medical image segmentation. In: Proceedings of the IEEE/CVF Winter Conference on Applications of Computer Vision, pp. 574–584 (2022)

10. Isensee, F., et al.: nnu-net: Self-adapting framework for u-net-based medical image segmentation. arXiv preprint arXiv:1809.10486 (2018)

11. Karimi, D., Dou, H., Warfield, S.K., Gholipour, A.: Deep learning with noisy labels: exploring techniques and remedies in medical image analysis. Med. Image Anal. 65, 101759 (2020)

12. Khanna, A., Londhe, N.D., Gupta, S., Semwal, A.: A deep residual u-net convolutional neural network for automated lung segmentation in computed tomography images. Biocybernet. Biomed. Eng. 40(3), 1314–1327 (2020)

13. Kim, M., Kwon, H., Wang, C., Kwak, S., Cho, M.: Relational self-attention: what's missing in attention for video understanding. Adv. Neural Inf. Process. Syst. 34 (2021)

14. Krizhevsky, A., Sutskever, I., Hinton, G.E.: Imagenet classification with deep convolutional neural networks. Commun. ACM 60(6), 84–90 (2017)

15. Landman, B., Xu, Z., Igelsias, J., Styner, M., Langerak, T., Klein, A.: Miccai multi-atlas labeling beyond the cranial vault-workshop and challenge. In: Proceedings of the MICCAI Multi-Atlas Labeling Beyond Cranial Vault-Workshop Challenge, vol. 5, p. 12 (2015)

16. Li, D., et al.: Involution: inverting the inherence of convolution for visual recognition. In: Proceedings of the IEEE/CVF Conference on Computer Vision and Pattern Recognition, pp. 12321–12330 (2021)

17. Li, X., Chen, H., Qi, X., Dou, Q., Fu, C.W., Heng, P.A.: H-denseunet: hybrid densely connected unet for liver and tumor segmentation from ct volumes. IEEE Trans. Med. Imaging 37(12), 2663–2674 (2018)

18. Malhotra, P., Gupta, S., Koundal, D., Zaguia, A., Enbeyle, W.: Deep neural networks for medical image segmentation. J. Healthc. Eng. 2022 (2022)

19. Malinin, A., et al.: Shifts: a dataset of real distributional shift across multiple large-scale tasks. arXiv preprint arXiv:2107.07455 (2021)
20. Oktay, O., et al.: Attention u-net: learning where to look for the pancreas. arXiv preprint arXiv:1804.03999 (2018)
21. Ronneberger, O., Fischer, P., Brox, T.: U-net: convolutional networks for biomedical image segmentation. In: Navab, N., Hornegger, J., Wells, W.M., Frangi, A.F. (eds.) MICCAI 2015. LNCS, vol. 9351, pp. 234–241. Springer, Cham (2015). https://doi.org/10.1007/978-3-319-24574-4_28
22. Roth, H.R., et al.: Hierarchical 3d fully convolutional networks for multi-organ segmentation. arXiv preprint arXiv:1704.06382 (2017)
23. Simonyan, K., Zisserman, A.: Very deep convolutional networks for large-scale image recognition. arXiv preprint arXiv:1409.1556 (2014)
24. Valanarasu, J.M.J., Patel, V.M.: Unext: MLP-based rapid medical image segmentation network. In: Medical Image Computing and Computer Assisted Intervention-MICCAI 2022, pp. 23–33. Springer, Cham (2022). https://doi.org/10.1007/978-3-031-16443-9_3
25. Yan, X., Tang, H., Sun, S., Ma, H., Kong, D., Xie, X.: After-unet: axial fusion transformer unet for medical image segmentation. In: Proceedings of the IEEE/CVF Winter Conference on Applications of Computer Vision, pp. 3971–3981 (2022)

Interpretability-Guided Data Augmentation for Robust Segmentation in Multi-centre Colonoscopy Data

Valentina Corbetta[1,2]([envelope]) [iD], Regina Beets-Tan[1,2] [iD], and Wilson Silva[1] [iD]

[1] Department of Radiology, The Netherlands Cancer Institute, Amsterdam,
The Netherlands
v.corbetta@nki.nl

[2] GROW School for Oncology and Developmental Biology, Maastricht University
Medical Center, Maastricht, The Netherlands

Abstract. Multi-centre colonoscopy images from various medical centres exhibit distinct complicating factors and overlays that impact the image content, contingent on the specific acquisition centre. Existing Deep Segmentation networks struggle to achieve adequate generalizability in such data sets, and the currently available data augmentation methods do not effectively address these sources of data variability. As a solution, we introduce an innovative data augmentation approach centred on interpretability saliency maps, aimed at enhancing the generalizability of Deep Learning models within the realm of multi-centre colonoscopy image segmentation. The proposed augmentation technique demonstrates increased robustness across different segmentation models and domains. Thorough testing on a publicly available multi-centre dataset for polyp detection demonstrates the effectiveness and versatility of our approach, which is observed both in quantitative and qualitative results. The code is publicly available at: https://github.com/nki-radiology/interpretability_augmentation.

1 Introduction

The adoption of Deep Learning (DL) techniques has significantly advanced medical image segmentation in recent years [4,12]. UNet and other U-shaped architectures have been pivotal in this revolution [11], remaining competitive even with the introduction of newer models.

However, when DL models are applied to unseen datasets acquired from different scanners or clinical centers, their performance at inference time declines noticeably [2,16]. This is due to *domain shifts*, caused by variations in data statistics between different clinical centers, resulting from varying patient populations, scanners, and scan settings [20,21]. These disparities in patient characteristics

Supplementary Information The online version contains supplementary material available at https://doi.org/10.1007/978-3-031-45673-2_33.

and imaging settings can significantly affect the model's ability to generalize effectively [10,21].

To further integrate DL models into clinical practice, it is crucial for them to be robust against these changes and demonstrate a high level of generalizability. The most straightforward approach to address domain shifts is by collecting and annotating as many varied samples as possible. Nevertheless, acquiring and labeling enough data to encompass real-world variation is prohibitively time-consuming and costly.

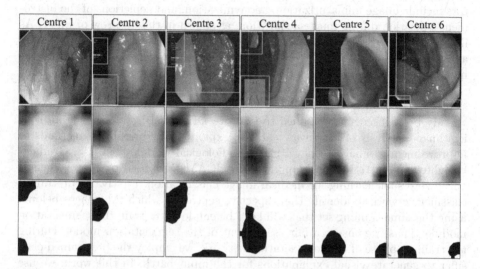

Fig. 1. Illustration of examples of the "extra"-anatomical content across the different centres in the PolypGen dataset and its impact on the GradCAM visualizations. The first row depicts the original image, the second and third row show the GradCAM visualizations and their binarization, respectively.

In recent years, Disentangled Representation Learning (DRL) has emerged as a promising solution to address the aforementioned limitations. This method encodes underlying variation factors into separate latent variables, capturing valuable information relevant to the task at hand. By adopting DRL, DL models gain increased robustness against domain shifts, reducing the need for a large number of meticulously labeled samples [13]. Various DRL models have been employed for segmentation in the context of multi-centre datasets, yielding state-of-the-art outcomes. One such model is the Spatial Decomposition Network (SDNet), which decomposes 2-dimensional (2D) medical images into spatial anatomical factors (content) and non-spatial modality factors (style) [3]. Expanding upon SDNet, Jiang et al. [8] have made additional advancements by further disentangling the pathology factor from the anatomy, particularly when the ground truth mask for anatomy is available. To further improve generalizability, Liu et al. [14] combined DRL with meta-learning, while Shin et al. [18]

have effectively disentangled intensity and non-intensity factors to enable domain adaptation in Computerised Tomography (CT) images.

Despite the significant advancements made by DRL methods in improving model generalizability, it is important to acknowledge that these methods assume that the shift introduced by unseen domains is embedded within the "style" features. However, this assumption does not always hold true, especially in scenarios like videos and images of colonoscopies, or other endoscopy applications. In such cases, various confounding factors affect the content of the images, depending on the domain, from hereon also referred to as *centre*, of acquisition. These factors include image miniaturization, anonymization, and depictions of the instrument's position during image acquisition, as shown in the first row of Fig. 1. As illustrated in Sect. 3, both traditional methods, like UNet and DeepLabV3+ [5], and DRL models like SDNet encounter challenges in generalizing to domains heavily characterized by this additional content that is unrelated to the anatomy.

To address these limitations, we propose an innovative data augmentation strategy based on interpretability techniques. Interpretability techniques have already been successfully applied to improve DL models' performance in medical image analysis tasks: Silva et al. [19] exploited interpretability methods to improve medical image retrieval in the radiological workflow; in [15], Gradientweighted Class Activation Mappings (GradCAMs) are used to improve generalized zero shot learning for medical image classification. Firstly, we pretrain a classifier network to identify the respective centres to which the images belong, using the same training set that will later be employed to train the segmentation module. Thus, we ensure a fair assessment of the segmentation model. During the training phase of the segmentation network, we employ the pre-trained classifier to generate visual explanations for the input batch. In this work, we use GradCAM [17] to produce the visual explanations, a widely adopted technique for visualizing and interpreting the decision-making process of Convolutional Neural Networks (CNNs) in a wide variety of computer vision tasks. By leveraging gradients of the predicted class, GradCAM assigns importance weights to different spatial locations within the last convolutional layer. This allows us to identify the regions in the input image that significantly contribute to the model's decision-making process. Figure 1, specifically the second row, showcases examples of the generated GradCAM visualizations for each centre. Notably, the classifier predominantly focuses on areas where the "extra"-anatomical content resides (indicated by darker regions). We binarize the generated GradCAMs, as depicted in the third row of Fig. 1, and multiply them with a probability p with the input to the segmentation network. Thus, this approach randomly blocks out the additional information, enabling the segmentation network to place greater emphasis on the anatomical regions.

The key contributions of our research can be summarized as follows:

- We introduce a novel data augmentation technique based on interpretability techniques. By incorporating visual explanations, we develop a robust technique that can be readily applied to different and multiple domains. This

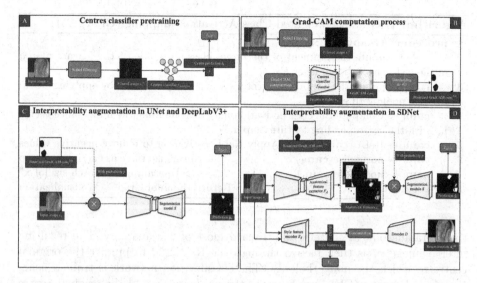

Fig. 2. Illustration of the proposed method. In **A**, the pretraining of the classifier is shown. **B** depicts the process of computing GradCAM visualizations. As for **C** and **D**, they demonstrate the incorporation of the interpretability-guided augmentation in the training phase of UNet and DeepLabV3+, and SDNet, respectively.

stands in contrast to standard augmentation techniques, which are limited in their ability to capture the variability of non-synthetic data [21].

- We apply and adapt the proposed methodology to both two conventional baseline models, UNet and DeepLabV3+ with ResNet101 as backbone, and a DRL model, specifically SDNet. This showcases the versatility of our augmentation strategy across different architectures.
- We conduct thorough testing of our method using an open-source multi-centre dataset, PolypGen [1], to demonstrate the robustness of our technique and its effectiveness in diverse domains.

2   Methodology

The proposed methodology is illustrated in Fig. 2, outlining the pretraining of the classifier module, the generation process of the GradCAM visualizations and their integration into the UNet, DeepLabV3+ and SDNet models.

2.1   Pretraining of Centres Classifier

For the classifier backbone, we employ a ResNet50 architecture [7] with pretraining on ImageNet [6]. The module is trained on the same training set used for subsequent training of the segmentation networks. Given an input image x_i, we initially apply Sobel filtering [9] to emphasize the edges characterizing

---

Algorithm 1. Gradient-weighted Class Activation Mapping (GradCAM)

1: **procedure** GRADCAM($\mathbf{x}_i^s$, $\mathbf{w_c}$)

2:     $\mathbf{g}$ ← Compute the gradient of the target class score c w.r.t the feature map a of the last convolutional layer l_c

3:     $\mathbf{i_w}$ ← Compute the importance vector of each feature channel by applying global average pooling to $\mathbf{g}$

4:     $\mathbf{cam_i}$ ← Compute the class activation map by combining the importance vector $\mathbf{i_w}$ with the corresponding feature map a

5:     $\mathbf{cam_i}$ ← ReLU($\mathbf{cam_i}$)   ▷ Apply ReLU activation to remove negative values

6:     $\mathbf{cam_i}$ ← normalize($\mathbf{cam_i}$)   ▷ Normalize the class activation map

7:     $\mathbf{cam_i}$ ← upsample($\mathbf{cam_i}$)   ▷ Upasample to match size of $\mathbf{x}_i^s$

8:     **return cam$_i$**   ▷ Return the final GradCAM visualization

---

the extra-anatomical content and discard most of the anatomy. The resulting filtered image x_i^s is then passed through the ResNet50 to predict the original centre to which the image belongs. The classification network is trained using the Cross-Entropy (CE) Loss between the predicted $\hat{c}_i$ and the original centre label c_i.

2.2   GradCAM Visualizations Generation Process

During the training process of the segmentation network, the pretrained weights w_c of the centres classifier are loaded and kept frozen. The classifier is then used to perform inference on the Sobel-filtered input image x_i^s. Subsequently, the GradCAM visualization cam_i is generated following the steps outlined in Algorithm 1. The GradCAM visualizations provide a coarse representation of the areas in the input image x_i^s that the classifier focused on to make its prediction. To ensure that we do not inadvertently block useful content for the downstream segmentation task, we binarize cam_i using a threshold $th = 0.5$ to obtain cam_i^{bin}. This ensures that the augmentation procedure described in the subsequent paragraphs masks only the most relevant "extra"-anatomical content.

2.3   Interpretability-Guided Data Augmentation

We will now provide a detailed explanation of how the GradCAM visualizations are utilized as an augmentation technique to enhance the robustness and generalizability of segmentation models.

UNet and DeepLabV3+. The integration of the GradCAM visualizations into the UNet and DeepLabV3+ training process is straightforward. With a probability p, we multiply the input image x_i by the corresponding GradCAM visualization cam_i^{bin}. We introduce a probability p for the multiplication step to mitigate the risk of covering important areas for the downstream task. Figure 3 provides two examples of augmented samples in the UNet and DeepLabV3+

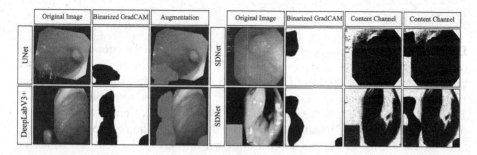

Fig. 3. Examples of augmented samples with our interpretability-guided augmentation, for UNet, DeepLabV3+ and SDNet.

training process. The models are trained by computing the Dice Loss between the predicted segmentation mask $\hat{y}_i$ and the ground truth y_i.

SDNet. To gain a better understanding of how interpretability-guided augmentation is integrated into the SDNet, it is necessary to provide a brief overview of the model's structure. Initially, the anatomy encoder $F_{anatomy}$ encodes input image x_i into a multi-channel spatial representation, the anatomical features z_i. It is important to note that z_i has shape $N \times H \times W$, where N represents a fixed number of channels (e.g. 8), while H and W correspond to the height and width of the original image, respectively. The modality encoder E_s uses factor z_i along with the input image x_i to produce the latent vector s_i, representing the style features. These two representations, s_i and z_i, are combined to reconstruct the input image through the decoder network D. The anatomical representation z_i is then fed into the segmentation network S to produce the segmentation mask $\hat{y}_i$. For a more detailed explanation of the SDNet and its associated losses, we refer the reader to the schematic in Fig. 2(D) and to the original paper [3]. To perform the interpretability-guided augmentation, we multiply the multi-channel anatomical representation z_i with the binary GradCAM visualization cam_i^{bin} using a probability p before feeding it as input to the segmentation module S. Figure 3 provides two examples of augmented samples in the SDNet training process, in particular we report the effect on only one of the N channels of z_i for synthesis purposes. We made the decision to apply the augmentation on the anatomy representation z_i and not on the input image x_i. This approach is intended to mimic the process used in the UNet and DeepLabv3+, where we directly manipulate the input to the module dedicated to the downstream segmentation task, while keeping the rest of the SDNet architecture intact.

3   Results

Dataset and Implementation Details. To evaluate the proposed methodology, we utilized the publicly available PolypGen dataset, which comprises colonoscopy data collected from 6 different centres, encompassing diverse patient populations. Our analysis focused on single frame samples, resulting in a total of 1537 images (Centre 1: 256, Centre 2: 301, Centre 3: 457, Centre 4: 227, Centre 5: 208, Centre 6: 88). Notably, as illustrated in Fig. 1, the centres exhibit substantial variability in image content, both within and across centres. To assess

Table 1. The comparison results of the proposed method with the corresponding baseline models. The best result concerning the specific baseline considered is highlighted in **bold**. In the Dice column, the value of probability p that produced the best result in terms of Dice score in the interpretability-guided augmentation is indicated in parentheses.

	Dice	Recall	Accuracy	Dice	Recall	Accuracy
Out-dist set	UNet			UNet interpretability augmentation		
centre 1	0.7257	0.7725	0.9589	**0.7353 (60%)**	0.7716	0.9587
centre 2	0.5762	**0.6454**	0,9440	**0.6062 (60%)**	0.5948	**0.9530**
centre 3	0.7054	0.6724	0.9619	**0.7470 (40%)**	**0.6966**	**0.9658**
centre 4	0.4223	**0.3804**	0.9594	**0.4519 (40%)**	0.3757	**0.9607**
centre 5	0.4725	0.4934	0.9581	**0.4893 (40%)**	**0.5369**	0.9583
centre 6	0.6423	0.5597	0.9593	**0.6574 (60%)**	**0.6132**	**0.9634**
	Dice	Recall	Accuracy	Dice	Recall	Accuracy
Out-dist set	Deeplabv3+			Deeplabv3+ interpretability augmentation		
centre 1	0.6003	**0.6284**	0.9482	**0.6155 (60%)**	0.6195	**0.9545**
centre 2	0.5498	**0.5358**	0.9398	**0.5679 (60%)**	0.5159	**0.9496**
centre 3	0.5696	0.5407	**0.9585**	**0.6314 (40%)**	**0.6745**	0.9561
centre 4	0.3424	**0.2471**	0.9483	**0.3592 (50%)**	0.2166	**0.9493**
centre 5	0.3867	0.4171	**0.9543**	**0.4046 (60%)**	**0.4479**	0.9533
centre 6	0.6268	0.5932	0.9631	**0.6342 (40%)**	**0.6126**	**0.9658**
	Dice	Recall	Accuracy	Dice	Recall	Accuracy
Out-dist set	SDNet			SDNet interpretability augmentation		
centre 1	0.7130	0.7583	0.9551	**0.7226 (40%)**	**0.7726**	**0.9575**
centre 2	0.5489	**0.5794**	0.9328	**0.5579 (50%)**	0.5482	**0.9464**
centre 3	0.7151	0.7082	**0.9620**	**0.7208 (40%)**	**0.7245**	0.9603
centre 4	**0.3981**	**0.3254**	**0.9591**	0.3841 (50%)	0.3360	0.9557
centre 5	0.4312	0.4293	0.9587	**0.4546 (40%)**	**0.4722**	0.9588
centre 6	0.6398	**0.6195**	0.9610	**0.6626 (60%)**	0.5977	**0.9621**

the generalizability of the models, we conducted 6 distinct experiments for each tested model: we trained and validated the models using 5 centres while reserving one centre as an out-of-distribution test set. Throughout the experiments, a patient-level split was applied for the in-distribution frames, with 80% of patients allocated for training, 10% for validation, and 10% for the in-distribution test set.

Standard preprocessing techniques were applied to the images, including resizing them to a dimension of 256×256 and normalizing the pixel values. Furthermore, additional augmentations were performed exclusively on the training set. These augmentations comprised rotation, horizontal and vertical flips (with a 50% probability), which were also applied to the ground truth and binary GradCAM visualizations. Additionally, colour jitter was applied to the images with a 30% probability. The models were trained for 300 epochs, except for the centres classifier which was trained just for 10 epochs, on an NVIDIA RTX^{TM} A6000 GPU, utilizing a batch size of 4 and a learning rate of 10^{-5}.

Results and Discussion. The experimental results are presented in Table 1, where we evaluate segmentation outcomes using as metrics the Dice Score (DSC), recall, and accuracy. To determine the best-performing augmented models based on DSC, we conduct a parameter study on the probability value p, as elaborated in the following paragraph, and subsequently compare them with the corresponding baseline models. Concerning the DSC, the models trained with interpretability-guided augmentation demonstrate superior performance in nearly all experiments. Particularly noteworthy is the substantial improvement in Centre 3 for the UNet, with an increase of 4.16%. Additionally, the SDNet exhibits an increment of 2.57% in Centre 5, and for DeepLabV3+, there is a significant 6.18% increase in Centre 3. The results also consistently demonstrate improvements in terms of accuracy and recall in nearly all experiments. Figure 4 presents several qualitative examples of the segmentation results. Notably, the masks obtained using our proposed methodology exhibit reduced noise levels, and our approach demonstrates greater performance in detecting smaller polyps.

Parameters Study. We delve deeper into the effectiveness of our proposed technique, conducting a detailed study of the parameters for the all the analyzed architectures. Our augmentation method was employed with probabilities of 40%, 50% and 60%. The results of this study are displayed in the Supplementary Material. The study proves that the fine-tuning of the probability value p within our augmentation approach plays a pivotal role in enhancing the models' generalizability. Indeed, when the suitable probability p is applied, the augmented architectures surpass the performance of the baseline models in nearly all the tests.

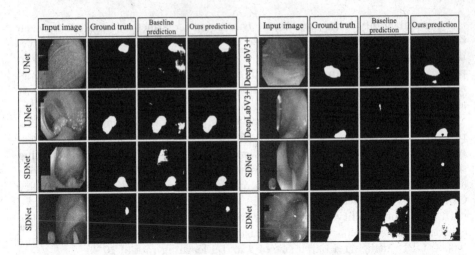

Fig. 4. Examples of qualitative results for the baselines and the interpretability-guided augmentation approach.

4   Conclusion

We have introduced an interpretability-guided augmentation technique aimed at improving the generalizability of DL models to unseen domains in colonoscopy segmentation tasks. We have demonstrated the strength and reliability of this proposed technique by successfully adapting it to three distinct architectures: UNet, SDNet and DeeplabV3+. Our future work will involve testing the proposed augmentation in low-data and semi-supervised settings, where DRL models, and in particular SDNet, significantly outperform conventional models. Moreover, we plan to investigate the adaptability of our methodology to other multi-centre endoscopy images, such as cystoscopy or laparoscopy, as they exhibit the same instrumentation-specific and User Interface (UI) overlays as colonoscopy data. Consequently, our method could improve models' generalizability to such data.

Acknowledgements. Research at the Netherlands Cancer Institute is supported by grants from the Dutch Cancer Society, the Dutch Ministry of Health, Welfare and Sport and private sectors.

References

1. Ali, S., et al.: A multi-centre polyp detection and segmentation dataset for generalisability assessment. Sci. Data **10**(1), 75 (2023)
2. Campello, V.M., et al.: Multi-centre, multi-vendor and multi-disease cardiac segmentation: the M&Ms challenge. IEEE Trans. Med. Imaging **40**(12), 3543–3554 (2021)
3. Chartsias, A., et al.: Disentangled representation learning in cardiac image analysis. Med. Image Anal. **58**, 101535 (2019)

4. Chen, C., et al.: Deep learning for cardiac image segmentation: a review. Front. Cardiovasc. Med. **7**, 25 (2020)

5. Chen, L.-C., Zhu, Y., Papandreou, G., Schroff, F., Adam, H.: Encoder-decoder with atrous separable convolution for semantic image segmentation. In: Ferrari, V., Hebert, M., Sminchisescu, C., Weiss, Y. (eds.) ECCV 2018. LNCS, vol. 11211, pp. 833–851. Springer, Cham (2018). https://doi.org/10.1007/978-3-030-01234-2_49

6. Deng, J., Dong, W., Socher, R., Li, L.J., Li, K., Fei-Fei, L.: ImageNet: a large-scale hierarchical image database. In: 2009 IEEE Conference on Computer Vision and Pattern Recognition, pp. 248–255. IEEE (2009)

7. He, K., Zhang, X., Ren, S., Sun, J.: Deep residual learning for image recognition. In: Proceedings of the IEEE Conference on Computer Vision and Pattern Recognition, pp. 770–778 (2016)

8. Jiang, H., et al.: Semi-supervised pathology segmentation with disentangled representations. In: Albarqouni, S., et al. (eds.) DART/DCL -2020. LNCS, vol. 12444, pp. 62–72. Springer, Cham (2020). https://doi.org/10.1007/978-3-030-60548-3_7

9. Kanopoulos, N., Vasanthavada, N., Baker, R.: Design of an image edge detection filter using the Sobel operator. IEEE J. Solid-State Circuits **23**(2), 358–367 (1988). https://doi.org/10.1109/4.996

10. Li, Y., Chen, J., Xie, X., Ma, K., Zheng, Y.: Self-loop uncertainty: a novel pseudo-label for semi-supervised medical image segmentation. In: Martel, A.L., et al. (eds.) MICCAI 2020. LNCS, vol. 12261, pp. 614–623. Springer, Cham (2020). https://doi.org/10.1007/978-3-030-59710-8_60

11. Liu, L., Cheng, J., Quan, Q., Wu, F.X., Wang, Y.P., Wang, J.: A survey on U-shaped networks in medical image segmentations. Neurocomputing **409**, 244–258 (2020)

12. Liu, X., Song, L., Liu, S., Zhang, Y.: A review of deep-learning-based medical image segmentation methods. Sustainability **13**(3), 1224 (2021)

13. Liu, X., Sanchez, P., Thermos, S., O'Neil, A.Q., Tsaftaris, S.A.: Learning disentangled representations in the imaging domain. Med. Image Anal. **80**, 102516 (2022)

14. Liu, X., Thermos, S., O'Neil, A., Tsaftaris, S.A.: Semi-supervised meta-learning with disentanglement for domain-generalised medical image segmentation. In: de Bruijne, M., et al. (eds.) MICCAI 2021. LNCS, vol. 12902, pp. 307–317. Springer, Cham (2021). https://doi.org/10.1007/978-3-030-87196-3_29

15. Mahapatra, D., Ge, Z., Reyes, M.: Self-supervised generalized zero shot learning for medical image classification using novel interpretable saliency maps. IEEE Trans. Med. Imaging **41**(9), 2443–2456 (2022). https://doi.org/10.1109/TMI.2022.3163232

16. Prados, F., et al.: Spinal cord grey matter segmentation challenge. Neuroimage **152**, 312–329 (2017)

17. Selvaraju, R.R., Cogswell, M., Das, A., Vedantam, R., Parikh, D., Batra, D.: Grad-CAM: visual explanations from deep networks via gradient-based localization. In: Proceedings of the IEEE International Conference on Computer Vision, pp. 618–626 (2017)

18. Shin, S.Y., Lee, S., Summers, R.M.: Unsupervised domain adaptation for small bowel segmentation using disentangled representation. In: de Bruijne, M., et al. (eds.) Medical Image Computing and Computer Assisted Intervention-MICCAI 2021: 24th International Conference, Strasbourg, France, 27 September–1 October 2021, Proceedings, Part III 24, vol. 12903, pp. 282–292. Springer, Cham (2021). https://doi.org/10.1007/978-3-030-87199-4_27

19. Silva, W., Poellinger, A., Cardoso, J.S., Reyes, M.: Interpretability-guided content-based medical image retrieval. In: Martel, A.L., et al. (eds.) MICCAI 2020. LNCS, vol. 12261, pp. 305–314. Springer, Cham (2020). https://doi.org/10.1007/978-3-030-59710-8_30

20. Tao, Q., et al.: Deep learning-based method for fully automatic quantification of left ventricle function from cine MR images: a multivendor, multicenter study. Radiology **290**(1), 81–88 (2019)

21. Zhang, L., et al.: Generalizing deep learning for medical image segmentation to unseen domains via deep stacked transformation. IEEE Trans. Med. Imaging **39**(7), 2531–2540 (2020)

Improving Automated Prostate Cancer Detection and Classification Accuracy with Multi-scale Cancer Information

Cynthia Xinran Li[1]([✉]), Indrani Bhattacharya[2], Sulaiman Vesal[3],
Sara Saunders[3][iD], Simon John Christoph Soerensen[3,4], Richard E. Fan[3],
Geoffrey A. Sonn[2,3], and Mirabela Rusu[2,3,5][iD]

[1] Institute for Computational and Mathematical Engineering, Stanford University,
Stanford, CA 94305, USA
xli0429@stanford.edu
[2] Department of Radiology, Stanford University School of Medicine, Stanford, CA
94305, USA
[3] Department of Urology, Stanford University School of Medicine, Stanford, CA
94305, USA
[4] Department of Epidemiology and Population Health, Stanford University School of
Medicine, Stanford, CA 94305, USA
[5] Department of Biomedical Data Science, Stanford University, Stanford, CA 94305,
USA

Abstract. Automated detection of prostate cancer via multi-parametric
Magnetic Resonance Imaging (mp-MRI) could help radiologists in the
detection and localization of cancer. Several existing deep learning-based
prostate cancer detection methods have high cancer detection sensitivity
but suffer from high rates of false positives and misclassification between
indolent (Gleason Pattern $= 3$) and aggressive (Gleason Pattern ≥ 4)
cancer. In this work, we propose a multi-scale Decision Prediction Module (DPM), a novel lightweight false-positive reduction module that can
be added to cancer detection models to reduce false positives, while maintaining high sensitivity. The module guides pixel-level predictions with
local context information inferred from multi-resolution coarse labels,
which are derived from ground truth pixel-level labels with patch-wise
calculation. The coarse label resolution varies from a quarter size and 16
times smaller, to a single label for the whole slice, indicating that the
slice is normal, indolent, or aggressive. We also propose a novel multi-
scale decision loss that supervises cancer prediction at each resolution.
Evaluated on an internal test set of 56 studies, our proposed model,
DecNet, which adds the DPM and multi-scale loss to the baseline model
SPCNet, significantly increases precision from 0.49 to 0.63 ($p \leq 0.005$ in
paired t-test) while keeping the same level of sensitivity (0.90) for clinically significant cancer predictions. Our model also significantly outperforms U-Net in sensitivity and Dice coefficient ($p \leq 0.05$ and $p \leq 0.005$,

---

Supplementary Information The online version contains supplementary material
available at https://doi.org/10.1007/978-3-031-45673-2_34.

X. Cao et al. (Eds.): MLMI 2023, LNCS 14348, pp. 341–350, 2024.
https://doi.org/10.1007/978-3-031-45673-2_34

respectively). As shown in the appendix, a similar trend was found when validating with an external dataset containing multi-vendor MRI exams. An ablation study on different label resolutions of the DPM shows that decision loss at all three scales achieves the best performance.

Keywords: prostate cancer · automated segmentation · false-positive reduction

1   Introduction

Prostate cancer is the most commonly diagnosed cancer in American men [13]. Magnetic Resonance Imaging (MRI) is increasingly used for the early detection of prostate cancer and subsequent biopsy and treatment planning. However, the interpretation of MR images is challenging due to the subtle differences between benign and cancerous tissue. Moreover, it is difficult for radiologists to separate indolent cancer from aggressive cancer correctly due to overlapping MRI features. As such, radiologists miss up to 34% of aggressive cancers and 81% of indolent cancers on MRI exams of men undergoing radical prostatectomy [1]. Radiologist interpretations of MRI also suffer from wide inter-reader variability [14].

Recent deep learning-based models have demonstrated prostate cancer detection sensitivities similar to experienced radiologists [3,5]. However, these models often also demonstrate high false positive rates, thus showing unsatisfying specificity for clinical deployment.

To improve cancer detection accuracy and specificity, some machine learning methods specifically integrate false positive reduction modules [11,15]. Yu et al. [15] used the true positive and false positive prediction patches from the detection network to train a classification network. However, this approach requires separate training for the detection and discrimination networks and is time-consuming. The false positive reduction module used by Saha et al. [11] considers patches of the input images and predicts their malignancy scores on a patch level. While the patch-level classification scheme in [11] can be trained jointly with the detection model, the false positive reduction module based on residual learning [4,6] is still heavyweight since it learns similar level of image details as the original image, and requires larger training sets. Moreover, although existing studies focused on reducing false positive predictions to improve clinically significant prostate cancer (csPCa), they did not consider simultaneously improving the classification and localization of aggressive and indolent cancer components in mixed lesions.

In this study, we propose a convolutional neural network model Decision Network for prostate cancer detection (DecNet), that combines a lightweight Decision Prediction Module (DPM) and a novel multi-scale decision loss with an existing baseline model to improve MRI-based prostate cancer detection. DPM regularizes predictions on multiple scales and effectively improves prediction accuracy. DPM leverages the ground truth pixel-level labels to calculate a series of coarse labels, which captures local to global information of the ground truth label. The proposed multi-scale decision loss adjusts the pixel-level predictions with the coarse labels. DPM can be easily attached to and trained with any deep learning-based cancer detection model. It is lightweight because it only requires parameters of 1.4% of the input volume. Moreover, DPM and multi-scale loss consider cancer subtypes, enabling distinguishing not only between normal and cancer tissue, but also between aggressive and indolent cancer subtypes. Model performances are evaluated on both lesion level and patient level.

2 Methods

2.1 Data

Our IRB-approved study includes 156 patients from Stanford University Medical Center comprised of 117 patients who underwent radical prostatectomy, and 39 patients without prostate cancer, confirmed by a negative biopsy. The MRI exam included T2-weighted (T2w), and Apparent Diffusion Coefficient (ADC) maps acquired using 3 T GE scanners.

The high-resolution whole-mount histopathology images of the patients that underwent radical prostatectomy, were registered [9] to the pre-surgery MRI, which enables the mapping of ground truth histopathology labels onto T2w images. ADC images are registered manually to T2 images with affine registration. T2w and ADC images are resampled to size of 224×224 with a pixel size of 0.29×0.29 mm^2. We standardized the T2w and ADC intensities within the prostate gland following [7] to ensure similar MRI intensity distribution among patients.

Ground truth labels included pixel-level Gleason patterns from a deep learning model [10] confirmed by pathologists with > 6 months of experience. We consider pixels of Gleason pattern ≥ 4 to be aggressive cancer (labeled 2) and Gleason pattern $= 3$ to be indolent cancer (labeled 1). Normal pixels are labeled 0. Pixels with overlapping aggressive and indolent cancer labels were considered aggressive. The prostate was outlined by image analysis experts with > 6 months of experience on T2w images.

2.2 DecNet

Our proposed model, DecNet has two parts: (1) a Cancer Detection Network, and (2) a Decision Prediction Module (DPM) (Fig. 1).

(1) **Cancer Detection Network:** The Cancer Detection Network of DecNet consists of a pixel-level prostate cancer semantic segmentation model inspired by Stanford Prostate Cancer Network (SPCNet) [12]. As shown in Fig. 1, DecNet uses a 2.5-dimensional input where three adjacent slices of T2w and ADC images are used to predict cancer on the center slice. T2w and ADC images are input into two separate branches for feature extraction, and subsequently

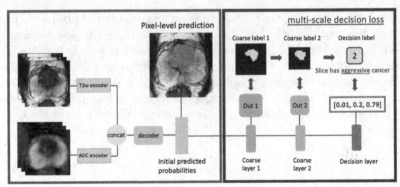

Cancer Detection Network   **Decision Prediction Module**

Fig. 1. Overview of DecNet. The cancer detection network has output from two branches of encoders concatenated together and passed through a decoder for the initial pixel-level probabilities of cancer. The Decision Prediction Module (DPM) contains three convolutional layers that predict fine to coarse image labels (layers and outputs are shown in corresponding colors). Labels generation process is shown in Fig. 2. A multi-scale decision loss was computed as the weighted sum of the losses of the three layers in DPM. The total loss combines multi-scale decision loss and the cross-entropy loss.

concatenated for the prediction of cancer probability maps. The resulting prediction is masked with the prostate gland segmentation to only backpropagate through pixels within the prostate. The loss function for pixel-level prediction is a weighted cross-entropy loss where each class is weighted by the inverse proportion of the class size across all labels in the training dataset

$$L_{pixel} = -\frac{1}{M} \sum_{i=1}^{N} N_{\hat{y}_i} \frac{\exp(y_{i,\hat{y}_i})}{\sum_{c=0}^{2} \exp(y_{i,c})},$$

where $\hat{y}_i \in \{0, 1, 2\}$ is the pixel in the ground truth label representing normal, indolent, and aggressive. y_i is the prediction of each pixel and is a vector with 3 elements. $N_{\hat{y}_i}$ refers to the number of pixels of class $\hat{y}_i$ across all images in the training data masked by the prostate gland segmentations, and $M = N_0 + N_1 + N_2$.

(2) Decision Prediction Module (DPM): We attach DPM to the last layer of the cancer detection network to improve prediction performance. DPM starts with 2 consecutive convolutional layers with kernel size 4 and stride 4, which sequentially encodes information in 4×4 and 16×16 patches from the initial prediction map of the cancer detection model. We supervise the output of each layer with low-resolution labels described below and shown in Fig. 2. Since our X-Y image size is 224×224, we can only apply 2 layers until the low-resolution label is too coarse to provide additional information (see Fig. 2 Coarse label 2). At the last layer of DPM, we perform a convolution with kernel size 14, generating a single prediction to encode the slice-level probability to be one of the three classes, i.e., normal, indolent, or aggressive.

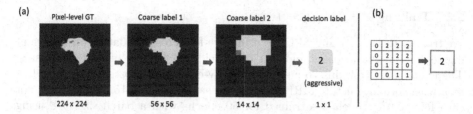

Fig. 2. Generation of coarse labels from ground truth (GT) pixel-level labels. (a): the process of coarse label generation, indolent (green) and aggressive (yellow) regions are kept. (b): Sample maximum calculation of a 4×4 patch towards the next coarse level. (Color figure online)

We acquire fine-to-coarse labels from ground truth pixel-level labels to add supervision at each level, encouraging the model to "make decisions" matching the ground truth label from local to global scales. The label generation process is shown in Fig. 2(a). Mimicking the convolution process of DPM, we divide the ground truth labels into grids of 4×4 and 16×16 patches. We compute the maximum value within each patch and assign it as the label at the coarser level. An example of maximum computation for a 4×4 patch is shown in Fig. 2(b). In the last step, we take the largest value within the ground truth label slice as the slice decision label. Taking the maximum ensures that cancer pixels inside the patch are kept throughout the process while normal patches remain normal. In the case that both indolent and aggressive pixels are present, the aggressiveness will be kept. Although normal regions can be contaminated by neighboring cancer pixels when the patch size is large, this information is kept in the finer labels and propagates through layers with our multi-scale loss function in Eq. 1.

Multi-scale Decision Loss: The multi-scale decision loss is defined as the weighted sum of losses of the 3 layers in DPM:

$$L_{decision} = \sum_{k=1}^{3} \alpha_k L_k, \tag{1}$$

where the loss of each layer L_k is the cross entropy loss between the layer output and the coarse label of the layer described above:

$$L_k = \sum_{i=1}^{N_k} \frac{\exp(d^k_{i,\hat{d}^k_i})}{\sum_{c=0}^{2} \exp(d^k_{i,c})}, \tag{2}$$

where N_k refers to the number of pixels at coarse level k, and $\hat{d}^k \in \{0, 1, 2\}$ represents the normal, indolent, or aggressive class in the coarse label at level k. With backpropagation, pixel-level prediction is adjusted with global to local cancer information, and the adjustments are performed locally with our strided convolution. We adapt the ratio between pixel-level loss and multi-scale decision loss with weight factor λ. The combined loss is $L_{total} = L_{pixel} + \lambda L_{decision}$.

2.3   Training

We train both DecNet and SPCNet using five-fold cross-validation on the train-ing set of 100 (65%) cases and test on a held-out test set of 56 (35%) cases. The training set contains 76 men with biopsy-confirmed prostate cancer and 24 men with normal prostate MRI, and the test set has 41 and 15, respectively. For each fold, both models are trained for 50 epochs with a batch size of 8 using Adam optimizer with learning rate 5×10^{-4}. The test set predictions are the average of predictions of each of the 5-fold models. In our training, we chose $\alpha_1 = \alpha_2 = \alpha_3 = 1$ to enforce equal importance in global to local information. We performed a grid search for λ and found $\lambda = 0.5$ has the best performance.

2.4   Evaluation

We evaluate the models on both lesion-level and patient-level. For lesion-level evaluation, we process predicted labels with the morphological closing operation to fill small holes in the prediction. Lesion candidates with a volume smaller than 250 mm^3 are dropped. A lesion prediction is considered a true positive if there is at least 10% overlap between the predicted lesion and the ground truth lesion. For assessing true negatives and false positives, we divide the prostate into 6 roughly equal regions (sextants) by dividing the prostate into left and right halves, and dividing each half into apex, mid, and base. A prostate sextant is considered negative if it contains less than 5% cancer pixels. If the model predicts at least 90% of a ground truth negative sextant as negative, the prediction is considered as true negative; otherwise, the prediction is a false positive prediction.

We perform the lesion-level evaluation in 2 ways: clinically significant cancer detection (defined as lesions containing > 1% aggressive voxels, i.e., Gleason Pat-tern ≥ 4), and indolent cancer detection (only indolent voxels). We compare the performance of DecNet with the SPCNet and a vanilla U-Net [8] and branched U-Net (br-U-Net) [2]. For patient-level evaluation, we evaluate the models' per-formances in distinguishing patients with and without cancer ((including both indolent and aggressive pixels, i.e., Gleason Patterns ≥ 3). A prediction is con-sidered true positive if at least one lesion is detected. For normal cases, the prediction is true negative if none of the 6 divided regions has a false posi-tively predicted lesion. Otherwise, the prediction is a false positive. For both evaluations, we compute several standard evaluation metrics based on the above criterion: PR-AUC, sensitivity, precision, Dice, and Accuracy.

Table 1. Lesion-level evaluation

Clinically Significant Cancer vs. Normal					
Model	PR-AUC	Sensitivity	Precision	Dice	Accuracy
br-U-Net	0.73 ± 0.39	0.54 ± 0.49	0.44 ± 0.47	0.22 ± 0.21	0.77 ± 0.23
U-Net	0.83 ± 0.32	0.74 ± 0.43	0.55 ± 0.42	0.29 ± 0.22	0.77 ± 0.23
SPCNet	0.75 ± 0.36	$\mathbf{0.94 \pm 0.22}$	0.49 ± 0.28	0.36 ± 0.19	0.64 ± 0.27
DecNet	$\mathbf{0.89 \pm 0.27}$	0.90 ± 0.28	$\mathbf{0.63 \pm 0.33}$	$\mathbf{0.42 \pm 0.22}$	$\mathbf{0.80 \pm 0.23}$
Indolent vs. Normal					
Model	PR-AUC	Sensitivity	Precision	Dice	Accuracy
br-U-Net	0.73 ± 0.37	0.25 ± 0.43	0.21 ± 0.39	0.09 ± 0.09	0.74 ± 0.18
U-Net	$\mathbf{0.82 \pm 0.31}$	0.38 ± 0.48	0.38 ± 0.48	0.11 ± 0.12	$\mathbf{0.80 \pm 0.19}$
SPCNet	0.75 ± 0.35	0.42 ± 0.45	0.43 ± 0.46	0.13 ± 0.12	0.77 ± 0.16
DecNet	0.74 ± 0.33	$\mathbf{0.64 \pm 0.45}$	$\mathbf{0.58 \pm 0.45}$	$\mathbf{0.18 \pm 0.14}$	$\mathbf{0.80 \pm 0.20}$

3 Results

3.1 Quantitative Results

In Table 1, for clinically significant cancer detection, SPCNet has the highest sensitivity. While DecNet has slightly lower but similar sensitivity compared to SPCNet, it has statistically significantly improved sensitivity from UNet ($p \leq 0.05$) in paired t-test. For precision, DecNet has significantly improved from 0.49 to 0.63 ($p \leq 0.005$) over SPCNet, and slightly improved from UNet. The Dice coefficient has also been statistically significantly improved from both UNet and SPCNet, with $p \leq 0.005$ in both paired t-tests. The average volume of predicted false positive voxels dropped from 1187 mm^3 to 461 mm^3, comparing DecNet and SPCNet when predicting clinically significant cancer. In both tasks, DecNet reduced false positives while retaining true positive predictions. DecNet also has the highest sensitivity, precision, and Dice in predicting indolent cancer alone, with accuracy close to SPCNet and UNet. This shows that DecNet detects more indolent lesions compared to SPCNet.

Table 2 details model performances of predicting cancer (indolent and aggressive) compared to SPCNet evaluated at the patient level. DecNet improves accuracy and F1-score, and better distinguishes between patients with and without cancer.

Table 2. Patient-level Evaluation Cancer vs. Normal

Model	Sensitivity	Specificity	Precision	NPV	F1-score	Accuracy
SPCNet	**0.94**	0.26	0.76	0.66	0.84	0.75
DecNet	0.92	**0.73**	**0.89**	**0.79**	**0.90**	**0.86**

3.2   Qualitative Results

In the first example in Fig. 3 Row 1, both DecNet and SPCNet detected the aggressive and indolent components within the lesion, while DecNet avoided the two false positives at the lower corners that SPCNet detected (pointed by orange arrows). In Fig. 3 Row 2, the example slice has only indolent cancer. DecNet successfully detected the lesion at the lower left and correctly identified it as indolent cancer, while SPCNet missed the lesion. Both models missed the small indolent lesion at the lower right. In the last row in Fig. 3, the slice does not show cancer. While both models predicted some false positive lesion on the slice, DecNet predicted an indolent lesion in a much smaller size, while SPCNet predicted both indolent and aggressive lesions on the slice with a much larger size.

4   Discussion and Conclusion

We presented DecNet, a prostate cancer detection network with a novel Decision Prediction Module (DPM) and multi-scale loss function, that improves prostate

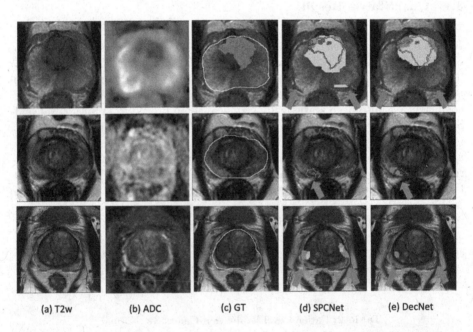

(a) T2w   (b) ADC   (c) GT   (d) SPCNet   (e) DecNet

Fig. 3. Qualitative performance comparison between DecNet and SPCNet in three different test set patients. (a) T2w image, (b) ADC image, and (c) T2w image overlaid with ground truth indolent (green) and aggressive (yellow) cancer labels. Prostate gland segmentation is outlined in white in (c). Indolent (green) and aggressive (yellow) cancer predictions in (d), (e), with ground truth labels outlined. (d) SPCNet predictions (e) DecNet predictions. Orange arrows point to detailed differences. (Color figure online)

cancer detection performance by reducing false positives while retaining true positive predictions. The DPM module is lightweight and can be trained end-to-end with no extra label or training. The proposed multi-scale decision loss encourages pixel-level predictions to combine information from neighboring pixels within the same image patch in a cascaded way.

We showed that DPM improves the performance over SPCNet, U-Net and branched U-Net in detecting both clinically significant cancer and indolent cancer. The evaluation of our internal dataset is limited since it contains MRI exams solely from GE scanners, so we also evaluated the effectiveness of DecNet with the multi-vendor, publicly available Prostate Imaging: Cancer AI (PI-CAI) dataset, containing 1500 MRI exams from Philips and Siemens scanners in Appendix.A. We saw DecNet outperforms SPCNet on this external dataset when evaluated using both PI-CAI metrics, and our own evaluation metrics. We have also performed an ablation study in Appendix.B on the layers of DPM, and we found using decision loss at all three scales achieves the best performance.

Falsely identified and classified prostate cancer affects treatment planning. Thus, it is crucial for automated methods to accurately detect and classify cancer aggressiveness to avoid unnecessary biopsy and overtreatment. By reducing false positive predictions and providing indolent and aggressive cancer segmentations with higher accuracy, DecNet improves automated cancer detection, localization, and aggressiveness classification.

Acknowledgements. This work was supported by the Departments of Radiology and Urology at Stanford University, and by the National Cancer Institute of the National Institutes of Health (R37CA260346). The content is solely the responsibility of the authors and does not necessarily represent the official views of the National Institutes of Health.

References

1. Johnson, D.C., et al.: Detection of individual prostate cancer foci via multiparametric magnetic resonance imaging. Eur. Urol. **75**(5), 712–720 (2019)
2. Bhattacharya, I., et al.: Selective identification and localization of indolent and aggressive prostate cancers via CorrSigNIA: an MRI-pathology correlation and deep learning framework. Med. Image Anal. **75**, 102288 (2022)
3. Cao, R., et al.: Joint prostate cancer detection and Gleason score prediction in mp-MRI via FocalNet. IEEE Trans. Med. Imaging **38**(11), 2496–2506 (2019). https://doi.org/10.1109/TMI.2019.2901928
4. He, K., Zhang, X., Ren, S., Sun, J.: Identity mappings in deep residual networks. In: Leibe, B., Matas, J., Sebe, N., Welling, M. (eds.) Computer Vision-ECCV 2016: 14th European Conference, Amsterdam, The Netherlands, 11–14 October 2016, Proceedings, Part IV 14, vol. 9908, pp. 630–645. Springer, Cham (2016). https://doi.org/10.1007/978-3-319-46493-0_38
5. Hosseinzadeh, M., Saha, A., Brand, P., Slootweg, I., de Rooij, M., Huisman, H.: Deep learning-assisted prostate cancer detection on bi-parametric MRI: minimum training data size requirements and effect of prior knowledge. Eur. Radiol. **32**, 2224–2234 (2021). https://doi.org/10.1007/s00330-021-08320-y

6. Hu, J., Shen, L., Sun, G.: Squeeze-and-excitation networks, pp. 7132–7141 (2018)
7. Nyúl, L., Udupa, J., Zhang, X.: New variants of a method of MRI scale standardization. IEEE Trans. Med. Imaging 19(2), 143–150 (2000). https://doi.org/10.1109/42.836373
8. Ronneberger, O., Fischer, P., Brox, T.: U-Net: convolutional networks for biomedical image segmentation. In: Navab, N., Hornegger, J., Wells, W.M., Frangi, A.F. (eds.) Medical Image Computing and Computer-Assisted Intervention - MICCAI 2015, vol. 9351, pp. 234–241. Springer, Cham (2015). https://doi.org/10.1007/978-3-319-24574-4_28
9. Rusu, M., et al.: Registration of presurgical MRI and histopathology images from radical prostatectomy via RAPSODI. Med. Phys. 47(9), 4177–4188 (2020)
10. Ryu, H.S., et al.: Automated Gleason scoring and tumor quantification in prostate core needle biopsy images using deep neural networks and its comparison with pathologist-based assessment. Cancers 11(12), 1860 (2019). https://doi.org/10.3390/cancers11121860
11. Saha, A., Hosseinzadeh, M., Huisman, H.: End-to-end prostate cancer detection in BPMRI via 3D CNNs: effects of attention mechanisms, clinical priori and decoupled false positive reduction. Med. Image Anal. 73, 102155 (2021). https://doi.org/10.1016/j.media.2021.102155, https://www.sciencedirect.com/science/article/pii/S1361841521002012
12. Seetharaman, A., et al.: Automated detection of aggressive and indolent prostate cancer on magnetic resonance imaging. Med. Phys. 48(6), 2960–2972 (2021)
13. Siegel, R.L., Miller, K.D., Fuchs, H.E., Jemal, A.: Cancer statistics, 2022. Cancer J. Clin. 72(1), 7–33 (2022)
14. Sonn, G.A., et al.: Prostate magnetic resonance imaging interpretation varies substantially across radiologists. Eur. Urol. Focus 5(4), 592–599 (2019). https://doi.org/10.1016/j.euf.2017.11.010, https://www.sciencedirect.com/science/article/pii/S2405456917302663
15. Yu, X., et al.: False positive reduction using multiscale contextual features for prostate cancer detection in multi-parametric MRI scans. In: 2020 IEEE 17th International Symposium on Biomedical Imaging (ISBI), pp. 1355–1359 (2020). https://doi.org/10.1109/ISBI45749.2020.9098338

Skin Lesion Segmentation Improved by Transformer-Based Networks with Inter-scale Dependency Modeling

Sania Eskandari[✉], Janet Lumpp, and Luis Sanchez Giraldo[✉]

Department of Electrical Engineering, University of Kentucky, Lexington, USA
{ses235,luis.sanchez}@uky.edu

Abstract. Melanoma, a dangerous type of skin cancer resulting from abnormal skin cell growth, can be treated if detected early. Various approaches using Fully Convolutional Networks (FCNs) have been proposed, with the U-Net architecture being prominent To aid in its diagnosis through automatic skin lesion segmentation. However, the symmetrical U-Net model's reliance on convolutional operations hinders its ability to capture long-range dependencies crucial for accurate medical image segmentation. Several Transformer-based U-Net topologies have recently been created to overcome this limitation by replacing CNN blocks with different Transformer modules to capture local and global representations. Furthermore, the U-shaped structure is hampered by semantic gaps between the encoder and decoder. This study intends to increase the network's feature re-usability by carefully building the skip connection path. Integrating an already calculated attention affinity within the skip connection path improves the typical concatenation process utilized in the conventional skip connection path. As a result, we propose a U-shaped hierarchical Transformer-based structure for skin lesion segmentation and an Inter-scale Context Fusion (ISCF) method that uses attention correlations in each stage of the encoder to adaptively combine the contexts from each stage to mitigate semantic gaps. The findings from two skin lesion segmentation benchmarks support the ISCF module's applicability and effectiveness. The code is publicly available at https://github.com/saniaesk/skin-lesion-segmentation.

Keywords: Deep learning · Transformer · Skin lesion segmentation · Inter-scale context fusion

1 Introduction

The skin comprises three layers: the epidermis, dermis, and hypodermis [13]. When exposed to ultraviolet radiation from the sun, the epidermis produces melanin, which can be produced at an abnormal rate if too many melanocytes are present. Malignant melanoma is a deadly form of skin cancer caused by the abnormal growth of melanocytes in the epidermis; in 2023, it was estimated that

© The Author(s), under exclusive license to Springer Nature Switzerland AG 2024
X. Cao et al. (Eds.): MLMI 2023, LNCS 14348, pp. 351–360, 2024.
https://doi.org/10.1007/978-3-031-45673-2_35

there would be 97,610 new cases of melanoma with a mortality rate of 8.18% [21]. The survival rate drops from 99% to 25% when melanoma is diagnosed at an advanced stage due to its aggressive nature [1,22]. Therefore, early diagnosis is crucial in reducing the number of deaths from this disease. Dermoscopy has been introduced to improve the diagnosis procedure; it is a non-invasive technique that produces lighted and improved pictures of skin patches. Dermatologists use this method to detect skin cancer, which was formerly done by visual examination and manual screening, which was ineffective and time-consuming [28]. Size, symmetry, boundary definition, and irregularity of lesion shape are essential for identifying skin cancer. Localization and delineation of lesions are required for both surgical excision and radiation treatment [22]. Manual delineation is a tedious and laborious task. Therefore, automatic segmentation becomes essential for developing pre and post-diagnosis processes in computer-aided diagnosis (CAD); however, automatic segmentation is challenging. Lighting and contrast difficulties, underlying inter-class similarities and intra-class variations, occlusions, artifacts, and various imaging tools impede automated skin lesion segmentation. The scarcity of large datasets with expert-generated ground-truth segmentation masks exacerbates the situation, hindering both model training and trustworthy assessment.

Before the Deep Learning (DL) era, most of the segmentation algorithms were based on hand-crafted and classical vision-based and conventional machine learning-based techniques. Celebi et al. [10] utilized adaptive thresholding, and in another study, [9], they investigated a region growing strategy, Erkol et al. [11] applied the active contour method, and Hwang et al. [15] proposed a hybrid segmentation pipeline with an unsupervised clustering remedy where a hierarchical k-means with a level set technique was used. The aforementioned algorithms rely on human-engineered features, which may be challenging to construct and frequently have low invariance and discriminative ability. As a result, the inadequacies of classic ML-based segmentation approaches emphasize the need for more advanced DL-based methods capable of handling complex data and producing more accurate results.

U-Net [19] is the de facto in segmentation tasks. This DL-based framework is a cornerstone in medical image segmentation. U-Net is a hierarchical encoder-decoder framework comprising successive convolution operations in the encoding path, which downsample the spatial resolution while embedding the input space in a high dimensional space. The encoder provides a highly semantic representation, which is gradually upsampled in the decoding path to recover the input's spatial dimensions. Skip connections between the encoder and decoder in this design are used to mitigate the loss of spatial information, which is vital for segmentation tasks. Due to the modular design of the U-Net, thousands of variants of this network have been introduced that alleviate any shortcomings of the U-Net [3], *e.g.*, U-Net++ [29], H-DenseUNet [16], Attention U-Net [18], *etc.*. U-Net++ [29] takes advantage of embedding nested U-Net structures in each stage by using a dense flow of semantic information from the encoder to the decoder with skip connections. Li et al. [16] replaced each naive convolutional encoder block with residual blocks besides using dense skip connections to extract more semantic information. However, applying successive convolution operations in dense structures still could not prevent the CNN-based U-shaped frameworks from having limited receptive fields. Attention U-Net [18] utilized the image-grid-based gating module that includes skip connections to let signals pass through and capture the gradient of relevant localization information from the encoder path before it merges with decoder features on the same scale. This strategy was the first seminal medical image segmentation study investigating the attention mechanism.

Transformer models in language translation tasks have been a huge success. This is related to its ability to calculate the self-affinity between the input tokens [23]. Dosovitskiy et al. [8] proposed a Vision Transformer (ViT) which implements the attention mechanism on an image by partitioning the input images into a 1D sequence of patches to address the lack of globality of convolution operations. Soon, the Vision Transformer's ability to capture long-range dependencies in encoding the object's shape information inspired several studies that utilize the ViT for various tasks such as classification [8] and segmentation [5]. However, due to the lack of intrinsic spatial inductive bias (impeding the ViT from capturing local representation) and the quadratic computational complexity of ViT (making ViT to be data hungry) with respect to the number of patches, the vanilla ViT performs poorly in dense prediction tasks like segmentation and object detection in comparison with the CNN models. Therefore, to mitigate the loss of local interactions within ViT, TransUNet [5], TransBTS [25], UCTransNet [24], and FAT-Net [26] successfully bridge the CNN and Transformer designs in hybrid models. Due to the U-shaped design's success [5,25,26] hierarchical CNN-Transformer models tried capturing long and local dependencies simultaneously. The main drawbacks of these methods are that they still suffer from a high number of parameters and are dependent on the pre-trained weights to perform competitively. UCTransNet [24] investigated the semantic gap between encoder and decoder by designing a new Transformer-based module over skip connections to fuse the multi-scale spatial semantic information, but their method still requires pre-training weights.

Thus, various studies, such as the Efficient Transformer [27] and the Swin Transformer [17], have explored minimizing this computational burden to make ViTs suited for segmentation tasks by delving into the inner structure of the Transformer's multi-head self-attention (MHSA) calculation or by changing the tokenization process. Swin-Unet [4] is a hierarchical U-shaped pure Transformer structure that successfully utilized a Linear Swin Transformer as a main counterpart to segment the abdominal computer tomography inputs. Due to the shifting window strategy in Swin blocks, Swin-Unet captures the contextual information locally and is heavily dependent on pre-training weights. Huang et al. [14] applied the Efficient Transformer from [27] for medical image segmentation as a pure Transformer design, namely MISSFormer. Efficient Transformer [27] utilizes the irreversible downsampling step after the patch embedding to lessen the computational complexity, but this method suffers from loss of spatial information.

Our Contribution – To address the aforementioned deficiencies, we propose a new pure Transformer-based U-shaped structure that utilizes the efficient attention mechanism by Shen et al. [20] to capture the global context in linear complexity without redundant context extraction. Moreover, to shield the U-Net-like structures from the semantic gaps between the encoder and decoder, we devised a new module that uses an already calculated attention correlation at various scales to fuse the attention information for better localization. Our contributions are as follows: ❶ A novel Transformer-based structure in a U-shaped framework to capture the global dependency in an efficient manner without the need for pre-training weights (see Fig. 1). ❷ The design of a new skip connection module that integrates the multi-scale attention maps to lessen the encoder-decoder gap rather than the plain copy-and-paste skip connection paradigm, namely **Inter-Scale Context Fusion (ISCF)**. It is noteworthy that this study is the extended abstract version of [12]. ❸ SOTA results on two public skin lesion segmentation datasets and publicly available implementation source code via GitHub.

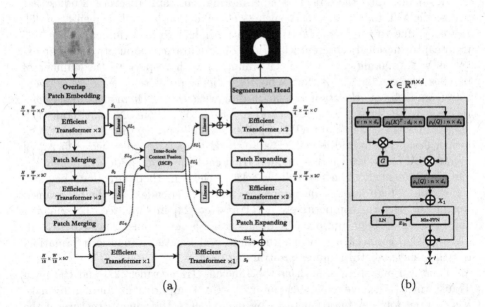

Fig. 1. (*a*) Proposed method. Our U-shaped structure comprises modified Efficient attention. (*b*) Our Efficient Transformer block.

2   Proposed Method

Our overall framework proposed in Fig. 1 is a convolution-free hierarchical U-shaped pure Transformer, designed for skin lesion segmentation. For an input image Im $\in \mathbb{R}^{H \times W \times C}$, where H, W, and C denote the spatial dimensions and channels, respectively, our structure, as in Fig. 1a, uses the patch merging and patch expanding strategies from [4,17]. The patch embedding module extracts overlapping patch tokens of size 4×4 from the embedded tokens ($X \in \mathbb{R}^{n \times d}$) and then passes them through an encoder module with three stacked encoder blocks. Each block has two consecutive Efficient Transformer blocks and a patch merging layer that reduces spatial dimension (by merging 2×2 tokens) while doubling the channel dimension. The decoder expands the tokens by a factor of two in each block and integrates the output of each patch-expanding layer with the features forwarded by the fusion of the skip connection from the parallel encoder layer using ISCF. This approach enables the network to obtain a hierarchical representation.

2.1 Efficient Attention

Let $\mathbf{Q}$, $\mathbf{K}$, and $\mathbf{V}$ denote the Query, Key, and Value matrices that are produced by the embedded tokens X in each stage, and d is the embedding dimension. The standard self-attention equation is given by:

$$S(\mathbf{Q}, \mathbf{K}, \mathbf{V}) = \text{softmax} \left(\frac{\mathbf{Q}\mathbf{K}^{\mathbf{T}}}{\sqrt{d}} \right) \mathbf{V}. \tag{1}$$

The standard self-attention mechanism has a quadratic computational complexity ($\mathcal{O}(n^2)$), which limits its applicability in high-resolution tasks. Shen et al. [20] proposed an approach called "Efficient Attention" that takes advantage of the fact that regular self-attention creates repetitive context matrix entries. They suggested a more efficient method for computing self-attention, as follows:

$$E(\mathbf{Q}, \mathbf{K}, \mathbf{V}) = \rho_{\mathbf{q}}(\mathbf{Q})(\rho_{\mathbf{k}}(\mathbf{K})^{\mathbf{T}} \mathbf{V}), \tag{2}$$

where ρ_q and ρ_k are **Softmax**, normalization functions for the Queries and Keys. Shifting the order of multiplication drastically decreases the computation complexity to $\mathcal{O}(d^2 n)$ when $d_v = d, d_k = \frac{d}{2}$, which is a typical setting (see Fig. 1b). In contrast to naive dot-product attention (self-attention), efficient attention does not first compute pairwise similarities between points. Instead, the keys are represented as d_k attention maps $\mathbf{k}^{\mathbf{T}}{}_j$, with j referring to position j in the input feature. To sum up, according to Fig. 1b, our Efficient Transformer block applies the following operations:

$$X_1 = E(\mathbf{Q}, \mathbf{K}, \mathbf{V}) + X \tag{3}$$
$$X' = \text{Mix-FFN}(\text{LN}(X_1)) + X_1, \tag{4}$$

where LN is a LayerNorm operation, and Mix-FFN is an enhanced Feed Forward Network (FFN) operation. It has been proven such a design can align features and make discriminative representations [14,27], that mixes a 3×3 convolution and an MLP into each FFN. This operation actually plays as the dynamic positional encoding for each stage's Efficient transformer.

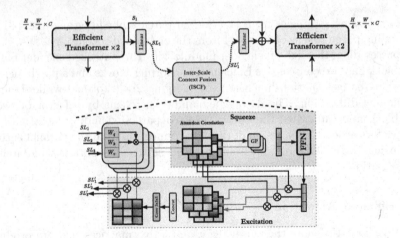

Fig. 2. An overview of the Inter-Scale Context Fusion (ISCF) module. This module applies to already calculated attention maps in each stage's Transformer block and produces a general context over stages to compensate for the inter and intra-semantic gaps within the U-shaped design to improve the performance.

2.2 Inter-scale Context Fusion

The ISCF module is displayed in Fig. 2. Instead of simply concatenating the features from the encoder and decoder layers, we devised a context fusion module to decrease the encoder-decoder semantic gap. Our proposed module not only can effectively provide spatial information to each decoder to recover fine-grained details when producing output masks, but also it does not require additional parameters for the model. Our proposed U-shaped structure is defined as a three stage multi-scale representation coupled with an ISCF module. Due to the hierarchical design of the structure, the attention maps' shape at each level differs from the next one. Therefore, we used a Linear layer in the first two stages two make the attention map sizes the same as the last stage. This operation is done at the output of the ISCF module to remap the attention maps to their original sizes. In the ISCF module, we utilize the Global Pooling (GP) operation to produce a single value for each stage's attention correlation and concatenate them, followed by an FFN to amalgamate the contribution of each global value into a scaling factor. Then each attention map applies the point-wise production with the corresponding scaling value and concatenates the resulting attention maps. Further, to adaptively fuse these global contexts to lessen the mentioned semantic gaps, a $3 \times 1 \times 1$ convolution is used. Finally, the resulting context fusion tensor adds to each plain skip connection from the encoder to the decoder to highlight the spatial localization for better segmentation results.

3   Experimental Setup

The PyTorch library was used to implement our proposed architecture and was run on a single RTX 3090 GPU. A batch size of 24 and an Adam solver with an empirically chosen learning rate of $1e - 4$ were used for a 100 epochs training. The loss function for the segmentation task was binary cross-entropy.

3.1 Datasets

ISIC 2017 – The ISIC 2017 dataset [7] comprises 2,000 skin dermoscopic images (cancer-positive and negative samples) with their corresponding annotations. We used 600 samples for the test set, 150 samples for data validation, and 1,250 samples for training. Each sample has an original size of 576×767 pixels. Images that are larger than 224×224 pixels are resized using the same pre-processing as [2].

ISIC 2018 – ISIC 2018 [6] provides independent datasets for the classification and segmentation tasks for the first time, with 2,594 training (20% melanomas, 72% nevi, and 8% seborrheic keratoses). Like prior approaches [2], we used 1,815 samples for training, 259 for validation, and 520 for testing. We downsize each sample image to 224×224 pixels from its original size of 2016×3024 pixels.

3.2 Quantitative and Qualitative Results

In Table 1, the quantitative results for our proposed method are displayed. We reported the performance of the model on the Dice score (DSC), sensitivity (SE), specificity (SP), and accuracy (ACC). Our results show that the proposed design can outperform SOTA methods without pre-training weights and having fewer parameters. In addition, Fig. 3 provides qualitative results that show the network performs well with respect to the ground truth segmentation and preserves the high-frequency details such as boundary information. It is evident that the boundary information with an Efficient Transformer and the ISCF module preserve high-frequency details effectively in comparison to the naive Swin U-Net [4] and highlights the efficacy of the ISCF module in compensating the information gap between encoder and decoder.

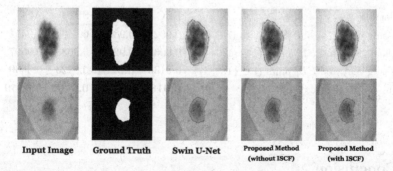

Input Image   **Ground Truth**   **Swin U-Net**   **Proposed Method (without ISCF)**   **Proposed Method (with ISCF)**

Fig. 3. Qualitative results on *ISIC 2018* dataset. Green represents the ground truth contour and blue corresponds to the prediction mask contour. (Color figure online)

Table 1. Performance comparison of the proposed method against SOTA approaches on the *ISIC 2017*, and *ISIC 2018* skin lesion segmentation tasks.

Methods	# Params(M)	ISIC 2017				ISIC 2018			
		DSC	SE	SP	ACC	DSC	SE	SP	ACC
U-Net [19]	14.8	0.8159	0.8172	0.9680	0.9164	0.8545	0.8800	0.9697	0.9404
Att U-Net [18]	34.88	0.8082	0.7998	0.9776	0.9145	0.8566	0.8674	**0.9863**	0.9376
TransUNet [5]	105.28	0.8123	0.8263	0.9577	0.9207	0.8499	0.8578	0.9653	0.9452
FAT-Net [26]	28.75	0.8500	0.8392	0.9725	0.9326	0.8903	0.9100	0.9699	0.9578
Swin U-Net [4]	82.3	0.9183	0.9142	**0.9798**	**0.9701**	0.8946	0.9056	0.9798	**0.9645**
Efficient Transformer (without ISCF)	22.31	0.8998	0.8834	0.9530	0.9578	0.8817	0.8534	0.9698	0.9519
Efficient Transformer (with ISCF)	**23.43**	**0.9257**	**0.9321**	0.9793	0.9698	**0.9136**	**0.9284**	0.9723	0.9630

3.3 Ablation Study

To investigate how various settings affect the performance of our model, we conducted ablation studies on the number of times our proposed attention strategy was used in skip connections, as well as on different input sizes. We examined the effects of incorporating our proposed attention module at 1/4, 1/8, and 1/16 resolution scales. We studied the impact on our model by selecting which pairs of scales SL_i and SL_i' for $i \in \{1, 2, 3\}$. As shown in Table 2, the segmentation performance improved as we increased the number of skip connection modules (scales), supporting our attention module's efficacy in capturing a rich representation. We also evaluated the impact of increasing the input size to 384×384 and found that while it led to slightly improved segmentation results, it also incurred a higher computational cost.

Table 2. Ablation study on the *ISIC 2018* skin lesion segmentation benchmark.

Setting	DSC	SE	SP	ACC
Using only SL_i and SL_i' pairs ($i \in \{1\}$)	0.9025	0.9305	0.9490	0.9451
Using SL_i and SL_i' pairs ($i \in \{1, 2\}$)	0.9065	0.9274	0.9683	0.9597
Using SL_i and SL_i' pairs ($i \in \{1, 2, 3\}$)	0.9136	**0.9284**	0.9723	0.9630
Input image size 384×384	**0.9189**	0.9265	**0.9799**	**0.9659**

4 Conclusion

The semantic gap between the encoder and decoder in a U-shaped Transformer-based network can be mitigated by carefully recalibrating the already calculated attention maps from each stage. In this study, not only do we address the hierarchical semantic gap drawback, but also, our ISCF module highlights the importance of spatial attention. ISCF module is a plug-and-play and computation-friendly module that can effectively be applied to any Transformer-based architecture. The qualitative and quantitative results endorse the applicability of the proposed module.

References

1. Aghdam, E.K., Azad, R., Zarvani, M., Merhof, D.: Attention Swin U-Net: cross-contextual attention mechanism for skin lesion segmentation. arXiv preprint arXiv:2210.16898 (2022)
2. Alom, M.Z., Hasan, M., Yakopcic, C., Taha, T.M., Asari, V.K.: Recurrent residual convolutional neural network based on U-Net (R2U-Net) for medical image segmentation. arXiv preprint arXiv:1802.06955 (2018)
3. Azad, R., et al.: Medical image segmentation review: the success of U-Net. arXiv preprint arXiv:2211.14830 (2022)
4. Cao, H., et al.: Swin-Unet: Unet-like pure transformer for medical image segmentation. In: Karlinsky, L., Michaeli, T., Nishino, K. (eds.) Computer Vision-ECCV 2022 Workshops: Tel Aviv, Israel, 23–27 October 2022, Proceedings, Part III, vol. 13803, pp. 205–218. Springer, Cham (2023). https://doi.org/10.1007/978-3-031-25066-8_9
5. Chen, J., et al.: TransUNet: transformers make strong encoders for medical image segmentation. arXiv preprint arXiv:2102.04306 (2021)
6. Codella, N., et al.: Skin lesion analysis toward melanoma detection 2018: a challenge hosted by the international skin imaging collaboration (ISIC). arXiv preprint arXiv:1902.03368 (2019)
7. Codella, N.C., et al.: Skin lesion analysis toward melanoma detection: a challenge at the 2017 international symposium on biomedical imaging (ISBI), hosted by the international skin imaging collaboration (ISIC). In: 2018 IEEE 15th International Symposium on Biomedical Imaging (ISBI 2018), pp. 168–172. IEEE (2018)
8. Dosovitskiy, A., et al.: An image is worth 16×16 words: transformers for image recognition at scale. In: International Conference on Learning Representations (2021). https://openreview.net/forum?id=YicbFdNTTy
9. Emre Celebi, M., Alp Aslandogan, Y., Stoecker, W.V., Iyatomi, H., Oka, H., Chen, X.: Unsupervised border detection in dermoscopy images. Skin Res. Technol. 13(4), 454–462 (2007)
10. Emre Celebi, M., Wen, Q., Hwang, S., Iyatomi, H., Schaefer, G.: Lesion border detection in dermoscopy images using ensembles of thresholding methods. Skin Res. Technol. 19(1), e252–e258 (2013)
11. Erkol, B., Moss, R.H., Joe Stanley, R., Stoecker, W.V., Hvatum, E.: Automatic lesion boundary detection in dermoscopy images using gradient vector flow snakes. Skin Res. Technol. 11(1), 17–26 (2005)
12. Eskandari, S., Lumpp, J.: Inter-scale dependency modeling for skin lesion segmentation with transformer-based networks. In: Medical Imaging with Deep Learning, Short Paper Track (2023). https://openreview.net/forum?id=JExQEfV5um
13. Gordon, R.: Skin cancer: an overview of epidemiology and risk factors. In: Seminars in Oncology Nursing, vol. 29, pp. 160–169. Elsevier (2013)

14. Huang, X., Deng, Z., Li, D., Yuan, X., Fu, Y.: MISSFormer: an effective transformer for 2D medical image segmentation. IEEE Trans. Med. Imaging **42**(5), 1484–1494 (2022)

15. Hwang, Y.N., Seo, M.J., Kim, S.M.: A segmentation of melanocytic skin lesions in dermoscopic and standard images using a hybrid two-stage approach. Biomed. Res. Int. **2021**, 1–19 (2021)

16. Li, X., Chen, H., Qi, X., Dou, Q., Fu, C.W., Heng, P.A.: H-DenseUNet: hybrid densely connected UNet for liver and tumor segmentation from CT volumes. IEEE Trans. Med. Imaging **37**(12), 2663–2674 (2018)

17. Liu, Z., et al.: Swin transformer: hierarchical vision transformer using shifted windows. In: Proceedings of the IEEE/CVF International Conference on Computer Vision, pp. 10012–10022 (2021)

18. Oktay, O., et al.: Attention U-Net: learning where to look for the pancreas. arXiv preprint arXiv:1804.03999 (2018)

19. Ronneberger, O., Fischer, P., Brox, T.: U-Net: convolutional networks for biomedical image segmentation. In: Navab, N., Hornegger, J., Wells, W.M., Frangi, A.F. (eds.) MICCAI 2015. LNCS, vol. 9351, pp. 234–241. Springer, Cham (2015). https://doi.org/10.1007/978-3-319-24574-4_28

20. Shen, Z., Zhang, M., Zhao, H., Yi, S., Li, H.: Efficient attention: attention with linear complexities. In: Proceedings of the IEEE/CVF Winter Conference on Applications of Computer Vision, pp. 3531–3539 (2021)

21. Siegel, R.L., Miller, K.D., Wagle, N.S., Jemal, A.: Cancer statistics, 2023. CA: Cancer J. Clin. **73**(1), 17–48 (2023). https://doi.org/10.3322/caac.21763, https://acsjournals.onlinelibrary.wiley.com/doi/abs/10.3322/caac.21763

22. Society, A.C.: Cancer facts and figures (2023). https://www.cancer.org/content/dam/cancer-org/research/cancer-facts-and-statistics/annual-cancer-facts-and-figures/2023/slideshow-2023-cancer-facts-and-figures.pptx. Accessed 10 Mar 2023

23. Vaswani, A., et al.: Attention is all you need. In: Advances in Neural Information Processing Systems, vol. 30 (2017)

24. Wang, H., Cao, P., Wang, J., Zaiane, O.R.: UCTransNet: rethinking the skip connections in U-Net from a channel-wise perspective with transformer. In: Proceedings of the AAAI Conference on Artificial Intelligence, vol. 36, pp. 2441–2449 (2022)

25. Wang, W., Chen, C., Ding, M., Yu, H., Zha, S., Li, J.: TransBTS: multimodal brain tumor segmentation using transformer. In: de Bruijne, M., et al. (eds.) Medical Image Computing and Computer Assisted Intervention-MICCAI 2021: 24th International Conference, Strasbourg, France, 27 September–1 October 2021, Proceedings, Part I 24, pp. 109–119. Springer, Cham (2021). https://doi.org/10.1007/978-3-030-87193-2_11

26. Wu, H., Chen, S., Chen, G., Wang, W., Lei, B., Wen, Z.: FAT-Net: feature adaptive transformers for automated skin lesion segmentation. Med. Image Anal. **76**, 102327 (2022)

27. Xie, E., Wang, W., Yu, Z., Anandkumar, A., Alvarez, J.M., Luo, P.: SegFormer: simple and efficient design for semantic segmentation with transformers. Adv. Neural. Inf. Process. Syst. **34**, 12077–12090 (2021)

28. Yu, L., Chen, H., Dou, Q., Qin, J., Heng, P.A.: Automated melanoma recognition in dermoscopy images via very deep residual networks. IEEE Trans. Med. Imaging **36**(4), 994–1004 (2016)

29. Zhou, Z., Rahman Siddiquee, M.M., Tajbakhsh, N., Liang, J.: UNet++: a nested U-Net architecture for medical image segmentation. In: Stoyanov, D., et al. (eds.) DLMIA/ML-CDS -2018. LNCS, vol. 11045, pp. 3–11. Springer, Cham (2018). https://doi.org/10.1007/978-3-030-00889-5_1

MagNET: Modality-Agnostic Network for Brain Tumor Segmentation and Characterization with Missing Modalities

Aishik Konwer[1](✉), Chao Chen[2], and Prateek Prasanna[2]

[1] Department of Computer Science, Stony Brook University, Stony Brook, NY, USA
akonwer@cs.stonybrook.edu
[2] Department of Biomedical Informatics, Stony Brook University, Stony Brook, NY, USA
prateek.prasanna@stonybrook.edu

Abstract. Multiple modalities provide complementary information in medical image segmentation tasks. However, in practice, not all modalities are available during inference. Missing modalities may affect the performance of segmentation and other downstream tasks like genomic biomarker prediction. Previous approaches either attempt a naive fusion of multi-modal features or synthesize missing modalities in the image or feature space. We propose an end-to-end modality-agnostic segmentation network (MagNET) to handle heterogeneous modality combinations, which is also utilized for radiogenomics classification. An attention-based fusion module is designed to generate a modality-agnostic tumor-aware representation. We design an adversarial training strategy to improve the quality of the representation. A missing-modality detector is used as a discriminator to push the encoded feature representation to mimic a full-modality setting. In addition, we introduce a loss function to maximize inter-modal correlations; this helps generate the modality-agnostic representation. MagNET significantly outperforms state-of-the-art segmentation and methylation status prediction methods under missing modality scenarios, as demonstrated on brain tumor datasets. •

Keywords: Missing modality · Fusion · Brain tumor Segmentation · Adversarial Learning

1 Introduction

Clinical Motivation. Gliomas, the most familiar type of brain tumor, have different grades associated with the growth rate. Higher grade gliomas are very

Supplementary Information The online version contains supplementary material available at https://doi.org/10.1007/978-3-031-45673-2_36.

aggressive and fatal in nature. Precise segmentation of brain tumor is necessary in quantifying the tumor development, that eventually aids treatment planning [14]. Additionally, proper segmentation of tumor subcompartments is a precursor to several downstream tasks, such as survival analysis or mutation prediction. Multiple 3D Magnetic resonance (MR) protocols (T1, T1c, T2, Flair) are used together to understand the underlying texture and spatial characteristics of brain tumor and its surroundings [1,3]. In this work, we refer to these protocols as modalities. Each such modality accentuates distinct sub-areas within the brain; for example, T1c focuses more on the core part of tumor, while Flair highlights the edema. This complementary information extracted from different modalities helps to segment the brain tumor into well-defined regions (enhancing tumor, edema, and the necrotic and non-enhancing tumor core). In the past half decade, deep learning based approaches [5,8,20,26,30] have found great success in brain tumor segmentation. These methods mostly performed well in the scenario when all four acquisition modalities are available as input (i.e. in the full modality setting). However, in clinical practice, only a subset of the treatment modalities are available owing to varied reasons such as: 1) image degradation, 2) patient-movement related artifacts, 3) erroneous acquisition settings, 4) allergy prone to contrast mediums and 5) brief scan times. Hence it is crucial to develop a modality-independent robust method that can achieve state-of-the-art segmentation results, and also aid mutation prediction [24] tasks in missing modality settings.

Technical Motivation. A potential solution in the missing modality setting is to train separate models with all possible modalities; this is computationally inefficient and, often, data intensive [27]. Another popular approach is to synthesize the missing modalities en route segmentation. The synthesis is usually performed using a conditional Generative Network (cGAN) [13,28] or feature enhanced generator [6,31], serially followed by decoder for segmentation. However, from training perspective, it is difficult for a model to converge when both 3D image Generator and 3D Segmentation network are implemented in the same pipeline. Moreover, it is an overkill to reconstruct the missing image modalities when the end goal is essentially segmentation. The aforementioned drawbacks paved the way for utilizing latent features to tackle missing-modality scenarios. In [9,18], the authors computed first and second order statistics (mean, variance) to generate the unified representation before performing segmentation. Chartsias et. al. [4] aimed to attain the fused representation by minimizing the inter-modal feature distances. However the statistical techniques that these methods employ yield sub-optimal performance due to variation in the underlying intensity distributions of different MR modalities. More recent works [11,16,23,27] have applied traditional domain adaptation and knowledge distillation techniques, to push the missing modality-based features towards resembling the complete modality ones. Most of these methods need to learn a combinatorial number of imputation models, they are not inherently modality-agnostic in nature, rather becoming "catch-all" models. Moreover, performance increase comes at the expense of training time and computational resources.

In this work, we propose a robust hetero-modal architecture, MagNET, to extract meaningful features using a multi-encoder uni-decoder framework. An attention-weighted fusion module efficiently unifies the representations from the corresponding levels (skip connections, bottleneck) of the various modality-specific encoders. More importantly, we introduce an adversarial learning technique that overcomes the unwanted need for generating missing modalities in the image or latent space. This is specifically achieved by employing an MLP-based multi-label classifier as the discriminator. The discriminator predicts the presence/absence of modalities in a fused latent representation by performing a binary classification for each modality. The discriminator and segmentation decoder are trained in a multi-task paradigm so that they can complement each other. The segmentation model weights are further finetuned and passed through a classifier to predict MGMT promoter methylation status under different missing scenarios.

To summarize, the three-fold major contributions of the paper are as follows: 1)We introduce domain adaptation-inspired adversarial learning to tackle missing modalities without synthesizing them in image space, 2)We are the first to propose a deep learning model to classify the MGMT promoter methylation status from incomplete MRI sequences, and 3)We implement a fusion strategy to generate a shared modality-agnostic embedding from multiple modalities, and incorporate a correlation loss to further enrich the embedding.

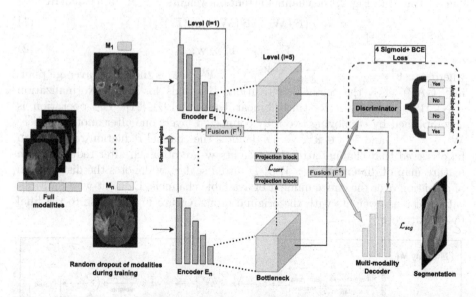

Fig. 1. Illustration of the proposed MagNET. After the modality encoders, the fusion module helps to obtain the unified representation from five different levels (the highest level and bottleneck are indicated here). Next, only the bottleneck embedding is used to train the discriminator. All five fused embeddings are used by the segmentation decoder. Once trained, the bottleneck feature is utilized to perform mutation prediction.

2   Methodology

We propose a multi-encoder uni-decoder segmentation framework, MagNET which is modality-agnostic during inference. Specifically, our framework includes a fusion module after the encoders, and a multi-label classifier based discriminator trained along with the decoder. MR modalities are manually dropped for each patient during training to simulate the scenario of missing modalities. Independent encoders E_n are trained to extract features from each available modality M_n where n = total number of modalities. These features F_n^l from the corresponding levels (l) of each encoder are concatenated and passed into a fusion module. The fused representation is eventually used as decoder input to perform segmentation. This representation is also trained in an adversarial manner using a discriminator that predicts absence of modalities. The overall framework is illustrated in Fig. 1.

2.1   Attention-Weighted Fusion

We aim to utilize multiple modalities (which vary in numbers per patient) and derive a common fused representation. This makes it feasible to be used later in training the discriminator and segmentation decoder. The generated features, $\mathbf{F}_n^l \in \mathbb{R}^{C \times H \times W \times D}$, from the lth level of E_n are added channel-wise to obtain a single feature at each level. For a particular level l, the summed feature $\mathbf{F}^l \in \mathbb{R}^{C \times H \times W \times D}$ includes C channels and feature maps of size $H \times W \times D$. As demonstrated in Fig. 2, the channel attention weights $\mathbf{w} \in \mathbb{R}^C$ are calculated as

$$\mathbf{w} = \sigma \left(\mathcal{B} \left(\mathbf{W}_2 \delta \left(\mathcal{B} \left(\mathbf{W}_1 (gap(\mathbf{F}^l)) \right) \right) \right) \right) \tag{1}$$

$$\mathbf{F}_{fused}^l = \mathbf{F}^l \otimes \mathbf{w}, \tag{2}$$

where $gap(\mathbf{F}^l) = \frac{1}{H \times W \times D} \sum_{i=1}^{H} \sum_{j=1}^{W} \sum_{k=1}^{D} \mathbf{F}^l_{[:,i,j,k]}$ is the global average pooling (GAP), σ is the Sigmoid activation, $\mathcal{B}$ stands for Batch Normalization (BN) [12], and δ is a Rectified Linear Unit (ReLU) [21]. The operation is accomplished by employing two fully connected layers one after another, where $\mathbf{W}_1 \in \mathbb{R}^{\frac{C}{rr} \times C}$ and $\mathbf{W}_2 \in \mathbb{R}^{C \times \frac{C}{rr}}$. rr denotes the channel reduction ratio. It can be observed that channel attention weights $\mathbf{w}$ are obtained after reducing each feature map of dimension $H \times W \times D$ into a scalar. $\mathbf{w}$ denotes the distribution of attention over the entire number of available channels. Element-wise multiplication of these weights with the original input feature $\mathbf{F}^l$ gives rise to the final fused feature $\mathbf{F}_{fused}^l$.

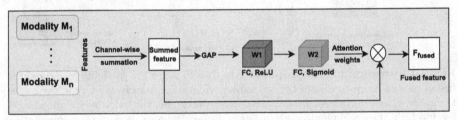

Fig. 2. Attention-weighted Fusion module

2.2 Adversarial Learning

Considering that the full modality contains richer information, we encourage the encoder outputs to mimic their representations, irrespective of the input combination. The segmentation encoders can be seen as a generator E. We then introduce an MLP-based multi-label classifier as our discriminator $\mathcal{D}$. The objective of $\mathcal{D}$ is to predict the absence/presence of modalities from the fused bottleneck embedding $\mathbf{F}^5_{fused}$. $\mathcal{D}$ utilizes Binary Cross-Entropy loss $\mathcal{L}_{BCE}$, and sigmoid activation to output n binary predictions $\hat{d}$, denoting whether a modality is available or not. While calculating the discriminator loss $\mathcal{L}_{\mathcal{D}}$ indicated below, the label of data used as ground truth (T_{actual}) is a vector of size n which reflects the true combination of modalities available at the input for that iteration. For example, assuming that $n = 4$ and only M_1 and M_2 are available, $T_{actual} = \{1, 1, 0, 0\}$.

The generator loss is a combination of segmentation loss and an adversarial loss used to train the generator to fool the discriminator. We consider a dummy ground truth variable T_{dummy}. In order to encourage the generator to encode representations that confuse or "fool" the discriminator into inferring that all modalities are present, we set $T_{dummy} = \{1, 1, 1, 1\}$, masquerading all generated representations as full modality representations. Thus $\mathcal{D}$ pushes the generator E to agnostically produce full modality representations.

$$\mathcal{L}_{\mathcal{D}} = \sum_{z=1}^{N} \mathcal{L}_{BCE}(\hat{d}_z, T_{actual_z}). \quad (3) \qquad \mathcal{L}_E = \lambda_1 \mathcal{L}_{seg} + \lambda_2 \sum_{z=1}^{N} \mathcal{L}_{BCE}(\hat{d}_z, T_{dummy_z}). \quad (4)$$

2.3 Segmentation Encoder-Decoder

We aim to perform volumetric segmentation on 3D brain MR images. Hence for MagNET, we used an architecture inspired by UNETR [8] for the encoder-decoder framework in our task. UNETR houses a single encoder-decoder combination that simultaneously accepts input from all n MR modalities. The input tensor has a dimension $H \times W \times D \times n$. However, we trained independent encoders for each of our available MR modalities, essentially setting up a multi-encoder uni-decoder structure. All the encoders have shared weights. In our case, the input tensor has a shape of $H \times W \times D \times 1$. The proposed fusion module in Sect. 2.1 made it feasible to design such an architecture, by fusing multi-encoder outputs into a single representation for the decoder.

Firstly, non-overlapping patches were extracted from 3D MR volumes and passed through a linear projection layer to form a 768-dimensional embedding. After obtaining this embedding, we exploit a stack of 12 Vision transformer layers [7,17] which includes multi-head self-attention and multi-layer perceptron as essential sub-modules. Following that, we extract the transformer output at $l = 5$ different levels (4 higher resolutions and 1 bottleneck) for each of the n available encoders and fuse them correspondingly via our fusion block. Next, we perform necessary convolutions and deconvolutions to make the representations compatible for concatenation with previous levels. The final output corresponding to the input MR resolution is passed through $1 \times 1 \times 1$ convolutional layer to perform voxel-wise semantic segmentation. The segmentation loss function $\mathcal{L}_{seg}$ is defined as follows: $\mathcal{L}_{seg}(G, P) = 1 - \frac{2}{V} \sum_{v=1}^{V} \frac{\sum_{u=1}^{U} G_{u,v} P_{u,v}}{\sum_{u=1}^{U} G_{u,v}^2 + \sum_{u=1}^{U} P_{u,v}^2} - \frac{1}{U} \sum_{u=1}^{U} \sum_{v=1}^{V} G_{u,v} \log P_{u,v}$. where V is the number of classes and U is the number of voxels. $P_{u,v}$ and $G_{u,v}$ refer to the predicted output and one-hot encoded ground truth for class v at voxel u, respectively.

2.4   Inter-modal Correlation Loss

We aim to maximize the correlation [15] between every pair of modalities for a particular patient. This will enable MagNET to learn different characteristics of the same tumor present in all modalities. Eventually we obtain better embeddings of the bottleneck layers to improve the quality of segmentations. Our proposed loss function is inspired by the Barlow Twins loss [29]. The loss was originally applied between distorted augmentations of the same image. Since the different MR modalities (T1, T1c, T2, Flair) are basically complementary views of the same brain (differentially highlighting tissue regions), we considered them as matching candidates. The bottleneck features generated by modality-specific encoders are passed through a projection layer. The projection layer begins with a GAP, followed by two consecutive FC+ReLU+BN blocks and finally terminated by another FC layer.

Let Z^X and Z^Y be the projection encoded outputs of two MR modalities M_X and M_Y of a given patient. A cross-correlation matrix ($\mathcal{C}$) is computed between Z^X, Z^Y. Barlow Twins loss, $\mathcal{L}_{corr}$, is then employed to push $\mathcal{C}$ towards resembling an identity matrix.

$$\mathcal{L}_{corr} = \sum_s (1 - \mathcal{C}_{ss})^2 + \lambda \sum_s \sum_{t \neq s} \mathcal{C}_{st}{}^2; \quad \mathcal{C}_{st} = \frac{\sum_b z_{b,s}^X z_{b,t}^Y}{\sqrt{\sum_b \left(z_{b,s}^X\right)^2} \sqrt{\sum_b \left(z_{b,t}^Y\right)^2}} \quad (5)$$

where $\sum_s (1 - \mathcal{C}_{ss})^2$ is an invariance term, $\sum_s \sum_{t \neq s} \mathcal{C}_{st}{}^2$ is a redundancy reduction term. $\lambda = 0.2$, s and t are the dimensions of two projection outputs. b denotes the index of pair-wise samples in a batch.

3   Experiment Design and Results

Dataset. The segmentation dataset utilized in our experiments comprises 285 training cases from BraTS2018 [2,19]. Radiogenomics classification is performed in a 5-fold cross-validation manner on BraTS2021 which contains 585 cases. It should be noted that there were no overlapping patients between the two cohorts, thus eliminating data leakage. All patients have $n = 4$ sequences of MR volumetric images: a) native (T1), b) post-contrast T1-weighted (T1c), c) T2-weighted

Table 1. Comparison with SOTA (DSC %) for the different combinations of available modalities on three nested tumor subregions (Complete, Core, Enhancing). Modalities present are denoted by •, the missing ones by o.

Modalities				Complete				Core				Enhancing			
F	T_1	T_{1c}	T_2	U-HeMIS	U-HVED	ACN	MagNET(↑)	U-HeMIS	U-HVED	ACN	MagNET(↑)	U-HeMIS	U-HVED	ACN	MagNET(↑)
o	o	o	•	80.9	79.8	85.5	**87.3**	57.2	54.6	67.9	**68.4**	25.6	22.8	42.9	**44.3**
o	o	•	o	61.5	53.6	80.5	**82.4**	65.2	59.5	84.1	**84.6**	62.0	57.6	78.0	**79.6**
o	•	o	o	57.6	49.5	79.3	**81.2**	37.3	33.9	71.1	**73.3**	10.1	8.6	41.5	**43.2**
•	o	o	o	52.4	84.3	87.3	**88.4**	26.0	57.9	67.7	**69.8**	11.7	23.8	42.7	**45.2**
o	o	•	•	82.4	81.3	86.4	**88.2**	76.6	73.9	84.4	**86.0**	67.8	67.8	75.6	**77.1**
o	•	•	o	68.4	64.2	80.0	**83.3**	72.4	67.5	84.5	**87.3**	66.2	61.1	75.2	**76.5**
•	•	o	o	64.6	85.7	87.4	**87.9**	41.1	61.1	71.3	**74.6**	10.7	27.9	43.7	**45.0**
o	•	o	•	82.4	81.5	85.5	**86.8**	60.9	56.2	73.2	**74.9**	32.3	24.2	47.3	**47.6**
•	o	o	•	82.9	87.5	**87.7**	87.6	57.6	62.7	71.6	**72.3**	30.2	32.3	**45.9**	45.4
•	o	•	o	68.9	85.9	88.2	**89.3**	71.4	75.0	83.3	**84.7**	66.1	68.3	77.4	**79.8**
•	•	•	o	72.3	86.7	88.9	**89.0**	76.0	77.0	84.2	**85.9**	68.5	68.6	76.1	**77.0**
•	•	o	•	83.4	88.0	88.3	**88.5**	60.3	63.1	67.8	**68.3**	31.0	32.3	42.0	**44.9**
•	•	o	•	83.8	88.0	88.3	**88.7**	77.5	76.7	82.8	**83.1**	68.7	68.9	75.9	**77.0**
o	•	•	•	83.9	82.3	86.9	**87.8**	78.9	75.2	84.6	**86.9**	69.9	67.7	76.1	**78.3**
•	•	•	•	84.7	88.4	89.2	**91.2**	79.4	77.7	85.1	**87.5**	70.2	69.0	77.0	**81.1**
Avg				74	79.11	85.96	87.17	62.52	6.8	77.57	79.17	46.06	46.72	61.15	62.8
Std				±11.03	±12.70	±3.30	±2.74	±16.53	±11.91	±7.41	±7.58	±24.65	±22.31	±16.95	±17.21

(T2), and d) T2 Fluid Attenuated Inversion Recovery (Flair). As part of pre-processing, the organizers skull-stripped the volumes and interpolated them to an isotropic 1 mm³ resolution. For a given patient, the four sequences have been co-registered to the same anatomical template. The segmentation classes include whole or complete tumor (WT), tumor core (TC), and enhancing tumor (ET).

Implementation Details. MagNET is implemented in Pytorch [22] with a 48 GB Nvidia Quadro RTX 8000 GPU. We randomly dropped modalities during training and pursued with an optimum combination of 40-20-20-20 ($n \in \{4,3,2,1\}$) over the entire training data. This denotes 40% cases were selected with full modality while 20% cases were reserved for each of tri-, bi-, uni- modal combinations. To utilize 3D information we performed volumetric segmentation with the 155 × 240 × 240 images. Before input to encoder they were resized to 128 × 128 × 128. An Adam optimiser is applied with initial learning rate 5e-4 that gets reduced by a factor of 2 every 50 epochs till a maximum of 250 epochs. Following the settings of ACN [27] we randomly split the dataset into training and validation sets by a ratio of 2:1 and use the same seed criteria. Batch size was kept as 1. λ_1 and λ_2 are experimentally set to 0.5. SOTA comparison results have been reported from [27]. We used Dice similarity coefficient (DSC ↑) (Table 1) and Hausdorff distance (HD ↓) (Table 1 in Supplementary) as metrics to evaluate the segmentation performance.

Comparison with SOTA. We compare our MagNET with three state-of-the-art methods: 1) 'latent fusion' based HeMIS [9], 2) U-HVED [6] that synthesizes missing modality, and 3) 'knowledge distillation' based ACN [27]. Table 1 shows that MagNET significantly outperforms HeMIS, U-HVED, ACN for almost all the possible combinations of missing modality. It may be observed that when

Table 2. Ablation results to show the importance of proposed modules on full modality inference setting.

Method	DSC(↑)			
	WT	TC	ET	Avg ± Std
Base	88.3	83.1	76.3	82.5 ± 6.01
Base + Adv	90.8	87.0	80.4	86.0 ± 5.26
Base + Adv + Corr	91.2	87.5	81.1	86.6 ± 5.10

Table 3. Prediction results of MGMT methylation status on BraTS2021.

Modalities				Accuracy		
F	T_1	$T_{1}c$	T_2	ResNet	Eff3DNet	**MagNET(↑)**
○	○	○	●	0.47	0.53	**0.60**
○	○	●	○	0.55	0.61	**0.65**
○	●	○	○	0.46	0.49	**0.59**
●	○	○	○	0.57	0.63	**0.67**
●	●	●	●	0.59	0.64	**0.70**

T1c is absent, there is a notable drop in performance for both core and enhancing tumor classes. Similar decrease in DSC can be seen for complete tumor when Flair is missing. This is expected since T1c is used to highlight the enhancing component of the tumor and FLAIR highlights the peritumoral edema showing the extent of the tumor habitat boundaries. Previous methods, HeMIS and U-HVED, struggled in these scenarios with low DSC scores (8%–57%). The proposed approach not only outperforms them but also ACN (≈ 2.4%–4.1%). In 14/15 situations of uni-, bi-, tri- and full modalities, MagNET surpasses SOTA. Decent quality segmentation maps are obtained even in missing modality scenarios, as viewed in Fig. 3 and Supplementary Fig. 1.

Ablation Studies. We conducted an ablation experiment in full modality inference settings to explore the efficacy of each proposed component, including the Adversarial classifier (Adv) and Correlation module (Corr). As a baseline, we started with our fusion-based multi-encoder segmentation network which only uses Dice loss $\mathcal{L}_{seg}$. Then we gradually zeroed into our final framework of MagNET. It can be observed from Table 2 that MagNET benefits significantly from latent adversarial training (+4.2%), while correlation module also boosts the performance (+0.7%). Results for uni-modal combinations in supplementary Table 2. We also conduct an ablation by varying the proportion of full modality data in training (supplementary Fig. 2).

Genomic Biomarker Prediction. The trained segmentation model was further finetuned for binary prediction of MGMT methylation status on 585 cases from BraTS2021. Table 3 indicates that in both uni-modal and full-modal inference settings, MGMT status prediction accuracy surpasses Efficient-3DNet [25] and ResNet50 [10] by 6% and 11%, respectively.

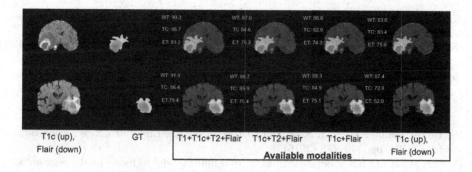

Fig. 3. Qualitative results. Red: NCR, NET; Yellow: ED; Blue: ET. The segmentations do not degrade sharply as additional modalities are dropped during inference phase. Even with single T1c or Flair modalities, we achieve decent segmentations. (Color figure online)

4   Conclusion

This paper presents a novel framework, MagNET, to perform brain tumor segmentation and characterization in real clinical scenarios where MRI sequences are often incomplete. We introduce the application of domain adaptation-inspired adversarial learning in this context, which discards the generation of missing modalities in image space. We are the first to validate radiogenomics classification for missing modality scenarios. Our model outperforms state-of-the-art approaches for both tumor segmentation and genomic biomarker prediction tasks on BraTS2018 and BraTS2021 datasets respectively. In the future, we will experiment on cross-modal datasets (CT, MRI) to improve generalizability.

References

1. Bakas, S., et al.: Advancing the cancer genome atlas glioma MRI collections with expert segmentation labels and radiomic features. Sci. Data **4**(1), 1–13 (2017)
2. Bakas, S., et al.: Identifying the best machine learning algorithms for brain tumor segmentation, progression assessment, and overall survival prediction in the brats challenge. arXiv preprint arXiv:1811.02629 (2018)
3. Bauer, S., Wiest, R., Nolte, L.P., Reyes, M.: A survey of MRI-based medical image analysis for brain tumor studies. Phys. Med. Biol. **58**(13), R97 (2013)
4. Chartsias, A., Joyce, T., Giuffrida, M.V., Tsaftaris, S.A.: Multimodal MR synthesis via modality-invariant latent representation. IEEE TMI **37**(3), 803–814 (2017)
5. Chen, S., Ding, C., Liu, M.: Dual-force convolutional neural networks for accurate brain tumor segmentation. Pattern Recogn. **88**, 90–100 (2019)
6. Dorent, R., Joutard, S., Modat, M., Ourselin, S., Vercauteren, T.: Hetero-modal variational encoder-decoder for joint modality completion and segmentation. In: Shen, D., et al. (eds.) MICCAI 2019. LNCS, vol. 11765, pp. 74–82. Springer, Cham (2019). https://doi.org/10.1007/978-3-030-32245-8_9

7. Dosovitskiy, A., et al.: An image is worth 16×16 words: transformers for image recognition at scale. arXiv preprint arXiv:2010.11929 (2020)

8. Havaei, M., et al.: Brain tumor segmentation with deep neural networks. Med. Image Anal. **35**, 18–31 (2017)

9. Havaei, M., Guizard, N., Chapados, N., Bengio, Y.: HeMIS: hetero-modal image segmentation. In: Ourselin, S., Joskowicz, L., Sabuncu, M.R., Unal, G., Wells, W. (eds.) MICCAI 2016. LNCS, vol. 9901, pp. 469–477. Springer, Cham (2016). https://doi.org/10.1007/978-3-319-46723-8_54

10. He, K., Zhang, X., Ren, S., Sun, J.: Deep residual learning for image recognition. In: CVPR, pp. 770–778 (2016)

11. Hu, M., et al.: Knowledge distillation from multi-modal to mono-modal segmentation networks. In: Martel, A.L., et al. (eds.) MICCAI 2020. LNCS, vol. 12261, pp. 772–781. Springer, Cham (2020). https://doi.org/10.1007/978-3-030-59710-8_75

12. Ioffe, S., Szegedy, C.: Batch normalization: accelerating deep network training by reducing internal covariate shift. In: ICML, pp. 448–456. PMLR (2015)

13. Islam, M., Wijethilake, N., Ren, H.: Glioblastoma multiforme prognosis: MRI missing modality generation, segmentation and radiogenomic survival prediction. Comput. Med. Imaging Graph. **91**, 101906 (2021)

14. Konwer, A., et al.: Predicting COVID-19 lung infiltrate progression on chest radiographs using spatio-temporal LSTM based encoder-decoder network. In: MIDL, pp. 384–398. PMLR (2021)

15. Konwer, A., et al.: Attention-based multi-scale gated recurrent encoder with novel correlation loss for COVID-19 progression prediction. In: de Bruijne, M., et al. (eds.) MICCAI 2021. LNCS, vol. 12905, pp. 824–833. Springer, Cham (2021). https://doi.org/10.1007/978-3-030-87240-3_79

16. Konwer, A., Hu, X., Xu, X., Bae, J., Chen, C., Prasanna, P.: Enhancing modality-agnostic representations via meta-learning for brain tumor segmentation. arXiv preprint arXiv:2302.04308 (2023)

17. Konwer, A., Xu, X., Bae, J., Chen, C., Prasanna, P.: Temporal context matters: enhancing single image prediction with disease progression representations. In: CVPR, pp. 18824–18835 (2022)

18. Lau, K., Adler, J., Sjölund, J.: A unified representation network for segmentation with missing modalities. arXiv preprint arXiv:1908.06683 (2019)

19. Menze, B.H., et al.: The multimodal brain tumor image segmentation benchmark (BRATS). IEEE TMI **34**(10), 1993–2024 (2014)

20. Myronenko, A.: 3D MRI brain tumor segmentation using autoencoder regularization. In: Crimi, A., Bakas, S., Kuijf, H., Keyvan, F., Reyes, M., van Walsum, T. (eds.) BrainLes 2018. LNCS, vol. 11384, pp. 311–320. Springer, Cham (2019). https://doi.org/10.1007/978-3-030-11726-9_28

21. Nair, V., Hinton, G.E.: Rectified linear units improve restricted Boltzmann machines. In: ICML (2010)

22. Paszke, A., et al.: Pytorch: An imperative style, high-performance deep learning library. In: Advances in Neural Information Processing Systems 32, pp. 8024–8035. Curran Associates, Inc. (2019). http://papers.neurips.cc/paper/9015-pytorch-an-imperative-style-high-performance-deep-learning-library.pdf

23. Shen, Y., Gao, M.: Brain tumor segmentation on MRI with missing modalities. In: Chung, A., Gee, J., Yushkevich, P., Bao, S. (eds.) IPMI 2019. LNCS, vol. 11492, pp. 417–428. Springer, Cham (2019). https://doi.org/10.1007/978-3-030-20351-1_32

24. Singh, G., et al.: Radiomics and radiogenomics in gliomas: a contemporary update. Br. J. Cancer **125**(5), 641–657 (2021)

25. Tan, M., Le, Q.: Efficientnet: rethinking model scaling for convolutional neural networks. In: ICML, pp. 6105–6114. PMLR (2019)
26. Wang, W., Chen, C., Ding, M., Yu, H., Zha, S., Li, J.: TransBTS: multimodal brain tumor segmentation using transformer. In: de Bruijne, M., et al. (eds.) MICCAI 2021. LNCS, vol. 12901, pp. 109–119. Springer, Cham (2021). https://doi.org/10.1007/978-3-030-87193-2_11
27. Wang, Y., et al.: ACN: adversarial co-training network for brain tumor segmentation with missing modalities. In: de Bruijne, M., et al. (eds.) MICCAI 2021. LNCS, vol. 12907, pp. 410–420. Springer, Cham (2021). https://doi.org/10.1007/978-3-030-87234-2_39
28. Yu, Z., Zhai, Y., Han, X., Peng, T., Zhang, X.-Y.: MouseGAN: GAN-Based multiple MRI modalities synthesis and segmentation for mouse brain structures. In: de Bruijne, M., et al. (eds.) MICCAI 2021. LNCS, vol. 12901, pp. 442–450. Springer, Cham (2021). https://doi.org/10.1007/978-3-030-87193-2_42
29. Zbontar, J., Jing, L., Misra, I., LeCun, Y., Deny, S.: Barlow twins: self-supervised learning via redundancy reduction. In: ICML, pp. 12310–12320. PMLR (2021)
30. Zhou, C., Ding, C., Wang, X., Lu, Z., Tao, D.: One-pass multi-task networks with cross-task guided attention for brain tumor segmentation. IEEE TIP **29**, 4516–4529 (2020)
31. Zhou, T., Canu, S., Vera, P., Ruan, S.: Feature-enhanced generation and multi-modality fusion based deep neural network for brain tumor segmentation with missing MR modalities. Neurocomputing **466**, 102–112 (2021)

Unsupervised Anomaly Detection in Medical Images Using Masked Diffusion Model

Hasan Iqbal[1]([✉])(iD), Umar Khalid[2](iD), Chen Chen[2](iD), and Jing Hua[1](iD)

[1] Department of Computer Science, Wayne State University, Detroit, MI, USA
hasan.iqbal.cs@wayne.edu
[2] Center for Research in Computer Vision, University of Central Florida, Orlando, FL, USA

Abstract. It can be challenging to identify brain MRI anomalies using supervised deep-learning techniques due to anatomical heterogeneity and the requirement for pixel-level labeling. Unsupervised anomaly detection approaches provide an alternative solution by relying only on sample-level labels of healthy brains to generate a desired representation to identify abnormalities at the pixel level. Although, generative models are crucial for generating such anatomically consistent representations of healthy brains, accurately generating the intricate anatomy of the human brain remains a challenge. In this study, we present a method called the masked-denoising diffusion probabilistic model (mDDPM), which introduces masking-based regularization to reframe the generation task of diffusion models. Specifically, we introduce Masked Image Modeling (MIM) and Masked Frequency Modeling (MFM) in our self-supervised approach that enables models to learn visual representations from unlabeled data. To the best of our knowledge, this is the first attempt to apply MFM in denoising diffusion probabilistic models (DDPMs) for medical applications. We evaluate our approach on datasets containing tumors and numerous sclerosis lesions and exhibit the superior performance of our unsupervised method as compared to the existing fully/weakly supervised baselines. Project website: https://mddpm.github.io/.

Keywords: Diffusion Models · Medical Imaging · Anomaly Detection

1 Introduction

Medical imaging (MI) systems play a crucial role in aiding radiologists in their diagnostic and decision-making processes. These systems provide medical imaging specialists with detailed visual information to detect abnormalities, make accurate diagnoses, plan treatments, and monitor patients. More recently, advanced machine learning techniques and image processing algorithms are being utilized to automate the medical diagnostic process. Among these techniques, deep learning models based on convolutional neural networks (CNNs) have exhibited significant achievements in accurately identifying anomalies in medical images [12,30]. However, supervised CNN approaches have inherent limitations, including the requirement for extensive expert-annotated training

---

H. Iqbal and U. Khalid—Equal Contribution.

X. Cao et al. (Eds.): MLMI 2023, LNCS 14348, pp. 372–381, 2024.
https://doi.org/10.1007/978-3-031-45673-2_37

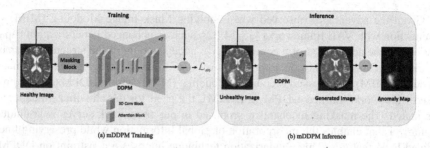

(a) mDDPM Training   (b) mDDPM Inference

Fig. 1. Schematic diagram of our framework. (**a**) During training, only healthy images are used, and no classifier guidance is required. The healthy image is passed through the **Masking Block** before feeding into the DDPM. The reconstruction loss is calculated between the original image and the image generated by the DDPM. Here, **Masking Block** plays the role of regularizer and eliminates the need for classifier guidance which is discussed in Sect. 2. Further, the schematic of the **Masking Block** is illustrated in Fig. 2. (**b**) During inference, DDPM considers the tumor in the unhealthy image as an augmented patch and eliminates it to generate a healthy image. The difference between the generated image and the given unhealthy image is then calculated to report the anomaly map. No masking mechanism is employed at inference.

data and the difficulty of learning from noisy or imbalanced data [11,16,17]. On the contrary, pixel-level annotations are not necessary for unsupervised anomaly detection (UAD) that uses only healthy examples for training.

In recent years, numerous architectures have been explored to investigate UAD for brain MRI anomaly detection. Autoencoders (AE) and variational autoencoders (VAE) have proven to be effective in training models and achieving efficient inference. However, their reconstructions often suffer from blurriness, limiting their effectiveness in UAD [3]. To address this limitation, researchers have focused on enhancing the understanding of image context by utilizing the spatial context through techniques such as spatial erasing [35] and leveraging 3D information [4,6]. Additionally, vector-quantized VAEs [26], adversarial autoencoders [7], and encoder activation maps [31] have been proposed to improve the image restoration quality. Generative adversarial networks (GANs) have emerged as an alternative to AE-based architectures for the task of UAD [29]. However, the unstable training nature of GANs poses challenges, and GANs often suffer from mode collapse and a lack of anatomical coherence [3,22].

Recently, denoising diffusion probabilistic models (DDPMs) [15] have shown promise for UAD in brain MRI [5,25,34]. In the context of DDPMs, the approach involves introducing noise to an input image and subsequently utilizing a trained model to eliminate the noise and estimate or reconstruct the original image [15]. Unlike most autoencoder-based methods, DDPMs retain spatial information in their hidden representations, which is crucial for the image generation process [27]. Recent works in medical imaging establish that they exhibit scalable and stable training properties and generate high-quality, sharp images with classifier guidance [25,26,28,33,34]. Further, [5,21,24] introduce patch-based DDPM (pDDPM) which offer better brain MRI reconstruction by incorporating global context information about individual brain structures and appearances while estimating individual patches.

Given the advantages observed when applying Mask Image Modeling (MIM) in conjunction with VAE frameworks [13,14,32], such as enhanced generalization capabilities and the acquisition of a comprehensive understanding of the structural characteristics of images, we introduce the first investigation into leveraging Masked Image Modeling (MIM) and Masked Frequency Modeling (MFM) within DDPMs. In our proposed framework, masked DDPM (mDDPM), the need for the classifier guidance is eliminated. The masking mechanism proposed in our framework serves as a unique regularizer that enables the incorporation of global information while preserving fine-grained local features. This regularization technique imposes a constraint on DDPM, ensuring the generation of a healthy image during inference, regardless of the input image characteristics. In this study, we focus on exploring three specific variants of masking-based regularization: *(i) image patch-masking (IPM)*, *(ii) frequency patch-masking (FPM)*, and *(iii) frequency patch-masking CutMix (FPM-CM)*. In the IPM approach, random pixel-level masks are applied to patches extracted from the original image before subjecting them to the diffusion process in DDPMs. The unmasked version of the same image is used as the reference for comparison. However, in the FPM approach, the image is first transformed into the frequency domain using the Fast Fourier Transform. Subsequently, patch-level masking is performed in the frequency domain as shown in Fig. 2a. The inverse Fourier Transform is then applied to obtain the reconstructed image, which is utilized to calculate the reconstruction loss. In FPM-CM, random patches are sampled from the augmented image generated through FPM and subsequently inserted at corresponding positions within the original clean image as shown in Fig. 2b. To evaluate the performance of our method, we use two publicly available datasets: BraTS21 [2], and MSLUB [20], and demonstrate a significant improvement ($p < 0.05$) in tumor segmentation performance.

2   Method

Given the potential occurrence of anomalies in any region of the brain during testing, we introduce data augmentation techniques that involve the insertion of random augmented patches into the healthy input image, $z \in \mathbb{R}^{C,W,H}$ with C channels, width W and height H, prior to the application of DDPM noise addition and removal. This approach allows us to generate a healthy image based on an unhealthy image during inference, facilitating the calculation of the anomaly map. The illustration of our approach can be seen in Fig. 1. We will be further discussing our unsupervised mDDPM approach and proposed masking strategies in this section.

2.1   Fourier Transform

We first briefly introduce the Discrete Fourier Transform (DFT) as it plays a crucial role in our mDDPM approach. Given a 2D signal $z \in \mathbb{R}^{W \times H}$, the corresponding 2D-DFT, a widely used signal analysis technique, can be defined as follows:

$$f(x,y) = \sum_{h=0}^{H-1} \sum_{w=0}^{W-1} z(h,w) e^{-j2\pi\left(\frac{xh}{H} + \frac{yw}{W}\right)}, \tag{1}$$

$z(h, w)$ denotes the signal located at position (h, w) in z, while x and y serve as indices representing the horizontal and vertical spatial frequencies in the Fourier spectrum. The inverse 2D DFT (2D-IDFT) is defined as:

$$F(h, w) = \frac{1}{HW} \sum_{x=0}^{H-1} \sum_{y=0}^{W-1} f(x, y) e^{j2\pi\left(\frac{xh}{H} + \frac{yw}{W}\right)}, \tag{2}$$

Both the DFT and IDFT can be efficiently computed using the Fast Fourier Transform (FFT) algorithm, as in [23].

In the context of medical images with various modalities, the Fourier Transform is applied independently to each channel. Additionally, previous works such as [8, 19, 32] have demonstrated that the high-frequency part of the Fourier spectrum contains detailed structural texture information, while the low-frequency part contains global information.

2.2 DDPMs

In DDPMs, the forward process involves gradually introducing noise to the input image z_0 according to a predefined schedule $\beta_1, ..., \beta_T$. The noise is sampled from a Gaussian distribution $\mathcal{N}(0, I)$, and at each time step t, the noisy image z_t is generated as follows:

$$z_t \sim q(z_t | z_0) = \mathcal{N}(\sqrt{\bar{\alpha}_t} z_0, (1 - \bar{\alpha}_t) I), \tag{3}$$

where $\bar{\alpha}_t = \prod_{s=0}^{t} (1 - \beta_t)$ and t is sampled from a uniform distribution. For $t = T$, the image becomes pure Gaussian noise $z_t = \varepsilon \sim \mathcal{N}(0, I)$, while for $t = 0$, z_t remains x_0.

In the denoising process, the objective is to reverse the forward process and reconstruct the original image z_0. The reconstruction is achieved by sampling z_0 from the joint distribution:

$$z_0 \sim p_\theta(z_t) \prod_{t=1}^{T} p_\theta(z_{t-1} | z_t), \tag{4}$$

where $p_\theta(z_{t-1} | z_t)$ is modeled as a Gaussian distribution $\mathcal{N}(\mu_\theta(z_t, t), \Sigma_\theta(z_t, t))$. The parameters μ_θ and Σ_θ are estimated by a neural network, and we use a U-Net architecture for this purpose. The covariance $\Sigma_\theta(z_t, t)$ is fixed as $\frac{1-\alpha_{t-1}}{1-\alpha_t} \beta_t I$, following the approach in [15]. In this work, we simplify the loss derivation by directly estimating the reconstruction $z_0' \sim p_\theta(z_0 | z_t, t)$ as in [5] and using the mean absolute error ($l1$-error) as the loss function:

$$\mathcal{L}_{rec} = |z_0 - z_0'|, \tag{5}$$

Instead of performing step-wise denoising for all time steps starting from $t = T$, as commonly done for sampling images with DDPMs, we directly estimate z_0' at a fixed time step t_{fix}.

2.3 Masked DDPMs

As stated above, we model mDDPM by introducing a Masking block before DDPM stage as shown in Fig. 1 that can be incorporated in three different forms during training, namely IPM, FPM, and FPM-CM.

Image Patch-Masking (IPM). In the IPM approach, random masks are applied at the pixel level to patches extracted from the original image. The masked image is then subjected to the diffusion process in DDPMs. In contrast, the unmasked version of the same image is used as a reference for comparison during training. For reference, the output of the IPM block can be observed in Fig. 2b.

During training, we sample N patch regions, $[p_1, p_2, ..., p_N]$ at random positions such that $\Sigma_{n=0}^{N} A_n < A_z$, where A_n is the area of patch p_n, and A_z is the area of image, z in pixel space. Let $M_p^I \in \mathbb{R}^{C,H,W}$ be a binary mask in the pixel domain that indicates which pixels overlap with the patches $[p_1, p_2, ..., p_N]$. Specifically, the pixels within each p_n are assigned a value of zero, while the pixels outside of p_n are assigned a value of one. We obtain z_M, by combining the original image z_0 with the masked region using element-wise multiplication:

$$z_M^I = z_0 \odot M_p^I, \tag{6}$$

Here, $\odot$ represents the Hadamard product. z_M is then fed to DDPM forward process.

In the backward process, the denoised image, denoted as $\tilde{z}_0$, is generated by the network as an estimate of the original image z_0. As we calculate the absolute difference between the original image z_0 and the denoised image $\tilde{z}_0$ during training, the objective function $\mathscr{L}_M = \mathscr{L}_{rec}$, where $\mathscr{L}_{rec}$ is defined in Eq. 5.

By applying random pixel-level masks to the patches, the IPM approach introduces a form of regularization that encourages the DDPMs to generate images that closely resemble the unmasked reference image.

Frequency Patch-Masking (FPM). The proposed FPM mechanism has been illustrated in Fig. 2a. FPM block mainly consists of three blocks, FFT-2D, frequency spectrum masking (FSM), and IFFT-2D. Here, the image undergoes a series of transformations in the frequency domain using the Fast Fourier Transform (FFT). The Fast Fourier Transform allows us to analyze the image in terms of different frequencies present within it. After the image is transformed into the frequency domain, patch-level masking is performed using a binary mask M_p^F similar to the one used in IPM. This

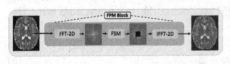
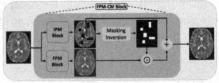

(a) Block diagram of FPM. Here, Fast Fourier Transform is performed on each slice, and then masking is performed on the frequency spectrum. The Inverse Fourier Transform of this masked spectrum outputs an augmented image of the input image.

(b) Block diagram of FPM-CM. Here, random patches are chosen from the augmented image obtained after FPM. These selected patches are then inserted at the corresponding positions in the original clean image to generated an augmented image.

Fig. 2. Proposed data augmentation techniques which are implemented in the Masking Block of the Fig. 1.

means that specific regions or patches within the frequency representation of the image are masked out. Masking in the frequency domain allows for selective modification of certain frequency components of the image while leaving others intact. Here, we don't specify the high-frequency or low-frequency masking, rather MFM is performed randomly. Once the patch-level masking is complete, the inverse Fourier Transform is applied to the modified frequency representation. This transforms the image back from the frequency domain to the spatial domain, reconstructing the modified image. The FPM block is mathematically formulated as,

$$z_M^F = IDFT(M_p^F \odot DFT(z_0)), \tag{7}$$

The reconstructed image is subsequently inputted into DDPM, utilizing the identical objective function as described earlier in Eq. 5.

Frequency Patch-Masking CutMix (FPM-CM). FPM-CM extends the application of FPM to apply patched augmentation in the original image as shown in Fig. 2b. In FPM-CutMix, the image is first passed through an FPM augmentation stage, and then patches are sampled from this augmented image. These patches are essentially small rectangular regions with varying sizes that capture specific features or information from the augmented image. After sampling the patches, they are inserted into the original clean image at corresponding positions. This means that the patches are placed in the same spatial locations within the original image as they were in the augmented image. By inserting the patches at corresponding positions, the intention is to transfer the modified features from the augmented image to the original clean image. It can be observed that these augmented patches behave as anomalies during training as the key idea is to use DDPM to generate the clean un-augmented image. Therefore such augmentation serves the purpose of unsupervised anomaly training. Assuming that we have obtained z_M^I, and z_M^F from IPM and FPM block respectively, FPM-CM can be mathematically written as,

$$z_M^{FC} = z_M^I + (z_M^F \odot \neg M_p^I), \tag{8}$$

Here, $\neg$ indicates binary inversion, and z_M^{FC} is the FPM-CM block output. We feed in DDPM with z_M^{FC} using the same objective function as used by other approaches mentioned above and stated in Eq. 5.

3 Experimental Evaluation

3.1 Implementation Details

For our experiments, we utilize the publicly available IXI dataset for training [1]. The IXI dataset comprises 560 pairs of T1 and T2-weighted brain MRI scans. To ensure robust evaluation, the IXI dataset is partitioned into eight folds, comprising 400/160 training/validation samples. To evaluate our approach, we employ two publicly available datasets: the Multimodal Brain Tumor Segmentation Challenge 2021 (BraTS21) dataset [2] and the multiple sclerosis dataset from the University Hospital of Ljubljana

(MSLUB) [20]. The BraTS21 dataset includes 1251 brain MRI scans with four different weightings (T1, T1-CE, T2, FLAIR). Following [5], it was divided into a validation
set of 100 samples and a test set of 1151 samples, both containing unhealthy scans. The
MSLUB dataset consists of brain MRI scans from 30 multiple sclerosis (MS) patients,
with each patient having T1, T2, and FLAIR-weighted scans. This dataset was split into
a validation set of 10 samples and a test set of 20 samples as in [5], all representing
unhealthy scans. Thus, our training, validation, and testing data consist of 400, 270, and
1171 samples respectively. In all our experimental setups, we exclusively employ T2-
weighted images extracted from the respective dataset and perform the pre-processing
such as affine transformation, skull stripping, and downsampling as in [5]. With the
specifically designed pre-processing techniques, we filtered out the regions belonging
to the foreground area so that Masking Block can only be applied to the foreground
pixel patches.

We assess the performance of our proposed method, mDDPM, in comparison to
various established baselines for UAD in brain MRI. These baselines include: *(i)* VAE
[3], *(ii)* Sequential VAE (SVAE) [4], *(iii)* denoising AE (DAE) [18], the GAN-based
(iv) f-AnoGAN [29], *(v)* DDPM [34], and *(iv)* patched DDPM (pDDPM), which feeds
patched input to the DDPM. We evaluate all baseline methods via in-house training
using their default parameters. For VAE, SVAE, and f-AnoGAN, we set the value of the
hyperparameter according to [5]. For DDPM, pDDPM, and mDDPM, we employ structured simplex noise instead of Gaussian noise as it better captures the natural frequency
distribution of MRI images [34]. We follow [5] for all other training and inference settings. By default, the models undergo training for 1600 epochs. During the training
phase, the volumes are processed in a slice-wise fashion, where slices are uniformly
sampled with replacement. However, during the testing phase, all slices are iterated
over to reconstruct the entire volume. Further, we conducted all our experiments with a
masking ratio randomly varying between 10%–90% of the whole foreground region.

3.2 Inference Criteria

During the training phase, all models are trained to minimize the $l1$ error between the
healthy input image and its corresponding reconstruction. At the test stage, we utilize
the reconstruction error as a pixel-wise anomaly score denoted as $\Lambda_S = |z_0 - z_0'|$, where
higher values correspond to larger reconstruction errors.

To enhance the quality of the anomaly maps, we employ commonly used post-
processing techniques [3,35]. Prior to binarizing Λ_S, we apply a median filter with a
kernel size of $K_M = 5$ to smooth the anomaly scores and perform brain mask erosion
for three iterations. After binarization and calculating threshold as in [5], we iteratively
calculate DICE scores [10] for different thresholds and select the threshold that yields
the highest average DICE score on the selected test set. Additionally, we record the
average Area Under the Precision-Recall Curve (AUPRC) [9] on the test set.

3.3 Results

The comparison of our proposed mDDPM with the baseline method is presented in
Table 1. It can be observed that our mDDPM outperforms all baseline approaches on

Table 1. Comparison of the models under consideration, with the best outcomes denoted in bold using DICE and AUPRC as evaluation metrics.

Model	BraTS21		MSLUB	
	DICE [%]	AUPRC [%]	DICE [%]	AUPRC [%]
VAE [3]	30.57 ± 1.67	28.47 ± 1.38	6.63 ± 0.12	5.01 ± 0.54
SVAE [4]	33.86 ± 0.19	33.53 ± 0.23	5.71 ± 0.48	5.05 ± 0.11
DAE [18]	36.85 ± 1.62	45.19 ± 1.35	3.67 ± 0.82	5.24 ± 0.53
f-AnoGAN [29]	24.44±2.28	22.52±2.37	4.29±1.02	4.09±0.79
DDPM [34]	40.82 ± 1.34	49.82 ± 1.13	8.52 ± 1.42	8.44 ± 1.54
pDDPM [5] + *fixed sampling* + $\mathscr{L}_p$	49.12 ± 1.27	53.98 ± 2.16	9.04 ± 0.66	9.23 ± 0.83
mDDPM (IPM)	52.16 ± 1.64	58.12 ± 1.56	10.39 ± 0.88	10.58 ± 0.92
mDDPM (FPM)	50.91 ± 1.28	56.27 ± 1.44	9.31 ± 0.46	9.51 ± 0.52
mDDPM (FPM-CM)	**53.02 ± 1.34**	**59.04 ± 1.26**	**10.71 ± 0.62**	**10.59 ± 0.57**

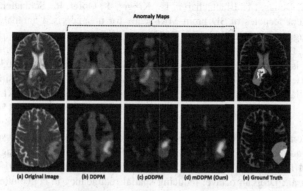

Fig. 3. In the comparison of the qualitative results between mDDPM (FPM-CM) and the baselines, we present the anomaly map comparisons for two samples. In the first sample, mDDPM demonstrates a more precise detection of the anomaly, exhibiting a final anomaly map without foreground noise. Similarly, in the second sample, mDDPM performs closest to the ground truth in terms of anomaly detection as shown in the results of the second row where we only included results for the FPM-CM variant of our method.

both datasets in terms of DICE [10] and AUPRC [9]. In terms of qualitative evaluation, we observe smaller reconstruction errors from mDDPM compared to patched DDPM [5] for healthy brain anatomy, as shown in Fig. 3. It can be observed that, mDDPM (FPM-CM) showcases a higher level of precision in detecting the anomaly, resulting in an anomaly map that is free from foreground noise.

4 Conclusion

This study introduces an approach for reconstructing the healthy brain anatomy using masked DDPM, which incorporates image-mask and frequency-mask regularization. Our method, known as mDDPM, surpasses established baselines, even with unsupervised training. However, a limitation of the proposed approach is the increased inference time associated with the diffusion architecture. To address this, future research could concentrate on enhancing the efficiency of the diffusion denoising process by leveraging spatial context more effectively. Further, we intend to explore the Masked Diffusion Transformer architecture in our future studies where we can incorporate a latent modeling scheme using masks to specifically improve the contextual relationship learning capabilities of DDPMs for object semantic parts within an image.

References

1. https://brain-development.org/ixi-dataset/
2. Baid, U., et al.: The RSNA-ASNR-MICCAI BraTS 2021 benchmark on brain tumor segmentation and radiogenomic classification. arXiv preprint arXiv:2107.02314 (2021)
3. Baur, C., Denner, S., Wiestler, B., Navab, N., Albarqouni, S.: Autoencoders for unsupervised anomaly segmentation in brain MR images: a comparative study. Med. Image Anal. **69**, 101952 (2021)
4. Behrendt, F., Bengs, M., Bhattacharya, D., Krüger, J., Opfer, R., Schlaefer, A.: Capturing inter-slice dependencies of 3D brain MRI-scans for unsupervised anomaly detection. In: Medical Imaging with Deep Learning (2022)
5. Behrendt, F., Bhattacharya, D., Krüger, J., Opfer, R., Schlaefer, A.: Patched diffusion models for unsupervised anomaly detection in brain MRI. arXiv preprint arXiv:2303.03758 (2023)
6. Bengs, M., Behrendt, F., Krüger, J., Opfer, R., Schlaefer, A.: Three-dimensional deep learning with spatial erasing for unsupervised anomaly segmentation in brain MRI. Int. J. Comput. Assist. Radiol. Surg. **16**(9), 1413–1423 (2021). https://doi.org/10.1007/s11548-021-02451-9
7. Chen, X., Konukoglu, E.: Unsupervised detection of lesions in brain MRI using constrained adversarial auto-encoders. In: International Conference on Medical Imaging with Deep Learning (MIDL). Proceedings of Machine Learning Research, PMLR (2018)
8. Chen, Y., et al.: Drop an octave: reducing spatial redundancy in convolutional neural networks with octave convolution. In: Proceedings of the IEEE/CVF International Conference on Computer Vision, pp. 3435–3444 (2019)
9. Davis, J., Goadrich, M.: The relationship between precision-recall and ROC curves. In: Proceedings of the 23rd International Conference on Machine Learning, pp. 233–240 (2006)
10. Dice, L.R.: Measures of the amount of ecologic association between species. Ecology **26**(3), 297–302 (1945)
11. Ellis, R.J., Sander, R.M., Limon, A.: Twelve key challenges in medical machine learning and solutions. Intell.-Based Med. **6**, 100068 (2022)
12. Fernando, T., Gammulle, H., Denman, S., Sridharan, S., Fookes, C.: Deep learning for medical anomaly detection - a survey. ACM Comput. Surv. **54**(7), 1–37 (2021). https://doi.org/10.1145/3464423
13. Gao, P., Ma, T., Li, H., Lin, Z., Dai, J., Qiao, Y.: ConvMAE: masked convolution meets masked autoencoders (2022)
14. He, K., Chen, X., Xie, S., Li, Y., Dollár, P., Girshick, R.: Masked autoencoders are scalable vision learners (2021)
15. Ho, J., Jain, A., Abbeel, P.: Denoising diffusion probabilistic models. In: Advances in Neural Information Processing Systems, vol. 33, pp. 6840–6851 (2020)
16. Johnson, J.M., Khoshgoftaar, T.M.: Survey on deep learning with class imbalance. J. Big Data **6**(1), 1–54 (2019). https://doi.org/10.1186/s40537-019-0192-5
17. Karimi, D., Dou, H., Warfield, S.K., Gholipour, A.: Deep learning with noisy labels: exploring techniques and remedies in medical image analysis. Med. Image Anal. **65**, 101759 (2020)

18. Kascenas, A., Pugeault, N., O'Neil, A.Q.: Denoising autoencoders for unsupervised anomaly detection in brain MRI. In: International Conference on Medical Imaging with Deep Learning (MIDL). Proceedings of Machine Learning Research, PMLR (2022)

19. Kauffmann, L., Ramanoël, S., Peyrin, C.: The neural bases of spatial frequency processing during scene perception. Front. Integr. Neurosci. **8**, 37 (2014)

20. Lesjak, Ž, et al.: A novel public MR image dataset of multiple sclerosis patients with lesion segmentations based on multi-rater consensus. Neuroinformatics **16**(1), 51–63 (2017). https://doi.org/10.1007/s12021-017-9348-7

21. Lugmayr, A., Danelljan, M., Romero, A., Yu, F., Timofte, R., Van Gool, L.: RePaint: inpainting using denoising diffusion probabilistic models. In: Proceedings of the IEEE/CVF Conference on Computer Vision and Pattern Recognition, pp. 11461–11471 (2022)

22. Nguyen, B., Feldman, A., Bethapudi, S., Jennings, A., Willcocks, C.G.: Unsupervised region-based anomaly detection in brain MRI with adversarial image inpainting. In: 2021 IEEE 18th International Symposium on Biomedical Imaging (ISBI), pp. 1127–1131. IEEE (2021)

23. Nussbaumer, H.J.: The fast Fourier transform. In: Fast Fourier Transform and Convolution Algorithms, pp. 80–111. Springer, Berlin, Heidelberg (1981). https://doi.org/10.1007/978-3-662-00551-4_4

24. Özdenizci, O., Legenstein, R.: Restoring vision in adverse weather conditions with patch-based denoising diffusion models. IEEE Trans. Pattern Anal. Mach. Intell. **45**(8), 10346–10357 (2023)

25. Pinaya, W.H., et al.: Fast unsupervised brain anomaly detection and segmentation with diffusion models. arXiv preprint arXiv:2206.03461 (2022)

26. Pinaya, W.H., et al.: Unsupervised brain imaging 3D anomaly detection and segmentation with transformers. Med. Image Anal. **79**, 102475 (2022)

27. Rombach, R., Blattmann, A., Lorenz, D., Esser, P., Ommer, B.: High-resolution image synthesis with latent diffusion models. In: Proceedings of the IEEE/CVF Conference on Computer Vision and Pattern Recognition, pp. 10684–10695 (2022)

28. Sanchez, P., Kascenas, A., Liu, X., O'Neil, A.Q., Tsaftaris, S.A.: What is healthy? generative counterfactual diffusion for lesion localization. In: Mukhopadhyay, A., Oksuz, I., Engelhardt, S., Zhu, D., Yuan, Y. (eds.) Deep Generative Models: Second MICCAI Workshop, DGM4MICCAI 2022, Held in Conjunction with MICCAI 2022, Singapore, September 22, 2022, Proceedings, pp. 34–44. Springer, Cham (2022). https://doi.org/10.1007/978-3-031-18576-2_4

29. Schlegl, T., Seeböck, P., Waldstein, S.M., Langs, G., Schmidt-Erfurth, U.: f-AnoGAN: fast unsupervised anomaly detection with generative adversarial networks. Med. Image Anal. **54**, 30–44 (2019)

30. Shen, D., Wu, G., Suk, H.I.: Deep learning in medical image analysis. Annu. Rev. Biomed. Eng. **19**, 221–248 (2017)

31. Silva-Rodríguez, J., Naranjo, V., Dolz, J.: Constrained unsupervised anomaly segmentation. Med. Image Anal. **80**, 102526 (2022)

32. Wang, W., et al.: FreMAE: Fourier transform meets masked autoencoders for medical image segmentation (2023)

33. Wolleb, J., Bieder, F., Sandkühler, R., Cattin, P.C.: Diffusion models for medical anomaly detection. arXiv preprint arXiv:2203.04306 (2022)

34. Wyatt, J., Leach, A., Schmon, S.M., Willcocks, C.G.: AnoDDPM: anomaly detection with denoising diffusion probabilistic models using simplex noise. In: Proceedings of the IEEE/CVF Conference on Computer Vision and Pattern Recognition, pp. 650–656 (2022)

35. Zimmerer, D., Kohl, S., Petersen, J., Isensee, F., Maier-Hein, K.: Context-encoding variational autoencoder for unsupervised anomaly detection. In: International Conference on Medical Imaging with Deep Learning-Extended Abstract Track (2019)

IA-GCN: Interpretable Attention Based Graph Convolutional Network for Disease Prediction

Anees Kazi[1,2,3(✉)], Soroush Farghadani[4,5], Iman Aganj[2,3], and Nassir Navab[1,6]

[1] Computer Aided Medical Procedures, Technical University of Munich, Munich, Germany
[2] Radiology Department, Martinos Center for Biomedical Imaging, Massachusetts General Hospital, Boston, USA
[3] Harvard Medical School, Boston, USA
akazi1@mgh.harvard.edu
[4] Sharif University of Technology, Tehran, Iran
[5] University of Toronto, Toronto, Canada
[6] Whiting School of Engineering, Johns Hopkins University, Baltimore, USA

Abstract. Interpretability in Graph Convolutional Networks (GCNs) has been explored to some extent in general in computer vision; yet, in the medical domain, it requires further examination. Most of the interpretability approaches for GCNs, especially in the medical domain, focus on interpreting the output of the model in a *post-hoc* fashion. In this paper, we propose an interpretable attention module (IAM) that explains the relevance of the input features to the classification task on a GNN Model. The model uses these interpretations to improve its performance. In a clinical scenario, such a model can assist the clinical experts in better decision-making for diagnosis and treatment planning. The main novelty lies in the IAM, which directly operates on input features. IAM learns the attention for each feature based on the unique interpretability-specific losses. We show the application of our model on two publicly available datasets, Tadpole and the UK Biobank (UKBB). For Tadpole we choose the task of disease classification, and for UKBB, age, and sex prediction. The proposed model achieves an increase in an average accuracy of 3.2% for Tadpole and 1.6% for UKBB sex and 2% for the UKBB age prediction task compared to the state-of-the-art. Further, we show exhaustive validation and clinical interpretation of our results.

Keywords: Interpretability · Graph Convolutional Network · Disease prediction

Equal contribution. This work was done when A. Kazi and S. Farghadani were affiliated to the Technical University of Munich.

Supplementary Information The online version contains supplementary material available at https://doi.org/10.1007/978-3-031-45673-2_38.

X. Cao et al. (Eds.): MLMI 2023, LNCS 14348, pp. 382–392, 2024.
https://doi.org/10.1007/978-3-031-45673-2_38

1   Introduction

Graph Convolutional Networks (GCNs) have shown great impact in the medical domain [2] such as brain imaging [5], ophthalmology [16], breast cancer [12] , and thorax disease diagnosis [18]. Recently many methodological advances have also been made, especially for medical tasks, such as dealing with missing [6]/imbalanced [8] data, out-of-sample extension [9], handling the multiple-graphs [28] and, graph learning [7] to name a few. In spite of their great success, GCNs are still less transparent than other models. Interpreting the model's outcome with respect to input (graph and node features) and task (loss) is essential. Interpretability techniques dealing with the analysis of GCNs have been gaining importance in the last couple of years [20]. GNNExplainer [30], for example, is one of the pioneer works in this direction. The paper proposes a *post hoc* technique to generate an explanation for the outcome of Graph Convolution (GC) based models with respect to the input graph and features. This is obtained by maximizing the mutual information between the pre-trained model output and the output with selected input sub-graph and features. Further conventional gradient-based and decomposition-based techniques have also been applied to GCN [4]. Another work [11] proposes a local interpretable model explanation for graphs. It uses a nonlinear feature selection method leveraging the Hilbert-Schmidt Independence Criterion. However, the method is computationally complex as it generates a nonlinear interpretable model.

Deploying non-interpretable Graph-based deep learning models in medicine could lead to incorrect diagnostic decisions [3]. Therefore, adopting interpretability in machine learning (ML) models is important, especially in healthcare [23]. Recently, the main efforts have been towards creating transparent and explainable ML models [26]. One recent work [14] proposes a post hoc approach similar to the GNNexplainer applied to digital pathology. Another method, Brainexplainer [19], proposes an ROI-selection pooling layer (R-pool) that highlights ROIs (nodes in the graph) important for the prediction of neurological disorders. Interpretability for GCNs is still an open challenge in the medical domain.

In this paper, we target the interpretability of GCNs and design a model capable of incorporating the interpretations in the form of attentions for better model performance. We propose an interpretability-specific loss that helps in increasing the confidence of the interpretability. We show that such an interpretation-based feature selection enhances the model performance. We also adapt a graph learning module from [7] to learn the latent population graph thus our model does not require a pre-computed graph. We show the superiority of the proposed model on 2 datasets for 3 tasks, Tadpole [21] for Alzheimer's disease prediction (three-class classification) and UK Biobank (UKBB) [22] for sex (2 classes) and age (4 classes) classification task. We provide several ablation tests and validations supporting our hypothesis that incorporating the interpretation with in the model is beneficial. In the following sections, we first provide mathematical details of the proposed method, then describe the experiments, conclude the paper by a discussion.

2   Method

Given the dataset $\mathbf{Z} = [\mathbf{X}, \mathbf{Y}]$, where $\mathbf{X} \in \mathbb{R}^{N \times D}$ is the feature matrix with dimension D for N subjects, with $\mathbf{Y} \in \mathbb{R}^N$, being the labels with c classes, the task is to classify each patient into the respective class from 1,2,...,c. To achieve this, we design an end-to-end model, mathematically defined as $\hat{y} = f_\theta(h_M(X), g_\phi(X))$. Here $\hat{y}$ is the model prediction, $f_\theta(.)$ is the classification module, h_M is the Interpretable Attention Module (IAM) designed to learn the attentions for features, and g_ϕ is the model to learn the latent graph. In the following paragraphs, we explain h_M, g_ϕ, the proposed loss, and the base model used for the classification task.

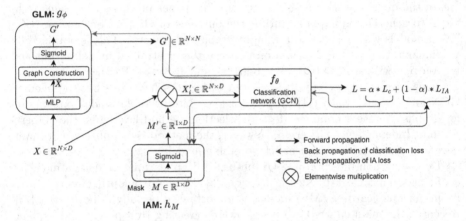

Fig. 1. IA-GCN consists of three main components: 1) Interpretable Attention Module (IAM): h_M, 2) Graph Learning Module (GLM):g_ϕ, and 3) Classification Module: f_θ. These are trained in an end-to-end fashion. In backpropagation, two loss functions are playing roles which are demonstrated in blue and red arrows.

2.1   Interpretable Attention Module (IAM): h_M

For h_M, we define a differentiable and continuous mask $\mathbf{M} \in \mathbb{R}^{1 \times D}$ that learns an attention coefficient for each feature element from D features. IAM can be mathematically defined as $\mathbf{x}'_i = h_M(\mathbf{x_i}) = \sigma(\mathbf{M}) \times \mathbf{x_i}$ where, $\mathbf{x}'_i \in \mathbb{R}^{1 \times D}$ is the masked output, σ is the sigmoid function. $\sigma(\mathbf{M})$ represents the learned mask $\mathbf{M}' \subset [0, 1]$. The mask $\mathbf{M}$ is continuous as the aim is to learn the interpretable attentions while the model is training. Conceptually, the corresponding weights in the mask $\mathbf{M}'$ should take a value close to zero when a particular feature is not significant towards the task. In effect, m_i corresponding to d_i may improve or deteriorate the model performance based on the importance of d_i towards the task. The proposed IAM is trained by a customized loss discussed in Sect. 2.4.

2.2   Graph Learning Module (GLM): g_ϕ

Inspired by DGM [7] we define our GLM mathematically denoted as g_ϕ. Given the input $\mathbf{X}$, GLM predicts an optimal graph $\mathbf{G}' \in \mathbb{R}^{N \times N}$ which is then used in the GCN for the classification, as shown in Fig. 1. GLM consists of 2 layered multilayer perceptron (MLP) followed by a graph construction step and a graph pruning step. MLP takes the feature matrix $\mathbf{X} \in \mathbb{R}^{N \times D}$ as input and produces $\hat{\mathbf{X}}$ embedding specific for the optimal latent graph as output. A fully connected graph is computed with continuous edge values (shown as graph construction in Fig. 1) using the Euclidean distance metric between the feature embedding $\hat{x}_i$ and $\hat{x}_j$ where $[\hat{x}_i, \hat{x}_j] \in \hat{\mathbf{X}}$. Sigmoid function is used for soft thresholding keeping the GLM differentiable. g'_{ij} is computed as $g'_{ij} = \frac{1}{1+e^{\left(t\|\hat{x}_i - \hat{x}_j\|_2 + T\right)}}$ with T being the threshold parameter and $t(> 0)$ the temperature parameter pushing values of g'_{ij} to either 0 or 1. Both t and T are optimized during training. Thus, $\mathbf{G}'$ is obtained.

2.3   Classification Module with Joint Optimization of GLM and IAM

As mentioned before, the primary goal is to classify each patient x_i into the respective class y_i. The classification model can be mathematically defined as $\hat{\mathbf{Y}} = f_\theta(\mathbf{X}', \mathbf{G}')$ where f_θ is the classification function with learnable parameters θ, $\mathbf{G}'$ the learned latent population graph structure, and $\mathbf{X}'$ the output of IAM. We define f_θ as a generic GCN targeted towards node classification. The whole model is trained end to end using a customized loss focusing more on interpretability. This loss is discussed below.

2.4   Interpretability-Focused Loss Functions

Empirically we observed that training the model with only softmax cross-entropy loss L_c was sub-optimal and, specifically, 1) the performance was not the best, 2) the mask learned average values for all the features reflecting uncertainty, and 3) unimportant features would take considerable weight in the mask. In order to optimize the whole network in an end-to-end fashion, we define the loss L as

$$L = (1 - \alpha)L_c + \alpha * (\alpha_1 * \sum_{i=0}^{D-1} -m'_i log_2(m'_i) + \alpha_2 * \sum_{i=0}^{D-1} m'_i) \qquad (1)$$

where L_c is the softmax cross-entropy loss, α is the weighting factor chosen experimentally, the next two terms being F_{MEL}: the feature mask entropy loss, and F_{MSL}: the feature mask size loss respectively with respective weights factors α_1 and α_2. F_{MEL} and F_{MSL} are used to regularize L_c. Firstly, $F_{MSL} = \sum_{i=0}^{D-1} m'_i$ lowers the sum of the values of individual m'_i. Otherwise, all the features would get the highest importance with all the $m's$ taking up the value 1. On the other hand, $F_{MEL} = \sum_{i=0}^{D-1} -m'_i log_2(m'_i)$ pushes the values away from 0.5, which makes the model more confident about the importance of the feature d_i. F_{MEL} and F_{MSL} are used only by h_M for the back propagation.

3   Experiments

Two publicly available datasets were used for three tasks. Tadpole [21] for Alzheim- er's disease prediction and UK Biobank (UKBB) [22] for age and sex prediction. The task in the **Tadpole** dataset was to classify 564 subjects into three categories (Normal, Mild Cognitive Impairment, Alzheimer's) that represent their clinical status. Each subject had 354 multi-modal features that included cognitive tests, MRI ROIs measures, PET imaging, DTI ROI measures, demographics, etc. On the other hand, the **UKBB** dataset consisted of 14,503 subjects with 440 features per individual, which were extracted from MRI and fMRI images. Two classification tasks were considered for this dataset: 1) sex prediction, 2) categorical age prediction. In the second task, subjects' ages were quantized into four decades as the classification targets. Table 1 shows the results of the classification task for both datasets. We performed an experiment with a linear classifier (LC) to see the complexity of the task as well as with the Chebyshev polynomial-based spectral-GCN [25], and Graph Attention Network (GAT) [27], which is a spatial method. We compared with these two methods as they require a pre-defined graph structure for the classification task, whereas our method and DGM [17] do not. Our reasoning behind learning the graph is that pre-computed/preprocessed graphs can be noisy, irrelevant to the task, or unavailable. Depending on the model, learning the population graph is much more clinically semantic. Unlike Spectral-GCN and GAT, DGCNN [29] constructs a KNN graph at each layer dynamically during training. This removes the requirement for a pre-computed graph. However, the method still lacks the ability to learn the latent graph. DGM [7] and the proposed method, on the other hand, do not require any graph structure to be defined and they only utilize the given features. **Implementation Details:** M is initialized either with Gaussian normal distribution or constant values. Experiments were performed using Google Colab with a Tesla T4 GPU with PyTorch 1.6. Number of epochs was 600. Same 10 folds with the train:test split of 90:10 were used in all the experiments. We used two MLP layers (16→8) for GLM and two Conv layers followed by a FC layer (32→16→# classes) for the classification network. ReLU was used as the activation function.

Classification Performance: For Tadpole, the proposed method performed best for all the three measures (Accuracy, AUC, F1). The overall lower F1-score indicates that the task was challenging due to the class imbalance ($\frac{2}{7}, \frac{4}{7}, \frac{1}{7}$) present in the dataset. The proposed IAM adds interpretable attention to features, which improves the model performance by 3.16% compared to the state-of-the-art (DGM). The low variance shows the stability of the proposed method. The UKBB was chosen due to its much larger dataset size. Sex prediction covers the challenge of larger dataset size, whereas age prediction deals with both large size and imbalance. The results are shown in Table 3. For sex prediction, our method shows superior performance and AUC reconfirms the consistency of the model's performance. For age prediction, results demonstrate that the overall

Table 1. Performance of the proposed method (mean ± STD) compared with several state-of-the-art and baseline methods on the Tadpole and UKBB dataset for classification.

Dataset	Task.	Method	Accuracy	AUC	F1
Tadpole	Disease	LC [10]	70.22 ± 06.32	80.26 ± 04.81	68.73 ± 06.70
		GCN [25]	81.00 ± 06.40	74.70 ± 04.32	78.4 ± 06.77
		GAT [27]	81.86 ± 05.80	91.76 ± 03.71	80.90 ± 05.80
		DGCNN [29]	84.59 ± 04.33	83.56 ± 04.11	82.87 ± 04.27
		DGM [7]	92.92 ± 02.50	97.16 ± 01.32	91.4 ± 03.32
		IA-GCN	**96.08 ± 02.49**	**98.6 ± 01.93**	**94.77 ± 04.05**
UKBB	Sex	LC	81.70 ± 01.64	90.05 ± 01.11	81.62 ± 01.62
		GCN [25]	83.70 ± 01.06	83.55 ± 00.83	83.63 ± 00.86
		DGCNN [29]	87.06 ± 02.89	90.05 ± 01.11	86.74 ± 02.82
		DGM [7]	90.67 ± 01.26	96.47 ± 00.66	90.65 ± 01.25
		IA-GCN	**92.32 ± 00.89**	**97.04 ± 00.59**	**92.25 ± 00.87**
UKBB	Age	LC	59.66 ± 01.17	80.26 ± 00.91	48.32 ± 03.35
		GCN [25]	55.55 ± 01.82	61.00 ± 02.70	40.68 ± 02.82
		DGCNN [29]	58.35 ± 00.91	76.82 ± 03.03	47.12 ± 03.95
		DGM [7]	63.62 ± 01.23	82.79 ± 01.14	50.23 ± 02.52
		IA-GCN	**65.64 ± 01.12**	**83.49 ± 01.04**	**51.73 ± 02.68**

task is much more challenging than the sex prediction. Lower F1-score shows the existence of class imbalance. Our method outperforms the DGM by 2.02% and 1.65% in accuracy for the sex and age task, respectively. Moreover, the performance trend of other comparative methods can be seen similar to be similar to Tadpole. The above results indicate that the incorporation of graph convolutions helps in better representation learning, resulting in more accurate classification. Further, GAT requires full data in one batch along with the affinity graph, which causes the out-of-memory issue in UKBB experiments. Moreover, DGCNN and DGM achieve higher accuracy compared to Spectral-GCN and GAT. This confirms our hypothesis that a pre-computed graph might not be optimal. Between DGCNN and DGM, the latter performs better, confirming that learning a graph is beneficial to the final task and for getting latent semantic graph as output.

Analysis of the Loss Function: Next, we investigated the contribution of all the loss terms, toward the optimization of the task. We report the accuracy of classification, the average attention for the top four features (Avg.4) selected by the model and other features (Avg.O). Table 2 (top) shows changes in the performance and the average of attention values (Avg.4 and Avg.O) with respect to α. The performance drops significantly with $\alpha = 0$. Best accuracy at $\alpha = 0.6$ shows that both loss terms are necessary for the optimal performance of the model. Avg.4 and Avg.O surge dramatically each time α increases. This proves the importance of F_{MEL} and F_{MSL} in shrinking the attention values of features.

In the second experiment shown in Table 2 (bottom), two specific cases were investigated. While α is at its optimum value of 0.6, the contribution of α_1 and α_2 was investigated, to show the contribution of F_{MEL} and F_{MSL} in the model. α_1 and α_2 were set to 0 respectively. F_{MEL} seems to have more importance in the certainty of interpretable attentions for important features. However, F_{MSL} pushes the attention values to 0 which helps us in distinguishing more and less important features. The combination of all three terms leads to the best performance as shown by optimal αs.

Interpretability: Here, we show validation experiments to prove the relevance of features selected by the IAM to the clinical task and model performance. We measured the classification performance by manually adding and removing the features from the input for two traditional methods of GCN [25] and DGM [7]. Table 3 presents experiments with different input features, including a) method trained traditionally on all available features, b) conventional feature selection technique using Ridge classifier applied to the input features at the preprocessing step, c) method trained conventionally with all input features except the features selected by the IAM, and d) model trained on only features selected by the proposed method. Overall, the feature selection approach (b and d) was advantageous, with the proposed IA-based feature selection (d)

Table 2. Performance of IA-GCN on the Tadpole dataset in different settings w.r.t α. Here, we show the model performance for classification. We report the accuracy of classification, the average attention for the top 4 features (Avg.4), and other features (Avg.O). We also show model performance when the values of α_1 and α_2 are changed.

α	Accuracy	Avg.4	Avg.O
0	57.00 ± 09.78	10^{-6}	10^{-6}
0.2	94.20 ± 03.44	0.12	0.002
0.4	95.10 ± 02.62	0.29	0.001
0.6	**96.10 ± 02.49**	**0.74**	**0.0**
0.8	95.80 ± 02.31	0.78	0.23
1.0	95.40 ± 02.32	0.82	0.42
$\alpha_1 = 0$	95.60 ± 02.44	0.54	0.13
$\alpha_2 = 0$	95.10 ± 03.69	0.86	0.26

performing the best. When models were trained with features other than the selected ones, their performance drastically dropped. Similar experiments were repeated on the UKBB dataset for age and sex classification using DGM, and the Ridge classifier with feature selection during preprocessing performed the best, indicating the necessity of feature selection. Further discussion of these results will be provided later in the limitations section.

Clinical Interpretation: In the Tadpole dataset, our model selects four cognitive features CDR, CDR at baseline, MMSE and MMSE at baseline. It is reported in the clinical literature that the cognitive measure of Clinical Dementia Rating Sum of Boxes (CDRSB) compares well with the global CDR score for dementia staging. Cognitive tests measure the decline in a straightforward and quantifiable way with the disease condition. Therefore, these are important in Alzheimer's disease prediction [13], in particular, CDRSB and the Mini-Mental

Table 3. Performance for the classification task. We show results for four baselines on GCN [25] for Tadpole and UKBB and DGM [7] for Tadpole with different input feature settings. ACC represents accuracy.

Data	Task	Method	Measure	a	b	c	d
Tadpole	Disease	GCN	ACC	77.4 ± 02.41	81.00 ± 06.40	74.50 ± 3.44	**82.4 ± 04.14**
			AUC	79.79 ± 04.75	74.70 ± 04.32	72.11 ± 08.24	**83.89 ± 09.06**
			F1	74.70 ± 05.32	78.4 ± 06.77	65.23 ± 08.46	**78.73 ± 07.60**
		DGM	ACC	89.2 ± 05.26	92.92 ± 02.50	79.70 ± 04.22	**95.09 ± 03.15**
			AUC	96.47 ± 02.47	97.16 ± 01.32	90.66 ± 02.64	**98.33 ± 02.07**
			F1	88.60 ± 05.32	91.4 ± 03.32	77.9 ± 6.38	**93.36 ± 03.28**
UKBB	Age	DGM	ACC	62.10 ± 01.45	**63.62 ± 01.23**	61.54 ± 01.83	59.45 ± 03.15
			AUC	76.57 ± 02.47	76.82 ± 03.03	**81.40 ± 04.73**	77.23 ± 02.17
			F1	46.80 ± 04.83	**50.23 ± 02.52**	47.31 ± 03.54	47.46 ± 03.19
	Sex	DGM	ACC	89.93 ± 01.3	**90.67 ± 01.26**	89.04 ± 01.84	87.46 ± 03.32
			AUC	95.83 ± 00.76	**96.47 ± 00.66**	95.02 ± 00.92	93.98 ± 02.43
			F1	89.83 ± 01.34	**90.65 ± 01.25**	89.01 ± 01.75	87.4 ± 03.23

State Examination (MMSE) [1, 24]. MMSE is the best-known and the most often used short screening tool for providing an overall measure of cognitive impairment in clinical, research, and community settings. Apart from cognitive tests, the Tadpole dataset includes other imaging features. We observed that the Pearson correlation coefficient with respect to the ground truth and the attention value computed by IAM are roughly linearly related. For the UKBB sex classification task, in the order of importance, our model selected volume features of peripheral cortical gray matter (normalized for (1) head size, (2) white matter, (3) brain, gray+white matter, (4) cortical gray matter, and (5) peripheral cortical gray matter) which is also supported by [15]. For age prediction, the most relevant features selected by our network were (1) volume of peripheral cortical gray matter, mean (2) MD and (3) L2 in fornix on FA skeleton, (4) mean L3 in anterior corona radiata on FA skeleton right and (5) mean L3 in anterior corona radiata on FA skeleton left which are also supported by [15]. For both the Tadpole and UKBB datasets, it is observed that the set of selected features are different depending on the task. Our interpretation of the model not selecting the MRI features in the Tadpole experiments is that attention is distributed over 314 features, which are indistinct compared to cognitive features. MRI features may nevertheless be more valuable when two scans taken over time between two visits to the hospital are compared to check the loss in volume (atrophy). However, in our case, we only considered scans at the baseline.

Limitations: The model fails in the case of UKBB in Table 3. The best performance is shown by feature selection by the Ridge classifier (b). Intuitively, UKBB is much large data with a difficult task. A more complex attention module design could be helpful. In the case of clinical interpretation for the age classification task, we observed a much larger set of features were given higher attention (Only the top 4 shown due to page limit), confirming the task complexity. Further, the values of αs are empirically chosen. They could be learned automatically.

4 Discussion and Conclusion

We developed a GCN-based model featuring an interpretable attention module (IAM) and a distinct loss function. The IAM learns feature attention and aids model training. Our experiments reveal strong feature correlations via IAM for Tadpole and UKBB sex classification, and our model outperforms state-of-the-art methods in disease and age classification.To address the issue of ignoring important features, we marginalized overall feature subsets and used a Monte Carlo estimate to sample from empirical marginal distribution for nodes during training. Our proposed method handles class imbalance well and achieved higher accuracy and F1-score than DGM in both tasks for UKBB. In terms of results, the method exceeded the highest accuracy by 3.5% for the disease classification task. Furthermore, the proposed method's F1-score was 3.2% higher than that of the state-of-the-art methods, which shows that it handles class imbalance well. For both tasks in UKBB, the accuracy and F1-score of the proposed method was 1.7% and 1.8% higher than the DGM method, respectively. For the UKBB age prediction task, we observed ~2% gain in accuracy and F1-score.

Acknowledgment. Anees Kazi's financial support was provided by BigPicture (IMI945358) from the Technical University of Munich during this project. Support for this research was partly provided by the National Institutes of Health (NIH), specifically the National Institute on Aging (RF1AG068261).

References

1. A-Rodriguez, I., et al.: Mini-mental state examination (MMSE) for the detection of Alzheimer's disease and other dementias in people with mild cognitive impairment (MCI). CDSR (3) (2015)
2. Ahmedt-Aristizabal, D., Armin, M.A., Denman, S., Fookes, C., Petersson, L.: Graph-based deep learning for medical diagnosis and analysis: past, present and future. Sensors **21**(14), 4758 (2021)
3. Amann, J., Blasimme, A., Vayena, E., Frey, D., Madai, V.I.: Explainability for artificial intelligence in healthcare: a multidisciplinary perspective. BMC Med. Inform. Decis. Mak. **20**(1), 1–9 (2020)
4. Baldassarre, F., Azizpour, H.: Explainability techniques for graph convolutional networks. arXiv preprint arXiv:1905.13686 (2019)
5. Bessadok, A., Mahjoub, M.A., Rekik, I.: Graph neural networks in network neuroscience. IEEE Trans. Pattern Anal. Mach. Intell. **45**(5), 5833–5848 (2023)
6. Chang, Y.W., et al.: Neural network training with highly incomplete medical datasets. Mach. Learn. Science. Technol. **3**(3), 035001 (2022)
7. Cosmo, L., Kazi, A., Ahmadi, S.-A., Navab, N., Bronstein, M.: Latent-graph learning for disease prediction. In: Martel, A.L., et al. (eds.) MICCAI 2020. LNCS, vol. 12262, pp. 643–653. Springer, Cham (2020). https://doi.org/10.1007/978-3-030-59713-9_62

8. Ghorbani, M., Kazi, A., Baghshah, M.S., Rabiee, H.R., Navab, N.: RA-GCN: graph convolutional network for disease prediction problems with imbalanced data. Media **75**, 102272 (2022)

9. Hamilton, W., Ying, Z., Leskovec, J.: Inductive representation learning on large graphs. In: Proceedings NIPS (2017)

10. Hoerl, A.E., Kennard, R.W.: Ridge regression: biased estimation for nonorthogonal problems. Technometrics **12**, 55–67 (1970)

11. Huang, Q., Yamada, M., Tian, Y., Singh, D., Yin, D., Chang, Y.: GraphLIME: local interpretable model explanations for graph neural networks. arXiv preprint arXiv:2001.06216 (2020)

12. Ibrahim, M., Henna, S., Cullen, G.: Multi-graph convolutional neural network for breast cancer multi-task classification. In: Longo, L., O'Reilly, R. (eds.) Artificial Intelligence and Cognitive Science: 30th Irish Conference, AICS 2022, Munster, Ireland, December 8–9, 2022, Revised Selected Papers, pp. 40–54. Springer, Cham (2023). https://doi.org/10.1007/978-3-031-26438-2_4

13. Jack, C.R., Holtzman, D.M.: Biomarker modeling of Alzheimer's disease. Neuron **80**(6), 1347–1358 (2013)

14. Jaume, G., et al.: Towards explainable graph representations in digital pathology. arXiv preprint arXiv:2007.00311 (2020)

15. Jiang, H., et al.: Predicting brain age of healthy adults based on structural MRI parcellation using convolutional neural networks. Front. Neurol. **10**, 1346 (2020)

16. Joshi, A., Sharma, K.K.: Graph deep network for optic disc and optic cup segmentation for glaucoma disease using retinal imaging. Phys. Eng. Sci. Med. **45**(3), 847–858 (2022). https://doi.org/10.1007/s13246-022-01154-y

17. Kazi, A., Cosmo, L., Ahmadi, S.-A., Navab, N., Bronstein, M.M.: Differentiable graph module (DGM) for graph convolutional networks. IEEE Trans. Pattern Anal. Mach. Intell. **45**(2), 1606–1617 (2023)

18. Lee, Y.W., Huang, S.K., Chang, R.F.: CheXGAT: a disease correlation-aware network for thorax disease diagnosis from chest x-ray images. Artif. Intell. Med. **132**, 102382 (2022)

19. Li, X., Duncan, J.: BrainGNN: Interpretable brain graph neural network for FRMI analysis. bioRxiv (2020)

20. Liu, N., Feng, Q., Hu, X.: Interpretability in graph neural networks. In: Graph Neural Networks: Foundations, Frontiers, and Applications, pp. 121–147. Springer, Singapore (2022). https://doi.org/10.1007/978-981-16-6054-2_7

21. Marinescu, R.V., et al.: TADPOLE Challenge: prediction of longitudinal evolution in Alzheimer's disease. arXiv preprint arXiv:1805.03909 (2018)

22. Miller, K.L., et al.: Multimodal population brain imaging in the UK biobank prospective epidemiological study. Nat. Neurosci. **19**(11), 1523–1536 (2016)

23. Molnar, C.: Interpretable Machine Learning. Lulu. com (2020)

24. O'Bryant, S.E., et al.: Staging dementia using clinical dementia rating scale sum of boxes scores: a Texas Alzheimer's research consortium study. Arch. Neurol. **65**(8), 1091–1095 (2008)

25. Parisot, S., et al.: Spectral graph convolutions for population-based disease prediction. In: Descoteaux, M., Maier-Hein, L., Franz, A., Jannin, P., Collins, D.L., Duchesne, S. (eds.) MICCAI 2017. LNCS, vol. 10435, pp. 177–185. Springer, Cham (2017). https://doi.org/10.1007/978-3-319-66179-7_21

26. Stiglic, G., Kocbek, P., Fijacko, N., Zitnik, M., Verbert, K., Cilar, L.: Interpretability of machine learning-based prediction models in healthcare. Wiley Interdisc. Rev.: Data Min. Knowl. Discovery **10**(5), e1379 (2020)

27. Velickovic, P., Cucurull, G., Casanova, A., Romero, A., Lio, P., Bengio, Y.: Graph attention networks. Stat. **1050**(20), 48510–48550 (2017)
28. Vivar, G., Kazi, A., Burwinkel, H., Zwergal, A., Navab, N., Ahmadi, S.A.: Simultaneous imputation and disease classification in incomplete medical datasets using multigraph geometric matrix completion (MGMC). arXiv preprint arXiv:2005.06935
29. Wang, Y., Sun, Y., Liu, Z., Sarma, S.E., Bronstein, M.M., Solomon, J.M.: Dynamic graph CNN for learning on point clouds. ACM (TOG) **38**(5) (2018)
30. Ying, Z., Bourgeois, D., You, J., Zitnik, M., Leskovec, J.: GNNExplainer: generating explanations for graph neural networks. In: NeurIPs, pp. 9244–9255 (2019)

Multi-modal Adapter for Medical Vision-and-Language Learning

Zheng Yu, Yanyuan Qiao, Yutong Xie, and Qi Wu[✉]

Australian Institute for Machine Learning, The University of Adelaide, Adelaide,
Australia
{zheng.yu,yanyuan.qiao,qi.wu01}@adelaide.edu.au

Abstract. Recently, medical vision-and-language learning has attracted
great attention from biomedical communities. Thanks to the development of large pre-trained models, the performances on these medical
multi-modal learning benchmarks have been greatly improved. However,
due to the rapid growth of the model size, full fine-tuning these large
pre-trained models has become costly in training and storing such huge
parameters for each downstream task. Thus, we propose a parameter-
efficient transfer learning method named **M**edical **M**ulti-**M**odal **Ad**apter
(M^3AD) to mediate this problem. We select the state-of-the-art M^3AE
model as our baseline, which is pre-trained on 30k medical image-text
pairs with multiple proxy tasks and has about 340M parameters. To be
specific, we first insert general adapters after multi-head attention layers and feed-forward layers in all transformer blocks of M^3AE. Then,
we specifically design a modality-fusion adapter that adopts multi-head
attention mechanisms and we insert them in the cross-modal encoder to
enhance the multi-modal interactions. Compared to full fine-tuning, we
freeze most parameters in M^3AE and only train these inserted adapters
with much smaller sizes. Extensive experimental results on three medical
visual question answering datasets and one medical multi-modal classification dataset demonstrate the effectiveness of our proposed method,
where M^3AD achieves competitive performances compared to full fine-
tuning with much fewer training parameters and memory consumption.

Keywords: Medical Vision-and-Language Learning ·
Parameter-Efficient Transfer Learning · Multi-Modal Adapter

1 Introduction

In the past few years, the research of medical vision-and-language learning
(MedVL) [7,15,16,24] has received great attention from the biomedical community. For example, given a medical image and a clinical question, the task
of medical visual question answering (MedVQA) [1,5,23,26,27] requires a multi-
modal model to answer the question correctly. Such MedVL tasks not only have
significance in clinical education, but also in practical applications. However, the

X. Cao et al. (Eds.): MLMI 2023, LNCS 14348, pp. 393–402, 2024.
https://doi.org/10.1007/978-3-031-45673-2_39

lack of well-annotated data by experts with medical-specific knowledge may hinder the further application of medical vision-and-language learning [21]. Recently, many works [8,13,17,19,20] of medical vision-and-language pre-trained models (MedVLP) have been proposed to overcome the shortage of annotated data. By pre-training the large MedVLP models on abundant medical image-report pairs with elaborately designed proxy tasks and fine-tuning the full pre-trained model for each downstream medical V&L task, the performance on these tasks has been significantly improved. Since the model sizes of these MedVLP are growing rapidly, this paradigm of pretraining-and-finetuning has become costly in even fully finetuning and storing a copy of the entire pre-trained model for each downstream MedVL task. Parameter-Efficient Transfer Learning (PETL) [9–11,18,22] has emerged as a recent research hotspot with promising potential in effectively fine-tuning large pre-trained models. Among multiple PETL methods, Adapter [10] is shown with superior in many general vision-and-language tasks.

Thus, in this work, we propose to adapt the PETL technique to MedVL with our creatively proposed Medical Multi-modal Adapter (M^3AD), by inserting the general adapters and the specifically designed modality-fusion adapters into the large medical vision-and-language pretrained model of M^3AE [3]. To be specific, the general adapters are bottleneck layers that recover the down-sampled features into the original dimension and fused these recovered features with the frozen knowledge implied in the pre-trained model. Furthermore, considering the importance of cross-modal feature interactions, we specifically designed a modality-fusion adapter based on the general adapter, which incorporates multi-head self-attention layers to further fuse the multi-modal features from a more comprehensive and global view. As shown in Fig. 1, the general adapters are inserted after each multi-head attention layer and each feed-forward layer in all the transformer blocks of M^3AE. While the modality-fusion adapters are inserted only after each transformer block in the cross-modal encoder, to enhance the interactions between the multi-modal features of images and texts. During the training process for downstream medical vision-and-language tasks, we freeze the native parameters of M^3AE and only keep the newly inserted adapters and the prediction heads trainable.

To evaluate the effectiveness of our proposed M^3AD, we conduct extensive experiments on three medical visual question answering tasks of VQA-RAD [14], SLAKE [16], and VQA-Med-19 [2], and one medical multi-modal classification task of MELINDA. Though only training a minimal set of about 4% parameters, M^3AD achieves comparable performances in all downstream MedVL tasks compared to full fine-tuning M^3AE and reduces about 20%-35% memory consumption. We also conduct the ablation study to analyze the impact of different components of M^3AD in efficiently tuning the large pre-trained model.

2   Proposed Method

As shown in Fig. 1, based on the large MedVLP of M^3AE, we first insert the general adapters after each multi-head attention layer including self-attention

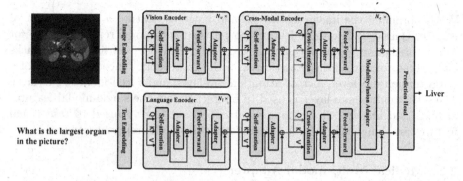

Fig. 1. Overview of the proposed M³AD. To efficiently tune M³AE for MedVQA, most parameters of M³AE are frozen and only the newly inserted adapters are kept trainable during training. Blue blocks represent the frozen parameters while Green blocks represent the trainable parameters during training for different downstream MedVL tasks. (Color figure online)

layer and cross-attention layer and each feed-forward layer in M³AE. Then, we specifically design a modality-fusion adapter and insert it in the cross-modal encoder of M³AE to enhance the multi-modal feature interactions. Compared to full fine-tuning that trains all parameters in the pre-trained model, we freeze most parameters of M³AE and only keep these inserted adapters trainable with the prediction head for different downstream MedVL tasks.

2.1   Baseline Model

We use the state-of-the-art large medical vision-and-language pre-trained model of M³AE [3] as our baseline, which is pre-trained on about 30k medical image-text pairs and has 340M parameters. As shown in Fig. 1, M³AE mainly consists of three parts: a vision encoder based on Vision Transformer [6] (ViT) to extract vision features, a language encoder based on BERT [4] to extract language features, and a cross-modal encoder to implement multi-modal interactions between the vision and language features. Different prediction heads are incorporated after the cross-modal encoder to predict answers for different downstream medical vision-and-language tasks.

Specifically, the input medical image is first reshaped into a sequence of flattened patches, which are projected by a linear projection layer and summed with the position embedding to generate visual embedding e_v. Then, e_v is passed through N_v transformer blocks consisting of a self-attention layer and a feed-forward layer to generate visual features h_v for multi-modal interaction. Meanwhile, the clinical text corresponding to the medical image is first tokenized by WordPiece [12] into a sequence of tokens, which are transformed into word embeddings by the linear projection layer and summed with the position embeddings to generate the language embedding e_w. Then, the language embedding e_w is passed through N_l transformer blocks that consist of a self-attention layer and a feed-forward layer to generate language features h_w for multi-modal interaction.

When obtaining the image feature h_v and language feature h_w, a cross-modal encoder is used to fuse these features to get the multi-modal representations for answer prediction. To be specific, the cross-modal encoder consists of two stacks of N_c Transformer blocks. Each block in the cross-modal encoder consists of a self-attention sub-layer, a cross-attention sub-layer, and a feed-forward sub-layer. The self-attention sub-layer is used to interact within the single-modal features of texts and images, while the cross-attention sub-layer is used to fuse visual features and language features to generate multi-modal representations.

2.2   Medical Multi-modal Adapter

As shown in Fig. 2, our proposed Medical Multi-Modal Adapter (M^3AD) mainly consists of two parts, the general adapter and the specifically designed modality-fusion adapter based on the general adapter. The details are illustrated as follows.

General Adapter. As shown in Fig. 2, the general adapter is a bottleneck layer that consists of a trainable linear down-projection with $W_{\text{down}} \in \mathbb{R}^{D_{\text{hidden}} \times D_{\text{mid}}}$, a non-linear activation function $\sigma(\cdot)$ and a learnable linear up-projection $W_{\text{up}} \in \mathbb{R}^{D_{\text{mid}} \times D_{\text{hidden}}}$. Given the input feature f_{in}, the general adapter first down-samples the input feature f_{in} into the bottleneck dimension of D_{mid}. Then, the general adapter up-samples the intermediate feature back into the D_{hidden} dimension as follows:

$$f_{\text{out}} = W_{\text{up}}^{\text{T}} \sigma(W_{\text{down}}^{\text{T}} f_{\text{in}}). \tag{1}$$

As shown in Fig. 1, we insert general adapters with trainable bottleneck layers after all multi-head attention layers and feed-forward layers of the transformer blocks in M^3AE. Specifically, given the output feature $\hat{f}_{\text{att}}$ of a multi-head attention layer, $\hat{f}_{\text{att}}$ is passed through the adapter and $f_{\text{out\_att}}$ is get. Then, $f_{\text{out\_att}}$ is summed with $\hat{f}_{\text{att}}$ in skip connection followed by the layer normalization to generate the final output feature f_{att}.

$$f_{\text{out\_att}} = W_{\text{up\_att}}^{\text{T}} \sigma(W_{\text{down\_att}}^{\text{T}} \hat{f}_{\text{att}}), \tag{2}$$

$$f_{\text{att}} = \text{LN}(f_{\text{out\_att}} + \hat{f}_{\text{att}}). \tag{3}$$

where $\text{LN}(\cdot)$ represents the layer normalization. For the feed-forward layers, the process is as same as the multi-head attention layer:

$$f_{\text{out\_ffn}} = W_{\text{up\_ffn}}^{\text{T}} \sigma(W_{\text{down\_ffn}}^{\text{T}} \hat{f}_{\text{ffn}}), \tag{4}$$

$$f_{\text{ffn}} = \text{LN}(f_{\text{out\_ffn}} + \hat{f}_{\text{ffn}}). \tag{5}$$

Modality-Fusion Adapter. Different from single-stream vision-and-language large pre-trained models that use unified transformer blocks consisting of multi-head self-attention layers and feed-forward layers to fuse multi-modal features,

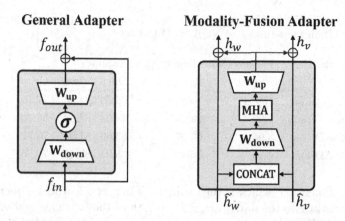

Fig. 2. Details of General Adapter and Modality-Fusion Adapter

M^3AE utilizes two-stream transformer blocks that respectively consist of multi-head self-attention layers, multi-head cross-attention layers, and feed-forward layers to implement multi-modal interactions. These cross-attention layers absorb cross-modal features from the counterpart modality by changing keys and values in the multi-head attention layers. This kind of multi-modal fusion may concentrate more on the internal modality while neglecting the global multi-modal fusion compared to the single-stream models, which have much more comprehensive multi-modal fusions from a global view. When most parameters in M^3AE are frozen, this self-focused two-stream architecture may bring performance degeneration even though we insert general adapters into all transformer blocks. Therefore, to bring a global multi-modal feature fusion to a frozen large pre-trained model, we specifically design a Modality-Fusion Adapter for M^3AE.

As shown in Fig. 2, the modality-fusion adapter mainly consists of a trainable linear down-projection with $W_{\mathrm{down}} \in \mathbb{R}^{D_{\mathrm{hidden}} \times D_{\mathrm{mid}}}$, a multi-head self-attention layer and a learnable linear up-projection $W_{\mathrm{up}} \in \mathbb{R}^{D_{\mathrm{mid}} \times D_{\mathrm{hidden}}}$. Different from inserting general adapters after all transformer layers in M^3AE, we only insert the modality-fusion adapter after each fusion block in the cross-modal encoder. To be specific, given the output language feature $\hat{h}_{\mathrm{w}}$ and the output visual feature $\hat{h}_{\mathrm{v}}$ from the two independent transformer blocks in the cross-modal encoder, we first concatenate $\hat{h}_{\mathrm{w}}$ with $\hat{h}_{\mathrm{v}}$ and get the global vision-and-language feature h_{wv}. Then, h_{wv} is passed through the linear down-projection W_{down} to generate $h_{\mathrm{wv}}^{\mathrm{down}}$. A comprehensive multi-modal fusion is implemented between all feature tokens in $h_{\mathrm{down}}^{\mathrm{wv}}$ by the multi-head self-attention layer and h_{fus} is obtained:

$$h_{\mathrm{wv}} = \mathrm{CONCAT}(\hat{h}_{\mathrm{w}}, \hat{h}_{\mathrm{v}}), \tag{6}$$

$$h_{\mathrm{wv}}^{\mathrm{down}} = W_{\mathrm{down}}^{\mathrm{T}} h_{\mathrm{wv}}, \tag{7}$$

$$h_{\mathrm{fus}} = \mathrm{MSA}(h_{\mathrm{wv}}^{\mathrm{down}}). \tag{8}$$

where $\mathrm{CONCAT}(\cdot)$ represents the process of concatenation and $\mathrm{MSA}(\cdot)$ represents the multi-head self-attention layer. Next, h_{fus} is up-sampled into $h_{\mathrm{fus}}^{\mathrm{up}}$ with

Table 1. Comparison of trainable parameters, memory consumption and performance between M^3AD and fine-tuning on VQA-RAD test set.

Methods	Params (%)	Memory (%)	Accuracy		
			Open	Closed	Overall
Fine-Tuning	100	100	65.36	84.93	77.16
Head-Tuning	0.90	22.14	48.04	62.50	56.76
M^3AD-Gen	3.96	66.97	63.69	81.80	74.72
M^3AD-Fus	4.16	76.43	65.92	83.46	76.50

the D_{hidden} dimension, which is then split back into two features sequences h_w^{fus} and h_w^{fus} according to the native sequence length of the language and vision feature tokens. At last, the fused language feature h_w^{fus} and the fused vision feature h_v^{fus}, which are summed with $\hat{h}_w$ with $\hat{h}_v$ followed by the layer normalization. This feed-forward pass can be formulated as follow:

$$h_{fus}^{up} = W_{up}^T h_{fus}, \tag{9}$$

$$h_w^{fus}, h_v^{fus} = \text{SPLIT}(h_{fus}^{up}), \tag{10}$$

$$h_w = \text{LN}(h_w^{fus} + \hat{h}_w), \tag{11}$$

$$h_v = \text{LN}(h_v^{fus} + \hat{h}_v). \tag{12}$$

where $\text{SPLIT}(\cdot)$ represents the process of split.

3   Experiments

3.1   Datasets

We conduct experiments on three medical visual question answering datasets including the VQA-RAD [14], SLAKE [16] and VQA-Med-19 [2] and a medical multi-modal classification dataset of MELINDA [25]. VQA-RAD has 315 medical images and 3,515 clinician-annotated question-answer pairs, which are classified into the open-ended category and closed-ended category. SLAKE-EN has 642 medical images and 7,033 open-ended and closed-ended clinical questions with the corresponding answers. VQA-Med-19 has 4,200 medical images and 15,292 question-answer pairs, which are classified into 4 categories: Modality, Plane, Organ and Abnormality. MELINDA is aimed at classifying biomedical experiment methods given the corresponding image-caption pairs, which have 5,371 paired samples.

3.2   Implementation Details

Most of the default settings of M^3AE are adopted for full fine-tuning and efficient tuning. For all experiments, the batch size is fixed to 64. Especially for efficient tuning, we adjust the default learning rates into a fixed number of 1e-4 with the

Table 2. Comparison between M^3AD and fine-tuning on SLAKE test set.

Methods	Params (%)	Memory (%)	Accuracy		
			Open	Closed	Overall
Fine-Tuning	100	100	80.93	86.54	83.13
Head-Tuning	0.78	22.36	69.15	70.43	69.65
M^3AD-Gen	3.85	67.24	77.67	83.38	80.30
M^3AD-Fus	4.04	76.89	80.00	87.26	82.85

Table 3. Comparison between M^3AD and fine-tuning on VQA-Med-19 test set.

Methods	Params (%)	Memory (%)	Accuracy				
			Plane	Organ	Modality	Abnormal	Overall
Fine-Tuning	100	100	81.60	72.80	80.80	11.20	61.60
Head-Tuning	1.44	22.92	72.00	71.20	71.20	8.00	55.60
M^3AD-Gen	4.48	67.78	82.40	75.20	80.00	8.80	61.60
M^3AD-Fus	4.68	77.46	81.60	76.00	84.80	8.80	62.80

default warmup steps. The bottleneck dimension of all insert adapters is set as 64, the hidden size of the Multi-head Self Attention layers in the modality-fusion adapters is set as 64 and the number of heads is set as 8. All experiments are conducted on a single NVIDIA 3090 GPU with 24GB VRAM.

3.3 Comparisons with Fine-Tuning

As shown in Table 1, 2, 3 and 4, we compare our proposed M^3AD to fine-tuning in different aspects from trainable parameters to memory consumption and performance on multiple four MedVL tasks. We report two settings of M^3AD: M^3AD-Gen which only inserts general adapters into M^3AE and M^3AD-Fus which inserts both the general adapters and the modality-fusion adapters. We also report the result of Head-Tuning which only trains the prediction heads of M^3AE without inserted adapters.

Compared to full fine-tuning, M^3AD efficiently trains the large pre-trained model by tuning only about 4% parameters of M^3AE and reducing about 20%-35% memory consumption. We can see that the efficient tuning only with general adapters has worse performance than full fine-tuning with non-trivial margins. When incorporating the specifically designed modality-fusion adapters, M^3AD successfully mediates the performance gap with full fine-tuning and even surpasses fine-tuning on the VQA-Med-19 test set. These promising results demonstrate the effectiveness of our proposed method with the creatively designed modality-fusion adapter for MedVL tasks.

Table 4. Comparison between M³AD and fine-tuning on MELINDA test set.

Method	Params(%)	Memory(%)	Accuracy
Fine-Tuning	100	100	76.54
Head-Tuning	0.68	21.42	57.44
M³AD-Gen	3.76	69.74	73.87
M³AD-Fus	3.95	80.83	76.30

Table 5. Ablation on different components of M³AD on VQA-RAD test set.

LAD	VAD	CAD	Accuracy		
			Open	Closed	Overall
✗	✗	✗	48.04	62.50	56.76
✓	✗	✗	61.45	79.78	72.51
✗	✓	✗	58.66	80.88	72.06
✗	✗	✓	60.34	81.25	72.95
✓	✓	✓	63.69	81.80	74.72

3.4 Ablation on Different Components

Following the practice of M³AE, we select the VQA-RAD test set to conduct ablation study on the contribution of different components during efficient tuning. As shown in Table 5, We take into account the results of only insert general adapters in the language encoder (LAD), only in the vision encoder (VAD) and only in the cross-modal encoder (CAD). We can see that each independent component is significant in the performance improvement while CAD and LAD contributes a lot more compared to VAD. When combing all these components, the final performance is further increased.

4   Conclusion

In this paper, we present the first study to adapt the parameter-efficient transfer learning technique to the large medical vision-and-language pre-trained model with our proposed medical multi-modal adapter (M³AD), which mainly consists of the general adapter and the specifically designed modality-fusion adapter. We conduct extensive experiments on multiple medical vision-and language downstream tasks, where M³AD achieves promising performance and reduces considerable memory consumption while only tuning a minimal set of parameters compared to full fine-tuning, demonstrating the effectiveness of our proposed method. However, though most parameters are frozen, due to the full feed-forward and backward propagation in training, the reduce of memory is not that huge as 50% or more, which is a common slight defect of all PETL methods. Thus, our future work will concentrate on combining M^3AD with other techniques that could further reduce the memory consumption while keep the competitive performance as well.

References

1. Abacha, A.B., Gayen, S., Lau, J., Rajaraman, S., Demner-Fushman, D.: Nlm at imageclef 2018 visual question answering in the medical domain. In: CLEF (Working Notes) (2018)
2. Abacha, A.B., Hasan, S.A., Datla, V., Liu, J., Demner-Fushman, D., Müller, H.: Vqa-med: overview of the medical visual question answering task at imageclef 2019. In: CLEF (2019)
3. Chen, Z., et al.: Multi-modal masked autoencoders for medical vision-and-language pre-training. In: MICCAI, vol. 13435, pp. 679–689 (2022). https://doi.org/10.1007/978-3-031-16443-9_65
4. Devlin, J., Chang, M.W., Lee, K., Toutanova, K.: Bert: pre-training of deep bidirectional transformers for language understanding. ArXiv abs/ arXiv: 1810.04805 (2019)
5. Do, T., Nguyen, B.X., Tjiputra, E., Tran, M., Tran, Q.D., Nguyen, A.: Multiple meta-model quantifying for medical visual question answering. In: de Bruijne, M., Cattin, P.C., Cotin, S., Padoy, N., Speidel, S., Zheng, Y., Essert, C. (eds.) MICCAI 2021. LNCS, vol. 12905, pp. 64–74. Springer, Cham (2021). https://doi.org/10.1007/978-3-030-87240-3_7
6. Dosovitskiy, A., et al.: An image is worth 16x16 words: transformers for image recognition at scale. ArXiv abs/ arXiv: 2010.11929 (2020)
7. Eslami, S., de Melo, G., Meinel, C.: Does CLIP benefit visual question answering in the medical domain as much as it does in the general domain? CoRR abs/ arXiv: 2112.13906 (2021)
8. Gong, H., Chen, G., Liu, S., Yu, Y., Li, G.: Cross-modal self-attention with multi-task pre-training for medical visual question answering. In: ICMR, pp. 456–460. ACM (2021)
9. He, R., et al.: On the effectiveness of adapter-based tuning for pretrained language model adaptation. In: Zong, C., Xia, F., Li, W., Navigli, R. (eds.) ACL/IJCNLP, pp. 2208–2222 (2021)
10. Houlsby, N., et al.: Parameter-efficient transfer learning for nlp. In: ICML, pp. 2790–2799 (2019)
11. Hu, E., et al.: Lora: low-rank adaptation of large language models. In: ICLR (2022)
12. Johnson, M., et al.: Google's multilingual neural machine translation system: Enabling zero-shot translation. Trans. Assoc. Comput. Linguistics **5**, 339–351 (2017)
13. Khare, Y., Bagal, V., Mathew, M., Devi, A., Priyakumar, U.D., Jawahar, C.V.: MMBERT: multimodal BERT pretraining for improved medical VQA. In: ISBI, pp. 1033–1036. IEEE (2021)
14. Lau, J., Gayen, S., Abacha, A.B., Demner-Fushman, D.: A dataset of clinically generated visual questions and answers about radiology images. Sci. Data **5** (2018)
15. Li, Y., Wang, H., Luo, Y.: A comparison of pre-trained vision-and-language models for multimodal representation learning across medical images and reports. In: Park, T., et al (eds.) IEEE International Conference on Bioinformatics and Biomedicine, BIBM 2020, Virtual Event, South Korea, 16–19 December 2020, pp. 1999–2004 (2020)

16. Liu, B., Zhan, L.M., Xu, L., Ma, L., Yang, Y., Wu, X.M.: Slake: a semantically-labeled knowledge-enhanced dataset for medical visual question answering. 2021 IEEE 18th International Symposium on Biomedical Imaging (ISBI), pp. 1650–1654 (2021)

17. Liu, B., Zhan, L.-M., Wu, X.-M.: Contrastive pre-training and representation distillation for medical visual question answering based on radiology images. In: de Bruijne, M., Cattin, P.C., Cotin, S., Padoy, N., Speidel, S., Zheng, Y., Essert, C. (eds.) MICCAI 2021. LNCS, vol. 12902, pp. 210–220. Springer, Cham (2021). https://doi.org/10.1007/978-3-030-87196-3_20

18. Mahabadi, R.K., Henderson, J., Ruder, S.: Compacter: efficient low-rank hypercomplex adapter layers. In: NeurIPS, pp. 1022–1035 (2021)

19. Moon, J.H., Lee, H., Shin, W., Choi, E.: Multi-modal understanding and generation for medical images and text via vision-language pre-training. CoRR (2021)

20. Moon, J.H., Lee, H., Shin, W., Kim, Y., Choi, E.: Multi-modal understanding and generation for medical images and text via vision-language pre-training. IEEE J. Biomed. Health Informatics **26**(12), 6070–6080 (2022)

21. Nguyen, B.D., Do, T.-T., Nguyen, B.X., Do, T., Tjiputra, E., Tran, Q.D.: Overcoming data limitation in medical visual question answering. In: Shen, D., et al. (eds.) MICCAI 2019. LNCS, vol. 11767, pp. 522–530. Springer, Cham (2019). https://doi.org/10.1007/978-3-030-32251-9_57

22. Pfeiffer, J., et al.: Adapterhub: a framework for adapting transformers. In: Proceedings of the 2020 Conference on Empirical Methods in Natural Language Processing: System Demonstrations, EMNLP 2020 - Demos, Online, 16–20 November 2020, pp. 46–54. Association for Computational Linguistics (2020)

23. Ren, F., Zhou, Y.: CGMVQA: a new classification and generative model for medical visual question answering. IEEE Access **8**, 50626–50636 (2020)

24. Subramanian, S., et al.: Medicat: a dataset of medical images, captions, and textual references. In: Cohn, T., He, Y., Liu, Y. (eds.) Findings of the Association for Computational Linguistics: EMNLP 2020, Online Event, 16–20 November 2020, pp. 2112–2120. Findings of ACL, Association for Computational Linguistics (2020)

25. Wu, T., Singh, S., Paul, S., Burns, G.A., Peng, N.: MELINDA: a multimodal dataset for biomedical experiment method classification. In: AAAI, pp. 14076–14084 (2021)

26. Yan, X., Li, L., Xie, C., Xiao, J., Gu, L.: Zhejiang university at imageclef 2019 visual question answering in the medical domain. In: CLEF (2019)

27. Zhan, L., Liu, B., Fan, L., Chen, J., Wu, X.: Medical visual question answering via conditional reasoning. In: ACM MM, pp. 2345–2354. ACM (2020)

Vector Quantized Multi-modal Guidance for Alzheimer's Disease Diagnosis Based on Feature Imputation

Yuanwang Zhang[1], Kaicong Sun[1], Yuxiao Liu[1], Zaixin Ou[1],
and Dinggang Shen[1,2,3]($\boxtimes$)

[1] School of Biomedical Engineering, ShanghaiTech University, Shanghai, China
dgshen@shanghaitech.edu.cn
[2] Shanghai United Imaging Intelligence Co., Ltd., Shanghai, China
[3] Shanghai Clinical Research and Trial Center, Shanghai, China

Abstract. Magnetic Resonance Imaging (MRI) and positron emission tomography (PET) are the most used imaging modalities for Alzheimer's disease (AD) diagnosis in clinics. Although PET can better capture AD-specific pathologies than MRI, it is less used compared with MRI due to high cost and radiation exposure. Imputing PET images from MRI is one way to bypass the issue of unavailable PET, but is challenging due to severe ill-posedness. Instead, we propose to directly impute classification-oriented PET features and combine them with real MRI to improve the overall performance of AD diagnosis. In order to more effectively impute PET features, we discretize the feature space by vector quantization and employ transformer to perform feature transition between MRI and PET. Our model is composed of three stages including codebook generation, mapping construction, and classifier enhancement based on combined features. We employ paired MRI-PET data during training to enhance the performance of MRI data during inference. Experimental results on ADNI dataset including 1346 subjects show a boost in classification performance of MRI without requiring PET. Our proposed method also outperforms other state-of-the-art data imputation methods.

Keywords: Discrete Learning · Multimodal Learning · Incomplete Modalities · Alzheimer's Disease · Classification · Feature Imputation

1 Introduction

Alzheimer's disease (AD) is one of the most common forms of dementia among older adults, which gradually causes irreversible brain damage and affects brain functions [1]. AD patients are known to have certain changes in the brain such as brain atrophy and abnormal glucose metabolism, which can be detected by structural Magnetic Resonance Imaging (sMRI) and Positron Emission Tomography (PET) imaging [2,3]. Studies have shown that PET has a higher diagnostic and prognostic performance on AD than MRI, as PET can better capture AD-specific pathologies [4–7]. Although PET is a superior modality for AD diagnosis,

© The Author(s), under exclusive license to Springer Nature Switzerland AG 2024
X. Cao et al. (Eds.): MLMI 2023, LNCS 14348, pp. 403–412, 2024.
https://doi.org/10.1007/978-3-031-45673-2_40

it is not always available in clinical settings due to high costs, radiation exposure risks, and limited access to imaging facilities. In fact, MRI is more widely used. Therefore, it will be advantageous if we could make use of pseudo-PET information for AD diagnosis in case that PET is not available.

To achieve this, a straight forward idea is to impute PET data from MRI data. In the literature, two primary data imputation methods have been proposed for AD diagnosis. The first category involves imputing hand-crafted region-of-interest (ROI) features extracted from volumetric images [8,9]. However, these hand-crafted features may not be sufficiently discriminative for AD diagnosis, so imputing them may not improve the final classification performance. The second category involves synthesizing voxel-wise images using generative models like VoxGAN [10,11] and cycleGAN [12–14]. While these synthesized images appear similar to the original MR images, it may bring limited contribution to the final classification task since AD primarily affects specific small regions of the brain, especially in the early stages. Additionally, GAN-based generative models require a large number of parameters and the training may be challenging since they have to build mappings for millions of voxels.

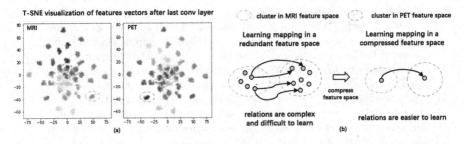

Fig. 1. (a) Visualization of encoded feature vectors of MRI and PET in continuous space, where feature vectors are clustered into separate groups. (b) Advantages of learning mapping in a sparse feature space, where overfitting issue could be largely alleviated, especially when training data is not easily available.

Rather than first imputing PET images and then extracting the features again for downstream classification tasks, directly imputing classification-oriented features is supposed to be more effective. A recent study [15] proposed a multimodal guidance framework that imputes the latent representations of the superior modality from the counterpart of the inferior modality. However, mapping from one modality to another can be difficult due to the redundancy of the continuous feature space, particularly when the training data is limited. To illustrate this redundancy, we follow [16] and visualize feature vectors after the last convolution layer of a classification network in Fig. 1 (a). These feature vectors are clustered into a small number of groups, implying that the entire feature space can be roughly represented using a limited number of latent vectors. Mapping between modalities can be easier in such a compressed sparse feature space because the model does not have to create mappings between many redundant feature vectors.

In this work, we propose a three-stage framework to effectively impute the latent representations of PET from MRI in a compressed feature space discretized by vector quantization [17,18], as shown in Fig. 2. The feature spaces of different modalities are compressed and regularized using corresponding classification-oriented codebooks with a finite number of feature vectors. The relation between the two feature spaces is then learned by a transformer to impute PET features. Finally, the imputed PET features are combined with original MRI features to enhance the final classification task. During training, we use existing paired MRI and PET data, while only MRI data is used during inference. Our imputing strategy has shown significant improvement in the classification performance of MRI without requiring PET.

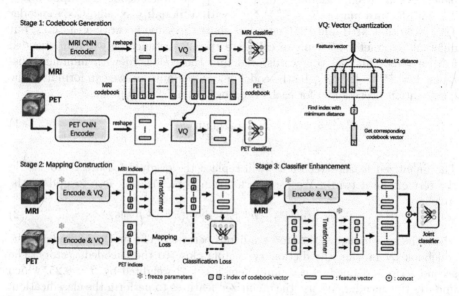

Fig. 2. Schematic illustration of the three stages of our proposed method. Stage 1: Generate AD-oriented codebooks of MRI and PET; Stage 2: Construct mapping between MRI and PET based on indices of codebooks using transformer; Stage 3: Enhance the AD classifier using combined features of MRI and PET.

2   Methods

As a matter of fact that the use of PET obtains better performance for AD diagnosis in clinics, we want to exploit pseudo-PET information in case that PET is unavailable in practice. Unlike the most existing methods which try to synthesize voxel-wise PET images, we propose to impute the PET features from MRI data in the discrete feature space compressed by vector quantization and employ index mapping from MRI to PET based on transformer. The overall model consists of three stages, as demonstrated in Fig. 2. We will provide detailed explanations in the following section.

Stage 1: Codebook Generation. We first build up individual codebook for MRI and PET based on vector quantization using a classifier. The main idea is to generate a codebook with finite number of discrete feature vectors which embodies complete attributes of subjects and serves as a dictionary. The benefit of discrete space instead of a continuous one is mainly that it compresses the information and enforces a regularization upon the latent space, while still containing most discriminative features [16,19]. Each vector in the codebook has a dimension of c, which is referred to as 'codeword' in the remaining of the paper. For a specific modality $m \in \{M, P\}$, we first encode the input image X^m into feature maps $h^m \in \mathbb{R}^{h \times w \times d \times c}$ with a modality-specific CNN encoder f_{enc}^m. Note that MRI and PET encoders share the same network structures, but different parameters. Then, we calculate the L2 distance between an encoded feature vector and all codewords, and find the codeword with minimum distance. Let $\mathcal{Z}^m = \{z_k^m\}_{k=1}^K$ be the codebook of modality m, we can formulate the quantization process $q(\cdot)$ for each encoding vector h_i^m as follows:

$$\hat{h}_i^m = q(h_i^m; \mathcal{Z}^m) = \operatorname*{argmin}_{z_k^m \in \mathcal{Z}^m} \|h_i^m - z_k^m\|_2, \tag{1}$$

The quantized feature vector $\hat{h}^m$ will replace the original feature vector h^m in the rest of the network. We use the following quantization loss to update the codebook:

$$\mathcal{L}_Q = \|sg(h^m) - \hat{h}^m\|_2 + \beta\|sg(\hat{h}^m) - h^m\|_2, \tag{2}$$

where the $sg()$ indicates the stop gradient operation. The first term updates the codebook by moving the dictionary vectors close to the encoded vectors. The second term is defined as a commitment loss [17] weighted by $\beta = 0.25$, which updates the encoder. We use the quantized features to perform the classification: $\hat{y} = f_{cls}^m(\hat{h}^m)$, where f_{cls}^m denotes the classifier of modality m. Then the overall loss for Stage 1 is the combination of classification loss (cross entropy between the prediction $\hat{y}$ and label y and quantization loss:

$$\mathcal{L}_{S1} = -\mathbb{E}_y \log \hat{y} + \lambda_1 \mathcal{L}_Q, \tag{3}$$

where λ_1 is set to 1 empirically. Note that the quantization process is non-differentiable, so we directly take the gradient of $\hat{h}^m$ for h^m in the backpropagation process.

Stage 2: Mapping Construction. Learning the mapping from MRI to PET in the feature space could be easier, especially when the feature space has a finite number of feature vectors. After replacing feature maps by a series of codewords from the codebook, we can use a sequence of indices of the codewords in the codebook to represent the corresponding feature vectors. The quantized MRI features is $\hat{h}^M = q(f_{enc}^M(X^M))$, which is in fact a sequence of codeword indices $\mathbf{s}^M$. Then the mapping from MRI features to PET features becomes the mapping from a sequence of indices to another one. We employ transformer to carry out the mapping $\mathcal{G}$, as transformer is good at learning long-range dependencies of sequence data. The generated PET indices are formulated as $\hat{\mathbf{s}}^P = \mathcal{G}(\mathbf{s}^M)$. Since the outputs of transformer are probabilities over all codewords, we consider it as a multi-class classification problem and employ cross entropy between real PET indices $\mathbf{s}^P$ and predicted PET indices $\hat{\mathbf{s}}^P$ as the mapping loss for Stage 2:

$$\mathcal{L}_{map} = -\mathbb{E}_{\mathbf{s}^P} \log \hat{\mathbf{s}}^P, \tag{4}$$

The generated PET indices will be retrieved and replaced by codewords of the PET codebook $\hat{h}^P = \mathcal{Z}_{\hat{\mathbf{s}}^P}$. In order to make the mapping effective for AD diagnosis, we further pass the generated PET features $\hat{h}^P$ into the PET classifier trained in Stage 1 and get a classification result $\hat{y} = f_{cls}^P(\hat{h}^P)$. We expect the generated PET features could mimic real PET features and also get good classification results. The overall loss for this stage is the combination of a classification loss and the mapping loss:

$$\mathcal{L}_{S2} = -\mathbb{E}_y \log \hat{y} + \lambda_2 \mathcal{L}_{map}, \tag{5}$$

where λ_2 is set to 1 empirically. Note that the lookup procedure is also non-differentiable. So, we use Gumbel Softmax and re-parametrization trick to make such discrete sampling operation differentiable, so that the whole model can be trained in an end-to-end manner. During this stage, only the transformer module is updated, and other modules are frozen.

Stage 3: Classifier Enhancement. Finally, we use real MRI features and synthesized PET features to train a joint classifier f_{cls}. Features from both modalities are concatenated together and input to a series of fully connected layer for final classification $\hat{y} = f_{cls}(\hat{h}^M \oplus \hat{h}^P)$. The overall loss for this stage is the classification loss:

$$\mathcal{L}_{S3} = -\mathbb{E}_y \log \hat{y}, \tag{6}$$

It is worthy to note that encoders, codebooks, and the transformer module (trained in Stage 1 and Stage 2) are frozen during Stage 3.

3 Experiments

3.1 Experiment Settings

Datasets. We utilized the Alzheimer's Disease Neuroimaging Initiative (ADNI) dataset [20] for our study. Baseline images are collected from ADNI-1/GO, ADNI-2 and ADNI-3. Subjects were categorized into four groups based on diagnostic information: (1) AD; (2) CN (Cognitive normal); (3) pMCI (progressive mild cognitive impairment), i.e., subjects diagnosed as mild cognitive impairment (MCI) at the baseline and progressing to AD within 36 months; (4) sMCI (stable mild cognitive impairment), i.e., subjects diagnosed as MCI at the baseline and not converting to AD within the follow-up time. The dataset includes a total of 1346 subjects with 304 AD, 366 CN, 186 pMCI and 490 sMCI.

Date Preprocessing. We follow the standard pipeline to preprocess the raw data from ADNI [7,14]. First, bias correction and skull stripping were performed on the MRI scans using FreeSurfer [21]. Second, we perform linear registration to align the PET scan to its corresponding MRI scan. Sequentially, MRI scans are registered to the MNI template using affine transform, and the corresponding PET scans are transformed using the same affine parameters. Finally, PET images are smoothed by a Gaussian kernel with a full width at half maximum of 6mm. The above-mentioned registration and smoothing operations are conducted using SPM12 [22]. After these procedures, we further remove some backgrounds, and MRI and PET images have the same spatial resolution of $116 \times 138 \times 116$ with a voxel size of $1.5 \times 1.5 \times 1.5mm^3$.

Evaluation Methods. We use 5-fold cross-validation to evaluate the models in the experiments, with a train-validation-test ratio of 3:1:1. Accuracy (ACC), F1-score (F1) and area under the curve (AUC) are used as evaluation metrics. The mean and standard deviation across all folds are reported on both AD diagnosis (AD vs CN) task and MCI conversion prediction (pMCI vs sMCI) task.

Implementation Details. For CNN encoders, we adopt a VGG-like network with 4 CNN blocks [23]. Each CNN block consists of two successive CNN layers with ReLU activation and batch normalization. A max-pooling layer is attached to each CNN block to downsample the feature maps. For the transformer model, we adopt the minGPT structure [24,25], which consists of 6 transformer blocks. The codebook size is set as 128, with a codeword dimension of 64. We set training batch size as 16 and use the Adam optimizer with a learning rate of 5×10^{-4} for Stage 1, and 10^{-4} for Stage 2 and 3. The models are trained for 60 epochs in Stage 2, and 40 epochs in Stage 2 and 3 on an NVIDIA Tesla A100 GPU. The code is implemented using PyTorch backend.

3.2 Results

In the experiments, we conducted ablation study as shown in Table 1, and quantitative comparison with state-of-the-art data imputation methods for both AD diagnosis and MCI conversion prediction task as shown in Table 2. We use the commonly used metrics ACC, F1 score, and AUC for quantitative assessment.

Effectiveness of PET Feature Imputation. After employing the proposed method to impute PET features and using only the imputed PET features as

Table 1. Ablation study including effectiveness of the use of imputed PET features and the use of vector quantization. M indicates real MRI, and P indicates real PET. $\mathcal{G}$(M) indicates imputed PET features using model $\mathcal{G}$. M + $\mathcal{G}$(M) indicates combining real MRI and imputed PET with feature concatenation.

	Modality	Methods	AD vs CN			pMCI vs sMCI		
			ACC ↑	F1 ↑	AUC ↑	ACC ↑	F1 ↑	AUC ↑
1	M	wo/VQ	$82.8_{\pm1.4}$	$80.4_{\pm1.9}$	$90.2_{\pm2.6}$	$69.9_{\pm4.9}$	$51.9_{\pm2.6}$	$78.0_{\pm4.3}$
2		w/VQ	$82.8_{\pm3.6}$	$80.9_{\pm3.7}$	$89.0_{\pm3.5}$	$71.0_{\pm4.5}$	$47.9_{\pm9.6}$	$78.8_{\pm4.7}$
3	P	wo/VQ	$89.7_{\pm1.9}$	$88.8_{\pm2.0}$	$95.9_{\pm1.5}$	$75.8_{\pm1.7}$	$57.1_{\pm4.5}$	$84.5_{\pm2.1}$
4		w/VQ	$89.4_{\pm1.2}$	$88.2_{\pm0.9}$	$95.8_{\pm1.1}$	$76.9_{\pm3.2}$	$59.3_{\pm3.7}$	$85.7_{\pm4.6}$
5	$\mathcal{G}$(M)	wo/VQ	$81.3_{\pm2.9}$	$77.4_{\pm3.0}$	$88.4_{\pm2.5}$	$64.4_{\pm4.4}$	$50.0_{\pm4.3}$	$72.1_{\pm3.3}$
6		w/VQ	$84.9_{\pm0.8}$	$81.7_{\pm1.9}$	$91.5_{\pm1.7}$	$71.7_{\pm4.0}$	$49.8_{\pm4.6}$	$76.9_{\pm4.5}$
7	M + $\mathcal{G}$(M)	wo/VQ	$84.0_{\pm1.6}$	$82.0_{\pm1.7}$	$90.8_{\pm2.7}$	$72.5_{\pm5.1}$	$51.9_{\pm3.8}$	$79.8_{\pm4.0}$
8		w/VQ	$87.7_{\pm1.8}$	$85.8_{\pm1.9}$	$92.8_{\pm2.1}$	$73.9_{\pm2.8}$	$54.5_{\pm1.9}$	$81.7_{\pm3.1}$

input (row 1 vs row 6), there is a slight improvement in performance for the AD vs CN task, while there is a slight degradation for the pMCI vs sMCI task. The learned features might not be sufficiently discriminative for the MCI conversion prediction task, which could be a reason for the reduced performance. However, combining the original MRI information and the imputed PET information resulted in a significant improvement in performance for both tasks (row 1 vs row 10), demonstrating the effectiveness of our proposed method.

Effectiveness of Feature Quantization. To demonstrate the effectiveness of vector quantization (VQ), we first conduct experiments to show that mapping the encoded features to a discrete feature space does not lead to a degradation in classification performance. We compare the performance of CNN with or without VQ, which is shown in Table 2 (row 1-4). We perform paired t-test on 5-fold results and find the p-value $p > 0.05$ for most metrics, indicating that the performance difference is not significant. These results suggest that we can compress the feature space while retaining most discriminative features using VQ. Then we show that building mapping from MRI to PET in such lower dimensional feature spaces is more effective. We compare the performance in Stage 2 (using $\mathcal{G}$(M)) and Stage 3 (using M + $\mathcal{G}$(M)) with or without VQ, which is shown in Table 2 (row 5-8). The results demonstrate that imputing PET in a discrete feature space can significantly improve classification performance.

Comparison with State-of-the-Art Methods. We also compare our proposed method with three other representative data imputation methods: 1) Vox-GAN [26], a conditional GAN that generates voxel-wise mapping from MRI to PET images; 2) FGAN [13], a disease-oriented GAN that synthesizes PET images by encouraging classification features of synthetic and real PET images to be consistent; and 3) Deep Guidance [15], a multimodal guidance framework that imputes latent representations of the superior modality (e.g., PET), using information from the inferior modality (e.g., MRI). The two GAN-based methods generate voxel-wise PET images, while Deep Guidance imputes PET features in a continuous feature space. We present the results of the comparison in Table 2 (rows 3-10). We observe that VoxGAN performs the worst compared to other methods, as it may contain dominant non-disease-specific information. FGAN partially addresses this problem by incorporating a feature consistency loss, but its performance improvement is still limited since it is challenging to learn the mapping of millions of voxels in intensity and extract disease-specific information. Although Deep Guidance performs the mapping in the feature space, the improvement of Deep Guidance is also limited due to the large redundancy of imputed features in a continuous feature space. Our proposed method achieves the most significant improvement for both tasks, benefiting from learning the mapping in a discrete feature space. Notably, the other three methods generate poor results when only the generated PET $\mathcal{G}(M)$ is used as input, while our method produces slightly improved results on the AD vs. CN task and competitive results on the pMCI vs. sMCI task.

Table 2. Quantitative comparison with other state-of-the-art data imputation methods in terms of ACC, F1-score, and AUC on ADNI dataset. The mapping model $\mathcal{G}$ is replaced using other methods.

	Modality	Methods	AD vs CN			pMCI vs sMCI		
			ACC ↑	F1 ↑	AUC ↑	ACC ↑	F1 ↑	AUC ↑
1	M	-	$82.8_{\pm1.4}$	$80.4_{\pm1.9}$	$90.2_{\pm2.6}$	$69.9_{\pm4.9}$	$51.9_{\pm2.6}$	$78.0_{\pm4.3}$
2	P	-	$89.7_{\pm1.9}$	$88.8_{\pm2.0}$	$95.9_{\pm1.5}$	$75.8_{\pm4.5}$	$57.1_{\pm4.5}$	$84.5_{\pm2.1}$
3	$\mathcal{G}(M)$	VoxGAN	$75.6_{\pm3.3}$	$73.3_{\pm3.8}$	$85.2_{\pm4.2}$	$63.9_{\pm8.9}$	$50.6_{\pm5.0}$	$67.8_{\pm11.2}$
4		FGAN	$82.8_{\pm2.1}$	$80.2_{\pm4.0}$	$90.0_{\pm2.0}$	$68.9_{\pm5.3}$	$\mathbf{50.7_{\pm5.5}}$	$75.6_{\pm4.3}$
5		DeepGuidance	$80.6_{\pm3.2}$	$75.9_{\pm3.6}$	$89.3_{\pm2.8}$	$70.3_{\pm4.1}$	$47.3_{\pm3.8}$	$76.4_{\pm4.2}$
6		Proposed	$\mathbf{84.9_{\pm0.8}}$	$\mathbf{81.7_{\pm1.9}}$	$\mathbf{91.5_{\pm1.7}}$	$\mathbf{71.7_{\pm4.0}}$	$49.8_{\pm4.6}$	$\mathbf{76.9_{\pm4.5}}$
7	M + $\mathcal{G}(M)$	VoxGAN	$78.9_{\pm3.3}$	$75.3_{\pm3.1}$	$86.0_{\pm3.7}$	$63.6_{\pm5.9}$	$51.3_{\pm6.8}$	$70.6_{\pm4.3}$
8		FGAN	$85.3_{\pm3.4}$	$83.4_{\pm4.2}$	$91.2_{\pm2.2}$	$72.6_{\pm1.8}$	$52.0_{\pm4.6}$	$80.4_{\pm2.5}$
9		DeepGuidance	$84.3_{\pm1.9}$	$82.5_{\pm2.2}$	$90.8_{\pm2.9}$	$71.7_{\pm4.0}$	$\mathbf{54.9_{\pm4.6}}$	$77.8_{\pm3.5}$
10		Proposed	$\mathbf{87.7_{\pm1.8}}$	$\mathbf{85.8_{\pm1.9}}$	$\mathbf{92.8_{\pm2.1}}$	$\mathbf{73.9_{\pm2.8}}$	$54.5_{\pm1.9}$	$\mathbf{81.7_{\pm3.1}}$

4   Conclusions

PET is a highly effective modality for AD identification and MCI conversion prediction, but its use is limited due to high cost and radiative exposure. Deep learning-based imputation of PET from MRI has emerged as a promising solution to this issue. Existing works usually impute voxel-wise PET images from MRI, which is a much more complex generative task and might not be necessarily effective for downstream classification tasks. In this work, we propose a novel approach that directly imputes classification-oriented PET features using a multi-modal guidance model. In particular, we exploit the discrete feature space constructed by vector quantization to more effectively constrain the solution space of feature imputation. In the experiments, we have shown that the proposed method can largely improve the performance when only unimodal data is available during inference. By learning the information pattern of PET more effectively in a discrete feature space, our method outperforms state-of-the-art data imputation methods. In the future, we plan to extend our framework to other PET modalities, such as $A\beta$-PET and Tau-PET, which are also considered gold standards for AD diagnosis.

References

1. Association, A.: 2019 Alzheimer's disease facts and figures. Alzheimer's Dementia **15**(3), 321–387 (2019)
2. Aisen, P.S., et al.: On the path to 2025: understanding the Alzheimer's disease continuum. Alzheimer's Res. Therapy **9**, 1–10 (2017)
3. Marcus, C., Mena, E., Subramaniam, R.M.: Brain PET in the diagnosis of Alzheimer's disease. Clin. Nucl. Med. **39**(10), e413 (2014)
4. Bloudek, L.M., Spackman, D.E., Blankenburg, M., Sullivan, S.D.: Review and meta-analysis of biomarkers and diagnostic imaging in Alzheimer's disease. J. Alzheimers Dis. **26**(4), 627–645 (2011)
5. Frisoni, G.B., et al.: Imaging markers for Alzheimer disease: which vs how. Neurology **81**(5), 487–500 (2013)
6. Narazani, M., Sarasua, I., Pölsterl, S., Lizarraga, A., Yakushev, I., Wachinger, C.: Is a PET all you need? a multi-modal study for Alzheimer's disease using 3D CNNs. In: Wang, L., Dou, Q., Fletcher, P.T., Speidel, S., Li, S. (eds.) MICCAI 2022, pp. 66–76. Springer (2022). https://doi.org/10.1007/978-3-031-16431-6_7
7. Pan, X., et al.: Multi-view separable pyramid network for AD prediction at MCI stage by 18 F-FDG brain PET imaging. IEEE Trans. Med. Imaging **40**(1), 81–92 (2020)
8. Thung, K.H., Wee, C.Y., Yap, P.T., Shen, D., Initiative, A.D.N., et al.: Neurodegenerative disease diagnosis using incomplete multi-modality data via matrix shrinkage and completion. Neuroimage **91**, 386–400 (2014)
9. Ning, Z., Xiao, Q., Feng, Q., Chen, W., Zhang, Y.: Relation-induced multi-modal shared representation learning for Alzheimer's disease diagnosis. IEEE Trans. Med. Imaging **40**(6), 1632–1645 (2021)
10. Kazeminia, S., et al.: GANs for medical image analysis. Artif. Intell. Med. **109**, 101938 (2020)
11. Lin, W., et al.: Bidirectional mapping of brain MRI and PET with 3D reversible GAN for the diagnosis of Alzheimer's disease. Front. Neurosci. **15**, 646013 (2021)
12. Pan, Y., Liu, M., Lian, C., Zhou, T., Xia, Y., Shen, D.: Synthesizing missing PET from MRI with cycle-consistent generative adversarial networks for Alzheimer's Disease diagnosis. In: Frangi, A.F., Schnabel, J.A., Davatzikos, C., Alberola-López, C., Fichtinger, G. (eds.) MICCAI 2018. LNCS, vol. 11072, pp. 455–463. Springer, Cham (2018). https://doi.org/10.1007/978-3-030-00931-1_52
13. Pan, Y., Liu, M., Lian, C., Xia, Y., Shen, D.: Disease-image specific generative adversarial network for brain disease diagnosis with incomplete multi-modal neuroimages. In: Shen, D., et al. (eds.) MICCAI 2019. LNCS, vol. 11766, pp. 137–145. Springer, Cham (2019). https://doi.org/10.1007/978-3-030-32248-9_16

14. Pan, Y., Liu, M., Xia, Y., Shen, D.: Disease-image-specific learning for diagnosis-oriented neuroimage synthesis with incomplete multi-modality data. IEEE TPAMI **44**(10), 6839–6853 (2021)

15. Mallya, M., Hamarneh, G.: Deep multimodal guidance for medical image classification. In: Wang, L., Dou, Q., Fletcher, P.T., Speidel, S., Li, S. (eds.) MICCAI 2022, pp. 298–308. Springer (2022). https://doi.org/10.1007/978-3-031-16449-1_29

16. Shen, Z., et al.: Collaborative quantization embeddings for intra-subject prostate mr image registration. In: Wang, L., Dou, Q., Fletcher, P.T., Speidel, S., Li, S. (eds.) MICCAI 2022, pp. 237–247. Springer (2022). https://doi.org/10.1007/978-3-031-16446-0_23

17. Van Den Oord, A., Vinyals, O., et al.: Neural discrete representation learning. In: Advances in Neural Information Processing Systems 30 (2017)

18. Razavi, A., Van den Oord, A., Vinyals, O.: Generating diverse high-fidelity images with VQ-VAE-2. In: Advances in Neural Information Processing Systems 32 (2019)

19. Santhirasekaram, A., Kori, A., Winkler, M., Rockall, A., Glocker, B.: Vector quantisation for robust segmentation. In: Wang, L., Dou, Q., Fletcher, P.T., Speidel, S., Li, S. (eds.) MICCAI 2022, pp. 663–672. Springer (2022). https://doi.org/10.1007/978-3-031-16440-8_63

20. Jack, C.R., Jr., et al.: The Alzheimer's disease neuroimaging initiative (ADNI): MRI methods. J. Magn. Reson. Imaging **27**(4), 685–691 (2008)

21. Fischl, B.: Freesurfer. Neuroimage **62**(2), 774–781 (2012)

22. Kurth, F., Gaser, C., Luders, E.: A 12-step user guide for analyzing voxel-wise gray matter asymmetries in statistical parametric mapping (SPM). Nat. Protoc. **10**(2), 293–304 (2015)

23. He, S., Feng, Y., Grant, P.E., Ou, Y.: Deep relation learning for regression and its application to brain age estimation. IEEE Trans. Med. Imaging **41**(9), 2304–2317 (2022)

24. Radford, A., et al.: Language models are unsupervised multitask learners. OpenAI Blog **1**(8), 9 (2019)

25. Esser, P., Rombach, R., Ommer, B.: Taming transformers for high-resolution image synthesis. In: Proceedings of the IEEE/CVF Conference on Computer Vision and Pattern Recognition (CVPR), pp. 12873–12883 (2021)

26. Isola, P., Zhu, J.Y., Zhou, T., Efros, A.A.: Image-to-image translation with conditional adversarial networks. In: Proceedings of the IEEE Conference on Computer Vision and Pattern Recognition (CVPR), pp. 1125–1134 (2017)

Finding-Aware Anatomical Tokens for Chest X-Ray Automated Reporting

Francesco Dalla Serra[1,2(✉)], Chaoyang Wang[1], Fani Deligianni[2],
Jeffrey Dalton[2], and Alison Q. O'Neil[1,3]

[1] Canon Medical Research Europe, Edinburgh, UK
[2] University of Glasgow, Glasgow, UK
francesco.dallaserra@mre.medical.canon
[3] University of Edinburgh, Edinburgh, UK

Abstract. The task of radiology reporting comprises describing and interpreting the medical findings in radiographic images, including description of their location and appearance. Automated approaches to radiology reporting require the image to be encoded into a suitable token representation for input to the language model. Previous methods commonly use convolutional neural networks to encode an image into a series of *image-level* feature map representations. However, the generated reports often exhibit realistic style but imperfect accuracy. Inspired by recent works for image captioning in the general domain in which each visual token corresponds to an object detected in an image, we investigate whether using local tokens corresponding to anatomical structures can improve the quality of the generated reports. We introduce a novel adaptation of Faster R-CNN in which *finding detection* is performed for the candidate bounding boxes extracted during anatomical structure localisation. We use the resulting bounding box feature representations as our set of *finding-aware* anatomical tokens. This encourages the extracted anatomical tokens to be informative about the findings they contain (required for the final task of radiology reporting). Evaluating on the MIMIC-CXR dataset [12,16,17] of chest X-Ray images, we show that task-aware anatomical tokens give state-of-the-art performance when integrated into an automated reporting pipeline, yielding generated reports with improved clinical accuracy.

Keywords: CXR · Automated Reporting · Anatomy Localisation ·
Findings Detection · Multimodal Transformer · Triples Representation

1 Introduction

A radiology report is a detailed text description and interpretation of the findings in a medical scan, including description of their anatomical location and

Supplementary Information The online version contains supplementary material available at https://doi.org/10.1007/978-3-031-45673-2_41.

X. Cao et al. (Eds.): MLMI 2023, LNCS 14348, pp. 413–423, 2024.
https://doi.org/10.1007/978-3-031-45673-2_41

appearance. For example, a Chest X-Ray (CXR) report may describe an opacity (a type of finding) in the left upper lung (the relevant anatomical location) which is diagnosed as a lung nodule (interpretation). The combination of a finding and its anatomical location influences both the diagnosis and the clinical treatment decision, since the same finding may have a different list of possible clinical diagnoses depending on the location.

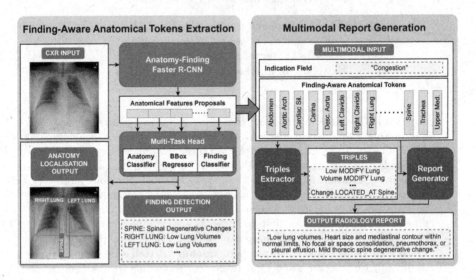

Fig. 1. Finding-aware anatomical tokens integrated into a multimodal CXR automated reporting pipeline. The CXR image and report are taken from the IU-Xray dataset [7].

Recent CXR automated reporting methods have adopted CNN-Transformer architectures, in which the CXR is encoded using Convolutional Neural Networks (CNNs) into global image-level features [13, 14] which are input to a Transformer language model [32] to generate the radiology report. However, the generated reports often exhibit realistic style but imperfect accuracy, for instance hallucinating additional findings or describing abnormal regions as normal. Inspired by recent image captioning works in the general domain [2, 19, 37] in which each visual token corresponds to an object detected in the input image, we investigate whether replacing the image-level tokens with local tokens – corresponding to anatomical structures – can improve the clinical accuracy of the generated reports. Our contributions are to:

1. Propose a novel multi-task Faster R-CNN [27] to extract *finding-aware anatomical tokens* by performing finding detection on the candidate bounding boxes identified during anatomical structure localisation. We ensure these tokens convey rich information by training the model on an extensive set of anatomy regions and associated findings from the Chest ImaGenome dataset [12, 34].

2. Integrate the extracted finding-aware anatomical tokens as the visual input in a state-of-the-art two-stage pipeline for radiology report generation [6]; this pipeline is multimodal, taking both image (CXR) and text (the corresponding text indication field) as inputs.
3. Demonstrate the benefit of using these tokens for CXR report generation through in-depth experiments on the MIMIC-CXR dataset.

2 Related Works

Automated Reporting. Previous works on CXR automated reporting have examined the model architecture [4,5], the use of additional loss functions [24], retrieval-based report generation [10,29], and grounding report generation with structured knowledge [6,35]. However, no specific focus has been given to the image encoding. Inspired by recent works in image captioning in the general domain [2,19,37], where each visual token corresponds to an object detected in an image, we propose to replace the image-level representations with local representations corresponding to anatomical structures detected in a CXR. To the best of our knowledge, only [30,33] have considered anatomical feature representations for CXR automated reporting. In [33], they extract anatomical features from an object detection model trained solely on the anatomy localisation task. In [30], they train the object detector through multiple steps – anatomy localisation, binary abnormality classification and region selection – and feed each anatomical region individually to the language model to generate one sentence at a time. This approach makes the simplistic assumption that one anatomical region is described in exactly one report sentence.

Finding Detection. Prior works have tackled the problem of finding detection in CXR images via weakly supervised approaches [36,38]. However, the design of these approaches does not allow the extraction of anatomy-specific vector representations, making them unsuited for our purpose. Agu et al. [1] proposed AnaXnet, comprising two modules trained independently: a standard Faster R-CNN trained to localise anatomical regions, and a Graph Convolutional Network (GCN) trained to classify the pathologies appearing in each anatomical region bounding box. This approach assumes that the finding information is present in the anatomical representations after the first stage of training.

3 Methods

We describe our method in two parts: (1) Finding-aware anatomical token extraction (Fig. 1, left) – a custom Faster R-CNN which is trained to jointly perform *anatomy localisation* and *finding detection*; and (2) Multimodal report generation (Fig. 1, right) – a two-step pipeline which is adapted to perform *triples extraction* and *report generation*, using the anatomical tokens extracted from the Faster R-CNN as the visual inputs for the multimodal Transformer backbone [32].

3.1 Finding-Aware Anatomical Token Extraction

Let us consider $A = \{a_n\}_{n=1}^N$ as the set of anatomical regions in a CXR and $F = \{f_m\}_{m=1}^M$ the set of findings we aim to detect. We define $f_{n,m} \in \{0,1\}$ indicating the absence or presence of the finding f_m in the anatomical region a_n, and $f_n = \{f_{n,m}\}_{m=1}^M$ as the set of findings in a_n. We define *anatomy localisation* as the task of predicting the top-left and bottom-right bounding box coordinates $c = (c_{x1}, c_{y1}, c_{x2}, c_{y2})$ of the anatomical regions A; and *finding detection* as the task of predicting the findings f_n at each location a_n.

We frame anatomy localisation as a general object detection task, employing the Faster R-CNN framework to compute the coordinates of the bounding boxes and the anatomical labels assigned to each of them. First, the image features are extracted from the CNN backbone, composed of a ResNet-50 [13] and a Feature Pyramid Network (FPN) [22]. Second, the multi-scale image features extracted from the FPN are passed to the Region Proposal Network (RPN) to generate the bounding box coordinates $c_k = (c_{k,x1}, c_{k,y1}, c_{k,x2}, c_{k,y2})$ for each proposal k and to the Region of Interest (RoI) pooling layer, designed to extract the respective fixed-length vector representation $l_k \in \mathbb{R}^{1024}$. Each proposal's local features l_k are then passed to a classification layer (*Anatomy Classifier*) to assign the anatomical label (a_k) and to a bounding box regressor layer to refine the coordinates. In parallel, we insert a multi-label classification head (*Findings Classifier*) – consisting of a single fully-connected layer with sigmoid activation functions – that classifies a set of findings for each proposal's local features.

During training, we use a multi-task loss comprising three terms: *anatomy classification loss*, *box regression loss*, and (multi-label) *finding classification loss*. Formally, for each predicted bounding box, this is computed as

$$\mathcal{L} = \mathcal{L}_{anatomy} + \mathcal{L}_{box} + \lambda \mathcal{L}_{finding}, \tag{1}$$

where $\mathcal{L}_{anatomy}$ and $\mathcal{L}_{box}$ correspond to the anatomy classification loss and the bounding box regression loss described in [11] and $L_{finding}$ is the finding classification loss that we introduce; λ is a balancing hyper-parameter set to $\lambda = 10^2$. We define

$$\mathcal{L}_{finding} = -\sum_{k=1}^K \sum_{m=1}^M w_m f_{k,m} \log(p_{k,m}) \tag{2}$$

a binary cross-entropy loss between the predicted probability $p_k = \{p_{k,m}\}_{m=1}^M$ of the k-th proposal and its associated ground truth $f_k = \{f_{k,m}\}_{m=1}^M$ (with $f_k = f_m$ if $a_k = a_m$). We class weight using $w_m = (1/\nu_m)^\alpha$, where ν_m is the frequency of the finding f_m in the training dataset and we empirically set α to 0.25.

At inference time, for each CXR image, we extract the finding-aware anatomical tokens $A_{tok} = \{l_n\}_{n=1}^N$, by selecting for each anatomical region the proposal with highest anatomical classification score and taking the associated latent vector representation l_n. Any non-detected regions are assigned a 1024-dimensional vector of zeros. A_{tok} is provided as input to the report generation model.

3.2 Multimodal Report Generation

We adopt the multimodal knowledge-grounded approach for automated report-ing on CXR images as proposed in [6]. Firstly, *triples extraction* is performed to extract structured information from a CXR image in the form of triples, given the indication field Ind as context. Secondly, *report generation* is performed to generate the radiology report from the triples with the CXR image and indica-tion field again provided as context.

Each step is treated as a sequence-to-sequence task; for this purpose, the triples are concatenated into a single text sequence (in the order they appear in the ground truth report) separated by the special [SEP] token to form Trp, and the visual tokens are concatenated in a fixed order of anatomical regions. Two multimodal encoder-decoder Transformers are employed as the Triples Extractor (TE) and Report Generator (RG). The overall approach is:

$$
\begin{aligned}
&\text{STEP 1} \quad Trp = TE(seg_1 = A_{tok}, seg_2 = Ind) \\
&\text{STEP 2} \quad R = RG(seg_1 = A_{tok}, seg_2 = Ind \ [\texttt{SEP}] \ Trp)
\end{aligned} \tag{3}
$$

where seg_1 and seg_2 are the two input segments which are themselves concate-nated at the input. In step 2, the indication field and the triples are merged into a single sequence of text by concatenating them, separated by the special [SEP] token. Similarly to [8], the input to a Transformer corresponds to the sum of the textual and visual *token embeddings*, the *positional embeddings*—to inform about the order of the tokens—and the *segment embeddings*—to discriminate between the two modalities.

4 Experimental Setup

4.1 Datasets and Metrics

We base our experiments on two open-source CXR imaging datasets, Chest ImaGenome [12,34] and MIMIC-CXR [12,16,17]. The MIMIC-CXR dataset comprises CXR image-report pairs and is used for the target task of report gen-eration. The Chest ImaGenome dataset is derived from MIMIC-CXR, extended with additional automatically extracted annotations for 242,072 anteroposte-rior and posteroanterior CXR images, which we use to train the finding-aware anatomical token extractor. We follow the same train/validation/test split as proposed in the Chest ImaGenome dataset. We extract the *Findings* section of each report as the target text[1]. For the textual input, we extract the *Indi-cation field* from each report.[2] We annotate the ground truth triples for each image-report pair following a semi-automated pipeline using RadGraph [15] and sciSpaCy [25], as described in [6].

To assess the quality of the generated reports, we compute Natural Language Generation (NLG) metrics: BLEU [26], ROUGE [21] and METEOR [3]. We fur-ther compute Clinical Efficiency (CE) metrics by applying the CheXbert labeller

---

[1] https://github.com/MIT-LCP/mimic-cxr/blob/master/txt/create_section_files.py.
[2] https://github.com/jacenkow/mmbt/blob/main/tools/mimic_cxr_preprocess.py.

[28] which extracts 14 findings to the ground truth and the generated reports, and evaluate F1, precision and recall scores. We repeat each experiment 3 times using different random seeds, reporting the average in our results.

4.2   Implementation

Finding-Aware Anatomical Token Extractor: We adapt the Faster R-CNN implementation from [20][3], by including the finding classifier. We initialise the network with weights pre-trained on the COCO dataset [23], then fine-tune it to localise 36 anatomical regions and to detect 71 findings within each region, as annotated in the Chest ImaGenome dataset. The CXR images are resized by matching the shorter dimension to 512 pixels (maintaining the original aspect ratio) and cropping to a resolution of 512×512 (random crop during training and centre crop during inference). We train the model for 25 epochs with a learning rate of 10^{-3}, decayed every 5 epochs by a factor of 0.8. We select the model with the highest finding detection performances for the validation set, measured by computing the AUROC score for each finding at each anatomical region.

Report Generator: We implement a vanilla Transformer encoder-decoder at each step of the automated reporting pipeline. Both the encoder and the decoder consist of 3 attention layers, each composed of 8 heads and 512 hidden units. All the parameters are randomly initialised. We train step 1 for 40 epochs, with the learning rate set to 10^{-4} and we decay it by a factor of 0.8 every 3 epochs; and step 2 for 20 epochs, with the same learning rate as step 1. During training, we follow [6] in masking out a proportion of the ground-truth triples (50%, determined empirically), while during inference we use the triples extracted at step 1. We select the model with the highest CE-F1 score on the validation set.

Baselines. We benchmark against other CXR automated reporting methods: R2Gen [5], R2GenCMN [4], $\mathcal{M}^2$ Tr.+fact$_{ENTNLI}$ [24], CNN+Two-Step [6] and RGRG [30]. All these methods (except RGRG) adopt a CNN-Transformer and have shown state-of-the-art performances in report generation on the MIMIC-CXR dataset. All reported values are re-computed using the original code based on the same data split and image resolution as our method, except for [30] who already used this data split and image resolution, therefore we cite their reported results. We keep remaining hyperparameters as the originally reported values.

5   Results

Overall results In Table 1, we benchmark against other state-of-the-art CXR automated reporting methods and compare with the A_{tok} integrated into the full pipeline versus a simpler approach of the report generator model only, *RG*, which

---

[3] https://pytorch.org/vision/main/models/generated/torchvision.models.detection. fasterrcnn_resnet50_fpn_v2.html.

generates the report directly from image and indication field (omitting triples extraction). The proposed finding-aware anatomical tokens integrated with a knowledge-grounded pipeline [6] generate reports with state-of-the-art fluency (NLG metrics) and clinical accuracy (CE metrics). Moreover, the superior results of our A_{tok} + RG approach compared to RGRG [30] suggests that providing the full set of anatomical tokens together, instead of separately, gives better results. The broader visual context is indeed necessary when describing findings that span multiple regions $e.g.$, assessing the position of a tube.

Table 1. Comparison of our proposed solution with previous approaches. TE = Triples Extractor, RG = Report Generator.

Method	NLG						CE		
	BL-1	BL-2	BL-3	BL-4	MTR	RG-L	F1	P	R
R2Gen [5]	0.381	0.248	0.174	0.130	0.152	0.314	0.431	0.511	0.395
R2GenCMN [4]	0.365	0.239	0.169	0.126	0.145	0.309	0.371	0.462	0.311
$\mathcal{M}^2$ Tr. + fact$_{\text{ENTNLI}}$ [24]	0.402	0.261	0.183	0.136	0.158	0.300	0.458	0.540	0.404
ResNet-101 + TE + RG [6]	0.468	0.343	0.271	0.223	0.200	0.390	0.477	0.556	0.418
RGRG [30]	0.400	0.266	0.187	0.135	0.168	-	0.461	0.475	0.447
A_{tok} + RG (ours)	0.422	0.324	0.265	0.225	0.201	**0.426**	0.515	0.579	0.464
A_{tok} + TE + RG (ours)	**0.490**	**0.363**	**0.288**	**0.237**	**0.213**	0.406	**0.537**	**0.585**	**0.496**

Table 2. Comparison of different visual input representations (ResNet-101 $vs.$ A_{tok}) using different pre-training supervision (ImageNet, Findings, Anatomy and Anatomy+Findings), integrated with the TE+RG two-step pipeline.

Visual Input	Supervision	NLG						CE		
		BL-1	BL-2	BL-3	BL-4	MTR	RG-L	F1	P	R
ResNet-101	ImageNet	0.468	0.343	0.271	0.223	0.200	0.390	0.477	0.556	0.418
ResNet-101	Findings	0.472	0.346	0.273	0.225	0.202	0.396	0.495	0.565	0.440
Naive A_{tok}	Anatomy	0.436	0.320	0.253	0.208	0.187	0.387	0.392	0.487	0.329
A_{tok}	Anatomy+Findings	**0.490**	**0.363**	**0.288**	**0.237**	**0.213**	**0.406**	**0.537**	**0.585**	**0.496**

Ablation Study. Table 2 shows the results of adopting different visual representations. Firstly, we use a CNN (**ResNet-101**) trained end-to-end with $TE+RG$ and initialised two ways: pre-trained on **ImageNet** versus pre-trained on the **Findings** labels of Chest ImaGenome. Secondly, we extract anatomical tokens ($\mathbf{A}_{tok}$) with different supervision of Faster R-CNN: anatomy localisation only (**Anatomy**) or anatomy localisation + finding detection (**Anatomy+Findings**). The results show the positive effect of including supervision with finding detection either when pre-training ResNet-101 or as an additional task for Faster R-CNN. Example reports are shown in Fig. 2.

GT Report	ResNet-101 + TE + RG	Naive A_{tok} + TE + RG	A_{tok} + TE + RG
The cardiac silhouette size is normal. There is mild calcification of the aortic knob.[1] The mediastinal and hilar contours are within normal limits. The pulmonary vascularity is normal. Lungs are clear. No pleural effusion focal consolidation or pneumothorax is present. There are multilevel degenerative changes in the thoracic spine.[2]	Heart size is normal. Atherosclerotic calcifications are noted at the aortic knob.[1] Mediastinal and hilar contours are otherwise unremarkable. Mild atherosclerotic calcifications are noted at the aortic knob.[1] Pulmonary vasculature is normal. Lungs are clear without focal consolidation. No pleural effusion or pneumothorax is present. There are no acute osseous abnormalities.	The heart size is normal. The mediastinal and hilar contours are unremarkable. The pulmonary vascularity is normal and the lungs are clear without focal consolidation. No pleural effusion or pneumothorax is present. There are mild degenerative changes in the thoracic spine.[2]	The heart size is normal. The aorta is mildly tortuous with atherosclerotic calcifications noted at the aortic knob.[1] The mediastinal and hilar contours are otherwise unremarkable and the pulmonary vascularity is not engorged. The lungs are clear without focal consolidation. No pleural effusion or pneumothorax is present. Mild degenerative changes are noted in the thoracic spine with mild loss of height of a mid thoracic vertebral body unchanged.[2]

Fig. 2. Example report generated with different visual representations. The ground truth (GT) report (left) is followed by the generated reports using ResNet-101, naive anatomical tokens, and finding-aware anatomical tokens. Correctly detected findings are coloured green and errors are coloured red; we number corresponding descriptions.

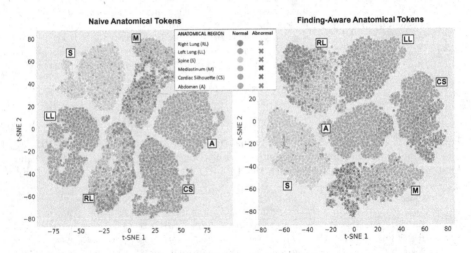

Fig. 3. T-SNE visualisation of normal and abnormal embeddings for a subset of visual tokens. Left: *naive anatomical token* embeddings extracted from Faster R-CNN trained solely on anatomy localisation. Right: *task-aware anatomical token* embeddings extracted from Faster R-CNN trained also on the finding detection task.

Anatomical Embedding Distributions. In Fig. 3, we visualise the impact of the finding detection task on the extracted anatomical tokens. To generate these plots, for 3000 randomly selected test set scans, we first perform principle component analysis [18] for dimensionality reduction of the token embeddings (from $\mathbb{R}^{1024}$ to $\mathbb{R}^{50}$), then use t-distributed stochastic neighbour embedding (t-SNE) [31], colour coding the extracted embeddings by their anatomical region and additionally categorising as *normal* or *abnormal* (a token is considered abnormal if at least one of the 71 findings is positively labeled). For most anatomical regions, the normal and abnormal groups are better separated by the finding-aware tokens, suggesting these tokens successfully transmit information about findings. We also compute the mean distance between normal and abnormal clusters using Fréchet Distance (mFD) [9], measuring mFD=8.80 (naive anatomical tokens) and mFD=78.67 (finding-aware anatomical tokens).

6   Conclusion

This work explores how to extract and integrate anatomical visual representations with language models, targeting the task of automated radiology reporting. We propose a novel multi-task Faster R-CNN adaptation that performs finding detection jointly with anatomy localisation, to extract *finding-aware anatomical tokens*. We then integrate these tokens as the visual input for a multimodal image+text report generation pipeline, showing that finding-aware anatomical tokens improve the fluency (NLG metrics) and clinical accuracy (CE metrics) of the generated reports, giving state-of-the-art results.

References

1. Agu, N.N., et al.: AnaXNet: anatomy aware multi-label finding classification in chest X-Ray. In: de Bruijne, M., et al. (eds.) MICCAI 2021. LNCS, vol. 12905, pp. 804–813. Springer, Cham (2021). https://doi.org/10.1007/978-3-030-87240-3_77
2. Anderson, P., et al.: Bottom-up and top-down attention for image captioning and visual question answering. In: CVPR (2018)
3. Banerjee, S., Lavie, A.: METEOR: an automatic metric for MT evaluation with improved correlation with human judgments. In: ACL Workshop on Intrinsic and Extrinsic Evaluation Measures for Machine Translation and/or Summarization, pp. 65–72 (2005)
4. Chen, Z., Shen, Y., Song, Y., Wan, X.: Cross-modal memory networks for radiology report generation. In: ACL-IJCNLP, pp. 5904–5914 (2021)
5. Chen, Z., Song, Y., Chang, T.-H., Wan, X.: Generating radiology reports via memory-driven transformer. In: EMNLP, pp. 1439–1449 (2020)
6. Dalla Serra, F., et al.: Multimodal generation of radiology reports using knowledge-grounded extraction of entities and relations. In: AACL-IJCNLP, pp. 615–624 (2022)
7. Demner-Fushman, D., et al.: Preparing a collection of radiology examinations for distribution and retrieval. JAMIA **23**(2), 304–10 (2016)
8. Devlin, J., Chang, M.-W., Lee, K., Toutanova, K.: BERT: pre-training of deep bidirectional transformers for language understanding. In: NAACL, pp. 4171–4186. Association for Computational Linguistics (2019)
9. Dowson, D., Landau, B.: The Fréchet distance between multivariate normal distributions. JMA **12**(3), 450–455 (1982)
10. Endo, M., Krishnan, R., Krishna, V., Ng, A.Y., Rajpurkar, P.: Retrieval-based chest x-ray report generation using a pre-trained contrastive language-image model. In: MLH, pp. 209–219 (2021)
11. Girshick, R.: Fast R-CNN. In: ICCV, pp. 1440–1448 (2015)

12. Goldberger, A.L., et al.: PhysioBank, PhysioToolkit, and PhysioNet: components of a new research resource for complex physiologic signals. Circulation 101(23), e215–e220 (2000)

13. He, K., Zhang, X., Ren, S., Sun, J.: Deep residual learning for image recognition. In: CVPR, pp. 770–778 (2016)

14. Huang, G., Liu, Z., Van Der Maaten, L., Weinberger, K.Q.: Densely connected convolutional networks. In: CVPR, pp. 4700–4708 (2017)

15. Jain, S., et al.: RadGraph: extracting clinical entities and relations from radiology reports. In: NeurIPS: Datasets and Benchmarks Track (Round 1) (2021)

16. Johnson, A.E., et al.: MIMIC-CXR, a de-identified publicly available database of chest radiographs with free-text reports. Sci. Data 6(1) (2019)

17. Johnson, A.E., et al.: MIMIC-CXR-JPG, a large publicly available database of labeled chest radiographs. arXiv preprint arXiv:1901.07042 (2019)

18. Jolliffe, I.: Principal component analysis. Springer Verlag, New York (2002). https://doi.org/10.1007/b98835

19. Li, X., et al.: OSCAR: object-semantics aligned pre-training for vision-language tasks. In: Vedaldi, A., Bischof, H., Brox, T., Frahm, J.-M. (eds.) ECCV 2020. LNCS, vol. 12375, pp. 121–137. Springer, Cham (2020). https://doi.org/10.1007/978-3-030-58577-8_8

20. Li, Y., Xie, S., Chen, X., Dollar, P., He, K., Girshick, R.: Benchmarking Detection Transfer Learning with Vision Transformers. arXiv preprint arXiv:2111.11429 (2021)

21. Lin, C.-Y.: ROUGE: a package for automatic evaluation of summaries. In: Text Summarization Branches Out, pp. 74–81 (2004)

22. Lin, T.-Y., Dollár, P., Girshick, R., He, K., Hariharan, B., Belongie, S.: Feature pyramid networks for object detection. In: CVPR, pp. 2117–2125 (2017)

23. Lin, T.-Y., et al.: Microsoft COCO: common objects in context. In: Fleet, D., Pajdla, T., Schiele, B., Tuytelaars, T. (eds.) ECCV 2014. LNCS, vol. 8693, pp. 740–755. Springer, Cham (2014). https://doi.org/10.1007/978-3-319-10602-1_48

24. Miura, Y., Zhang, Y., Tsai, E., Langlotz, C., Jurafsky, D.: Improving factual completeness and consistency of image-to-text radiology report generation. In: NAACL, pp. 5288–5304 (2021)

25. Neumann, M., King, D., Beltagy, I., Ammar, W.: ScispaCy: fast and robust models for biomedical natural language processing. In: BioNLP Workshop and Shared Task, pp. 319–327 (2019)

26. Papineni, K., Roukos, S., Ward, T., Zhu, W.-J.: BLEU: a method for automatic evaluation of machine translation. In: ACL, pp. 311–318 (2002)

27. Ren, S., He, K., Girshick, R., Sun, J.: Faster R-CNN: towards real-time object detection with region proposal networks. In: NIPS 28 (2015)

28. Smit, A., Jain, S., Rajpurkar, P., Pareek, A., Ng, A.Y., Lungren, M.: Combining automatic labelers and expert annotations for accurate radiology report labeling using BERT. In: EMNLP, pp. 1500–1519 (2020)

29. Syeda-Mahmood, T., et al.: Chest X-Ray report generation through fine-grained label learning. In: Martel, A.L., et al. (eds.) MICCAI 2020. LNCS, vol. 12262, pp. 561–571. Springer, Cham (2020). https://doi.org/10.1007/978-3-030-59713-9_54

30. Tanida, T., Müller, P., Kaissis, G., Rueckert, D.: Interactive and explainable region-guided radiology report generation. In: CVPR, pp. 7433–7442 (2023)

31. Van der Maaten, L., Hinton, G.: Visualizing data using t-SNE. JMLR 9(11) (2008)

32. Vaswani, A., et al.: Attention is all you need. In: NIPS 30 (2017)

33. Wang, Y., Wang, K., Liu, X., Gao, T., Zhang, J., Wang, G.: Self adaptive global-local feature enhancement for radiology report generation. arXiv preprint arXiv:2211.11380 (2022)
34. Wu, J.T., et al.: Chest ImaGenome dataset for clinical reasoning. In: NeurIPS: Datasets and Benchmarks Track (Round 2) (2021)
35. Yang, S., Wu, X., Ge, S., Zhou, S.K., Xiao, L.: Knowledge matters: chest radiology report generation with general and specific knowledge. In: MIA (2022)
36. Yu, K., Ghosh, S., Liu, Z., Deible, C., Batmanghelich, K.: Anatomy-guided weakly-supervised abnormality localization in chest X-rays. In: MICCAI (2022). https://doi.org/10.1007/978-3-031-16443-9_63
37. Zhang, P., et al.: VinVL: revisiting visual representations in vision-language models. In: CVPR, pp. 5579–5588 (2021)
38. Zhu, X., et al.: PCAN: pixel-wise classification and attention network for thoracic disease classification and weakly supervised localization. CMIG **102**, 102137 (2022)

Dual-Stream Model with Brain Metrics and Images for MRI-Based Fetal Brain Age Estimation

Shengxian Chen[1], Xin Zhang[1(✉)], Ruiyan Fang[1], Wenhao Zhang[1], He Zhang[2], Chaoxiang Yang[3], and Gang Li[4]

[1] School of Electronic and Information Engineering, South China University of Technology, Guangzhou, China
eexinzhang@scut.edu.cn

[2] Department of Radiology, Obstetrics and Gynecology Hospital, Fudan University, Shanghai, China

[3] Guangdong Women And Children Hospital, Guangzhou, China

[4] Department of Radiology and Biomedical Research Imaging Center, University of North Carolina at Chapel Hill, Chapel Hill, USA
gang_li@med.unc.edu

Abstract. The disparity between chronological age and estimated brain age from images is a significant indicator of abnormalities in brain development. However, MRI-based brain age estimation still encounters considerable challenges due to the unpredictable movement of the fetus and maternal abdominal motions, leading to fetal brain MRI scans of extremely low quality. In this work, we propose a novel deep learning-based dual-stream fetal brain age estimation framework, involving brain metrics and images. Given a stack of MRI data, we first locate and segment out brain regions of every slice. Since brain metrics are highly correlated with age, we introduce four brain metrics into the model. To enhance the representational capacity of these metrics in space, we design them as vector-based discrete spatial metrics(DSM). Then we design the 3D-FetalNet and DSM-Encoder to extract visual and metric features respectively. Additionally, we apply the Global and local regression to enable the model to learn various patterns across different age ranges. We evaluate our model on a fetal brain MRI dataset with 238 subjects and reach the age estimation error of 0.75 weeks. Our proposed method achieves state-of-the-art results compared with other models.

Keywords: MRI-based fetal brain age estimation · Brain metrics · Discrete spatial metircs · DSM-Encoder · 3D-FetalNet

1 Introduction

The discrepancy between the predicted brain age and chronological age can be an indicator of accelerated brain aging or delayed brain development [5,6]. It can

X. Cao et al. (Eds.): MLMI 2023, LNCS 14348, pp. 424–433, 2024.
https://doi.org/10.1007/978-3-031-45673-2_42

be a useful tool to characterize early brain development during the fetal stage and improve prenatal examinations given the limited diagnostic tools available for assessing the fetal brains. In-utero Magnetic Resonance Imaging (MRI) is increasingly recognized as an important tool to aid in prenatal diagnosis because it provides finer details of fetal brain anatomy and improves the identification of fetal brain abnormalities compared to ultrasound screening [13,19]. However, assessment of fetal brain MRI images remains difficult. In particular, although advanced fast imaging sequences can capture slices within seconds, unpredictable fetal and maternal abdominal movements can result in motion artifacts within and between slices, leading to generally low quality of obtained MRI images [4, 21]. Therefore, we need to explore methods that can overcome these challenges.

Routine radiological examinations heavily depend on physicians' subjective and potentially inaccurate visual assessments of the fetal brain. However, recent advancements in deep learning techniques have introduced numerous useful models for MRI-based fetal brain age estimation [18,23]. These models quantitatively assess developmental status by linking brain anatomical patterns to developmental timelines. J. Cheng et al. [17] proposes a 3D convolutional network called TSAN to accurately predict the brain age of healthy individuals from neuroimaging data with improved performance. Nevertheless, since it is developed based on adult brains, the effectiveness is not satisfactory when applied to the fetal brain. L. Shen et al. [23] combined the ResNet [14] architecture with the attention mechanism [26] to automatically localize the brain region, but it is an implicit constraint for the network to focus on the brain region, which means it cannot guarantee the attention mask exactly pay full attention to the brain region. L. Liao et al. [18] extracted one slice from the 3D images stack to predict fetal brain age, this will lose the global information of the whole brain.

In this work, we combine brain metrics and brain images to estimate fetal brain age. Our contributions can be summarized as follows:

- **Metric stream (M-stream).** We incorporate brain metrics into the model by discrete spatial metrics (DSM). Each metric is designed as a discrete spatial vector based on the 3D slice characteristics of MRI data. Additionally, we develop the DSM-Encoder to explore the spatial variations and extract features. Overall, the incorporation of brain metrics in our model enhances its expressive power and improves the interpretability of the results.
- **Image stream (I-stream).** We design a novel 3D convolutional network 3D-FetalNet with the multilayer AC block [7] and SE block [15] for extracting visual features from brain regions. This network is capable of eliminating noise interference and extracting diverse and effective features.
- **Metric stream and Image stream with Global and local regression (MI-stream-GL).** To estimate the fetal brain age using brain metrics information and brain visual features simultaneously, we combine the M-stream and I-stream into a dual stream(MI-stream) model. Additionally, to enable the model to learn various patterns across different age ranges, we apply Global and local regression(GL-regression), resulting in the MI-stream-GL model. Finally, through ablation experiments, we validate the significant con-

tribution of all our proposed techniques to the performance of fetal brain age estimation. Additionally, our MI-stream-GL model achieves state-of-the-art results compared with other models.

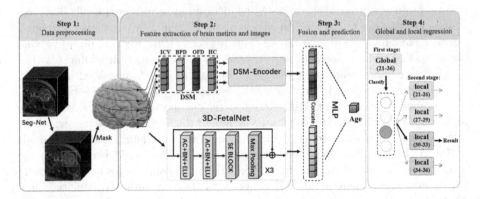

Fig. 1. The frame of MI-stream-GL model with brain metrics and images for MRI-based fetal brain age estimation.

2   Methodology

In this section, we will provide detailed information about our MI-stream-GL model design. As illustrated in Fig. 1, our proposed framework comprises four main steps: data preprocessing, feature extraction, fusion and prediction, and Global and local regression. The feature extraction step is composed of two streams: the M-stream and the I-stream. Additionally, the GL-regression component consists of one global regressor and four local regressors.

2.1   Data Preprocessing

To further eliminate the background noise, pay full attention to the brain region, and compute brain metrics, we adopt a segmentation baseline Seg-Net [9] to extract the brain region from the whole image. Subsequently, all segmented brain slices are utilized to generate input features for the M-stream and the I-stream. Specifically, we apply the 0–1 mask to extract the brain region and compute the DSM, which are the inputs of the M-stream and the I-stream, respectively.

2.2   M-Stream: Metric Stream for Learning Brain Metrics Information by Discrete Spatial Metrics

In existing hospital diagnostic protocols, doctors typically measure key brain metrics and construct equations to estimate fetal brain age [1,11,25,28]. We integrate primary brain metrics into our model, including Intracranial Volume

(ICV), Biparietal Diameter (BPD), Occipitofrontal Diameter (OFD), and Head
Circumference (HC). These brain metrics play a crucial role in assessing fetal
brain development and maturity. Our proprietary algorithm calculates these
metrics from segmented brain regions. Each brain metric is designed as a discrete
spatial vector based on the 3D slice characteristics of MRI data. Additionally,
we have developed the DSM-Encoder, which utilizes RNN and MLP models to
explore the spatial information of DSM, extract relevant features, and reduce
dimensions. Overall, the incorporation of brain metrics in our model enhances
its expressive power and improves the interpretability of the results.

Extracting Brain Metrics. To enhance the representational capacity of these
metrics in spatial analysis, we incorporate the 3D slice characteristics of MRI
data to represent the ICV, BPD, OFD, and HC as $n \times 1$ vectors $[s_1, ..., s_i, ..., s_n]$,
$[w_1, ..., w_i, ..., w_n]$, $[l_1, ..., l_i, ..., l_n]$ and $[p_1, ..., p_i, ..., p_n]$, respectively.

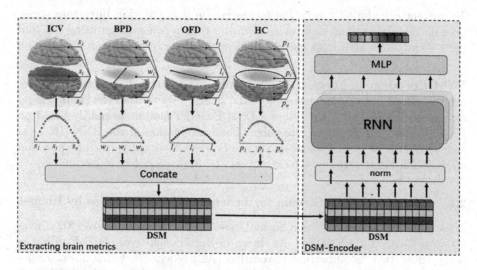

Fig. 2. Details of DSM and DSM-Encoder.

- **ICV:** ICV stands for Intracranial Volume, which refers to the total volume
  of the fetal cranium, including brain tissue, ventricles, meninges, and other
  intracranial structures [3]. To compute s_i from the segmented brain region is
  approximated as counting the number of pixels in the irregular region.
- **BPD:** BPD refers to the diameter between the parietal bones on both sides
  of the fetal head [16]. To compute w_i from the segmented brain region is
  approximated as finding the minimum circumscribed circle of the irregular
  region. The calculation method uses Welzl's algorithm [27], which iteratively
  updates the center and radius of a circle to recursively construct the minimum
  enclosing circle that contains all the given points.

- **OFD:** OFD refers to the diameter from the occipital bone to the frontal bone of the fetal head [20]. To compute l_i from the segmented brain region is approximated as finding the maximum inscribed circle of the irregular region. The calculation steps are as follows: first perform edge detection on the region, then use the Hough Circle Transform to detect circular contours, and finally, within the detected circular contours, find the largest circle, which represents the maximum inscribed circle.

- **HC:** HC refers to the measurement of the circumference or the distance around the head of a fetus [29]. To compute p_i from the segmented brain region is approximated as finding the perimeter of the irregular region. The calculation method used is the Ramer-Douglas-Peucker algorithm [8], which is a piecewise linear approximation algorithm used to approximate curves for a set of discrete points. In the case of calculating the perimeter of an irregular region, this algorithm approximates the contour by a series of line segments to estimate its total length.

To prevent calculation errors due to inaccurate segmentation in certain slices, we will identify outliers and revise them by taking the average using left and right slicing. The MRI data used is in the axial direction and the n is set to 30.

DSM-Encoder. In M-stream, we not only aim to extract numerical features from the DSM but also further explore the spatial variations of the DSM. Therefore, as shown in Fig. 2, we design a DSM-Encoder module primarily based on RNN [10] and MLP [22] architectures. The introduction of the RNN [10] aims to investigate the spatial variations of the DSM. Additionally, the incorporation of MLP [22] helps with feature extraction and dimensionality reduction.

2.3   I-Stream: Image Stream for Learning Visual Features by Images

Inspired by DenseNets [12] and ScaledDense [17], we propose a novel 3D convolution net, called 3D-FetalNet. As shown in Fig. 2, each layer of 3D-FetalNet is designed by two Asymmetric Convolution (AC) blocks [7], batch norm, Exponential Linear Unit (Elu) activation function, Squeeze-and-Excitation (SE) block [15] and max pooling. The original AC block in [7] was designed for natural images with three 2D Conv. Our AC block for MRI is designed with four 3D Conv in order to take the entire 3D images stack as input to guarantee the global information of the whole brain. The locations and directions of the fetal brain are randomly variable. So as to selectively emphasize informative features and suppress useless features, we add a SE block after the AC block. To accommodate the small-sized dataset, we only use a three-layer structure for 3D-FetalNet.

2.4   Fusion and Prediction

We directly concatenate the features learned by the M-stream and I-stream to integrate brain metrics information and brain visual features. Then, we use a two-layer MLP to further fuse and regressively predict fetal age.

As for the loss function, MAE provides greater stability and is less affected by outliers. On the other hand, MSE converges faster. Therefore, we fuse them in Eq. (1) as the loss function. The N is batch size and the α is set to 1.

$$Loss = L_{MAE} + \alpha L_{MSE} \tag{1}$$

2.5   Global and Local Regression

To accurately estimate fetal brain age, a network should be capable of learning various patterns across different age ranges. In fetal brain age estimation, the development process exhibits nonlinearity because each age group has different developmental characteristics. In the early stages of pregnancy, the growth rates of gray matter, subcortical gray matter, and white matter all peak, resulting in an increase in volume [2]. In the middle to later stages, the brain's surface begins to fold, leading to increased folding complexity and surface area.

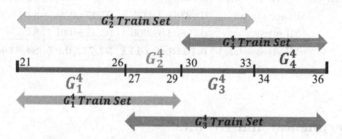

Fig. 3. Details of grouping and training set partitioning in Global and local regression.

A global regressor, used for the entire age range, should be able to learn these different developmental patterns. On the other hand, a local regressor would be more effective for a specific age range if it were trained with images within that range only. Training would also be easier because only the patterns within the smaller range need to be learned.

We observed that the fetal brains in the age range of 21–26 exhibit similar distribution patterns, as do those in the age range of 27–29, 30–33, and 34–36, respectively. Therefore, we employed a global regressor and four local regressors. In Fig. 3, we divided the entire age range into four age groups, G_1^4 (range: 21–26), G_2^4 (range: 27–29), G_3^4 (range: 30–33), and G_4^4 (range: 34–36).

During the prediction process, in the first stage, the global regressor (range: 21–36) is used to classify the fetal brain and determine which local subgroup it belongs to. In the second stage, the local regressor specific to that subgroup is invoked to accurately estimate the gestational age, which yields the final result.

However, in the first stage, the global regressor tends to misclassify data at the boundaries of the groups and incorrectly assign them to adjacent groups. To address this, when training the local regressor in the second stage, as shown in Fig. 3, we include not only the data from the target group but also data from the neighboring groups. For instance, the train Set for the G_1^4 group includes data ranging from 21 to 29. This ensures that even if there are classification errors in the first stage, the local regressor in the second stage is still able to accurately estimate the fetal brain age.

Table 1. Fetal brain age estimation results (mean±std) of all comparison methods on FetalBrain-SH dataset. (bold: best; underline: runner-up)

	Method	MAE(weeks)↓	R2(%)↑	CS(1.5)(%)↑
Baselines	VGG-reg [24]	0.8703±0.0280	89.59±2.13	73.55±0.64
	ResNet-reg [14]	0.8502±0.0189	89.94±1.64	81.98±1.17
	SENet-reg [15]	0.8348±0.0201	90.58±1.70	77.59±2.33
Related work	TSAN [17]	0.8459±0.0206	87.09±2.63	82.08±1.42
	DLVGG-reg [18]	0.8096±0.0153	91.43±1.04	83.43±1.27
Our methods	Metric stream	0.8333±0.0188	90.79±1.29	82.03±1.24
	Image stream	0.8278±0.0235	91.47±0.84	82.44±0.96
	MI-stream	0.7980±0.0206	93.41±0.91	83.67±0.74
	MI-stream-GL	**0.7476±0.0171**	**94.77±0.97**	**84.81+0.91**

3   Experimental and Results

3.1   Dataset and Implementation Details

We obtain a total of 238 fetal MRI scans (range:21–36 weeks) from the Obstetrics and Gynecology Hospital of Fudan University. The dataset is named "FetalBrain-SH". Prior to inclusion, each MRI stack undergoes meticulous manual examination by an expert to ensure normal fetal brain development.

To evaluate the performance of our approach, we compare it with several methods and perform ablation experiments to analyze the individual contributions of the various modules we designed. Specifically, we benchmark our method against DLVGG-reg [18], a framework for fetal brain age estimation. We compare our algorithm with TSAN [12], estimating the age of adult brains [17]. Additionally, we also compared our model with three baseline models, namely VGG-reg [24], ResNet-reg [14], and SeNet-reg [15], respectively. These age estimation algorithms are evaluated by 3 times 5-fold cross-validation with 3 criteria, i.e., MAE, correlation coefficient (R2), and the CS(1.5). The CS(1.5) is the proportion of cases with an MAE less than 1.5.

3.2   Results

Ablation Experiment. As shown in Table 1, in the "Our Method" section, we conducted several ablation experiments to further evaluate the effectiveness of our proposed method. These experiments included variations such as using only the I-stream, using only the M-stream, and performing GL-regression. We can draw the following conclusions: **(1)** Using either the I-stream or the M-stream alone yields good performance in estimating brain age. **(2)** The performance of the MI-stream approach is better than that of both the M-stream and I-stream methods. **(3)** Performing GL-regression on the model resulted in a 6% decrease in MAE. These findings demonstrate that all the modules proposed by us are reasonable and effective.

Performance Analysis. We first report the overall results of age estimation achieved by our MI-stream-GL model and related alternative methods on the FetalBrain-SH dataset in Table 1. We have the following interesting observations when comparing these statistics. First, the performance of exclusively utilizing I-stream with 3D-FetalNet outperforms other baseline models, highlighting the exceptional success of the 3D-FetalNet backbone in extracting image features. Second, when we combine the M-stream and I-stream into the dual-stream architecture, its performance surpasses all baselines and related work. Third, if we combine M-stream, I-stream, and GL-regression, its performance achieves state-of-the-art results compared with other methods. We can see a high correlation between the estimated brain age and the ground truth as shown in Fig. 4.

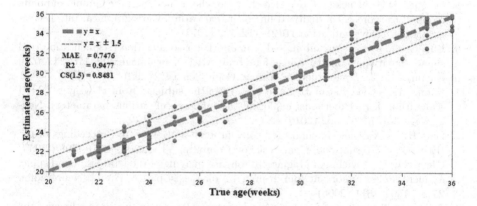

Fig. 4. Correlations between the estimated brain age and the ground truth of the testing data acquired by FetalBrain-SH of 238 samples.

4   Conclusion

In this paper, we developed a dual-stream network MI-stream-GL for MRI-based fetal brain age estimation, which can provide quantitative tools for prenatal diagnosis. MI-stream-GL consists of M-stream, I-stream, and GL-regression. In M-stream and I-stream, we design the DSM-Encoder and 3D-FetalNet to extract metric and visual features respectively. Experiments on the collected dataset demonstrated that our model achieved state-of-the-art results. In the future, we will further expand and develop our methods by adding multi-view MRI and improving image quality to achieve better performance.

References

1. Beheshti, I., Maikusa, N., Matsuda, H.: The accuracy of t1-weighted voxel-wise and region-wise metrics for brain age estimation. Comput. Methods Programs Biomed. **214**, 106585 (2021)
2. Bethlehem, R.A.I., et al.: Brain charts for the human lifespan. Nature **604**(7906), 525–533 (2022)
3. Brown, G.L., et al.: J. Neurol. Neurosurg. (2003)
4. Brugger, P., et al.: Methods of fetal MR: beyond t2-weighted imaging. Eur. J. Radiol. **57**, 172–181 (2006)
5. Cole, J.H., Franke, K.: Predicting age using neuroimaging: innovative brain ageing biomarkers. Trends Neurosci. **40**(12), 681–690 (2017)
6. Cole, J.H., et al.: Prediction of brain age suggests accelerated atrophy after traumatic brain injury. Ann. Neurol. **77**(4), 571–581 (2015)
7. Ding, X., Guo, Y., et al, X.: ACNet: strengthening the kernel skeletons for powerful CNN via asymmetric convolution blocks. In: Proceedings of the IEEE/CVF International Conference on Computer Vision, pp. 1911–1920 (2019)
8. Douglas, D.H., Peucker, T.K.: Algorithms for the reduction of the number of points required to represent a digitized line or its caricature. Cartographica: Int. J. Geograph. Inf. Geovisualization **10**(2), 112–122 (1973)
9. Ebner, M., et al.: An automated framework for localization, segmentation and super-resolution reconstruction of fetal brain MRI. NeuroImage **206**, 116324 (2020)
10. Elman, J.L.: Finding structure in time. Cogn. Sci. **14**(2), 179–211 (1990)
11. Franke, K., et al.: Estimating the age of healthy subjects from t1-weighted MRI scans using kernel methods: exploring the influence of various parameters. Neuroimage **50**(3), 883–892 (2010)
12. Gao, H., et al.: Densely connected convolutional networks. In: Proceedings of the IEEE/CVF International Conference on Computer Vision, pp. 4700–4708 (2017)
13. Glenn, O., Barkovich, A.: Magnetic resonance imaging of the fetal brain and spine: an increasingly important tool in prenatal diagnosis, part 1. Am. J. Neuroradiol. **27**(8), 1604–1611 (2006)
14. He, K., et al.: Deep residual learning for image recognition. In: Proceedings of the IEEE/CVF Iinternational Conference on Computer Vision, pp. 770–778 (2016)
15. Hu, J., Shen, L., Sun, G.: Squeeze-and-excitation networks. In: Proceedings of the IEEE/CVF International Conference on Computer Vision, pp. 7132–7141 (2018)
16. Işık, Ş, Büyüktiryaki, M., Şimşek, G.K., Kutman, H.G.K., Canpolat, F.E.: Relationship between biparietal diameter/ventricular ratio and neurodevelopmental outcomes in non-handicapped very preterm infants. Child's Nerv. Syst. **37**, 1121–1126 (2020)
17. Cheng, J., et al.: Brain age estimation from MRI using cascade networks with ranking loss. IEEE Trans. Med. Imaging **40**(12), 3400–3412 (2021)
18. Liao, L., et al.: Multi-branch deformable convolutional neural network with label distribution learning for fetal brain age prediction. In: 2020 IEEE 17th International Symposium on Biomedical Imaging (ISBI), pp. 424–427 (2020)

19. Malinger, G., et al.: Fetal brain imaging: a comparison between magnetic resonance imaging and dedicated neurosonography. Ultrasound in Obstetrics Gynecology: Official J. Int. Soc. Ultrasound Obstetrics Gynecol. **23**(4), 333–340 (2004)

20. Persson, P., Weldner, B.M.: Normal range growth curves for fetal biparietal diameter, occipito frontal diameter, mean abdominal diameters and femur length. Acta Obstetricia et Gynecologica Scandinavica 65 (1986)

21. Prayer, D., et al.: Fetal MRI: techniques and protocols. Pediatric Radiol. **34**, 685–693 (2004)

22. Rumelhart, D.E., Hinton, G.E., Williams, R.J.: Learning representations by back-propagating errors. Nature **323**(6088), 533–536 (1986)

23. Shen, L., et al.: Deep learning with attention to predict gestational age of the fetal brain. arXiv (2018)

24. Simonyan, K., Zisserman, A.: Very deep convolutional networks for large-scale image recognition. CoRR abs/1409.1556 (2014)

25. Van, P.V., et al.: Assessment of brain two-dimensional metrics in infants born preterm at term equivalent age: correlation of ultrasound scans with magnetic resonance imaging. Front. Pediatrics 10 (2022)

26. Vaswani, A., et al.: Attention is all you need. Adv. Neural. Inf. Process. Syst. **30**, 5998–6008 (2017)

27. Welzl, E.: Smallest enclosing disks (balls and ellipsoids). In: Maurer, H. (ed.) New Results and New Trends in Computer Science. LNCS, vol. 555, pp. 359–370. Springer, Heidelberg (1991). https://doi.org/10.1007/BFb0038202

28. Whitmore, L.B., Weston, S.J., Mills, K.L.: Brainage as a measure of maturation during early adolescence. bioRxiv (2023)

29. Zhang, J., Petitjean, C., Lopez, P., Ainouz, S.: Direct estimation of fetal head circumference from ultrasound images based on regression CNN. In: International Conference on Medical Imaging with Deep Learning (2020)

PECon: Contrastive Pretraining to Enhance Feature Alignment Between CT and EHR Data for Improved Pulmonary Embolism Diagnosis

Santosh Sanjeev[1]([✉]) [ID], Salwa K. Al Khatib[1] [ID], Mai A. Shaaban[1] [ID],
Ibrahim Almakky[1] [ID], Vijay Ram Papineni[2] [ID], and Mohammad Yaqub[1] [ID]

[1] Mohamed Bin Zayed University of Artificial Intelligence, Abu Dhabi, UAE
{santosh.sanjeev,salwa.khatib,mai.kassem,ibrahim.almakky,
mohammad.yaqub}@mbzuai.ac.ae
[2] Sheikh Shakbout Medical City, Abu Dhabi, UAE
vpapineni@ssmc.ae

Abstract. Previous deep learning efforts have focused on improving the performance of Pulmonary Embolism (PE) diagnosis from Computed Tomography (CT) scans using Convolutional Neural Networks (CNN). However, the features from CT scans alone are not always sufficient for the diagnosis of PE. CT scans along with electronic heath records (EHR) can provide a better insight into the patient's condition and can lead to more accurate PE diagnosis. In this paper, we propose **P**ulmonary **E**mbolism Detection using **Con**trastive Learning (**PECon**), a supervised contrastive pretraining strategy that employs both the patient's CT scans as well as the EHR data, aiming to enhance the alignment of feature representations between the two modalities and leverage information to improve the PE diagnosis. In order to achieve this, we make use of the class labels and pull the sample features of the same class together, while pushing away those of the other class. Results show that the proposed work outperforms the existing techniques and achieves state-of-the-art performance on the RadFusion dataset with an F1-score of 0.913, accuracy of 0.90 and an AUROC of 0.943. Furthermore, we also explore the explainability of our approach in comparison to other methods. Our code is publicly available at https://github.com/BioMedIA-MBZUAI/PECon.

Keywords: Contrastive learning · Multimodal data · Pulmonary Embolism · CT scans

1 Introduction

Pulmonary Embolism (PE) is an acute cardiovascular disorder considered the third most common cause of cardiovascular death after coronary artery disease

Supplementary Information The online version contains supplementary material available at https://doi.org/10.1007/978-3-031-45673-2_43.

X. Cao et al. (Eds.): MLMI 2023, LNCS 14348, pp. 434–443, 2024.
https://doi.org/10.1007/978-3-031-45673-2_43

and stroke. Despite advances in diagnosis and treatment over the past 30 years, PE has high early mortality rates [1], with nearly $100k$ to $200k$ deaths in the US each year [2]. Unfortunately, individuals diagnosed with PE frequently encounter a long delay before receiving a diagnosis, and approximately 25% of patients are initially misdiagnosed [3,4]. Hence, it is crucial to provide radiologists and clinicians with tools that can help them with the diagnosis.

Prior to deep learning, many efforts focused on using traditional feature extraction methods for PE diagnosis from Computed Tomography (CT) scans [5,6]. However, more recently research efforts have investigated improving the performance of PE diagnosis using deep Convolutional Neural Networks (CNNs) [7,8] and attention mechanisms [9]. More specifically, PENet [10] was introduced as a 3D CNN designed to detect PE using multiple CT slices. Such use of 3D convolutions allows the network to consider information from multiple slices when making predictions, which is important for diagnosing PE, as its presence is not limited to a single CT slice.

Features from CT scans alone could be insufficient for PE diagnosis. Therefore, CT scans along with the patient's Electronic Health Records (EHR) data can provide a better insight into the patient's condition and can help improve the diagnosis of PE. Previous studies have shown improved performance when combining demographic and clinical data with medical imaging data for various medical conditions such as Alzheimer's disease and skin cancer [11,12]. However, multimodal fusion is a non-trivial task, which can have varying results based on the fusion approach. Therefore, Huang et al. [13] compared different multimodal fusion approaches that combine inputs from both CT scans and EHR data to diagnose the presence of PE and explore optimal data selection and fusion strategies. Their results show that multimodal end-to-end deep learning models combining imaging and EHR provide better discrimination of abnormalities than using either modality independently. To this effect, Zhou et al. [14] released RadFusion, a multimodal dataset containing both EHR data and high-resolution CT scans labeled for PE. They assess the fairness properties across different subgroups and results suggest that integrating EHR data with medical images can improve classification performance and robustness without introducing large disparities between population groups.

Contrastive learning has been previously employed in medical imaging. Recently, Zhang et al. [15] made use of available pairs of images and text usually in medical contexts, and proposed Contrastive VIsual Representation Learning from Text (ConVIRT) [15]. The approach involves pretraining medical image encoders with paired text via a bidirectional contrastive objective. Nevertheless, their text encoder employs a general-purpose lexicon, resulting in issues with unknown words when processing a medical text. Although the division of terms into word pieces helps alleviate this problem, it results in the fragmentation of common biomedical terminology, leading to suboptimal results [16,17]. Furthermore, [18] has shown that constrastive learning between two images (same modality) in a fully-supervised setting can outperform the self-supervised contrastive pretraining by leveraging the label information. In this paper, we propose

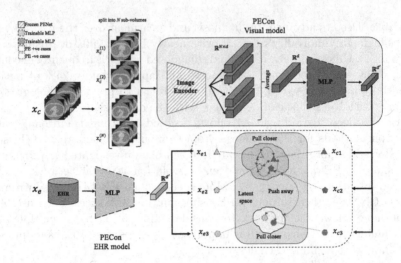

Fig. 1. PECon: a self-supervised contrastive pretraining method for improved PE diagnosis. The framework consists of a visual branch with a frozen image encoder backbone and a trainable MLP (top part) and an EHR branch with a trainable MLP (bottom left part). The output representations of both MLPs are later used for supervised contrastive loss (bottom right part). (x_{c1}, x_{e1}), (x_{c2}, x_{e2}), (x_{c3}, x_{e3}) are the CT and EHR embeddings of 3 patients in a given batch.

Pulmonary **E**mbolism Detection using **Con**trastive Learning (**PECon**), a novel fully supervised contrastive pretraining framework between 3D CT scans and EHR data. PECon uses a supervised contrastive learning objective to train a neural network to encode CT scans and EHR into a joint embedding space. In this space, we establish cross-modal alignment, through a fully supervised contrastive (SC) pretraining aiming to maximise the feature alignment between CT and EHR data. To the best of our knowledge we are the first to introduce a SC pretraining using different modalities (EHR,CT) in medical domain to maximize alignment between the feature representations. Unlike [18], which uses augmentations to get 2 views of the same image(single modality), we use CT and EHR as 2 views(multimodal) of the same patient. We demonstrate the effectiveness of our approach on a real-world dataset of PE patients, where we show that our pretrained model significantly outperforms state-of-the-art baselines. The main contributions of this work can be summarized as follows:

- We propose a supervised contrastive pretraining method that better aligns the image and EHR embeddings thereby achieving enhanced alignment between the features of both modalities.
- We demonstrate how the proposed method enhances the performance of multimodal PE detection and achieves state-of-the-art results on the RadFusion [14] dataset.

2   Methodology

In this section, we discuss the details of PECon, the proposed pretraining method illustrated in Fig. 1, and the fine-tuning stage that follows. The pretraining stage involves a supervised contrastive learning step, where, for each anchor feature extracted from a CT scan volume or an EHR record, all other features having the same label as this feature in the batch are *pulled* to it, and all those having the opposite label are *pushed* away. As such, given a pair of a CT scan and corresponding EHR data (x_c, x_e) belonging to the same patient, our goal is to learn a function $M(x_c, x_e) = \hat{y}_{x_c, x_e}$ that uses this paired input to approximate y, the ground truth PE diagnosis of patient. We set a pre-text task whose objective is to learn an image projection head f_c and an EHR projection head f_e that respectively map x_c and x_e to d'-dimensional vectors z. More precisely, projection head f_c takes as input a d-dimensional embedding, which is the result of averaging $N \times d$-dimensional embeddings from an image encoder. The image encoder is used to generate a d-dimensional embedding for each of the N subvolumes belonging to a single CT scan $(x_c^{(1)}, x_c^{(2)}, .. , x_c^{(N)}$ as shown in Fig. 1). For a given input batch of data, [18] applies data augmentation twice and obtains two copies of the batch. Alternatively, in PECon, considering a batch containing B pairs of (x_c, x_e), the training objective of PECon is a supervised contrastive loss with the following expression:

$$\mathcal{L}_{PECon} = \sum_{i=1}^{2B} \frac{-1}{|P(i)|} \sum_{p \in P(i)} log \frac{exp(\langle z_i . z_p \rangle / \tau)}{\sum_{a \in A(i)} exp(\langle z_i . z_a \rangle / \tau)} \tag{1}$$

where $A(i)$ is the set of all z features extracted from x_c and x_e samples in the batch distinct from feature i, $P(i) \equiv \{p \in A(i) : y_p = y_i\}$ is the set of features within the batch having the same ground truth label as feature i, $|.|$ is the set cardinality, and $\tau \in \mathcal{R}^+$ is a temperature parameter.

The visual projection head f_c and the EHR projection head f_e are then separately fine-tuned for classification by minimizing the cross-entropy loss. For this, the multimodal prediction is given by:

$$\hat{y}_{x_c, x_e} = \lambda \hat{y}_{x_c} + (1 - \lambda) \hat{y}_{x_e} \tag{2}$$

where λ is a modality weighting hyper-parameter $\in [0, 1]$, which can be used to set the influence each modality model will have on the overall multimodal classification.

3   Experimental Setup

3.1   Dataset

In this work, we use the RadFusion dataset [14], which consists of both CT scans and EHR data of the patients with and without PE. It consists of $1,837$ axial CTPA exams from 1794 patients captured with $1.25mm$ spacing. The number of negative PE cases is $1,111$, while the number of positive PE cases is 726. The EHR data consists of demographic features, vitals such as systolic and diastolic blood pressure, body temperature, 641 unique inpatient and outpatient medications, 141 unique diagnosis groupings, and laboratory tests of around 22 categories. This information is represented with a binary presence/absence as well as the latest value of the test. EHR in the dataset is structured (categorical, numerical) without text. PE cases are mainly of 3 types: central, segmental and subsegmental based on the location of PE within the arterial branches. The number of cases of central, segmental and subsegmental PE are 257, 387 and 52 respectively. For a fair comparison, we followed the standard split of Radfusion [14] which was, training, validation, and testing splits of 80%, 10%, 10% respectively. The scans are normalised between $[-1, 1]$ and the EHR data is normalised between $[0, 1]$.

3.2   Implementation Details

Pretraining stage. Pretraining is carried out in an end-to-end fashion with an SGD optimizer, using an initial learning rate (LR) of 0.1, $\tau = 0.8$, a batch size of 128, and for 100 epochs on a single NVIDIA RTX A6000 GPU. We select the weights of the epoch with the lowest validation loss for later fine-tuning. As for data augmentation, the 3D scans are randomly (1) cropped along their widths and heights to 192×192, (2) jittered up to 8 slices along the depth axis, and (3) rotated up to $15°$, as is done in [10]. The image encoder architecture we adopt is PeNet [10]. The PENet we use is an end to end 3D CNN model consisting of 4 encoder blocks primarily constructed from 3D convolutions with skip connections and squeeze-and-excitation blocks. The PENet backbone is initialized with weights trained on RadFusion, pretrained on Kinetics dataset [19] and it remains frozen during the pretraining. Input to the PENet models are subvolumes consisting of 24 slices, where PENet outputs a 2048 vector for each subvolume. The feature vector input to the visual projection head is an average of the 2048 embeddings generated for each subvolume. The image projection head is an MLP consisting of 2 hidden layers with 512 and 256 neurons respectively outputting a feature vector of size 128. Similarly, the EHR projection head consists of 1 hidden layer with 128 neurons. Image size and number of slices were chosen based on the PENet [10], which provided the best results and allowed for a fair comparison. The modality weighting parameter was selected empirically by testing with different values and choosing the best performing on validation set. We explored different embedding sizes (128,256,512) with performance stagnation after 128. Similarly other hyper-parameters have been experimentally selected. **Fine-tuning stage.** In the fine-tuning stage, the visual projection head and the EHR projection head are separately fine-tuned on the RadFusion dataset. The imaging model is fine-tuned for 25 epochs with an SGD optimizer, using a LR of 0.01 and a stepLR scheduler with a step size of 20. The EHR model is fine-tuned for 25 epochs using SGD optimizer with a LR of 0.1 and a stepLR scheduler with a step size of 10. We report results for a λ of 0.375 and the ablation is shown in the Supplementary.

4   Results and Discussion

Table 1. Performance Comparison of the imaging-only, EHR-only and multimodal fusion models with the state-of-the-art

Modality	Include subsegmental	Models	Accuracy	F1 score	AUROC
CT	✓	PENet [10]	0.689	0.677	**0.796**
		PECon Visual model	**0.726**	**0.752**	0.775
EHR		ElasticNet [14]	0.837	0.850	**0.922**
		PECon EHR model	**0.858**	**0.872**	0.922
Multimodal		Radfusion [14]	0.890	0.902	**0.946**
		PECon Multimodal	**0.900**	**0.913**	0.943
CT	✗	PENet [10]	**0.759**	0.735	**0.842**
		PECon Visual model	**0.759**	**0.766**	0.817
EHR		ElasticNet [14]	0.877	0.877	**0.932**
		PECon EHR model	**0.883**	**0.883**	0.930
Multimodal		Radfusion [14]	0.895	0.895	**0.962**
		PECon Multimodal	**0.914**	**0.918**	0.961

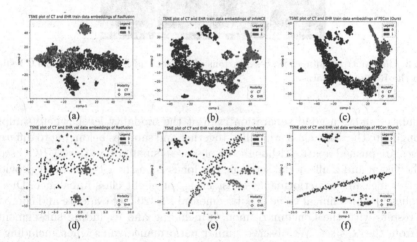

Fig. 2. Illustration of TSNE plots. (a) Train set embeddings of Radfusion, (b) Train set embeddings of InfoNCE, (c) Train set embeddings of PECon (Ours), (d) Val set embeddings of Radfusion, (e) Val set embeddings of InfoNCE and (f) Val set of embeddings of PECon

We consistently use the RadFusion test set to evaluate our approach and compare it with existing approaches. Table 1 compares the performance of our approach with the benchmark models. We observe that PECon achieves state-of-the-art performance on CT, EHR, as well as the multimodal setting. The boost in multimodal performance demonstrates the effectiveness of the approach in improving class separability in the latent space. Further, the improvement on the individual

Table 2. Ablation studies on the inclusion of subsegmental cases in the pre-training and finetuning stages.

Pre-training	Finetuning	Accuracy	F1 score	AUROC
✓	✓	0.889	0.905	0.931
✓	✗	0.878	0.893	0.941
✗	✓	0.873	0.887	0.920
✗	✗	**0.900**	**0.913**	**0.943**

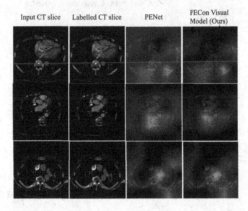

Fig. 3. Class activation map representations of PENet and PECon on sample CT slices from the RadFusion dataset.

modalities shows a good separation even at the modality level, where samples belonging to the same class are pulled together and samples belonging to different classes are pushed apart. Although, we observe a small dip in the AUROC score in the visual and multimodal settings, our focus is on the F1-score as it provides a better measure of performance considering the slight class imbalance. Due to the questionable clinical value of subsegmental PE [20], we evaluate and compare the results with the benchmark models including and excluding subsegmental PE from the test set. We observe similar performance gains when including or excluding the subsegmental cases from the test set. However, excluding sub-segmental cases from the training data for PECon during both pretraining and fine-tuning stages results in the best performance. We test the effect of excluding the sub-segmental cases in either the pretraining or the fine-tuning stages, while still including them in the evaluation. Results in Table 2 show that excluding samples from both stages achieves the best results.

In Fig. 2, we visualize and compare the TSNE plots of Radfusion and PECon (Ours) for the train and validation sets. We also visualize results of PECon with InfoNCE [21]. We observe that pretraining using a contrastive loss enhances the feature alignment between CT scan and EHR features, thereby making the data more separable and forming clusters. On the other hand, the RadFusion train and

Table 3. Ablation studies on the contrastive learning strategy

Contrastive Learning strategy	Accuracy	F1 score	AUROC
ConViRT (InfoNCE) [15]	0.892±0.003	0.907±0.002	0.938±0.001
Barlow Twins [23]	0.889±0.005	0.904±0.003	0.930±0.004
PECon	**0.898±0.003**	**0.912±0.002**	**0.939±0.003**

validation do not have a good separation. Furthermore, the use of supervision in PECon pulled together sample features of the same class while pushing away the features of different classes. This is lacking in the case of InfoNCE since it is a fully self-supervised method that is agnostic to the class labels of the features. To enhance the explainability of our approach, we use Grad-CAM [22] to visualize class activation maps of three CT slices pertaining to positive scans from the RadFusion dataset using PENet and PECon in Fig. 3. In the case of PENet, some attention is put on irrelevant regions around the lungs, whereas PECon imaging model consistently demonstrates better and more precise localization of the PE. This shows that combining the two modalities using our approach enables the visual model to learn more precise features from the input volumes. We further compare our pretraining stage with other contrastive learning methods. Results in Table 3 show that PECon achieves better performance. This demonstrates that the inclusion of supervision with contrastive learning is more effective for PE classification compared to self-supervised pretraining methods. The hyper-parameters for each method are reported in the Supplementary.

5   Conclusion and Future Work

In this work, we present (**PECon**), a supervised contrastive pretraining frame-work for multimodal prediction of PE to enhance the alignment between the CT and EHR latent representations. To accomplish this, we start from an anchor feature extracted from CT scans and EHR data and pull features from the same class to it and push away those from the other class. We trained and validated our approach on the RadFusion dataset, and report state-of-the-art results. Additionally, we show that our model attends to more precise and localized regions in the CT scans through class activation maps.

The limitations of this work include testing on other datasets to assess the generalizability of our approach. Furthermore, our study was limited to the impact of the contrastive pretraining loss on the late fusion multimodal approach, where in the future it would be important to study this impact on early and joint fusion approaches. Further, attention techniques will be explored for feature aggregation.

References

1. Bělohlávek, J., Dytrych, V., Linhart, A.: Pulmonary embolism, part I: epidemiology, risk factors and risk stratification, pathophysiology, clinical presentation, diagnosis and nonthrombotic pulmonary embolism. Exp. Clin. Cardiol. **18**(2), 129–138 (2013)
2. Tarbox, A., Swaroop, M.: Pulmonary embolism. Int. J. Crit. Illn. Inj. Sci. **3**(1), 69–72 (2013). https://doi.org/10.4103/2229-5151.109427, https://www.ijciis.org/text.asp?2013/3/1/69/109427
3. Hendriksen, J.M.T., et al.: Clinical characteristics associated with diagnostic delay of pulmonary embolism in primary care: a retrospective observational study. BMJ Open **7**(3), e012789 (2017). https://doi.org/10.1136/bmjopen-2016-012789
4. Alonso-Martínez, J.L., Sánchez, F.J., Echezarreta, M.A.: Delay and misdiagnosis in sub-massive and non-massive acute pulmonary embolism. Eur. J. Intern. Med. **21**, 278–282 (2010). https://doi.org/10.1016/J.EJIM.2010.04.005, https://pubmed.ncbi.nlm.nih.gov/20603035/
5. Masutani, Y., MacMahon, H., Doi, K.: Computerized detection of pulmonary embolism in spiral CT angiography based on volumetric image analysis. IEEE Trans. Med. Imaging **21**(12), 1517–1523 (2002). https://doi.org/10.1109/TMI.2002.806586
6. Liang, J., Bi, J.: Computer aided detection of pulmonary embolism with tobogganing and mutiple instance classification in CT pulmonary angiography. In: Karssemeijer, N., Lelieveldt, B. (eds.) IPMI 2007. LNCS, vol. 4584, pp. 630–641. Springer, Heidelberg (2007). https://doi.org/10.1007/978-3-540-73273-0_52
7. Lin, Y., et al.: Automated pulmonary embolism detection from CTPA images using an end-to-end convolutional neural network. In: Shen, D. (ed.) MICCAI 2019. LNCS, vol. 11767, pp. 280–288. Springer, Cham (2019). https://doi.org/10.1007/978-3-030-32251-9_31
8. Khachnaoui, H., Agrébi, M., Halouani, S., Khlifa, N.: Deep learning for automatic pulmonary embolism identification using CTA images. In: 2022 6th International Conference on Advanced Technologies for Signal and Image Processing (ATSIP), pp. 1–6 (2022). https://doi.org/10.1109/ATSIP55956.2022.9805929
9. Suman, S., et al.: Attention based CNN-LSTM network for pulmonary embolism prediction on chest computed tomography pulmonary angiograms (2021). https://arxiv.org/abs/2107.06276
10. Huang, S.C., et al.: PENet-a scalable deep-learning model for automated diagnosis of pulmonary embolism using volumetric CT imaging. NPJ Digital Med. **3**(1), 61 (2020). https://doi.org/10.1038/s41746-020-0266-y
11. Li, H., Fan, Y.: Early prediction of Alzheimer's disease dementia based on baseline hippocampal MRI and 1-year follow-up cognitive measures using deep recurrent neural networks. In: Proceedings. IEEE International Symposium on Biomedical Imaging 2019, pp. 368–371 (2019). https://doi.org/10.1109/ISBI.2019.8759397
12. Kawahara, J., Daneshvar, S., Argenziano, G., Hamarneh, G.: Seven-point checklist and skin lesion classification using multitask multimodal neural nets. IEEE J. Biomed. Health Inform. **23**(2), 538–546 (2019). https://doi.org/10.1109/JBHI.2018.2824327
13. Huang, S.C., Pareek, A., Zamanian, R., Banerjee, I., Lungren, M.P.: Multimodal fusion with deep neural networks for leveraging CT imaging and electronic health record: a case-study in pulmonary embolism detection. Sci. Rep. **10**(1), 1–9 (2020)

14. Zhou, Y., et al.: RadFusion: benchmarking performance and fairness for multimodal pulmonary embolism detection from CT and EHR (2021). https://arxiv.org/abs/2111.11665

15. Zhang, Y., Jiang, H., Miura, Y., Manning, C.D., Langlotz, C.P.: Contrastive learning of medical visual representations from paired images and text. In: Machine Learning for Healthcare Conference, pp. 2–25. PMLR (2022)

16. Zhang, S., et al.: Large-scale domain-specific pretraining for biomedical vision-language processing. arXiv preprint arXiv:2303.00915 (2023)

17. Gu, Y., et al.: Domain-specific language model pretraining for biomedical natural language processing. ACM Trans. Comput. Healthc. (HEALTH) **3**(1), 1–23 (2021)

18. Khosla, P., et al.: Supervised contrastive learning. Adv. Neural. Inf. Process. Syst. **33**, 18661–18673 (2020)

19. Carreira, J., Noland, E., Banki-Horvath, A., Hillier, C., Zisserman, A.: A short note about kinetics-600 (2018). arXiv:1808.01340

20. Albrecht, M.H., et al.: State-of-the-art pulmonary CT angiography for acute pulmonary embolism. Am. J. Roentgenol. **208**(3), 495–504 (2016). https://doi.org/10.2214/AJR.16.17202

21. Oord, A.V.D., Li, Y., Vinyals, O.: Representation learning with contrastive predictive coding. arXiv preprint arXiv:1807.03748 (2018)

22. Selvaraju, R.R., Cogswell, M., Das, A., Vedantam, R., Parikh, D., Batra, D.: Grad-CAM: visual explanations from deep networks via gradient-based localization. In: Proceedings of the IEEE International Conference on Computer Vision, pp. 618–626 (2017)

23. Zbontar, J., Jing, L., Misra, I., LeCun, Y., Deny, S.: Barlow Twins: self-supervised learning via redundancy reduction (2021). https://arxiv.org/abs/2103.03230

Exploring the Transfer Learning Capabilities of CLIP in Domain Generalization for Diabetic Retinopathy

Sanoojan Baliah[✉][iD], Fadillah A. Maani[iD], Santosh Sanjeev[iD],
and Muhammad Haris Khan[iD]

Mohamed bin Zayed University of Artificial Intelligence, Abu Dhabi, UAE
{sanoojan.baliah,fadillah.maani,santosh.sanjeev,
muhammad.haris}@mbzuai.ac.ae

Abstract. Diabetic Retinopathy (DR), a leading cause of vision impairment, requires early detection and treatment. Developing robust AI models for DR classification holds substantial potential, but a key challenge is ensuring their generalization in unfamiliar domains with varying data distributions. To address this, our paper investigates cross-domain generalization, also known as domain generalization (DG), within the context of DR classification. DG, a challenging problem in the medical domain, is complicated by the difficulty of gathering labeled data across different domains, such as patient demographics and disease stages. Some recent studies have shown the effectiveness of using CLIP to handle the DG problem in natural images. In this study, we investigate CLIP's transfer learning capabilities and its potential for cross-domain generalization in diabetic retinopathy (DR) classification. We carry out comprehensive experiments to assess the efficacy and potential of CLIP in addressing DG for DR classification. Further, we introduce a multi-modal fine-tuning strategy named Context Optimization with Learnable Visual Tokens (CoOpLVT), which enhances context optimization by conditioning on visual features. Our findings demonstrate that the proposed method increases the F1-score by 1.8% over the baseline, thus underlining its promise for effective DG in DR classification. Our code is publicly available at https://github.com/Sanoojan/CLIP-DRDG.

1 Introduction

Deep learning (DL) has become the standard for Computer Vision tasks, achieving state-of-the-art performance. However, DL models often suffer significant performance degradation when training and testing distributions differ [4,5], including in medical imaging, where diverse patient characteristics and scanners present challenges in deploying DL solutions globally. Additionally, data safety

Supplementary Information The online version contains supplementary material available at https://doi.org/10.1007/978-3-031-45673-2_44.

and privacy concerns limit access to data in medical centers. In medical imaging, it is intuitive to consider different medical centers as different domains. Thus, the research on domain generalization (DG) [13,20,28,42] can be a solution to widely mitigate major health problems.

Diabetic Retinopathy (DR) is a major global health concern, leading to blindness as a complication of diabetes [19]. Relying solely on clinicians for early prevention is impractical due to the complexity and variability in diagnosing DR. To address this, researchers have attempted to develop automatic systems for accurate DR diagnosis from fundus images [10,15,33]. Typically, DR stage diagnosis involves identifying specific lesions, which requires expert annotation for each image. To tackle this issue, several works have focused on deep learning-based DR classification [2,6,37], categorizing DR into five distinct groups [19]: no DR, mild, moderate, severe, and proliferative.

ImageNet pre-trained models benefit DR classification [2] indicating that models trained on natural images can still benefit the downstream DR classification task although the domain is very different. The CLIP model, trained on 400 million image-text pairs using contrastive learning, exhibits excellent multi-modality and zero-shot performance on diverse natural image datasets [30,36]. CLIP's success extends to tasks beyond classification, including object detection and video classification, owing to its joint vision and text contrastive learning approach. CLIP's success in zero-shot learning and transfer learning for natural tasks has inspired researchers to explore its potential in medical domains [12,26,35]. However, little to no work has been done on evaluating CLIP's performance for DR classification due to the lack of meaningful medical reports for each image sample. This study aims to investigate the potential of adopting CLIP methodology for DR in the domain generalization experiment setting.

Motivated by CLIP's generalizable capability in natural images, this work investigates CLIP performance for DR grading in the DG experiment setting by adopting CLIP with various strategies. Our key contributions are, **1.** We conduct an extensive analysis of CLIP multi-modalities for DR grading in the DG setting with various strategies for adapting CLIP architecture. **2.** We investigate various text modalities in the CLIP architecture. We show that combining the CLIP visual encoder and BioBERT [23] text encoder pre-trained on biomedical domain corpora results in a better performance. **3.** We propose **Context Optimization with Learnable Visual Tokens (CoOpLVT)**, a CLIP-based architecture that leverages image-conditioned prompt tuning with visual backbone fine-tuning. We also demonstrate how our proposed method performs better in terms of F1 score.

2   Related Work

Domain Generalization (DG): The earliest study on DG, known as empirical risk minimization (ERM) [34], aimed to minimize errors across source domains. Variants like multi-task autoencoders [13], maximum mean discrepancy (MMD) constraints [28], and contrastive learning approaches [11,21,27] have been proposed. DomainBed [14] introduced a fair evaluation protocol for DG, showing competitive performance for ERM. SWAD [8] proposed stochastic weight averaging to achieve flatter minima for DG.

Domain Generalization in Medical Imaging: In medical imaging analysis, diverse data distributions from different sources often lead to reduced model performance in new environments. However, few studies address the DG problem in medical imaging. One approach, presented in [24], utilized task-specific augmentations to enhance data diversity and employed episodic learning. Another work from [25] captured the shareable information by leveraging variational encoding to learn a representative feature space through linear-dependency regularization. Recently, for diabetic retinopathy (DR) classification, the DRGen method [3] combined the SWAD approach [8] for achieving flatter minima and the Fishr technique [32] for regularization to match gradient variations across domains.

Large Scale Models in Medical Domain: BioBERT [23] and ClinicalBERT [16] are two examples of domain-specific language models that have been pretrained on large-scale biomedical and clinical text datasets, respectively. The pre-training process involves fine-tuning a preexisting BERT model on a large biomedical or clinical text corpus, allowing the model to learn domain-specific features and terminology.

CLIP and its Adoption in DG: CLIP [30] has achieved significant success in image-text pre-training and supports various downstream tasks in computer vision and natural language processing. It has been applied to DG problems in image classification, demonstrated in [7,38], and [29]. In the medical domain, CLIP has been explored for segmentation [26] and Medical Visual Question Answering (MedVQA) [12]. However, adapting vision-text pre-training to the medical domain is challenging due to limited datasets and subtle differences within medical domains. [17,39], and [35] address these challenges with contrastive learning strategies, but there are no works that fully explore CLIP's domain generalizability in the medical domain.

Prompt Engineering: Recent works like [40,41] propose methods for generating effective prompts to enhance vision-language models, demonstrating improved performance on benchmark datasets. [40] introduces dynamic prompts that adapt to each instance, making them less sensitive to class shift and achieving better results in visual question-answering and image captioning tasks compared to fixed and learned prompts.

3   Methodology

The DG problem requires learning from multiple source domains or seldom a single source, and evaluating on an unseen target domain. Since target and source data distribution are different, the goal is to learn cross-domain generalizable features. Specifically, we characterize each domain d by $\mathcal{D}^d = \{(\mathbf{x}_i^d, y_i^d)\}_{d=1}^n$ that consists of samples drawn from i.i.d. with a probability $\mathcal{P}(\mathcal{X}^d, \mathcal{Y}^d)$, where $\mathbf{x} \in \mathcal{X}$ is an image with $C \times H \times W$ shape and $y \in \mathcal{Y} = \{1, 2, .., K\}$ is an associated class label. In multi-source domain generalization, a model is trained using data from multiple domains, i.e., the number of training domains $(d_{tr}) > 1$. We sample b

instances from every training domain for each step, making the total batch size of $B = b \times d_{tr}$ [14]. We target the task of classification, and the model can be represented as $\mathcal{F} = w \circ f_v$, which maps the input images to the target label space, where $f_v : \mathbf{x} \to h$ is the visual encoder and $w : h \to y$ is the classifier. Then, we evaluate the model generalization on data sampled from $\mathcal{P}(\mathcal{X}^{d_{te}}, \mathcal{Y}^{d_{te}})$, where d_{te} represents the target domain.

3.1   CLIP Adoption Techniques

Empirical Risk Minimization (ERM): ERM [34] uses the Cross Entropy (CE) loss over all samples in mini-batch to optimize a predictor. Under the DG settings, ERM simply aggregates data from all source domains and searches for the optimal predictor $\mathcal{F}$ by minimizing the empirical risk $\frac{1}{B}\sum_{i=1}^{B}\mathcal{L}_{\text{CE}}(\mathcal{F}(\mathbf{x}_i), y_i)$. A recent work [14] employed ERM for DG problem and showed competitive performance in DG datasets under a fair evaluation. Owing to the increasing adoption of Vision Transformers (ViT) for various Computer Vision tasks, there is a growing interest in benchmarking their DG performance. This naturally suggests the usage of ViT as the feature encoder f in the ERM pipeline, namely ERM-ViT.

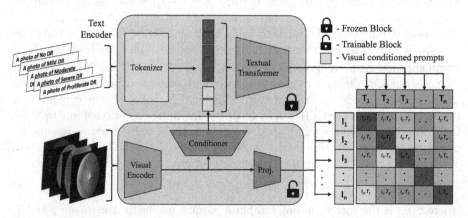

Fig. 1. Overall architecture of our proposed Context Optimization with Learnable Visual Tokens (**CoOpLVT**).

Linear Probing: Given the prominent generalizability of CLIP pre-trained models on fine-tuning [22] and zero-shot [36] settings, a rather simple alternative of adopting CLIP to medical imaging can be performed by retraining the classifier w while keeping the CLIP backbone f_v frozen.

Naive Multi-modal Fine-Tuning: We train the CLIP's visual encoder on the similarities between visual features V and text features T by applying cross-entropy loss, i.e. $-\frac{1}{B}\sum_{j=1}^{B} y_j log(\sigma(\langle V, T \rangle))$ where $\langle , \rangle$ symbolizes dot product. Note that it is a simple fine-tuning strategy under the zero-shot setting. More specifically, given K number of classes, we generate K text prompts (denoted as S) based on the class names following the CLIP's text template, e.g. "a photo of a {class name}". These text prompts are then encoded into textual features $T = f_t(S) \in \mathbb{R}^{S \times C_f}$, where f_t is the text encoder and C_f is the output feature dimension. For each optimization step, we sample B number of images (minibatch size) and obtain B visual features using CLIP visual encoder as $V = f_v(\boldsymbol{x}) \in \mathbb{R}^{B \times C_f}$. For architecture diagrams of ERM-ViT, Linear Probing and Naive CLIP refer to the supplementary material.

Further, we choose to only optimize CLIP's visual encoder while keeping the text encoder fixed. Training the full network might be crucial as medical domain labels are quite different from the natural image cases, however, this creates a challenge of high GPU memory consumption. To this end, we propose our fine-tuning strategy next, which considers both modalities while training, unlike fixed textual feature representation in the above approaches.

3.2 Context Optimization with Learnable Visual Tokens

Apart from the analysis of the CLIP with general fine-tuning approaches for DR, inspired from CoCoOp [40], we introduce a multi-modal fine-tuning strategy, namely **CoOpLVT**, which trains the visual model and a conditioner to interact with textual transformer. To formally introduce the approach, let us decompose the visual feature V as $V = f_v(\boldsymbol{x}) = p \circ f_I(\boldsymbol{x})$, where p is a linear projector often used in multi-modal models to bring both modalities' features to the same dimension, and f_I is the immediate higher dimension (d^I) feature from the CLIP visual encoder. Also, we denote the conditioner $\mathcal{G} : \mathbb{R}^{d^I} \rightarrow \mathbb{R}^{d'}$ an MLP network.

Since our motivation is to incorporate context-aware text prompts to interact on the textual encoder, we concatenate additional N_p tokens to the original text tokens created from the CLIP's text generation template mentioned in the Naive CLIP approach. As shown in Fig. 1, to generate these context-aware tokens, we introduce the learnable $\mathcal{G}$, where $d' = N_p \times d^T$. Here, d^T is the feature dimension of textual tokens. Thus, the text feature for an image j can be generated as,

$$T^j = f_t([S[y_j], \mathcal{G}(f_I(x^j))]) \tag{1}$$

where $S[y_j]$ is the corresponding template prompt tokens for the image j and f_t is a transformer-based text encoder that enables us to concatenate any number of context-aware tokens which can interact with the frozen tokens in self-attention. Unlike CoCoOp, we train the visual encoder as the CLIP has not seen any DR data. With the use of context-aware text tokens, we create different text features for all the images in the batch, i.e. all the images have different visual and textual feature pairs. This further allows us to formulate the loss as in the original CLIP contrastive loss [31], even on the downstream classification task with limited class settings. Thus, our proposed loss can be formulated as,

$$\mathcal{L}_{\text{contrastive}}(V, T) = \underbrace{-\frac{1}{B} \sum_{j=1}^{B} j \log \left(\sigma \left\langle V^j, T \right\rangle \right)}_{\mathcal{L}_{\text{text}}} \underbrace{-\frac{1}{B} \sum_{j=1}^{B} j \log \left(\sigma \left\langle V, T^j \right\rangle \right)}_{\mathcal{L}_{\text{visual}}} \tag{2}$$

where $\mathcal{L}_{visual}$, and $\mathcal{L}_{text}$ are the logits when we consider the similarity matrix in visual and text direction respectively.

4 Experiments

Table 1. Multi-Source DG Results(F1-score). Bold numbers mean best and underlined are the second best, and () denotes the standard deviation. CLIP-V, CLIP-VB, B(p), and B(s) denote CLIP visual-encoder, CLIP visual-encoder with BioBert text-encoder, BioBert pre-trained on PubMed, and BioBert pre-trained on SQuAD, respectively.

Strategy	Algorithm	F1-score				
		APTOS	EyePACS	Messidor	Messidor2	Avg
ERM	ResNet50	28.6 (0.8)	29.3 (0.4)	45.8 (0.9)	51.3 (0.7)	38.8 (0.5)
	ViT	24.0 (1.6)	30.9 (1.0)	46.6 (0.3)	53.4 0.6)	38.7 (0.3)
Zero-shot	CLIP	3.4 (0.1)	4.4 (0.0)	4.0 (0.1)	2.2 (0.1)	3.5 (0.0)
	CLIP-VB(p)	14.6 (0.0)	18.4 (0.0)	15.9 (0.1)	12.7 (0.1)	15.4 (0.0)
	CLIP-VB(s)	11.6 (0.1)	17.9 (0.0)	12.7 (0.1)	14.7 (0.1)	14.2 (0.0)
Linear probing	CLIP-V	13.0 (0.2)	14.0 (2.5)	12.4 (0.0)	14.2 (0.6)	13.4 (0.5)
Naive CLIP	CLIP	26.0 (1.0)	30.7 (1.0)	**47.7 (0.4)**	53.2 (0.5)	39.4 (0,7)
	CLIP-VB(p)	26.5 (0.6)	31.6 (0.7)	46.5 (0.5)	**53.5 (0.3)**	39.5 (0.3)
CoOpLVT (ours)	CLIP	**31.9 (2.6)**	**32.2 (0.3)**	46.2 (0.7)	51.6 (0.3)	**40.5 (0.5)**

Datasets: Following the work of [3], we validate the effectiveness of our approach on four datasets, EyePACs [18], APTOS [1], Messidor and Messidor-2 [9] which serve as 4 domains in the DG setting. EyePACs is the largest dataset with 88702 fundus images, while APTOS, Messidor, and Messidor-2 have 3657, 1200, and 1744 fundus images, respectively which are classified into 5 classes: no DR, mild, moderate, severe, and proliferative DR [19]. However, Messidor does not contain any proliferative samples. Following DRGen [3], we use images of size 224 × 224, with the random flip, random grayscaling, color jittering, random rotation, translation, and Gaussian blur augmentations.

Implementation and Training/Testing Details: For a fair comparison, we follow the default domainbed [14] settings throughout all the experiments. Specifically, we experiment with the 80:20 training, validation splits. We use the batch size (b) of 32 per domain making the total batch size 96, the learning rate of $5e-05$ for Imagenet trained models and the $5e-06$ for CLIP trained models, and no weight decay. We use the AdamW optimizer with the default Pytorch params. We train and validate the models on the source domains while the unseen target domain is used for testing only. We report accuracy by repeating the same experiment 3 times with different trial seeds (on two V100 GPUs, implemented in Pytorch). We strictly follow, not to access the test domain as a validation set (i.e. non-oracle model selection).

Metrics: Two widely-used image classification metrics are used to evaluate our experiments: accuracy and F1-score. We include accuracy as it has been used in

Table 2. Single Source domain generalization results (f1 score) .

Algorithm	Train Dataset	APTOS	EyePACS	Messidor	Messidor2	Avg
Naive CLIP	APTOS	–	26.2	13.4	22.8	20.7
CoOpLVT (Ours)	APTOS	–	**28.1**	**14.5**	**25.5**	**22.7**
Naive CLIP	EyePACS	**42.9**	–	30.1	44.9	39.3
CoOpLVT (Ours)	EyePACS	40.7	–	**34.9**	**48.4**	**41.3**
Naive CLIP	Messidor	20.7	20.3	-	37.3	26.1
CoOpLVT (Ours)	Messidor	**23.6**	**22.4**	–	**38.9**	**28.3**
Naive CLIP	Messidor2	33.9	29.3	**47.9**	–	37.0
CoOpLVT (Ours)	Messidor2	**38.4**	**29.7**	47.1	–	**38.4**

previous research on DG for DR (See supplementary), allowing us to benchmark our experiments with other works. However, given the highly imbalanced class distribution in DR, it is intuitive to choose the F1-score as the primary metric since it considers both precision and recall. Thus, in this work, we aim to explore and improve CLIP performance on DR with DG setting based on the F1-score.

Experiments. We conduct experiments to explore various strategies for adopting the CLIP model for tackling the DR problem in the DG setting. (1) We start with utilizing a pre-trained CLIP visual encoder as a DR classification model trained with the ERM algorithm. (2) Inspired by the decent performance of CLIP zero-shot in natural images, we inspect CLIP zero-shot performance on DR with two sets of prompts and several text encoders. (3) Linear probing is also investigated with CLIP visual encoder, as it is capable of extracting distinguishing features for general classification problems. (4) We experiment with naive multi-modal fine-tuning. (5) We evaluate our proposed **CoOpLVT** through various ablation studies. Table 1 provides the results of the main experiments, and we further discuss the experiments in Sect. 5.

5   Results and Discussion

Initially, we compare the effectiveness of CLIP pre-trained models with the Imagenet pre-trained models and also fine-tune the full network using the ERM-ViT approach (see Table 1). We explore the zero-shot capabilities of CLIP and also conduct other experiments changing the text-encoders. The zero-shot performance of CLIP is very poor as CLIP was trained only on the natural domain and has not seen much of medical-related data. We replace the CLIP's text encoder with BioBERT pre-trained on PubMed and SQuAD. We observe a boost in the performance of the models in both accuracy and F1-score. This is due to the fact that BioBERT was trained on large scale medical data and has a very good zero-shot performance on medical-related tasks.

Multi Source Results: Linear probing for CLIP gives 56.4% average accuracy and 13.4% average F1-score, suggesting the CLIP pre-trained weights, without

Table 3. Ablation study on zero-shot CLIP performance. **Prompt I**: *"a photo of a {c}"* where c belongs to [*No DR, mild DR, moderate DR, severe DR, proliferative DR*]. **Prompt II**: similar to Prompt I, with *"a photo of a {c}"* where c belongs to [*No Diabetic Retinopathy, mild Diabetic Retinopathy, moderate Diabetic Retinopathy, severe Diabetic Retinopathy, proliferative Diabetic Retinopathy*].

Model	CLIP		CLIP-VB(p)		CLIP-VB(s)	
Prompt	I	II	I	II	I	II
F1-Score	1.2 (0.0)	3.5 (0.0)	**15.4 (0.0)**	8.1 (0.0)	14.2 (0.0)	7.9 (0.1)
Acc	3.1 (0.1)	4.9 (0.1)	44.3 (0.1)	16.4 (0.0)	**47.9 (0.1)**	19.6 (0.1)

fine-tuning, cannot be applied to DR as the CLIP is trained mostly on natural images. Further, when train with the Naive-CLIP approach by replacing the text-encoder with BioBERT [23] pre-trained on PubMed we obtain 39.4% and 39.5% average F1-score respectively. In Naive-CLIP approach the prompts are fixed and it is a very challenging task to design the prompts as they are task-specific and require domain expertise. To overcome this problem, we design CoOpLVT which generates the best performance with an average F1-score of 40.5%.

We also conduct ablation studies to understand the effect of using different prompt designs (see Table 3). From the results we observe that slight changes in the prompts can have a huge effect on the F1-score and accuracy. We further conduct different experiments by freezing and fine-tuning the visual encoder, varying the number of MLP layers as well as the number of induced visual tokens (can be seen in supplementary material). We observed that increasing the number of MLP layers helps in improving the performance whereas increasing the number of induced tokens causes a drop in the performance.

Single Source DG Results: We show the comparison of single source DG results (see Table 2) between Naive CLIP and CoOpLVT (Top two performing strategies in multi-source DG). CoOpLVT consistently provides a better F1 score than Naive CLIP. **Attention maps:** Further, the analysis of attention maps (see supplementary) reveals that the CLIP model can find DR lesions in the eye in most cases.

6   Conclusion

In this work, we explore the effectiveness of the CLIP pre-trained model and its generalizability through various sets of experiments. To our knowledge, we are the first to investigate the DG performance of the CLIP model in medical imaging, especially in DR. We investigate and analyze the performance of zero-shot as well as fine-tuned settings. We investigate the effectiveness of the CLIP pre-trained model and its generalizability through various transfer learning techniques. In addition, we propose a multi-modal fine-tuning strategy named CoOpLVT to suit DR data. With extensive experiments, we showed the capabilities of CLIP for domain generalization in DR and demonstrated that our proposed approach results in a better F1-score by 1.8% compared to the baseline performance.

References

1. APTOS: APTOS 2019 Blindness Detection. https://www.kaggle.com/competitions/aptos2019-blindness-detection/data (2019)
2. Asiri, N., Hussain, M., Al Adel, F., Alzaidi, N.: Deep learning based computer-aided diagnosis systems for diabetic retinopathy: a survey. Artif. Intell. Med. 99 (2019). https://doi.org/10.1016/j.artmed.2019.07.009
3. Atwany, M., Yaqub, M.: DRGen: domain generalization in diabetic retinopathy classification. In: MICCAI 2022: Proceedings, Part II. pp. 635–644. Springer, Cham (2022). https://doi.org/10.1007/978-3-031-16434-7_61
4. Ben-David, S., Blitzer, J., Crammer, K., Kulesza, A., Pereira, F.C., Vaughan, J.W.: A theory of learning from different domains. Mach. Learn. **79**, 151–175 (2010)
5. Ben-David, S., Blitzer, J., Crammer, K., Pereira, F.: Analysis of representations for domain adaptation. In: Advances in Neural Information Processing Systems 19 (2006)
6. Bodapati, J.D., Shaik, N.S., Naralasetti, V.: Composite deep neural network with gated-attention mechanism for diabetic retinopathy severity classification. J. Ambient. Intell. Humaniz. Comput. **12**(10), 9825–9839 (2021)
7. Bose, S., Fini, E., Jha, A., Singha, M., Banerjee, B., Ricci, E.: StyLIP: multi-scale style-conditioned prompt learning for clip-based domain generalization (2023)
8. Cha, J., et al.: SWAD: domain generalization by seeking flat minima. In: NeurIPS 34 (2021)
9. Decencière, E., et al.: Feedback on a publicly distributed image database: the Messidor database. Image Anal. Stereol. **33**(3), 231–234 (2014). https://doi.org/10.5566/ias.1155
10. Dosovitskiy, A., et al.: An image is worth 16×16 words: transformers for image recognition at scale. arXiv preprint arXiv:2010.11929 (2020)
11. Dou, Q., de Castro, D.C., Kamnitsas, K., Glocker, B.: Domain generalization via model-agnostic learning of semantic features. In: NeurIPS, pp. 6450–6461 (2019)
12. Eslami, S., de Melo, G., Meinel, C.: Does clip benefit visual question answering in the medical domain as much as it does in the general domain? (2021)
13. Ghifary, M., Bastiaan Kleijn, W., Zhang, M., Balduzzi, D.: Domain generalization for object recognition with multi-task autoencoders. In: ICCV (2015)
14. Gulrajani, I., Lopez-Paz, D.: In search of lost domain generalization. ArXiv:2007.01434 (2021)
15. He, K., Zhang, X., Ren, S., Sun, J.: Deep residual learning for image recognition. In: CVPR, pp. 770–778 (2016)
16. Huang, K., Altosaar, J., Ranganath, R.: ClinicalBERT: modeling clinical notes and predicting hospital readmission (2020)
17. Huang, S.C., Shen, L., Lungren, M.P., Yeung, S.: Gloria: a multimodal global-local representation learning framework for label-efficient medical image recognition. In: ICCV, pp. 3942–3951 (2021)
18. Kaggle: diabetic retinopathy detection. https://www.kaggle.com/c/diabetic-retinopathy-detection. Accessed 28 Jan 2023
19. Kempen, J.H., et al.: The prevalence of diabetic retinopathy among adults in the united states. Archives of Ophthalmology (Chicago, Ill.: 1960) (2004)

20. Khan, M.H., Zaidi, T., Khan, S., Khan, F.S.: Mode-guided feature augmentation for domain generalization. In: Proceedings of British Machine Vision Conference (2021)
21. Kim, D., Yoo, Y., Park, S., Kim, J., Lee, J.: SelfReg: self-supervised contrastive regularization for domain generalization. In: ICCV, pp. 9619–9628 (2021)
22. Kumar, A., Raghunathan, A., Jones, R.M., Ma, T., Liang, P.: Fine-tuning can distort pretrained features and underperform out-of-distribution. In: ICLR (2022)
23. Lee, J., et al.: BioBERT: a pre-trained biomedical language representation model for biomedical text mining. Bioinformatics **36**(4), 1234–1240 (2019)
24. Li, C., et al.: Domain generalization on medical imaging classification using episodic training with task augmentation. Comput. Biol. Med. **141**, 105144 (2022)
25. Li, H., Wang, Y., Wan, R., Wang, S., Li, T.Q., Kot, A.: Domain generalization for medical imaging classification with linear-dependency regularization. In: NeurIPS (2020)
26. Liu, J., et al.: Clip-driven universal model for organ segmentation and tumor detection (2023)
27. Motiian, S., Piccirilli, M., Adjeroh, D.A., Doretto, G.: Unified deep supervised domain adaptation and generalization. In: ICCV, pp. 5715–5725 (2017)
28. Muandet, K., Balduzzi, D., Schölkopf, B.: Domain generalization via invariant feature representation. In: ICML (2013)
29. Niu, H., Li, H., Zhao, F., Li, B.: Domain-unified prompt representations for source-free domain generalization (2023)
30. Radford, A., et al.: Learning transferable visual models from natural language supervision. In: ICML, pp. 8748–8763. PMLR (2021)
31. Radford, A., et al.: Language models are unsupervised multitask learners. OpenAI blog **1**(8), 9 (2019)
32. Rame, A., Dancette, C., Cord, M.: Fishr: Invariant gradient variances for out-of-distribution generalization. In: ICML. PMLR (2022)
33. Touvron, H., Cord, M., Douze, M., Massa, F., Sablayrolles, A., Jégou, H.: Training data-efficient image transformers & distillation through attention. In: ICML (2021)
34. Vapnik, V.: The Nature of Statistical Learning Theory. Springer science & business media (1999). https://doi.org/10.1007/978-1-4757-3264-1
35. Wang, Z., Wu, Z., Agarwal, D., Sun, J.: MedCLIP: contrastive learning from unpaired medical images and text (2022)
36. Wortsman, M., et al.: Robust fine-tuning of zero-shot models. CoRR abs/2109.01903 (2021). https://arxiv.org/abs/2109.01903
37. Wu, Z., et al.: Coarse-to-fine classification for diabetic retinopathy grading using convolutional neural network. In: Artificial Intelligence in Medicine 108 (2020)
38. Zhang, X., Gu, S.S., Matsuo, Y., Iwasawa, Y.: Domain prompt learning for efficiently adapting clip to unseen domains (2022)
39. Zhang, Y., Jiang, H., Miura, Y., Manning, C.D., Langlotz, C.P.: Contrastive learning of medical visual representations from paired images and text (2022)
40. Zhou, K., Yang, J., Loy, C.C., Liu, Z.: Conditional prompt learning for vision-language models. In: CVPR, pp. 16816–16825 (2022)
41. Zhou, K., Yang, J,, Loy, C.C., Liu, Z.: Learning to prompt for vision-language models. Int. J. Comput. Vis. **130**(9), 2337–2348 (2022)
42. Zhou, K., Yang, Y., Hospedales, T., Xiang, T.: Learning to generate novel domains for domain generalization. In: Vedaldi, A., Bischof, H., Brox, T., Frahm, J.-M. (eds.) ECCV 2020. LNCS, vol. 12361, pp. 561–578. Springer, Cham (2020). https://doi.org/10.1007/978-3-030-58517-4_33

More from Less: Self-supervised Knowledge Distillation for Routine Histopathology Data

Lucas Farndale[1,2,3,4](✉) ⓘ, Robert Insall[1,2,5] ⓘ, and Ke Yuan[1,2,3] ⓘ

[1] School of Cancer Sciences, University of Glasgow, Glasgow, Scotland, UK
{lucas.farndale,robert.insall,ke.yuan}@glasgow.ac.uk
[2] Cancer Research UK Beatson Institute, Glasgow, Scotland, UK
[3] School of Computing Science, University of Glasgow, Glasgow, Scotland, UK
[4] School of Mathematics and Statistics, University of Glasgow, Glasgow, Scotland, UK
[5] Division of Biosciences, University College London, London, UK

Abstract. Medical imaging technologies are generating increasingly large amounts of high-quality, information-dense data. Despite the progress, practical use of advanced imaging technologies for research and diagnosis remains limited by cost and availability, so more information-sparse data such as H&E stains are relied on in practice. The study of diseased tissue would greatly benefit from methods which can leverage these information-dense data to extract more value from routine, information-sparse data. Using self-supervised learning (SSL), we demonstrate that it is possible to distil knowledge during training from information-dense data into models which only require information-sparse data for inference. This improves downstream classification accuracy on information-sparse data, making it comparable with the fully-supervised baseline. We find substantial effects on the learned representations, and pairing with relevant data can be used to extract desirable features without the arduous process of manual labelling. This approach enables the design of models which require only routine images, but contain insights from state-of-the-art data, allowing better use of the available resources.

Keywords: Representation Learning · Colon Cancer · Multi-Modality

1 Introduction

The complexity and amount of information generated by medical imaging technologies is constantly increasing. Developments such as spatial -omics, multiplex immunohistochemistry and super-resolution microscopy are continuously

---

Supplementary Information The online version contains supplementary material available at https://doi.org/10.1007/978-3-031-45673-2_45.

enabling greater insights into mechanisms of disease, but adoption of such technologies is prohibitively expensive for large cohorts or routine use. It is therefore highly desirable to develop methods which distil knowledge from these exceptionally dense and information-rich data into more accessible and affordable routine imaging models, so clinicians and researchers can obtain the most diagnostic information possible from the data available to them.

Typically, knowledge distillation focuses on distilling from a (possibly pretrained) larger teacher model into a smaller student model [7]. This is usually achieved using a self-supervised joint-embedding architecture, where two models are trained as parallel branches to output the same representations [7], so the smaller model can be more easily deployed in practice on the same dataset without sacrificing accuracy. This approach is ideal for digital pathology in which complete images are impractically large and often initially viewed at low magnification. Both knowledge distillation [11] and SSL [16] have been shown to improve performance on histopathology imaging tasks, including when used in tandem [4].

Existing approaches usually require that the data sources used in training are both available during inference, which is severely limiting where at least one source of data is not available, is more expensive, or is more computationally demanding. There are a limited number of SSL architectures which can train two encoders with different inputs concurrently, such as CLIP [17], VSE++ [6], Barlow Twins [20], SimCLR [3] and VICReg [1], which has been shown to outperform VSE++ and Barlow Twins on multi-modal data [1].

In this work, we focus on knowledge distillation from models of *information-dense* datasets (datasets rich in accessible information, e.g. high-resolution pathology images) into a models of *information-sparse* datasets (datasets containing little or obfuscated information, e.g. low-resolution images). We make the following contributions:

- We find that knowledge distillation significantly improves downstream classification accuracy on information-sparse data, comparable to a supervised baseline;
- We show that this training process results in measurably different representations compared to standard self-supervised and supervised training, and that the paired data significantly changes the areas of the image focused on by the model;
- We show the clinical utility of this method, by presenting a use-case where pan-cytokeratin (pan-CK) stained immunofluorescence (IF) imaging is used to train a better performing model for brightfield hematoxylin and eosin (H&E) stains.

2   Methods

2.1   Experimental Design

We use the self-supervised methods VICReg [1] and SimCLR [3] to distil knowledge from information-dense data to information-sparse data. As shown in Fig. 1,

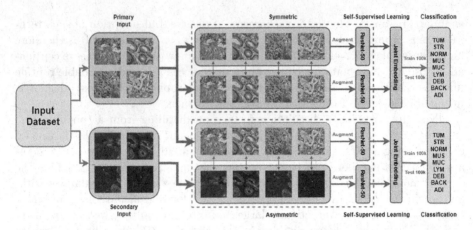

Fig. 1. Knowledge distillation between information-dense and information-sparse data. The upper branch shows a symmetric model, while the lower branch shows an asymmetric model. Images are preprocessed (augmentations detailed in Table S1), paired and passed to the self-suprevised model for training. The trained encoders are then used for downstream classification tasks to assess performance.

the model takes pairs of images as input, one for each branch. We refer to a pair consisting of two copies of the same image as a *symmetric* pair, and a pair consisting of distinct images as an *asymmetric* pair. For example, the primary input in an asymmetric pair might always be a lower resolution than the secondary input, while in a symmetric pair, both inputs would be the same resolution.

H&E/IF Distillation. We first demonstrate the efficacy of asymmetric training with a clinically-relevant use-case, with models trained on the SHIFT dataset [2], which contains 256×256 px patches from H&E stains, which are washed, restained with pan-CK and re-imaged. Pan-CK stains tumour cells, making it easier to differentiate tissue types. The trained models are evaluated on the standard tissue classification task using the NCT dataset [12], with label distributions shown in Table S2.

Contextual Distillation. Patches are usually created of an arbitrary, small size, making it difficult to detect patterns that may be obvious in larger patches. We show the effects of pairing inputs which can utilise surrounding contextual information, such as the area surrounding the border of a patch. We use the NCT dataset [12] to create two examples for original patches paired with: i) patches centre-cropped to 112×112 px and zero-padded back to 224×224 px, ii) patches downsampled to 7×7 px and resized to 224×224 px. Padding/resizing is to ensure consistency between encoders in asymmetric/symmetric models.

Table 1. Classification results from linear probing on the NCT dataset evaluation set [12] for models trained on the SHIFT dataset [2].

Asymmetric	Shared Weights	VICReg Accuracy	SimCLR Accuracy
✓		**0.8760**	0.8290
✓	✓	0.7934	0.8229
		0.8184	0.8417
	✓	0.8452	0.8375

Nuclear Segmentation Distillation. To enable finer-grained analysis with a toy example, we synthetically construct an information-sparse dataset by using HoVer-Net [9] to create a nuclear segmentation mask for each image, coloured by their predicted cell types (Background, Neoplastic, Inflammatory, Connective, Dead, Non-Neoplastic Epithelial). We repeat these 1D masks 3 times along channel 3 to ensure consistency of input size between models. Synthetic masks were used due to the limited number of manually annotated masks available, as SSL typically requires large datasets, as demonstrated in Table S3, and due to the lack of datasets with both nuclear segmentations and tissue type labels. These images contain significantly less relevant information for tissue type classification, as the majority of fine-grained morphological information in the image is lost, and the HoVer-Net segmentations are imperfect. The masks do not introduce additional information, as by definition all information which is present in these nuclear masks is present in the original images. In Table S6 we show that the method is robust to the use of manually annotated or synthetic data for evaluation.

We evaluate models' performance on the same tissue classification task as above for both the H&E patches and the masks, and we also evaluate performance on predicting the most common cell type in each image, to illustrate how learned representations transfer between tasks. For comparison, we include supervised models trained on tissue or cell classification before freezing weights, re-training the classifier head and evaluating on the other task. For both symmetric and asymmetric models, we present results where the same encoder is applied to both branches ('shared weights') and where there are distinct encoders used for each branch, with their weights updated independently. Having distinct encoders can marginally hurt performance [1], and this has to be balanced against the benefits of asymmetric learning.

2.2 Training

In our experiments[1], all models used a ResNet-50 encoder [10] with output size 2048, and expanders were composed of three dense layers with ReLU activations

---

[1] Code is available at https://github.com/lucasfarndale/More-From-Less.

Table 2. Classification accuracy for three toy examples for VICReg, SimCLR and supervised baselines. Bold indicates best performance on task for given architecture (VICReg/SimCLR). Tissue/Cell indicates tissue/cell classification task respectively. Tasks use the information-sparse input (Crop/Pad, Downsample, Mask) and model for inference except for the H&E results, which are paired with the masks for asymmetric architectures.

Input			Crop/Pad	Downsample	Mask		H&E	
Model	Asymmetric	Shared Weights	Tissue	Tissue	Tissue	Cell	Tissue	Cell
VICReg	✓		0.8297	**0.7743**	**0.5809**	**0.8650**	**0.8979**	**0.8127**
	✓	✓	**0.8334**	0.7730	0.3120	0.6110	0.8725	0.7312
			0.7369	0.7383	0.5249	0.8051	0.8419	0.6682
		✓	0.7845	0.6926	0.5338	0.8158	0.8855	0.6904
SimCLR	✓		0.8416	0.7505	**0.5600**	**0.8540**	0.8850	**0.8038**
	✓	✓	**0.8499**	**0.7780**	0.3582	0.6053	0.8824	0.7619
			0.7902	0.7243	0.5323	0.8151	0.8687	0.6816
		✓	0.8060	0.7241	0.5454	0.8235	**0.9075**	0.6914
Supervised	-	-	0.8966	0.7656	0.5909	0.9365	0.9451	0.8205
Supervised (Transfer)	-	-	-	-	0.5345	0.9103	0.9314	0.7045

and output size 8192. Encoder ablations are detailed in Table S4. Models are trained using Tensorflow 2.9.1 from a random initialisation for 100 epochs, with a batch size of 128. An Adam optimiser [13] with a warm-up cosine learning rate [8] was used, with a maximum learning rate of 10^{-4}, and warm-up period of 1/10th of an epoch. Training one model takes 8 h on an Nvidia A6000 GPU. For consistency with the original implementations, loss function parameters are $\lambda = 25, \mu = 25, \nu = 1$ for VICReg, and temperature $t = 0.5$ for SimCLR. With the encoder's weights frozen, we assess model performance by training a classifier (dense layer with softmax) for 100 epochs using an Adam optimiser and learning rate of 10^{-3}. We produce supervised baseline comparisons using an equivalent encoder and softmax classifier head, following the same training protocol.

3   Results

3.1   H&E-IF Distillation

As shown in Table 1, we observe that, for VICReg, the classification performance with asymmetric training with shared weights is significantly better than with symmetric training, while, for SimCLR, asymmetry results in worse performance. It has been demonstrated that VICReg outperforms other self-supervised methods on tasks with different architectures on each branch, as each branch is regularised independently [1], which is not the case with SimCLR, possibly explaining its lower performance with substantially different inputs to each branch. We observe that sharing weights is restrictive for asymmetric models, leading to lower performance, as the encoder must learn to process both H&E and IF images. Without sharing weights, the model on each branch can specialise on its input modality.

We also demonstrate in Table S5 that the model considerably outperforms models trained by predicting the IF stain directly from the H&E stain using a U-Net architecture [18]. Direct image-image translation retains primarily fine grained features which are not useful for the downstream task, while self-supervision leads to the extraction of more abstract coarse-grained features.

3.2 Contextual Distillation

We observe in Table 2 that asymmetric training leads to considerably more accurate classifications than symmetric training, showing that introducing contextual information into patch models improves performance. In contrast to other examples, there appears to be little negative effect from sharing weights, as the semantic content of the cropped image is entirely shared with the original. Note that asymmetric training significantly outperforms symmetric and even supervised training for the downsampled images. Notably, for all examples in Table 2 asymmetry leads to better performance for SimCLR, in contrast to the results for H&E/IF distillation.

3.3 Nuclear Segmentation Distillation

In all cases, asymmetric models considerably outperform symmetric models at predictions for masks, as being paired with H&E patches makes detection of patterns relating to different tissue types easier. Although the accuracy of the tissue classification task is lower for the masks than for the images, it is considerably better than random (0.1111). Despite comparable performance to symmetric models at tissue classification on H&E patches, asymmetric models perform considerably better at cell classification, as being paired with the nuclear mask forces asymmetric models to extract features relevant to nuclei. It appears that the asymmetry causes models to learn more features irrelevant to tissue classification, but which are essential for cell classification. Similarly, supervised models trained on cell classification from H&E patches transfer well to tissue classification, while the reverse is not true, as we observe a supervision collapse [5].

Figure S1 shows that the model performs fairly accurately on classes which typically contain at least one nucleus, but accurate classifications of most debris, background and adipose are impossible as the masks are simply matrices of zeros.

3.4 Comparing Layer Output Similarities

To further investigate the difference among learned representations, we use GradCAM [19] and centered kernel alignment (CKA) [14] to analyse the outputs of each convolutional layer. The differences between layers are demonstrated by the qualitative GradCAM analysis in Fig. 2. For H&E patches (Fig. 2b), Grad-CAM shows that the nuclei almost always black, meaning the symmetric model

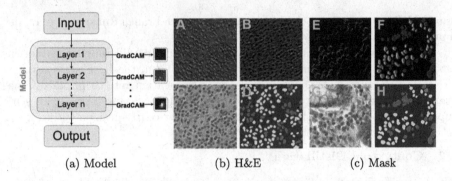

(a) Model   (b) H&E   (c) Mask

Fig. 2. (a) Schematic showing where in the model GradCAM images are obtained from, with samples generated from a (b) H&E patch, and (c) nuclear mask. (A, E) asymmetric GradCAM images, (B ,F) symmetric GradCAM images, (C, G) original tissue patches, (D, H) nuclear segmentation masks.

ignores sub-nuclear features such as nucleoli. This could explain the lower performance at cell classification, as these areas are heavily focused on by the asymmetric model. For the masks (Fig. 2c), symmetrically trained models focus more on coarse-grained morphological features, while asymmetrically trained models focus more on fine-grained features. Figure S2 shows that this difference is seen throughout all layers of the model.

We next use CKA, a measure of the similarity of the representations, to show that asymmetric, symmetric and supervised models obtain considerably different representations. Figures 3 and S3 show that asymmetric and symmetric models are highly dissimilar to each other, with the differences increasing throughout the model, and resulting in substantially different representations. The similarity score between asymmetric/asymmetric pairs is high, while similarity between symmetric/symmetric pairs is high only for SimCLR. This suggests that, with VICReg, there is a larger feature space which can be learned without the restriction imposed by asymmetry, leading to more variability between symmetric models.

4   Discussion and Conclusions

Our results show that asymmetric training can improve the performance of models on downstream tasks by extracting more relevant features from information-sparse inputs. Each branch in a joint-embedding architecture utilises the output of the other branch as a supervisory signal, meaning both branches extract approximately the same features from their input. The asymmetric pairing forces the information-sparse encoder to find patterns in the data which correlate with those found in the information-dense images, while possibly disregarding patterns which are easier to detect. This means that subtler features can be found which are harder to detect in the information-sparse input. In the symmetric

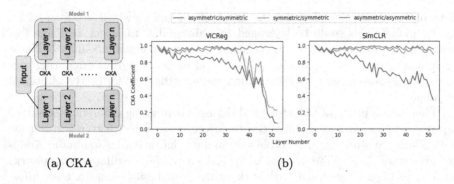

(a) CKA   (b)

Fig. 3. (a) Schematic showing the relationships between layers where centered kernel alignment (CKA) is calculated, (b) CKA analysis of internal representations for H&E/IF distillation, averaged over pairs from 10 asymmetric and 10 symmetric models. Higher values indicate more similarity between layer outputs.

case, the model is less able to focus on these weakly-signalled features, as they may not be as easily distinguished from noise.

Despite the objective forcing embeddings to be as similar as possible, we still observe significant differences in the classification accuracy of each branch, particularly in the nuclear mask example. It has been shown that the use of projection heads improves representation quality [3,15]; we conjecture that the projection head filters out information irrelevant to the opposite branch. This would explain the high classification accuracy observed in the asymmetrically trained H&E encoders for the nuclear mask example, which remains comparable with symmetric original image encoders despite the significantly lower performance of the paired mask encoders.

4.1 Asymmetric Learning Obtains Measurably Different Representations

In asymmetric models, we observe both qualitative and quantitative differences in the representations of both information-dense and information-sparse inputs compared to symmetric models. Intuitively, this is because knowledge distillation is bidirectional, so while knowledge is distilled from the information-dense model into the information-sparse model, there is also knowledge distilled from the information-sparse model to the information-dense model.

Our CKA analysis, shown in Figs. 3 and S3, further quantifies the difference between the layers of symmetric, asymmetric and supervised models. Models with different initialisations become very similar to other models with the same training regime, but significantly different to models with a different training regime, for both information-sparse and -dense data. We conclude that the representations of asymmetrically trained models contain different features from those of symmetrically trained models. This is corroborated by our qualitative GradCAM analysis in Fig. 2, where we demonstrate that the choice of paired data has a significant effect on the areas of the image which are focused on by the model. This could be leveraged to train models to target desirable features such as nucleoli without the arduous process of manual labelling. Utilising asymmetric training approaches, models could be developed for evaluating routine data, such as processing H&E stains, without the need for costly additional staining.

There are a plethora of histological datasets containing different modalities, inputs, or views, and asymmetry can be leveraged to detect features in routine data which are otherwise detectable only in data that is harder to obtain. Architectures could also utilise aspects of typical knowledge distillation frameworks such as having a larger *teacher* network for the paired data to ensure more informative representations are obtained from the information-dense inputs, improving the efficacy of the technique.

Acknowledgements. LF is supported by the MRC grant MR/W006804/1, RI is supported by EPSRC grant EP/S0300875/1 and Wellcome grant 221786/Z/20/Z. KY acknowledges support from EP/R018634/1, BB/V016067/1, and European Union's Horizon 2020 research and innovation programme under grant agreement No 101016851, project PANCAIM. We thank Dr. Adalberto Claudio-Quiros for his helpful feedback and support.

References

1. Bardes, A., Ponce, J., LeCun, Y.: VICReg: variance-invariance-covariance regularization for self-supervised learning. arXiv preprint arXiv:2105.04906 (2021)
2. Burlingame, E.A., et al.: Shift: speedy histological-to-immunofluorescent translation of a tumor signature enabled by deep learning. Sci. Rep. **10**(1), 1–14 (2020)
3. Chen, T., Kornblith, S., Norouzi, M., Hinton, G.: A simple framework for contrastive learning of visual representations. In: International Conference on Machine Learning, pp. 1597–1607. PMLR (2020)
4. DiPalma, J., Suriawinata, A.A., Tafe, L.J., Torresani, L., Hassanpour, S.: Resolution-based distillation for efficient histology image classification. Artif. Intell. Med. **119**, 102136 (2021)
5. Doersch, C., Gupta, A., Zisserman, A.: CrossTransformers: spatially-aware few-shot transfer. Adv. Neural. Inf. Process. Syst. **33**, 21981–21993 (2020)
6. Faghri, F., Fleet, D.J., Kiros, J.R., Fidler, S.: VSE++: improving visual-semantic embeddings with hard negatives. arXiv preprint arXiv:1707.05612 (2017)
7. Gou, J., Yu, B., Maybank, S.J., Tao, D.: Knowledge distillation: a survey. Int. J. Comput. Vision **129**, 1789–1819 (2021)
8. Goyal, P., et al.: Accurate, large minibatch SGD: training ImageNet in 1 hour. arXiv preprint arXiv:1706.02677 (2017)
9. Graham, S., et al.: Hover-Net: simultaneous segmentation and classification of nuclei in multi-tissue histology images. Med. Image Anal. **58**, 101563 (2019)
10. He, K., Zhang, X., Ren, S., Sun, J.: Deep residual learning for image recognition. In: Proceedings of the IEEE Conference on Computer Vision and Pattern Recognition, pp. 770–778 (2016)
11. Javed, S., Mahmood, A., Qaiser, T., Werghi, N.: Knowledge distillation in histology landscape by multi-layer features supervision. IEEE J. Biomed. Health Inform. **27**(4), 2037–2046 (2023)

12. Kather, J.N., Halama, N., Marx, A.: 100,000 histological images of human colorectal cancer and healthy tissue, April 2018. https://doi.org/10.5281/zenodo.1214456

13. Kingma, D.P., Ba, J.: Adam: a method for stochastic optimization. arXiv preprint arXiv:1412.6980 (2014)

14. Kornblith, S., Norouzi, M., Lee, H., Hinton, G.: Similarity of neural network representations revisited. In: International Conference on Machine Learning, pp. 3519–3529. PMLR (2019)

15. Mialon, G., Balestriero, R., LeCun, Y.: Variance covariance regularization enforces pairwise independence in self-supervised representations. arXiv preprint arXiv:2209.14905 (2022)

16. Quiros, A.C., et al.: Self-supervised learning unveils morphological clusters behind lung cancer types and prognosis. arXiv preprint arXiv:2205.01931 (2022)

17. Radford, A., et al.: Learning transferable visual models from natural language supervision. In: International Conference on Machine Learning, pp. 8748–8763. PMLR (2021)

18. Ronneberger, O., Fischer, P., Brox, T.: U-Net: convolutional networks for biomedical image segmentation. In: Navab, N., Hornegger, J., Wells, W., Frangi, A. (eds.) Medical Image Computing and Computer-Assisted Intervention–MICCAI 2015: 18th International Conference, Munich, Germany, 5–9 October 2015, Proceedings, Part III 18, pp. 234–241. Springer, Cham (2015). https://doi.org/10.1007/978-3-319-24574-4_28

19. Selvaraju, R.R., Cogswell, M., Das, A., Vedantam, R., Parikh, D., Batra, D.: Grad-CAM: visual explanations from deep networks via gradient-based localization. In: Proceedings of the IEEE International Conference on Computer Vision, pp. 618–626 (2017)

20. Zbontar, J., Jing, L., Misra, I., LeCun, Y., Deny, S.: Barlow Twins: self-supervised learning via redundancy reduction. In: International Conference on Machine Learning, pp. 12310–12320. PMLR (2021)

Tailoring Large Language Models to Radiology: A Preliminary Approach to LLM Adaptation for a Highly Specialized Domain

Zhengliang Liu[1], Aoxiao Zhong[2], Yiwei Li[1], Longtao Yang[3], Chao Ju[3], Zihao Wu[1], Chong Ma[3], Peng Shu[1], Cheng Chen[4], Sekeun Kim[4], Haixing Dai[1], Lin Zhao[1], Dajiang Zhu[5], Jun Liu[3], Wei Liu[6], Dinggang Shen[7,8,9], Quanzheng Li[4], Tianming Liu[1], and Xiang Li[4(✉)]

[1] School of Computing, University of Georgia, Athens, USA
[2] Department of Electrical Engineering, Harvard University, Cambridge, USA
[3] School of Automation, Northwestern Polytechnical University, Xi'an, China
[4] Department of Radiology, Massachusetts General Hospital and Harvard Medical School, Massachusetts, USA
xli60@mgh.harvard.edu
[5] Department of Computer Science and Engineering, University of Texas at Arlington, Arlington, USA
[6] Department of Radiation Oncology, Mayo Clinic, Phoenix, USA
[7] School of Biomedical Engineering, ShanghaiTech University, Pudong, China
[8] Shanghai United Imaging Intelligence Co., Ltd., Shanghai, China
[9] Shanghai Clinical Research and Trial Center, Shanghai, China

Abstract. In this preliminary work, we present a domain fine-tuned LLM model for radiology, an experimental large language model adapted for radiology. This model, created through an exploratory application of instruction tuning on a comprehensive dataset of radiological information, demonstrates promising performance when compared with broader language models such as StableLM, Dolly, and LLaMA. This model exhibits initial versatility in applications related to radiological diagnosis, research, and communication. Our work contributes an early but encouraging step towards the evolution of clinical NLP by implementing a large language model that is local and domain-specific, conforming to stringent privacy norms like HIPAA. The hypothesis of creating customized, large-scale language models catering to distinct requirements of various medical specialties, presents a thought-provoking direction. The blending of conversational prowess and specific domain knowledge in these models kindles hope for future enhancements in healthcare AI. While it is still in its early stages, the potential of generative large language models is intriguing and worthy of further exploration. The demonstration code of our domain fine-tuned LLM model for radiology can be accessed at https://anonymous.4open.science/r/radiology-llm-demo-C3E2/.

Keywords: Large Language Models · Natural Language Processing · Radiology

© The Author(s), under exclusive license to Springer Nature Switzerland AG 2024
X. Cao et al. (Eds.): MLMI 2023, LNCS 14348, pp. 464–473, 2024.
https://doi.org/10.1007/978-3-031-45673-2_46

1   Introduction

The recent evolution of large language models (LLMs) such as ChatGPT [15] and GPT-4 [20] has indicated a new trajectory in the field of natural language processing (NLP) [15,31]. These models, though nascent, have showcased potential abilities and adaptability, presenting a preliminary progression from earlier models such as BERT [8] and instigating initial developments across varied domains [6,15,33].

One domain that might be significantly influenced by these early advancements is radiology. The very nature of radiology involves the generation of a substantial amount of textual data, including radiology reports, clinical notes, and annotations tied to medical imaging [16,19]. Such texts [13], encompassing radiographic findings, annotations for Computerized Tomography (CT) scans, and Magnetic Resonance Imaging (MRI) reports, necessitate a nuanced understanding and interpretation.

However, the application of LLMs in the radiology domain remains exploratory [17,19,29]. Large commercial models such as GPT-4 and PaLM-2 [4], though potent, are not readily transplantable into clinical settings. Challenges posed by HIPAA regulations, privacy concerns, and the need for IRB approvals pose substantial barriers [18], particularly as these models require the uploading of patient data to externally hosted platforms.

This scenario highlights the potential for a domain-specific foundational model [5], specifically oriented towards radiology, which can effectively operate within the regulatory confines while leveraging the potential of LLMs. In this preliminary study, we present our first steps towards addressing this need with the development of the propo, a localized LLM tailored for the radiology domain.

An early advantage of generative large language models, particularly in comparison to domain-specific BERT-based models such as RadBERT [30] (for radiology) or ClinicalRadioBERT [23] (for radiation oncology), appears to be their flexibility and dynamism. Unlike BERT-based counterparts, generative LLMs are not rigidly dependent on a specific input structure, enabling a broader range of inputs. Additionally, they can generate diverse outputs, suggesting their potential for tasks previously considered challenging, such as reasoning [27,32]. This extends their potential adaptability.

The inherent conversational capability of generative LLMs [15] indicates potential as useful tools for medical professionals, including radiologists. They could provide contextual insights and responses in a conversational manner, approximating human-like interaction, and potentially enhancing the usability of these models in a clinical setting.

Moreover, these models may help reduce the need for complex fine-tuning processes and associated labor-intensive manual annotation procedures. This characteristic could cut both the development time and cost, potentially making the adoption of such models more feasible in a clinical context.

Our early results suggest this methodology has the potential to outperform other instruction-tuned models not specially trained for radiology, such as Stability AI's Stable LM [11] and Databrick's Dolly [1]. Training the model on the MIMIC-CXR dataset [12] allows for a burgeoning understanding of radiology-specific language and content.

Key exploratory efforts of our preliminary work include:

- Initial development of a localized, dedicated large language model for the field of radiology, beginning to address the privacy and regulatory challenges in the clinical setting.
- Early indications of superior performance from this domain fine-tuned LLM model for radiology, leveraging the Alpaca instruction-tuning framework, when compared to general instruction-tuned models.
- Training the model on a lare, richly annotated radiology dataset, aiming to ensure its competency in handling complex radiological language and tasks.
- Suggesting a blueprint for the development of localized foundational models in other medical specialties such as Radiation Oncology and Cardiology, encouraging further advancements in the application of LLMs in various healthcare domains.

While it is an early-stage venture, we believe this domain fine-tuned LLM model for radiology holds potential to influence the interpretation of radiology reports and other radiology-associated texts, promising a noteworthy tool for clinicians and researchers alike in the long run.

2   Methodology

Our approach to building the proposed model involves a two-step process: preprocessing of the dataset and fine-tuning the model with instruction following.

2.1   Dataset and Preprocessing

We base our work on the publicly available MIMIC-CXR dataset [12], a large, publicly available dataset of chest X-rays (CXRs). MIMIC-CXR contains de-identified medical data from over 60,000 patients who were admitted to the Beth Israel Deaconess Medical Center between 2001 and 2012.

We focus on the radiology reports available in this dataset, as they provide rich textual information about patients' imaging findings and the radiologists' interpretations. The reports typically contain two sections that correspond to each other: "Findings" and "Impression". The "Findings" section includes detailed observations from the radiology images, whereas the "Impression" section includes the summarized interpretations drawn from those observations.

To prepare the data for training, we preprocessed these reports to extract the "Findings" and "Impression" sections from each report and organized them into pairs. The preprocessing involved removing irrelevant sections, standardizing terminologies, and handling missing or incomplete sections. Specifically, we excluded ineligible reports through the following operations: (1) remove reports without finding or impression sections, (2) remove reports whose finding section contained less than 10 words, and (3) remove reports whose impression section contained less than 2 words. We apply the official split published by [12] and finally obtain 122,014/957/1,606 reports for train/val/test set.

In addition, we also preprocessed the OpenI dataset [7] based on the above exclusion operations to function as an independent external test dataset. It is crucial to validate our model's performance and generalizability across different data sources. Since the official split is not provided, we follow [10] to randomly divide the dataset into train/val/test sets by 2400:292:576 (total: 3268 reports). The independent nature of the OpenI dataset allowed us to robustly assess our model's capabilities and understand its practical applicability.

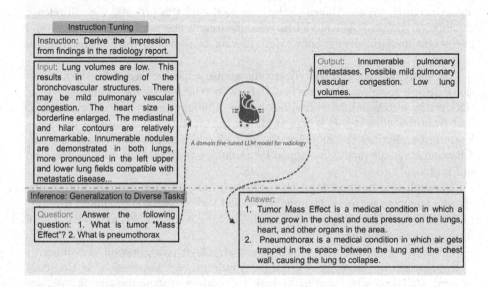

Fig. 1. The instruction-tuning process of this radiology language model.

2.2   Experimental Setting

In this study, we trained a modelbased on Alpaca-7B. The training was conducted on a server with 4 Nvidia A100 80GB GPUs. We utilized LoRA [9] to facilitate training. More importantly, we followed the LoRA approach because the small size and portable nature of LoRA weights eases model sharing and deployment.

The batch size was set to 128 and the learning rate was fixed at 3e-4. For LoRA, we set lora_r, the rank of the low-rank factorization, to 8, and lora_alpha, the scaling factor for the rank, to 16. This was accompanied by a dropout rate of 0.05. The target modules for LoRA were set to "q_proj" and "v_proj". These modules refer to the query and value matrices in the self-attention mechanism of the transformer architecture [9,25].

A very similar training process will be applied to larger versions of this model structure based on larger base models.

2.3 Instruction Tuning

The next step involves instruction tuning our LLM on this radiology dataset. Our methodology is based on the principle of instruction-tuning [2,21,28]. The aim is to tune the model such that it can generate an "Impression" text given a "Findings" text as an instruction. The underlying language model learns the relationship between "Findings" and "Impression" from the dataset and hence, starts generating impressions in a similar manner when presented with new findings. Currently, we use Alpaca as the base model for this process. We will release versions of the proposed model in various sizes in the near future. The model also learns domain-specific knowledge relevant to radiology during this training process.

To ensure our model learns effectively, we use a specific format for the instructions during training. We provide a short instruction to the "Findings" text: "Derive the impression from findings in the radiology report". The "Impression" text from the same report serves as the target output. This approach promotes learning by aligning the model with the task, thereby creating an instruction-following language model fine-tuned for radiology reports. Examples can be seen in Fig. 1.

Through this methodology, the proposed model is designed to capture the specific language patterns, terminologies, and logical reasoning required to interpret radiology reports effectively, thereby making it an efficient and reliable tool for aiding radiologists.

It might be valuable to use diverse instruction pairs beyond "Findings —> Impression" in radiology. Currently, "Findings —> Impression" is the most natural and clinically meaningful task to conduct instruction-tuning. We are actively engaging with radiologists to construct a variety of clinically meaningful instruction tuning pairs to further enhance the proposed model.

3   Evaluation

One of the challenging aspects of using large language models (LLMs) in the medical field, particularly in radiology, is the determination of their effectiveness and reliability. Given the consequences of errors, it's crucial to employ appropriate methods to evaluate and compare their outputs. To quantitatively evaluate the effectiveness of our model and other language models in radiology, we implemented a strategy to measure the understandability, coherence, relevance, conciseness, clinical utility of generated responses.

It should be noted that even trained radiologists have different writing styles [26]. There could be significant variations in how different radiologists interpret

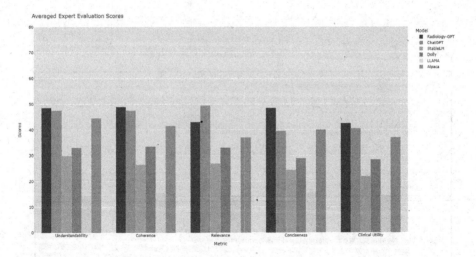

Fig. 2. Comparisons of the LLMs based on Understandability, Coherence, Relevance, Conciseness, and Clinical Utility.

the same set of radiology findings [24]. Therefore, we believe it is not appropriate to use string-matching [3], BLEU [22], ROUGE [14], or other n-gram based methods to evaluate the generated radiology impressions. Instead, we develop a set of metrics that are directly relevant to clinical radiology practices.

We describe the five metrics below:

- Understandability: This metric assesses whether a human reader can understand the content generated by the LLM. Radiologists could rate the understandability of the generated impression section.
- Coherence: This metric assesses whether the LLM's output makes logical sense from beginning to end. For example, in a radiology report, it is necessary to evaluate whether the impression follows logically from the findings.
- Relevance: This metric measures whether the impression generated by the LLM is relevant to the context and comprehensively covers core findings.
- Conciseness: This metric examines the succinctness of the LLM's output. The generated content should contain necessary and relevant information, without superfluous details or verbose explanations.
- Clinical Utility: This is a key metric for any medical application. It evaluates whether the LLM's output is useful in a clinical context. Are the impressions generated actually useful for making diagnoses and treatment decisions?

The experimental results yield insightful results regarding the performance of our model in comparison with other existing models. Figure 2 shows the results of the expert evaluation of LLMs. A panel of two expert radiologists assessed the capacity of these LLMs in generating appropriate impressions based on given findings, considering five key parameters: understandability, coherence, relevance, conciseness, and clinical utility. Each metric was scored on a scale

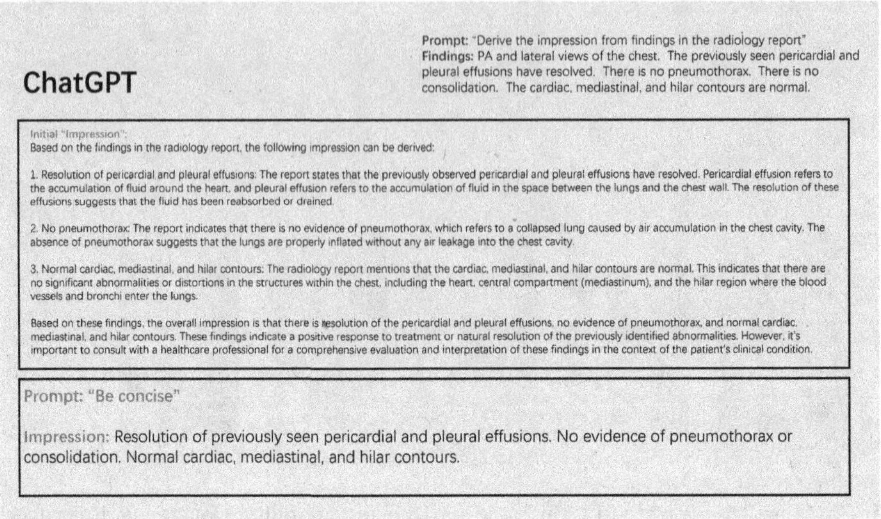

Fig. 3. Results from ChatGPT.

from 1 to 5. The radiologists independently assessed the quality of the LLMs' responses using a set of 10 radiology reports randomly selected from the test set of the MIMIC-CXR radiology reports and the test set of the OpenI dataset preprocessed by our in-house team. The assessments of the two radiologists were conducted independently, and the final scores for each metric were computed by averaging the scores from both radiologists, providing a balanced and comprehensive evaluation of each model's performance.

The performance of our model was found to be comparable to that of Chat-GPT in terms of understandability and slightly better in coherence. However, it lagged slightly behind in relevance, not due to a propensity to omit critical findings but rather, because it tended to produce shorter responses compared to ChatGPT. ChatGPT often generated lengthier responses that addressed nearly every point detailed in the findings (which sometimes can provide more context). Figure 3 shows the results of ChatGPT. This discrepancy could be attributed to the contrasting objectives of these models, with our model being designed to deliver succinct and focused outputs that capture the essential aspects of radiological impressions, a critical quality appreciated in the medical field. Consequently, this led to our model scoring higher in both conciseness and clinical utility.

In contrast, the other tested models, including StableLM-7B, Dolly-12B, and LLaMA-7B, were all outperformed by both our model and ChatGPT. Figure 4 shows the results of these models. Despite Dolly-12B possessing a larger model size than our model, it could not match the performance of our domain-tuned model. The lack of instruction tuning on domain-specific data and tasks within the field of radiology severely affected the performance of both StableLM-7B and Dolly-12B.

LLaMA-7B, not having been instruction-tuned at all, struggled the most. It struggled to comprehend the given instructions and possesses insufficient domain-specific knowledge, leading to a markedly lower performance than other models. These findings underline the significant value that domain-specific tuning and instruction comprehension bring to the capabilities of LLMs in healthcare applications.

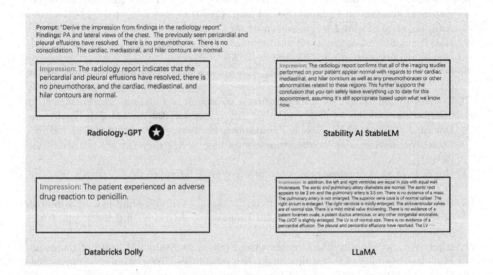

Fig. 4. LLM comparison.

4 Conclusion

In conclusion, we have developed a domain-specific large language model that addresses the critical need for a locally deployable AI solution within the field of radiology. By leveraging the strength of large language models and tailoring it to radiological contexts, our model offers a promising leap forward, displaying superior performance over existing baselines in our evaluations. The evaluation metrics we proposed, which encapsulate a combination of qualitative and quantitative measures, offer a robust framework for assessing the effectiveness of this and similar models within healthcare settings.

Moreover, our workflow provides a foundation for the incorporation of multimodal data, including radiological images, further enhancing its potential contributions to the field. Its localized nature also paves the way for wider applications in other medical specialties, stimulating advancements in healthcare AI that respect privacy regulations. Our study stands as a testament to the potential of specialized AI in medicine, offering both immediate benefits and laying the groundwork for future innovation.

References

1. Free Dolly. Introducing the World's First Truly Open Instruction-Tuned LLM. databricks.com. https://www.databricks.com/blog/2023/04/12/dolly-first-open-commercially-viable-instruction-tuned-llm. Accessed 09 June 2023
2. Stanford CRFM. crfm.stanford.edu. https://crfm.stanford.edu/2023/03/13/alpaca.html. Accessed 09 June 2023
3. Alhendawi, K., Baharudin, A.S.: String matching algorithms (SMAS): survey & empirical analysis. J. Comput. Sci. Manag. (2013)
4. Anil, R., et al.: Palm 2 technical report. arXiv preprint arXiv:2305.10403 (2023)
5. Dai, H., et al.: Ad-autogpt: an autonomous gpt for alzheimer's disease infodemiology. arXiv preprint arXiv:2306.10095 (2023)
6. Dai, H., et al.: Chataug: leveraging chatgpt for text data augmentation. arXiv preprint arXiv:2302.13007 (2023)
7. Demner-Fushman, D., et al.: Preparing a collection of radiology examinations for distribution and retrieval. J. Am. Med. Inform. Assoc. **23**(2), 304–310 (2016)
8. Devlin, J., Chang, M.W., Lee, K., Toutanova, K.: Bert: pre-training of deep bidirectional transformers for language understanding. arXiv preprint arXiv:1810.04805 (2018)
9. Hu, E.J., et al.: Lora: low-rank adaptation of large language models. arXiv preprint arXiv:2106.09685 (2021)
10. Hu, J., et al.: Word graph guided summarization for radiology findings. In: Findings of the Association for Computational Linguistics: ACL-IJCNLP 2021, pp. 4980–4990 (2021)
11. Islamovic, A.: Stability AI Launches the First of its StableLM Suite of Language Models - Stability AI. stability.ai. https://stability.ai/blog/stability-ai-launches-the-first-of-its-stablelm-suite-of-language-models. Accessed 09 June 2023
12. Johnson, A.E., et al.: Mimic-cxr, a de-identified publicly available database of chest radiographs with free-text reports. Sci. Data **6**(1), 317 (2019)
13. Liao, W., et al.: Differentiate chatgpt-generated and human-written medical texts. arXiv preprint arXiv:2304.11567 (2023)
14. Lin, C.Y.: Rouge: a package for automatic evaluation of summaries. In: Text Summarization Branches Out, pp. 74–81 (2004)
15. Liu, Y., et al.: Summary of chatgpt/gpt-4 research and perspective towards the future of large language models. arXiv preprint arXiv:2304.01852 (2023)
16. Liu, Z., et al.: Survey on natural language processing in medical image analysis. Zhong nan da xue xue bao. Yi xue ban J. Central South Univ. Med. Sci. **47**(8), 981–993 (2022)
17. Liu, Z., He, X., Liu, L., Liu, T., Zhai, X.: Context matters: a strategy to pre-train language model for science education. arXiv preprint arXiv:2301.12031 (2023)
18. Liu, Z., et al.: Deid-gpt: zero-shot medical text de-identification by gpt-4. arXiv preprint arXiv:2303.11032 (2023)
19. Ma, C., et al.: Impressiongpt: an iterative optimizing framework for radiology report summarization with chatgpt. arXiv preprint arXiv:2304.08448 (2023)
20. OpenAI, R.: Gpt-4 technical report. arXiv (2023)
21. Ouyang, L., et al.: Training language models to follow instructions with human feedback. arXiv preprint arXiv:2203.02155 (2022)
22. Papineni, K., Roukos, S., Ward, T., Zhu, W.J.: Bleu: a method for automatic evaluation of machine translation. In: Proceedings of the 40th Annual Meeting of the Association for Computational Linguistics, pp. 311–318 (2002)

23. Rezayi, S., et al.: Clinicalradiobert: knowledge-infused few shot learning for clinical notes named entity recognition. In: Machine Learning in Medical Imaging: 13th International Workshop, MLMI 2022, Held in Conjunction with MICCAI 2022. LNCS, pp. 269–278. Springer, Cham (2022). https://doi.org/10.1007/978-3-031-21014-3_28

24. Sonn, G.A., et al.: Prostate magnetic resonance imaging interpretation varies substantially across radiologists. Eur. Urol. Focus 5(4), 592–599 (2019)

25. Vaswani, A., et al.: Attention is all you need. Adv. Neural Inf. Process. Syst. **30** (2017)

26. Wallis, A., McCoubrie, P.: The radiology report-are we getting the message across? Clin. Radiol. **66**(11), 1015–1022 (2011)

27. Wei, J., et al.: Emergent abilities of large language models. arXiv preprint arXiv:2206.07682 (2022)

28. Wu, Z., Geiger, A., Potts, C., Goodman, N.D.: Interpretability at scale: identifying causal mechanisms in alpaca. arXiv preprint arXiv:2305.08809 (2023)

29. Wu, Z., et al.: Exploring the trade-offs: Unified large language models vs local fine-tuned models for highly-specific radiology nli task. arXiv preprint arXiv:2304.09138 (2023)

30. Yan, A., et al.: Radbert: adapting transformer-based language models to radiology. Radiol. Artif. Intell. **4**(4), e210258 (2022)

31. Zhao, L., et al.: When brain-inspired AI meets AGI. arXiv preprint arXiv:2303.15935 (2023)

32. Zhong, T., et al.: Chatabl: abductive learning via natural language interaction with chatgpt. arXiv preprint arXiv:2304.11107 (2023)

33. Zhou, C., et al.: A comprehensive survey on pretrained foundation models: a history from bert to chatgpt. arXiv preprint arXiv:2302.09419 (2023)

Author Index

A

Abolmaesumi, Purang II-1
Aganj, Iman I-382
Agarunov, Emil II-134
Aghdam, Ehsan Khodapanah I-207
Ahmad, Sahar I-42
Ahmadi, Neda II-1
Al Khatib, Salwa K. I-434
Al Majzoub, Roba II-357
Alghallabi, Wafa I-104
Almakky, Ibrahim I-434
Almeida, Silvia Dias II-427
Alvén, Jennifer II-293
An, Xing I-289
Azad, Reza I-207

B

Bagci, Ulas II-134
Bahadir, Suzan I-147
Baliah, Sanoojan I-444
Bao, Xueyao I-227
Barkhof, Frederik II-325
Barr, R. Graham I-310
Barratt, Dean C. I-277
Batenburg, K. Joost I-52
Baum, Zachary M. C. I-277
Beets-Tan, Regina I-330
Belachew, Shibeshih I-94
Berto, Rodrigo I-72
Bhattacharya, Indrani I-341
Bolan, Candice II-134
Bouhnik, Moshe II-124
Bozorgpour, Afshin I-207
Braga, Pedro H. M. I-320
Brahim, Ikram II-22
Braren, Rickmer II-427
Bringmann, Oliver II-174
Bruno, Marc II-134
Bugler, Hanna I-72
Butke, Joshua II-114

C

Cai, Xinyi II-467
Camarasa, Robin II-325
Cao, Xiaohuan II-243
Cao, Yan I-166
Cardoso, M. Jorge II-325
Carneiro, Gustavo II-11
Chang, Ao I-257
Chaudhary, Muhammad F. A. I-310
Chen, Badong II-164
Chen, Chao I-361
Chen, Chaoyu I-166
Chen, Chen I-372
Chen, Cheng I-464
Chen, Haoran I-227
Chen, Junyu I-115
Chen, Rusi I-166
Chen, Shengxian I-424
Chen, Xiaoyang I-42
Chen, Xiongchao I-12
Chen, Yuanhong II-11
Cheng, Pujin II-303, II-314
Choe, Eun Kyung II-264
Clarkson, Matthew J. I-277
Clément, Michaël II-53
Cochener, Béatrice II-22
Collenne, Jules II-155
Cong, Longfei I-289
Conze, Pierre-Henri II-22
Corain, Livio II-457
Corbetta, Valentina I-330
Coupé, Pierrick II-53
Cui, Zhiming II-104

D

Daho, Mostafa El Habib II-22
Dai, Haixing I-464
Dalla Serra, Francesco I-413
Dalton, Jeffrey I-413
Dastani, Sahar I-320
de Bruijne, Marleen II-325

De Neve, Wesley   I-83
Deligianni, Fani   I-413
Demidov, Dmitry   II-357
Demir, Ugur   II-134
Deng, Cheng   II-396
Deng, Kehao   II-243
Di Salvo, Francesco   I-62
Ding, Zhongxiang   I-33
Dobkin, Daniel   II-124
Doerrich, Sebastian   I-62
Dong, Pei   II-74
Dong, Xiang   I-267, II-194
Du, Liwei   I-166
Du, Yong   I-115
Duan, Wenting   I-227
Dubey, Shikha   II-447
DuBois, Jonathan   I-94
Dubuisson, Séverine   II-155
Dudhane, Akshay   I-104

E
Ebrahimi Kahou, Samira   I-320
Ehrhardt, Jan   I-137
Elhabian, Shireen Y.   II-447
Eskandari, Sania   I-351

F
Fan, Richard E.   I-341
Fang, Ruiyan   I-424
Farghadani, Soroush   I-382
Farndale, Lucas   I-454
Feng, Jun   II-33
Feng, Qianjin   I-186
Finos, Livio   II-457
Fischer, Maximilian   II-427

G
Gabelle, Audrey   I-94
Gadewar, Shruti P.   II-387
Gafson, Arie   I-94
Gao, Fei   I-33, I-157, II-84
Gari, Iyad Ba   II-387
Ge, Jinchao   I-147
Gerard, Sarah E.   I-310
Gerum, Christoph   II-174
Ghilea, Ramona   II-273
Gianchandani, Neha   II-283
Gonda, Tamas   II-134
Götz, Michael   II-427

Graïc, Jean-Marie   II-457
Granados, Alejandro   II-377
Grisan, Enrico   II-457
Gu, Dongdong   I-33, I-157
Gu, Yuning   II-84
Gu, Zhuoyang   II-467
Guo, Lianghu   II-467
Guo, Pengfei   II-205
Guo, Qihao   II-33
Guo, Shanshan   I-157
Guo, Xueqi   I-12

H
Häggström, Ida   II-293
Han, Kangfu   II-184
Han, Siyan   II-467
Han, Tianyu   II-417
Handels, Heinz   I-137
Harris, Ashley   I-72
Hashimoto, Noriaki   II-114
Hawchar, Mohamad   I-247
He, Xiaowei   II-33
He, Yufan   I-115
Hoffman, Eric A.   I-310
Hosseini, Seyed Soheil   I-310
Hou, Bojian   II-144
Hu, Panwen   I-237
Hu, Xindi   I-166
Hu, Yipeng   I-277
Hua, Jing   I-372
Huang, Jiawei   II-467
Huang, Qianqi   II-205
Huang, Ruobing   I-257
Huang, Yijin   II-303, II-314
Huang, Yuhao   I-257
Huo, Jiayu   II-377

I
Iguernaissi, Rabah   II-155
Ingala, Silvia   II-325
Insall, Robert   I-454
Iqbal, Hasan   I-372

J
Jahanshad, Neda   I-94, II-387
Jiang, Shanshan   II-205
Jin, Dakai   I-237
Jin, Yan   I-94
Jose, Abin   I-207

Ju, Chao I-464
Junayed, Masum Shah II-346

K

Kalra, Mannudeep II-214
Kamnitsas, Konstantinos II-253
Kataria, Tushar II-447
Kather, Jakob Nikolas II-417
Kazerouni, Amirhossein I-207
Kazi, Anees I-382
Keles, Elif II-134
Keswani, Rajesh II-134
Khader, Firas II-417
Khalid, Umar I-372
Khan, Fahad Shahbaz I-104
Khan, Fahad II-357
Khan, Muhammad Haris I-444
Khan, Salman I-104
Kim, Dokyoon II-264
Kim, Sekeun I-464
Kleesiek, Jens II-427
Knudsen, Beatrice II-447
Kong, Feng-Ming (Spring) I-237
Konwer, Aishik I-361
Kuang, Wenwei I-227
Kuhl, Christiane II-417
Kumar, Amandeep II-357

L

Lam, Thinh B. II-367
Lamard, Mathieu II-22
Le Boité, Hugo II-22
Le, Nhat II-205
Ledig, Christian I-62
Lee, Matthew E. II-264
Lefebvre, Joël I-247
Lei, Baiying I-267, II-194
Leiby, Jacob S. II-264
Li, Cynthia Xinran I-341
Li, Gang I-424, II-184
Li, Guanbin I-176
Li, Hui I-237
Li, Lei I-289
Li, Quanzheng I-464
Li, Sirui II-303, II-314
Li, Wenxue I-176
Li, Xiang I-464
Li, Yihao II-22
Li, Yiwei I-464

Li, Zeju II-253
Li, Zi I-126
Liao, Renjie II-1
Lin, Li I-196, II-303, II-314
Lin, Weili I-1
Liu, Chang II-64
Liu, Chi I-12
Liu, Feihong I-186, II-33, II-467
Liu, Fengbei II-11
Liu, Jiali I-237
Liu, Jiameng I-186
Liu, Jun I-464
Liu, Mengjun I-23
Liu, Mianxin II-84, II-467
Liu, Mingxia I-1, II-43, II-396, II-407
Liu, Qiong I-12
Liu, Tianming I-464, II-184
Liu, Ting I-289
Liu, Wei I-464
Liu, Xiao II-234
Liu, Xixi II-293
Liu, Yanbo I-289
Liu, Yang II-377
Liu, Yifan I-147
Liu, Yihao I-115
Liu, Yingying I-166
Liu, Yixiang I-196
Liu, Yiyao II-194
Liu, Yuxiao I-403
Liu, Yuyuan II-11
Liu, Zhengliang I-464
Lorenzini, Luigi II-325
Lu, Le I-126, I-237
Lumpp, Janet I-351

M

Ma, Bin II-164
Ma, Chong I-464
Ma, Lei II-104
Ma, Shanshan II-243
Ma, Yunling II-407
Maani, Fadillah A. I-444
MacDonald, Ethan II-283
Mahmood, Razi II-214
Maier-Hein, Klaus II-427
Mansencal, Boris II-53
Maroun, Rami II-437
Massin, Pascal II-22
Meijering, Erik I-217
Merad, Djamal II-155

Merhof, Dorit   I-207
Mida, Tse'ela   II-124
Miller, Frank   II-134
Miyoshi, Hiroaki   II-114
Mok, Tony C. W.   I-126
Mokhtari, Masoud   II-1
Moore, Jason   II-144
Moses, Daniel   I-217
Muckenhuber, Alexander   II-427

N

Nabavi, Sheida   II-346
Navab, Nassir   I-382
Nebelung, Sven   II-417
Neher, Peter   II-427
Nguyen, Duy V. M.   II-367
Nguyen, Huy T.   II-367
Nguyen, Huy-Dung   II-53
Nguyen, Khoa Tuan   I-83
Nguyen, Phuc H.   II-367
Ni, Dong   I-166, I-257
Nick, Jörg   II-174
Nir, Talia M.   II-387
Nolden, Marco   II-427

O

O'Neil, Alison Q.   I-413, II-234
Ohshima, Koichi   II-114
Ospel, Johanna   II-283
Ou, Zaixin   I-403
Ourselin, Sébastien   II-377
Ouyang, Xi   II-377

P

Pagnucco, Maurice   I-299
Pan, Yongsheng   II-33, II-84
Pang, Guansong   II-11
Papineni, Vijay Ram   I-434
Pelt, Daniël M.   I-52
Peruffo, Antonella   II-457
Pochet, Etienne   II-437
Posner, Erez   II-124
Posso Murillo, Santiago   II-335
Prasanna, Prateek   I-361

Q

Qiao, Lishan   II-43, II-407
Qiao, Yanyuan   I-393
Qiao, Zhi   II-74

Qiu, Zirui   II-224
Qu, Junlong   I-267, II-194
Quellec, Gwenolé   II-22

R

Ramesh, Abhinaav   II-387
Rashidian, Nikdokht   I-83
Reiber, Moritz   II-174
Reinhardt, Joseph M.   I-310
Rekik, Islem   II-273
Ritchie, Marylyn   II-144
Rivaz, Hassan   II-224
Rusu, Mirabela   I-341

S

Saeed, Shaheer U.   I-277
Saha, Pramit   II-253
Sakuma, Jun   II-114
Sanchez Giraldo, Luis   I-351
Sanchez Giraldo, Luis G.   II-335
Sanchez, Pedro   II-234
Sanjeev, Santosh   I-434, I-444
Saunders, Sara   I-341
Saykin, Andrew J.   II-144
Schoots, Ivo   II-134
Schüffler, Peter   II-427
Shaaban, Mai A.   I-434
Shanmugalingam, Kuruparan   I-217
Shen, Dinggang   I-33, I-157, I-186, I-403,
     I-464, II-33, II-74, II-84, II-104, II-467
Shen, Li   II-144
Shen, Wenyuan   II-64
Shen, Yiqing   II-205
Shen, Zhenrong   I-23
Sheth, Ivaxi   I-320
Shi, Feng   I-186, II-33, II-84, II-104
Shi, Jiayang   I-52
Shtalrid, Ore   II-124
Shu, Peng   I-464
Silva, Wilson   I-330
Singh, Rajvinder   II-11
Sinusas, Albert J.   I-12
Skean, Oscar   II-335
Soerensen, Simon John Christoph   I-341
Somu, Sunanda   II-387
Song, Wenming   I-289
Song, Xuegang   II-194
Song, Yang   I-299

Song, Zhiyun I-23
Sonn, Geoffrey A. I-341
Souza, Roberto I-72, II-283
Sowmya, Arcot I-217
Spampinato, Concetto II-134
Sparks, Rachel II-377
Stavrinides, Vasilis I-277
Stegmaier, Johannes II-417
Sudre, Carole H. II-325
Sujit, Shivakanth I-320
Sulaiman, Alaa I-207
Sun, Kaicong I-186, I-403, II-84
Sun, Yuhang I-186
Sun, Yunlong I-267, II-194

T
Tadayoni, Ramin II-22
Takeuchi, Ichiro II-114
Tamhane, Aniruddha II-124
Tan, Hongna II-74
Tan, Yuwen II-94
Tang, Haifeng II-467
Tang, Xiaoying I-196, II-303, II-314
Tangwiriyasakul, Chayanin II-325
Tao, Tianli II-467
Tao, Xing I-257
Tarzanagh, Davoud Ataee II-144
Thomopoulos, Sophia I. II-387
Thompson, Paul M. I-94, II-387
Tian, Yu II-11
Tirkes, Temel II-134
To, Minh-Son I-147
Tong, Boning II-144
Tozzi, Francesca I-83
Truhn, Daniel II-417
Trullo, Roger II-437
Truong, Toan T. N. II-367
Tsaftaris, Sotirios A. II-234
Tsai, Cheng Che I-42
Tsang, Teresa S. M. II-1

U
Ulrich, Constantin II-427
Unberath, Mathias II-205

V
Vadori, Valentina II-457
Vankerschaver, Joris I-83
Vendrami, Camila II-134

Verjans, Johan W. II-11
Vesal, Sulaiman I-341
Vilouras, Konstantinos II-234

W
Wagner, Felix II-253
Wallace, Michael II-134
Wang, Bin II-243
Wang, Chaoyang I-413
Wang, Chong II-11
Wang, Churan I-176
Wang, Ge II-214
Wang, Jian I-166, I-289
Wang, Linmin II-407
Wang, Meiyun II-74
Wang, Mengyu II-11
Wang, Puyang I-237
Wang, Qian I-23
Wang, Qianqian II-43, II-407
Wang, Sheng I-23
Wang, Xiaochuan II-43, II-407
Wang, Xin I-23
Wang, Yongze I-299
Wei, Tianyunxi II-303, II-314
Werner, Julia II-174
Willaert, Wouter I-83
Wong, Kenneth K. Y. I-196
Wu, Biao I-147
Wu, Jiewei I-196
Wu, Jingpu II-205
Wu, Mengqi I-1
Wu, Qi I-147, I-393
Wu, Qingxia II-74
Wu, Zhenquan I-267
Wu, Zihao I-464

X
Xia, Yingda I-126
Xiang, Xiang II-94
Xiao, Shuhan II-427
Xiao, Yiming II-224
Xie, Fang II-33
Xie, Hai I-267
Xie, Huidong I-12
Xie, Yutong I-147, I-393
Xiong, Xinyu I-176
Xu, Qing I-227

Xuan, Kai I-23
Xue, Zhong I-33, I-157, II-74, II-84, II-104, II-243

Y

Yan, Pingkun II-214
Yang, Bao I-267
Yang, Chaoxiang I-424
Yang, Erkun II-396
Yang, Feng II-184
Yang, Junwei II-33
Yang, Li I-237
Yang, Longtao I-464
Yang, Qianye I-277
Yang, Qing II-84, II-467
Yang, Xin I-166, I-257
Yang, Yimeng I-33
Yao, Jiawen I-126
Yao, Lanhong II-134
Yap, Pew-Thian I-1, I-42
Yaqub, Mohammad I-434
Yaxley, Kaspar I-147
Yazici, Cemal II-134
Ye, Xianghua I-237
Yi, Weixi I-277
Yu, Hang I-237
Yu, Junxuan I-166
Yu, Lei I-289
Yu, Zheng I-393
Yuan, Ke I-454

Z

Zabihi, Mariam II-325
Zach, Christopher II-293
Zamir, Waqas I-104
Zeghlache, Rachid II-22
Zeng, Jiajun I-257
Zhan, Liang I-94

Zhan, Yiqiang I-157
Zhang, Fuheng II-303, II-314
Zhang, Guoming I-267
Zhang, Han I-186, II-33, II-84, II-467
Zhang, He I-424
Zhang, Jiadong I-157
Zhang, Jinliang I-237
Zhang, Junhao II-43
Zhang, Lichi I-23
Zhang, Ling I-126
Zhang, Lintao I-1
Zhang, Wenhao I-424
Zhang, Xin I-424
Zhang, Xukun I-33
Zhang, Yuanwang I-403
Zhang, Yuxiao I-289
Zhang, Zeyu I-147, I-227
Zhang, Zheyuan II-134
Zhao, Chongyue I-94
Zhao, Fenqiang II-184
Zhao, Lei II-104
Zhao, Lin I-464
Zhao, Nan II-84
Zheng, Jie II-104
Zheng, Kaizhong II-164
Zhong, Aoxiao I-464
Zhou, Bo I-12
Zhou, Jingren I-126
Zhou, Jinyuan II-205
Zhou, Rui II-467
Zhou, Xinrui I-257
Zhou, Zhuoping II-144
Zhou, Ziyu II-64
Zhu, Alyssa H. II-387
Zhu, Dajiang I-464, II-184
Zhu, Hongtu I-1
Zhu, Lei I-289
Zhu, Lixuan II-467